Fundamentals of Kalman Filtering: A Practical Approach

Fundamentals of Kalman Filtering: A Practical Approach

Paul Zarchan and Howard Musoff
Charles Stark Draper Laboratory, Inc.
Cambridge, Massachusetts

Volume 190
PROGRESS IN
ASTRONAUTICS AND AERONAUTICS

Paul Zarchan, Editor-in-Chief
Charles Stark Draper Laboratory, Inc.
Cambridge, Massachusetts

Published by the
American Institute of Aeronautics and Astronautics, Inc.
1801 Alexander Bell Drive, Reston, Virginia 20191-4344

D

629.4015'1954

ZAR

To

Maxine, Adina, Ari, and Ronit

P.Z.

Wally, Sandy, Jay, Stephanie, Charlie, Scott, Cindy, Danielle, and Adam

H.M.

Preface

It has been four decades since Kalman introduced his systematic approach to linear filtering based on the method of least-squares (Kalman, R. E., "A New Approach to Linear Filtering and Prediction Problems," *Journal of Basic Engineering*, Vol. 82, No. 1, March 1960, pp. 35–46). Although his original journal article was difficult to read and understand, the results of the paper were applied immediately in many different fields by individuals with a variety of backgrounds because the filtering algorithm actually worked and was easy to implement on a digital computer. People were able to apply Kalman filtering without necessarily understanding or caring about the intricacies of its derivation. Because of the ease of implementation of the original recursive digital Kalman filter, engineers and scientists were able to find out immediately that this new filtering technique was often much better than existing filtering techniques in terms of performance. Both performance improvements and ease of implementation rather than analytical elegance made the Kalman filter popular in the world of applications. However, the Kalman filter was usually much more computationally expensive than existing filtering techniques, which was an issue in many cases for the primitive computers that were available at that time. In addition to improved performance, this new filtering technique also provided a systematic approach to many problems, which was also an improvement over some of the ad hoc schemes of the day. Today, because of the popularity and proliferation of Kalman filtering, many individuals either do not know (or care) about any other filtering techniques. Some actually believe that no filtering took place before 1960.

With the possible exception of the fast Fourier transform, Kalman filtering is probably the most important algorithmic technique ever devised. Papers on the subject have been filling numerous journals for decades. However, Kalman filtering is one of those rare topics that is not only popular in academic journals but also has a history of being rich in practical applications. Kalman filtering has been used in applications that include providing estimates for navigating the Apollo spacecraft, predicting short-term stock market fluctuations, and estimating user location with relatively inexpensive hand-held global positioning system (GPS) receivers.

The purpose of this text is not to make additional theoretical contributions in the world of Kalman filtering but is simply to show the reader how actually to build Kalman filters by example. It is the authors' belief that the best way of learning is by doing. Unlike other texts on Kalman filtering, which devote most of their time to derivations of the filter and the theoretical background in understanding the derivations, this text does not even bother to derive the filter. After all, the filter has been in use for 40 years and being an expert at derivations usually has nothing to do with getting a filter to work. Instead the Kalman-filtering equations are simply explained, and the text devotes its time to applying Kalman filtering to actual problems.

Numerous simplified, but nontrivial, real-world examples are presented in detail, showing the many ways in which Kalman filters can be designed. Sometimes mistakes are introduced intentionally to the initial designs to show the interested reader what happens when the filter is not working properly. Rarely in real life is a Kalman filter working after the first try. In fact, it usually takes many tries just to get the filter to fail (i.e., even getting the code to compile and give ridiculous answers is a challenge)! Therefore, we intentionally take the reader through part of that realistic iteration process. It is hoped that readers with varied learning styles will find the text's practical approach to Kalman filtering to be both useful and refreshing.

The text also spends a great deal of time in setting up a problem before the Kalman filter is actually formulated or designed. This is done to give the reader an intuitive feel for the problem being addressed. The time spent understanding the problem will always be important in later determining if the Kalman filter is performing as expected and if the resultant answers make sense.

Often the hardest part in Kalman filtering is the subject that no one talks about—setting up the problem. This is analogous to the quote from the recent engineering graduate who, upon arriving in industry, enthusiastically says, "Here I am, present me with your differential equations!" As the naive engineering graduate soon found out, problems in the real world are frequently not clear and are subject to many interpretations. Real problems are seldom presented in the form of differential equations, and they usually do not have unique solutions. Therefore, we will often do problems in several different ways to show the reader that there is no one way of tackling a problem. Each approach will have advantages and disadvantages, and often the filter designer will actually do the problem several different ways before determining the best approach for the application at hand.

The text only makes use of the fundamental techniques that were introduced by Kalman four decades ago. There have been many different implementations, variations, and innovations of the original Kalman filter, which were developed for real applications because of computer hardware constraints. However, most of those approaches are ignored in this text simply because computer hardware has improved so dramatically since the 1960s that many of those innovative techniques are no longer required. Using double-precision arithmetic on today's 32-bit desktop or flight computers usually negates the need for such heroic techniques as square-root filtering. It is our belief that discussing such techniques in an introductory text such as this will only confuse readers and possibly discourage them from ever attempting to build a Kalman filter because of its perceived complexity. This does not mean that Kalman filters always work on modern computers, but most of the time their failure to work is not associated with subtle numerical problems. Usually the filter's inability to work properly has more to do with mundane issues, such as how the filter was formulated or how design parameters were chosen. Simple programming mistakes rather than computer round-off errors often account for a majority of the initial errors encountered in debugging a Kalman filter on today's computers.

An attempt has been made to make this book as readable as possible. To make life easier for the reader, all of the background material required for under-standing the text's concepts is presented in the first chapter. In addition, numerous examples are presented in each of the chapters in order to illustrate all of the concepts presented in the book. It is our hope that the text will be understandable and useful to readers with different backgrounds.

Chapter 1 introduces all of the numerical techniques that are used throughout the text. Examples of matrix manipulation, which are necessary for implementing the Kalman filter, are presented, and subroutines automating their implementation in computer programs are provided. These subroutines are used in the computer programs scattered throughout the text. Numerical methods for integrating differential equations are discussed not only because they are important for solving our mathematical models of the real world but also because they are useful for projecting states forward in the Kalman filter. A brief discussion of random phenomenon, as applied to understanding Monte Carlo analysis and interpreting noise-driven simulation results, is presented. Finally state-space notation, which is required for modeling the real world in formulating a Kalman filter, is simply introduced as a technique for representing differential equations.

Chapter 2 introduces Gauss's method of least-squares as applied to measurements that can be described by a polynomial. All of the filtering techniques of the text are based on this classical technique. The least-squares method introduced in Chapter 2 is a batch processing technique in that all the data must first be collected before estimates can be made. The various properties of the method of least squares when applied to measurements that are polynomial signals corrupted by noise are demonstrated by numerous examples. Finally, formulas that can be used to solve for the coefficients of different least-squares polynomial fits to measurement data are presented.

Chapter 3 shows how the batch processing method of least squares can be made recursive. The recursive least-squares filter is ideal for computer implementation because estimates can be provided as soon as measurements are taken. Formulas used for the implementation of different order recursive least-squares filters are presented and demonstrated to be equivalent to the batch processing method of least squares presented in Chapter 2. Formulas are also presented that can be used to predict the errors in the estimate of a least-squares filter as a result of both measurement noise and truncation error. All of the formulas presented in this chapter are empirically validated by numerous simulation experiments.

Chapter 4 presents the general equations for a discrete Kalman filter. It is shown, via simulation experiment, that if the measurement data can be represented by a polynomial signal corrupted by zero mean Gaussian noise the discrete Kalman filter and recursive least-squares filter are identical. Several examples are presented illustrating the various possible implementations of a discrete Kalman filter for a tracking problem involving a high-speed falling object. It is shown in this chapter that superior estimation performance can be attained if the Kalman filter can make use of a priori information. Finally, it is demonstrated that the addition of process noise to the Kalman filter can sometimes be used as a tool for preventing filter divergence.

Chapter 5 considers problems in which the real world cannot be represented by a polynomial. Various polynomial and nonpolynomial Kalman filters are designed to illustrate various important concepts. The importance of having an accurate fundamental matrix for state propagation, but not for filter gain computation, is demonstrated via several experiments. It is also demonstrated in this chapter that if it is impossible to obtain an exact fundamental matrix then the addition of process noise to the filter can sometimes be used as a method for improving performance.

Chapter 6 introduces the equations for the continuous Kalman filter. Although the continuous Kalman filter is usually not implemented, it can be used as a vehicle for understanding the discrete Kalman filter from a classical point of view. The chapter first demonstrates, via simulation experiment, how the continuous and discrete polynomial Kalman filters are related. Steady-state solutions are then derived for the Kalman gains and covariance matrix elements of the continuous polynomial Kalman filter. These solutions can not only be used to predict performance of the filter but can also be used to derive transfer functions of the filter. The chapter shows how the bandwidth of the steady-state polynomial Kalman filter is related to the ratio of the process and measurement noise spectral densities.

Theoretically, the Kalman filter can only be used if the real world can be described by a set of linear differential equations. In practice, the model describing the real world is often nonlinear, and other techniques must be used if we desire to filter data and estimate states. Chapter 7 introduces the equations for the extended Kalman filter. A numerical example is presented involving estimation of the position and velocity of a high-speed falling object in the presence of drag. The example is similar to the one from Chapter 4, except the addition of drag, which is assumed to be known, makes the problem nonlinear. The example chosen illustrates the operation and characteristics of the extended Kalman filter. Usually the extended Kalman filter is more fragile than a linear Kalman filter. Examples are chosen to illustrate some of the things that can go wrong with the extended Kalman filter and some engineering fixes that can be used to improve filter performance.

Chapter 8 adds one more degree of complication to the high-speed falling object problem in that drag is now assumed to be unknown. In the new problem we have to estimate the position, velocity, and ballistic coefficient (or amount of drag) of the object. Two different extended Kalman filters, each of which has different states, are designed to illustrate filter design and performance tradeoffs. The use of a linear polynomial Kalman filter is also explored in this chapter for this nonlinear problem.

A projectile tracking problem is introduced in Chapter 9 to illustrate how different measurements from the same sensor can be incorporated into an extended Kalman filter. An extended Kalman filter is first designed in Cartesian coordinates, where the model of the real world is linear but the measurement model is nonlinear. Next, an alternative extended Kalman filter is designed in polar coordinates, where the equations modeling the real world are nonlinear but the measurement equations are linear. It is shown for the example considered that both filters performed equally well and it was really a matter of taste as to which one was preferred. It is also demonstrated that, for this example, various types of linear polynomial Kalman filters can also be built that have near identical performance to that of the extended Kalman filters. The advantage of the linear filter is that it is more robust, as is illustrated by an experiment in which the various linear and nonlinear filters are improperly initialized.

In all of the cases considered up until Chapter 10, the Kalman filter always worked. Chapter 10 chooses the example of building a Kalman filter to estimate the frequency of a sinusoid based on measurements of a sinusoidal signal contaminated by noise. This chapter demonstrates that simply choosing states and building an extended Kalman filter does not guarantee that it will actually work as expected if programmed correctly. Sometimes, some states that are being estimated are not observable. Numerous experiments are conducted, and various extended Kalman filters are designed, each of which has different states, to highlight some of the issues.

Chapter 11 considers a filtering example based on a simplified two-dimensional version of the GPS. Various extended Kalman-filtering options, including no filtering at all, are explored for determining the location of a receiver based on noisy range measurements from a satellite pair whose location is always known. The importance of satellite geometry is illustrated, and it is shown why an extended Kalman filter is better than a simple second-order recursive least-squares filter for this problem. The utility of using one or two satellites for this two-dimensional tracking problem is also explored. The value of adding states and process noise to the extended Kalman filter in tracking a receiver located on a vehicle traveling along an erratic path is also demonstrated.

Chapter 12 extends the simplified two-dimensional GPS example of the previous chapter by considering the case in which the range measurements from the receiver to the satellites are contaminated by a bias. It is demonstrated that if the satellite is causing the bias in the range measurement we can estimate the bias with an extended Kalman filter by taking range measurements from one satellite to a receiver whose location is precisely known. It is also shown that if the receiver, whose location is unknown, is causing the bias, then three satellites are required in two dimensions for an extended Kalman filter to estimate both the bias and receiver location.

Chapter 13 introduces the linearized Kalman filter. This type of Kalman filter is popular in applications where accurate nominal trajectory information is available on an object being tracked (i.e., satellite tracking). The equations for the linearized Kalman filter are stated, and two different numerical examples, taken from preceding chapters, illustrating their implementation are presented. It is demonstrated that when the nominal trajectory information, upon which the linearized Kalman filter is based, is accurate, wonderful estimates can be obtained from the linearized Kalman filter. However, it is also shown, via simulation experiments, that when the nominal trajectory information is inaccurate, the estimates from the linearized Kalman filter deteriorate. In fact, it is also demonstrated that sometimes significant errors in the nominal trajectory can render the linearized Kalman filter worthless.

Finally, Chapter 14 discusses various miscellaneous issues that were not covered in earlier chapters. First, Chapter 14 shows how filter performance degrades with decreasing signal-to-noise ratio and why the initialization of the filter states and the initial covariance matrix can be very important in reducing estimation errors under conditions of low signal-to-noise ratio. Second, the chapter shows how the performance of a polynomial Kalman filter behaves as a function of filter order when only a few measurements are available. It is demonstrated that, in general, lower-order filters provide better estimates more rapidly than higher-order filters. Third, the chapter shows how the Kalman-filter residual and its theoretical bounds can also be used as an indication of whether or not the filter is working properly. The residual test is important because it can be used both in simulations and in the real world. Fourth, an example is chosen in which a filter is intentionally designed where some states are not observable. Some nontheoretical but practical tests are suggested that can be used to determine the nonobservabilty of the states in question. Finally, the chapter introduces the reader to the concept of aiding. A simplified example is chosen in which there are two sensors, both of which do not work satisfactorily by themselves, but when combined work well under adverse circumstances.

Computer code, written in FORTRAN, accompanies all of the examples presented in the text in order to provide the interested reader with a vehicle for duplicating the examples presented and exploring issues beyond the scope of the text. The FORTRAN language was chosen for the code simply because that is the language most familiar to the authors (i.e., we are not trying to convert anyone). The computer listings will also clarify some issues or fine points, which may not be apparent from the equations or discussion of particular examples. The FORTRAN code also appears on a CD that is formatted for both Macintosh- and IBM-compatible personal computers. The CD is included with the text to save the interested reader precious time by not having to key in the code. For those with religious opposition to the FORTRAN language, the same code has also been converted on the CD to MATLAB®. Much of the FORTRAN code presented in the text can be abbreviated significantly by use of this very popular and powerful language. For those on a limited budget, the same code has also been converted on the CD to True BASIC. This low-cost interpreter (approximately $40) not only runs on Macintosh- and IBM-compatible computers but also on workstations.

It is our hope that the text will remove much of the mystique surrounding Kalman filtering and make the subject matter more accessible to readers with varied backgrounds and learning styles. Hopefully some readers and who have been turned off or even terrified of Kalman filters because of their perceived complexity will now be able to apply this popular filtering technique to their own problems. We believe that the text's pragmatic and nonintimidating approach to Kalman filtering will not only be of help to those trying to learn about this important subject but also may even be of value to some filtering experts.

Paul Zarchan and Howard Musoff
September 2000

Acknowledgments

Special thanks go to Norman Josephy, Professor of Mathematical Sciences at Bentley College, whose kind and constructive review of the first several chapters provided us with useful feedback and influenced other parts of the text. Informative conversations with Charles Stark Draper Laboratory, Inc., technical staff members Matthew Bottkol and Darold Riegsecker on important issues concerning extended Kalman filtering influenced several of the examples used. Without the superb technical environment of C.S. Draper Labs, the idea for such a text would not have been possible. We would also like to thank Rodger Williams of AIAA for helping us move this project forward as rapidly as possible.

Table of Contents

Chapter 1

Numerical Basics

Introduction

I N THIS chapter we will try to discuss all of the numerical techniques used throughout the text. We will begin with a discussion of vectors and matrices and illustrate various operations that we will need to know when we apply the Kalman-filtering equations. Next, we will show how two different numerical integration techniques can be used to solve both linear and nonlinear differential equations. The numerical integration techniques are necessary when we must integrate differential equations representing the real world in simulations made for evaluating the performance of Kalman filters. In addition, numerical integration techniques are sometimes required to propagate states from nonlinear differential equations. Next, we will review the basic concepts used in representing random phenomena. These techniques will be important in modeling measurement and process noise, which may be used to represent the real world and will also be used to add some uncertainty to the measurement inputs entering the Kalman filter. These concepts will also be useful in understanding how to apply Monte Carlo techniques for evaluating Kalman-filter performance. State-space notation will be introduced as a shorthand notation for representing differential equations. Models of the real world must be placed in state-space notation before Kalman filtering can be applied. Finally, methods for computing the fundamental matrix will be presented. The fundamental matrix is required before a Kalman filter can be put into digital form.

Simple Vector Operations[1]

An array of elements arranged in a column is known as a column vector. The number of elements in the column vector is known as the dimension of the column vector. For example, an n-dimensional column vector x can be expressed as

$$x = \begin{bmatrix} x_1 \\ x_2 \\ x_3 \\ \vdots \\ x_n \end{bmatrix}$$

1

where $x_1, x_2, x_3, \ldots, x_n$ are the elements of the column vector. An example of a three-dimensional column vector r is given by

$$r = \begin{bmatrix} 5 \\ 7 \\ 2 \end{bmatrix}$$

Two vectors x and y can be added or subtracted by either adding or subtracting each of their elements or

$$x + y = \begin{bmatrix} x_1 \\ x_2 \\ x_3 \\ \vdots \\ x_n \end{bmatrix} + \begin{bmatrix} y_1 \\ y_2 \\ y_3 \\ \vdots \\ y_n \end{bmatrix} = \begin{bmatrix} x_1 + y_1 \\ x_2 + y_2 \\ x_3 + y_3 \\ \vdots \\ x_n + y_n \end{bmatrix}$$

$$x - y = \begin{bmatrix} x_1 \\ x_2 \\ x_3 \\ \vdots \\ x_n \end{bmatrix} - \begin{bmatrix} y_1 \\ y_2 \\ y_3 \\ \vdots \\ y_n \end{bmatrix} = \begin{bmatrix} x_1 - y_1 \\ x_2 - y_2 \\ x_3 - y_3 \\ \vdots \\ x_n - y_n \end{bmatrix}$$

As a numerical example, consider the case where r is the three-dimensional column vector previously defined and s is another three-dimensional column vector given by

$$s = \begin{bmatrix} -2 \\ 6 \\ 4 \end{bmatrix}$$

Then the addition of r and s is

$$r + s = \begin{bmatrix} 5 \\ 7 \\ 2 \end{bmatrix} + \begin{bmatrix} -2 \\ 6 \\ 4 \end{bmatrix} = \begin{bmatrix} 3 \\ 13 \\ 6 \end{bmatrix}$$

whereas the subtraction of r and s yields

$$r - s = \begin{bmatrix} 5 \\ 7 \\ 2 \end{bmatrix} - \begin{bmatrix} -2 \\ 6 \\ 4 \end{bmatrix} = \begin{bmatrix} 7 \\ 1 \\ -2 \end{bmatrix}$$

The transpose of a column vector is a row vector. For example, if

$$x = \begin{bmatrix} x_1 \\ x_2 \\ x_3 \\ \vdots \\ x_n \end{bmatrix}$$

then the transpose of x or x^T is given by

$$x^T = \begin{bmatrix} x_1 \\ x_2 \\ x_3 \\ \vdots \\ x_n \end{bmatrix}^T = [x_1 \quad x_2 \quad x_3 \quad \cdots \quad x_n]$$

Similarly, if z was a row vector given by

$$z = [z_1 \quad z_2 \quad z_3 \quad \cdots \quad z_n]$$

then its transpose would be a column vector or

$$z^T = \begin{bmatrix} z_1 \\ z_2 \\ z_3 \\ \vdots \\ z_n \end{bmatrix}$$

Using our preceding example, we can see that the transpose of column vector r is given by

$$r^T = [5 \quad 7 \quad 2]$$

Simple Matrix Operations[1]

A matrix is an array of elements consisting of m rows and n columns. In this case the dimension of the matrix is denoted $m \times n$. For example, the $m \times n$ A matrix is given by

$$A = \begin{bmatrix} a_{11} & a_{12} & a_{13} & \cdots & a_{1n} \\ a_{21} & a_{22} & a_{23} & \cdots & a_{2n} \\ a_{31} & a_{32} & a_{33} & \cdots & a_{3n} \\ \vdots & \vdots & \vdots & \cdots & \vdots \\ a_{m1} & a_{m2} & a_{m3} & \cdots & a_{mn} \end{bmatrix}$$

An m-dimensional column vector can also be thought of as an $m \times 1$ matrix, whereas an m-dimensional row vector can be thought of as a $1 \times m$ matrix. If the number of rows equals the number of columns in a matrix, the matrix is known as a square matrix. An example of a 3×3 square matrix is

$$R = \begin{bmatrix} -1 & 6 & 2 \\ 3 & 4 & -5 \\ 7 & 2 & 8 \end{bmatrix}$$

Sometimes we refer to the diagonal elements of a square matrix. In the case of the preceding square matrix R, the diagonal elements are $-1, 4$, and 8.

Matrix addition and subtraction are only defined when the matrices involved have the same dimensions. In this case we simply add or subtract corresponding elements. Let us consider the 3×2 matrices S and T given by

$$S = \begin{bmatrix} 2 & -6 \\ 1 & 5 \\ -2 & 3 \end{bmatrix}$$

$$T = \begin{bmatrix} 9 & 1 \\ -7 & 2 \\ 5 & 8 \end{bmatrix}$$

Then the addition of S and T is given by

$$S + T = \begin{bmatrix} 2 & -6 \\ 1 & 5 \\ -2 & 3 \end{bmatrix} + \begin{bmatrix} 9 & 1 \\ -7 & 2 \\ 5 & 8 \end{bmatrix} = \begin{bmatrix} 11 & -5 \\ -6 & 7 \\ 3 & 11 \end{bmatrix}$$

All of the listings presented in this text are written in FORTRAN 77. Although this language may not be popular with computer language gourmets, it is ideal for number crunching. The language was chosen because the authors are most familiar with it and because it yields programs that run very quickly because FORTRAN is compiled. As was mentioned in the preface, all of the code has also been converted to the more popular MATLAB® language and to True BASIC. FORTRAN, MATLAB®, and True BASIC source code language versions of all of the text's programs are provided with this text on a CD that can be read by both IBM- and Macintosh-compatible personal computers.

We can write a FORTRAN program to get the same answers as the preceding example. Listing 1.1 presents the program that performs the matrix addition and uses the matrix addition subroutine MATADD that is highlighted in bold. The statement calling this subroutine is also highlighted in bold. With this calling statement we simply specify the two matrices to be added along with their dimensions (i.e., 3,2 in this example means 3 rows and 2 columns). When the matrices are added, their sum is placed in matrix U, whose elements are printed out using the WRITE statement, and yields the same answers as just stated for

Listing 1.1   Program that contains subroutine to perform matrix addition

```
IMPLICIT REAL*8 (A-H)
IMPLICIT REAL*8 (O-Z)
REAL*8 S(3,2),T(3,2),U(3,2)
S(1,1)=2.
S(1,2)=-6.
S(2,1)=1.
S(2,2)=5.
S(3,1)=-2.
S(3,2)=3.
T(1,1)=9.
T(1,2)=1.
T(2,1)=-7.
T(2,2)=2.
T(3,1)=5.
T(3,2)=8.
CALL MATADD(S,3,2,T,U)
WRITE(9,*)U(1,1),U(1,2)
WRITE(9,*)U(2,1),U(2,2)
WRITE(9,*)U(3,1),U(3,2)
PAUSE
END

SUBROUTINE MATADD(A,IROW,ICOL,B,C)
IMPLICIT REAL*8 (A-H)
IMPLICIT REAL*8 (O-Z)
REAL*8 A(IROW,ICOL),B(IROW,ICOL),C(IROW,ICOL)
DO 120 I=1,IROW
DO 120 J=1,ICOL
      C(I,J)=A(I,J)+B(I,J)
120 CONTINUE
RETURN
END
```

$S + T$. The subroutine MATADD will be used in programs throughout the text. We can also see from the listing that everything is in double precision (i.e., IMPLICIT REAL*8). Although double precision arithmetic is not required for matrix addition, we will use it to avoid potential numerical problems in Kalman filtering. Therefore, all of the programs presented in this text will use double precision arithmetic.

If we want to subtract T from S, we get

$$S - T = \begin{bmatrix} 2 & -6 \\ 1 & 5 \\ -2 & 3 \end{bmatrix} - \begin{bmatrix} 9 & 1 \\ -7 & 2 \\ 5 & 8 \end{bmatrix} = \begin{bmatrix} -7 & -7 \\ 8 & 3 \\ -7 & -5 \end{bmatrix}$$

We can also write a FORTRAN program to get the same answers as the preceding matrix subtraction example. Listing 1.2 presents the program that performs the matrix subtraction and uses the matrix subtraction subroutine MATSUB that is highlighted in bold. The statement calling this subroutine is also highlighted in bold. With this calling statement we simply specify the two matrices to be subtracted along with their dimensions. When the matrices are subtracted the difference is placed in matrix U, which is printed out and yields the same answers as above for $S - T$. The subroutine MATSUB will also be used in programs throughout the text.

When taking the transpose of a matrix, we simply interchange the rows and columns of the matrix (i.e., first row of original matrix becomes first column of

Listing 1.2   Program that contains subroutine to perform matrix subtraction

```
        IMPLICIT REAL*8 (A-H)
        IMPLICIT REAL*8 (O-Z)
        REAL*8  S(3,2),T(3,2),U(3,2)
        S(1,1)=2.
        S(1,2)=-6.
        S(2,1)=1.
        S(2,2)=5.
        S(3,1)=-2.
        S(3,2)=3.
        T(1,1)=9.
        T(1,2)=1.
        T(2,1)=-7.
        T(2,2)=2.
        T(3,1)=5.
        T(3,2)=8.
        CALL  MATSUB(S,3,2,T,U)
        WRITE(9,*)U(1,1),U(1,2)
        WRITE(9,*)U(2,1),U(2,2)
        WRITE(9,*)U(3,1),U(3,2)
        PAUSE
        END

        SUBROUTINE  MATSUB(A,IROW,ICOL,B,C)
        IMPLICIT REAL*8 (A-H)
        IMPLICIT REAL*8 (O-Z)
        REAL*8  A(IROW,ICOL),B(IROW,ICOL),C(IROW,ICOL)
        DO 120  I=1,IROW
        DO 120  J=1,ICOL
                C(I,J)=A(I,J)-B(I,J)
120     CONTINUE
        RETURN
        END
```

transposed matrix, second row of original matrix becomes second column of transposed matrix, etc.). For example the transpose of S is

$$S^T = \begin{bmatrix} 2 & -6 \\ 1 & 5 \\ -2 & 3 \end{bmatrix}^T = \begin{bmatrix} 2 & 1 & -2 \\ -6 & 5 & 3 \end{bmatrix}$$

We can also write a program to get the same answers as the preceding matrix transpose example. Listing 1.3 presents the FORTRAN program that performs the matrix transpose and uses the matrix transpose subroutine MATTRN, which is highlighted in bold. In calling this subroutine we simply specify the matrix to be transposed along with its dimensions. When the matrix is transposed, the results are placed in matrix ST, which is printed out and yields the same answers as just stated for S^T. The dimensions of matrix ST are 2×3, whereas the dimensions of matrix S are 3×2. The subroutine MATTRN will also be used in programs throughout the text.

Listing 1.3 Program that contains subroutine to take a matrix transpose

```
IMPLICIT REAL*8 (A-H)
IMPLICIT REAL*8 (O-Z)
REAL*8 S(3,2),ST(2,3)
S(1,1)=2.
S(1,2)=-6.
S(2,1)=1.
S(2,2)=5.
S(3,1)=-2.
S(3,2)=3.
CALL MATTRN(S,3,2,ST)
WRITE(9,*)ST(1,1),ST(1,2),ST(1,3)
WRITE(9,*)ST(2,1),ST(2,2),ST(2,3)
PAUSE
END

SUBROUTINE MATTRN(A,IROW,ICOL,AT)
IMPLICIT REAL*8 (A-H)
IMPLICIT REAL*8 (O-Z)
REAL*8 A(IROW,ICOL),AT(ICOL,IROW)
DO 105 I=1,IROW
DO 105 J=1,ICOL
AT(J,I)=A(I,J)
105 CONTINUE
RETURN
END
```

A square matrix A is said to be symmetric if the matrix equals its transpose or

$$A = A^T$$

In other words, the matrix

$$\begin{bmatrix} 1 & 2 & 3 \\ 2 & 5 & 6 \\ 3 & 6 & 9 \end{bmatrix}$$

is symmetric because

$$\begin{bmatrix} 1 & 2 & 3 \\ 2 & 5 & 6 \\ 3 & 6 & 9 \end{bmatrix}^T = \begin{bmatrix} 1 & 2 & 3 \\ 2 & 5 & 6 \\ 3 & 6 & 9 \end{bmatrix}$$

At the beginning of this section, we defined a general matrix A with m rows and n columns. The matrix A can only multiply a matrix B if the matrix B has n rows and q columns. If the condition is satisfied, then we multiply each element of the rows of matrix A with each element of the columns of matrix B, resulting in a new matrix with m rows and q columns. The method is easier to illustrate by example than to describe theoretically. For example, we can multiply the matrices already defined R (i.e., 3×3 matrix) with S (i.e., 3×2 matrix) and get

$$
\begin{aligned}
RS &= \begin{bmatrix} -1 & 6 & 2 \\ 3 & 4 & -5 \\ 7 & 2 & 8 \end{bmatrix} \begin{bmatrix} 2 & -6 \\ 1 & 5 \\ -2 & 3 \end{bmatrix} \\
&= \begin{bmatrix} -1*2 + 6*1 + 2*(-2) & -1*(-6) + 6*5 + 2*3 \\ 3*2 + 4*1 - 5*(-2) & 3*(-6) + 4*5 - 5*3 \\ 7*2 + 2*1 + 8*(-2) & 7*(-6) + 2*5 + 8*3 \end{bmatrix} \\
&= \begin{bmatrix} 0 & 42 \\ 20 & -13 \\ 0 & -8 \end{bmatrix}
\end{aligned}
$$

The preceding example also illustrates that when a 3×3 matrix multiplies a 3×2 matrix, we get a 3×2 matrix.

We can also write a FORTRAN program to get the same answers as the preceding matrix multiplication example. Listing 1.4 presents the program that performs the matrix multiplication and uses the matrix multiply subroutine MATMUL that is highlighted in bold. In calling this subroutine we simply specify the two matrices R and S to be multiplied along with their dimensions. When the matrices are multiplied, the product is placed in matrix RS. The subroutine MATMUL will also be used in programs throughout the text.

We can also multiply a matrix by a scalar. In that case we simply multiply each element of the matrix by the scalar. For example, if the matrix S is given by

$$S = \begin{bmatrix} 8 & 2 & -1 \\ 3 & 7 & 6 \\ 4 & 1 & 7 \end{bmatrix}$$

we want to multiply it by the constant or scalar 10 to get the new matrix T or

$$T = 10S = 10 \begin{bmatrix} 8 & 2 & -1 \\ 3 & 7 & 6 \\ 4 & 1 & 7 \end{bmatrix}$$

After element-by-element multiplication, we obtain

$$T = \begin{bmatrix} 80 & 20 & -10 \\ 30 & 70 & 60 \\ 40 & 10 & 70 \end{bmatrix}$$

We can also write a FORTRAN program to get the same answers as the preceding scalar matrix multiply example. Listing 1.5 presents the program that performs the scalar matrix multiplication and uses the matrix multiply subroutine MATSCA that is highlighted in bold. In calling this subroutine we simply specify the matrix to be multiplied S, along with the dimensions and the constant XK. When each of the matrix elements and scalar are multiplied, the product is placed in matrix T. The subroutine MATSCA will also be used, when required, in a few of the programs in the text.

The identity matrix is simply a square matrix whose diagonal elements are unity and off diagonal elements are zero. For example, the 2×2 identity matrix is

$$\begin{bmatrix} 1 & 0 \\ 0 & 1 \end{bmatrix}$$

whereas the 3×3 identity matrix is given by

$$\begin{bmatrix} 1 & 0 & 0 \\ 0 & 1 & 0 \\ 0 & 0 & 1 \end{bmatrix}$$

The inverse of a square matrix A is denoted A^{-1}. When a matrix is multiplied by its inverse, we get the identity matrix. In other words,

$$AA^{-1} = I$$

where I is the identity matrix. The inverse of a scalar is the reciprocal of the scalar. Taking the inverse of a matrix involves concepts such as determinants and

**Listing 1.4   Program that contains subroutine to perform
matrix multiplication**

```
IMPLICIT  REAL*8  (A-H)
IMPLICIT  REAL*8  (O-Z)
REAL*8  S(3,2),R(3,3),RS(3,2)
S(1,1)=2.
S(1,2)=-6.
S(2,1)=1.
S(2,2)=5.
S(3,1)=-2.
S(3,2)=3.
R(1,1)=-1.
R(1,2)=6.
R(1,3)=2.
R(2,1)=3.
R(2,2)=4.
R(2,3)=-5.
R(3,1)=7.
R(3,2)=2.
R(3,3)=8.
CALL  MATMUL(R,3,3,S,3,2,RS)
WRITE(9,*)RS(1,1),RS(1,2)
WRITE(9,*)RS(2,1),RS(2,2)
WRITE(9,*)RS(3,1),RS(3,2)
PAUSE
END

SUBROUTINE  MATMUL(A,IROW,ICOL,B,JROW,JCOL,C)
IMPLICIT  REAL*8  (A-H)
IMPLICIT  REAL*8  (O-Z)
REAL*8  A(IROW,ICOL),B(JROW,JCOL),C(IROW,JCOL)
DO  110  I=1,IROW
DO  110  J=1,JCOL
        C(I,J)=0.
        DO  110  K=1,ICOL
            C(I,J)=  C(I,J)+A(I,K)*B(K,J)
110  CONTINUE
    RETURN
    END
```

cofactors. Because in this text we will only be interested the inverses of square matrices with small dimensions, we will simply present the formulas for the inverses without any derivation. Readers who are interested in how these formulas were derived are referred to any of the numerous texts on linear algebra. In more advanced applications of Kalman filtering, where it may be necessary to take the inverse of matrices with large dimensions, numerical techniques for taking the inverses rather than the formulas are used. The formulas for the matrix inverse will only be important for those using the FORTRAN language (i.e.,

Listing 1.5   Program that contains subroutine to perform a scalar matrix multiplication

```
IMPLICIT REAL*8 (A-H)
IMPLICIT REAL*8 (O-Z)
REAL*8 S(3,3),T(3,3)
XK=10.
S(1,1)=8.
S(1,2)=2.
S(1,3)=-1.
S(2,1)=3.
S(2,2)=7
S(2,3)=6.
S(3,1)=4.
S(3,2)=1.
S(3,3)=7.
CALL MATSCA(S,3,3,XK,T)
WRITE(9,*)T(1,1),T(1,2),T(1,3)
WRITE(9,*)T(2,1),T(2,2),T(2,3)
WRITE(9,*)T(3,1),T(3,2),T(3,3)
PAUSE
END

SUBROUTINE MATSCA(A,IROW,ICOL,B,C)
IMPLICIT REAL*8 (A-H)
IMPLICIT REAL*8 (O-Z)
REAL*8 A(IROW,ICOL),C(IROW,ICOL)
DO 110 I=1,IROW
DO 110 J=1,ICOL
    C(I,J)=B*A(I,J)
110 CONTINUE
RETURN
END
```

FORTRAN 77). Users of both MATLAB® and True BASIC can use powerful statements to automatically take the inverse of a matrix of any dimension.

If A is a 2×2 matrix denoted by

$$A = \begin{bmatrix} a & b \\ c & d \end{bmatrix}$$

the inverse of A can be shown to be[2]

$$A^{-1} = \frac{1}{ad-bc} \begin{bmatrix} d & -b \\ -c & a \end{bmatrix}$$

If $ad = bc$, the matrix does not have an inverse. As a test of the preceding formula, let us consider the case where A is a 2×2 matrix given by

$$A = \begin{bmatrix} 2 & -4 \\ 1 & 3 \end{bmatrix}$$

Therefore, the inverse of A according to the formula for a 2×2 matrix is

$$A^{-1} = \frac{1}{2*3 - (-4)*1} \begin{bmatrix} 3 & 4 \\ -1 & 2 \end{bmatrix} = \frac{1}{10} \begin{bmatrix} 3 & 4 \\ -1 & 2 \end{bmatrix} = \begin{bmatrix} 0.3 & 0.4 \\ -0.1 & 0.2 \end{bmatrix}$$

As a check, we can multiply A^{-1} with A and get the identity matrix or

$$\begin{aligned} A^{-1}A &= \begin{bmatrix} 0.3 & 0.4 \\ -0.1 & 0.2 \end{bmatrix} \begin{bmatrix} 2 & -4 \\ 1 & 3 \end{bmatrix} \\ &= \begin{bmatrix} 0.3*2 + 0.4*1 & 0.3*(-4) + 0.4*3 \\ -0.1*2 + 0.2*1 & -0.1*(-4) + 0.2*3 \end{bmatrix} = \begin{bmatrix} 1 & 0 \\ 0 & 1 \end{bmatrix} \end{aligned}$$

If A is a 3×3 matrix given by

$$A = \begin{bmatrix} a & b & c \\ d & e & f \\ g & h & i \end{bmatrix}$$

the inverse of this matrix is given by[2]

$$A^{-1} = \frac{1}{aei + bfg + cdh - ceg - bdi - afh} \begin{bmatrix} ei - fh & ch - bi & bf - ec \\ gf - di & ai - gc & dc - af \\ dh - ge & gb - ah & ae - bd \end{bmatrix}$$

Again, if the denominator of the fraction multiplying the matrix in the preceding equation is zero, the inverse does not exist. As a test of the preceding formula, let us consider the case where A is a 3×3 matrix given by

$$A = \begin{bmatrix} 1 & 2 & 3 \\ 4 & 5 & 6 \\ 7 & 8 & 10 \end{bmatrix}$$

Therefore, we can find the inverse of A according to the preceding formula for a 3×3 matrix. First, we find the scalar multiplying the matrix as

$$\begin{aligned} &\frac{1}{aei + bfg + cdh - ceg - bdi - afh} \\ &= \frac{1}{1*5*10 + 2*6*7 + 3*4*8 - 3*5*7 - 2*4*10 - 1*6*8} \\ &= \frac{-1}{3} \end{aligned}$$

The matrix itself is given by

$$
\begin{bmatrix}
ei - fh & ch - bi & bf - ec \\
gf - di & ai - gc & dc - af \\
dh - ge & gb - ah & ae - bd
\end{bmatrix}
$$

$$
= \begin{bmatrix}
5*10 - 6*8 & 3*8 - 2*10 & 2*6 - 5*3 \\
7*6 - 4*10 & 1*10 - 7*3 & 4*3 - 1*6 \\
4*8 - 7*5 & 7*2 - 1*8 & 1*5 - 2*4
\end{bmatrix}
$$

$$
= \begin{bmatrix}
2 & 4 & -3 \\
2 & -11 & 6 \\
-3 & 6 & -3
\end{bmatrix}
$$

Therefore, the inverse of A can be computed as

$$
A^{-1} = \frac{-1}{3}\begin{bmatrix}
2 & 4 & -3 \\
2 & -11 & 6 \\
-3 & 6 & -3
\end{bmatrix} = \begin{bmatrix}
-2/3 & -4/3 & 1 \\
-2/3 & 11/3 & -2 \\
1 & -2 & 1
\end{bmatrix}
$$

As a check, we can multiply A^{-1} with A and get

$$
A^{-1}A = \begin{bmatrix}
-2/3 & -4/3 & 1 \\
-2/3 & 11/3 & -2 \\
1 & -2 & 1
\end{bmatrix}\begin{bmatrix}
1 & 2 & 3 \\
4 & 5 & 6 \\
7 & 8 & 10
\end{bmatrix}
$$

$$
= \begin{bmatrix}
-2/3 - 16/3 + 7 & -4/3 - 20/3 + 8 & -2 - 8 + 10 \\
-2/3 + 44/3 - 14 & -4/3 + 55/3 - 16 & -2 + 22 - 20 \\
1 - 8 + 7 & 2 - 10 + 8 & 3 - 12 + 10
\end{bmatrix}
$$

which is the identity matrix or

$$
A^{-1}A = \begin{bmatrix}
1 & 0 & 0 \\
0 & 1 & 0 \\
0 & 0 & 1
\end{bmatrix}
$$

As was already mentioned, there are a number of numerical techniques that can be used to invert square matrices of a higher order than three. Matrix inversion is part of the MATLAB® and True BASIC languages. Therefore, when using these languages, matrix inversion of any order can be achieved by using a one-line statement.

Numerical Integration of Differential Equations

Throughout this text we will be simulating both linear and nonlinear ordinary differential equations. Because, in general, these equations have no closed-form solutions, it will be necessary to resort to numerical integration techniques to solve or simulate these equations. Many numerical integration techniques exist for solving differential equations. Let us start with the simplest—Euler[3] integration.

Euler integration is a numerical integration technique that will work on most of the examples in this text. Consider a first-order differential equation of the form

$$\dot{x} = f(x, t)$$

We know from calculus that the definition of a derivative of a function can be approximated by

$$\dot{x} = f(x, t) = \frac{x(t + h) - x(t)}{h} = \frac{x_k - x_{k-1}}{h}$$

for small h. Rearranging terms yields

$$x_k = x_{k-1} + hf(x, t)$$

The preceding equation is known as Euler integration. With Euler integration the value of the state x at the next integration interval h is related to the earlier value of x plus a term proportional to the derivative evaluated at time t. For each recursion, time is incremented by the integration interval h. To start the numerical integration process, we need an initial value of x. This value comes from the initial conditions required for the differential equations.

We learned in calculus that integrating a function is equivalent to finding the area under the function when it is plotted. Euler integration is the simplest approximation available for computing the area under the curve represented by $f(x, t)$. It represents the sum of the areas of small rectangles having the height $f(x, t)$ and width h, as can be seen in Fig. 1.1. Of course using small rectangles to approximate the curve will yield integration errors, but these errors can be made smaller by decreasing the width of each of the rectangles. Decreasing the width of the rectangles is equivalent to making the integration step size smaller.

The integration step size h must be small enough to yield answers of sufficient accuracy. A simple test, commonly practiced among engineers, is to find the appropriate integration step size by experiment. As a rule of thumb, the initial step size is chosen to be several times smaller than the smallest time constant in the system under consideration. The step size is then halved to see if the answers change significantly. If the new answers are approximately the same, the larger integration step size is used to avoid excessive computer running time. If the answers change substantially, then the integration interval is again halved, and the process is repeated.

To see how the Euler integration technique can be used to solve a differential equation, let us make up a problem for which we know the solution. If the solution to the differential equation (i.e., we have not yet made up the differential equation) for x is given by the sinusoid

$$x = \sin \omega t$$

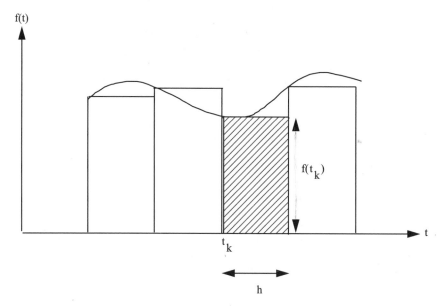

Fig. 1.1   Finding the area under a curve using rectangles is equivalent to Euler integration.

then differentiating once yields

$$\dot{x} = \omega \cos \omega t$$

and differentiating once again yields

$$\ddot{x} = -\omega^2 \sin \omega t$$

Recognizing that the last equation can be expressed in terms of the first, we obtain

$$\ddot{x} = -\omega^2 x$$

which is a linear second-order differential equation. The initial conditions for this differential equation can be obtained by setting time to zero in the first two equations (i.e., for x and its derivative), yielding

$$x(0) = 0$$
$$\dot{x}(0) = \omega$$

Again, we know in advance that the solution to the second-order linear differential equation with the two preceding initial conditions is simply

$$x = \sin \omega t$$

Let us see if the correct solution can also be obtained by numerical integration.

To check the preceding theoretical closed-form solution for x, a simulation involving numerical integration was written. The simulation of the second-order differential equation using the Euler integration numerical technique appears in Listing 1.6. First the second-order differential equation is presented. We solve for the first derivative of x, and next we solve for x. Time is then incremented by the integration interval h, and we repeat the process until time is greater than 10 s. We print out both the theoretical and simulated (i.e., obtained by Euler integration) values of x every 0.1 s.

The nominal case of Listing 1.6 was run, and the results appear in Fig. 1.2. Here we can see that for the integration step size chosen (i.e., $H = 0.01$) the Euler numerical integration technique accurately integrates the second-order differential equation because the exact and numerical solutions agree. As was already mentioned, because often we do not have exact solutions, we usually determine the correct integration interval to be used by experiment.

Sometimes the price paid for using Euler integration is less accuracy, compared to other methods of numerical integration, for a given integration

Listing 1.6   Simulation of second-order differential equation using Euler numerical integration

```
IMPLICIT  REAL*8(A-H)
IMPLICIT  REAL*8(O-Z)
OPEN(1,STATUS='UNKNOWN',FILE='DATFIL')
W=2.
T=0.
S=0.
X=0.
XD=W
H=.01
WHILE(T < =10.)
        S=S+H
        XDD=-W*W*X
        XD=XD+H*XDD
        X=X+H*XD
        T=T+H
        IF(S>=.09999)THEN
               S=0.
               XTHEORY=SIN(W*T)
               WRITE(9,*)T,X,XTHEORY
               WRITE(1,*)T,X,XTHEORY
        ENDIF
END  DO
PAUSE
CLOSE(1)
END
```

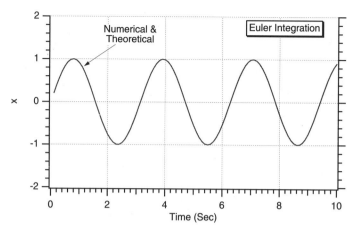

Fig. 1.2 Euler numerical technique accurately integrates second-order differential equation.

interval. Being prudent and using small integration step sizes is one way of getting accurate answers with Euler integration. However, sometimes Euler integration does not work at all, no matter how small the integration step size. For this reason we will now consider the more accurate second-order Runge–Kutta[4] numerical integration technique.

The second-order Runge–Kutta numerical integration procedure is easy to state. Given a first-order differential equation of the form

$$\dot{x} = f(x, t)$$

where t is time, we seek to find a recursive relationship for x as a function of time. With the second-order Runge–Kutta numerical technique the value of x at the next integration interval h is given by

$$x_k = x_{k-1} + 0.5h[f(x, t) + f(x, t + h)]$$

where the subscript $k - 1$ represents the last interval and k represents the new interval. From the preceding expression we can see that the new value of x is simply the old value of x plus a term proportional to the derivative evaluated at time t and another term with the derivative evaluated at time $t + h$. To start the integration, we need an initial value of x. As was already mentioned, this value comes from the initial conditions required for the differential equations.

If we look at Fig. 1.3, we can see that the second-order Runge–Kutta numerical integration is also a method for finding the area under a curve. However, rather than using small rectangles to approximate the curve we use small trapezoids. Because the area of a trapezoid is one-half the base times the sum of the two heights, we can see that the preceding integration formula for the second-order Runge–Kutta technique is really trapezoidal integration. Figure 1.3

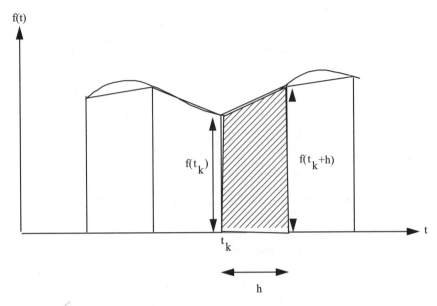

Fig. 1.3 Finding the area under a curve using trapezoids is equivalent to second-order Runge–Kutta integration.

also indicates that the area under the curve is better approximated by a trapezoid than by a rectangle. Therefore, the second-order Runge–Kutta method will be more accurate than Euler integration for a given integration step size.

A listing of the simulation, using the second-order Runge–Kutta integration technique, appears in Listing 1.7. We can see from Listing 1.7 that the differential equation or derivative information appears twice and is highlighted in bold. We evaluate the differential equation twice during the integration interval: once to evaluate the derivative at time t and once to evaluate the derivative at time $t + h$. We can also see from Listing 1.7 that every 0.1 s we print out the numerical solution to the differential equation X with the closed-form solution XTHEORY. In this particular example the natural frequency ω of the differential equation is 2 rad/s (i.e., $W = 2$).

The nominal case of Listing 1.7 was run, and the results appear in Fig. 1.4. Here we can see that for the integration step size chosen (i.e., $H = 0.01$) the second-order Runge–Kutta numerical integration technique accurately integrates the second-order differential equation because the actual and numerical solutions agree.

We will use the second-order Runge–Kutta technique throughout the text because it is simple to understand, easy to program, and, most importantly, yields accurate answers for all of the examples presented in this text. Therefore, all of the subsequent programs in this text involving numerical integration will have the same structure as Listing 1.7.

**Listing 1.7   Simulation of differential equation
using second-order Runge–Kutta numerical
integration ·**

```
IMPLICIT  REAL*8(A-H)
IMPLICIT  REAL*8(O-Z)
OPEN(1,STATUS='UNKNOWN',FILE='DATFIL')
W=2.
T=0.
S=0.
X=0.
XD=W
H=.01
WHILE(T<=10.)
        S=S+H
        XOLD=X
        XDOLD=XD
        XDD=-W*W*X
        X=X+H*XD
        XD=XD+H*XDD
        T=T+H
        XDD=-W*W*X
        X=.5*(XOLD+X+H*XD)
        XD=.5*(XDOLD+XD+H*XDD)
        IF(S>=.09999)THEN
                S=0.
                XTHEORY=SIN(W*T)
                WRITE(9,*)T,X,XTHEORY
                WRITE(1,*)T,X,XTHEORY
        ENDIF
END  DO
PAUSE
CLOSE(1)
END
```

Noise and Random Variables[2,5]

In this section we will start by defining some important quantities related to random variables. Because random variables have unknown specific values, they are usually quantified according to their statistical properties. One of the most important statistical properties of any random function x is its probability density function $p(x)$. This function is defined such that

$$p(x) \geq 0$$

and

$$\int_{-\infty}^{\infty} p(x)\,\mathrm{d}x = 1$$

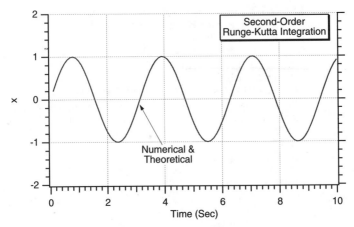

Fig. 1.4 Second-order Runge–Kutta numerical technique accurately integrates differential equation.

This means that there is a probability that x will occur, and it is certain that the value of x is somewhere between plus and minus infinity. The probability that x is between a and b can be expressed in terms of the probability density function as

$$\text{Prob}(a \leq x \leq b) = \int_a^b p(x)\, dx$$

Probability also can be viewed as the number of occurrences of a particular event divided by the total number of events under consideration.

Another important quantity related to random variables is the distribution function. A distribution function $P(x)$ is the probability that a random variable is less than or equal to x. Therefore, if the probability density function is known, the distribution function can be found by integration as

$$P(x) = \int_{-\infty}^x p(u)\, du$$

The mean or expected value of x is defined by

$$m = E(x) = \int_{-\infty}^\infty x p(x)\, dx$$

Therefore, the mean can also be thought of as the first moment of x. We can also think of the mean value of x as the sum (integral) of all values of x, each being weighted by its probability of occurrence. It can be shown that the expectation of the sum is the sum of the expectations, or

$$E(x_1 + x_2 + \cdots + x_n) = E(x_1) + E(x_2) + \cdots + E(x_n)$$

The second moment or mean squared value of x is defined as

$$E(x^2) = \int_{-\infty}^{\infty} x^2 p(x)\, dx$$

Therefore, the root mean square or rms of x can be obtained by taking the square root of the preceding equation, or

$$rms = \sqrt{E(x^2)}$$

The variance of x, σ^2, is defined as the expected squared deviation of x from its mean value. Mathematically, the variance can be expressed as

$$\sigma^2 = E\{[x - E(x)]^2\} = E(x^2) - E^2(x)$$

We can see that the variance is the difference between the mean squared value of x and the square of the mean of x. If we have independent random variables $x_1, x_2, \ldots, x_n$, then the variance of the sum can be shown to be the sum of the variances, or

$$\sigma^2 = \sigma_1^2 + \sigma_2^2 + \cdots + \sigma_n^2$$

The square root of the variance σ is also known as the standard deviation. In general, the rms value and standard deviation are not the same unless the random process under consideration has a zero mean.

An example of a probability density function is the uniform distribution, which is depicted in Fig. 1.5. With this probability density function all values of x between a and b are equally likely to occur.

An important practical example of a function that computes random values distributed according to the uniform distribution, which should be familiar to any engineer who has programmed on a personal computer, is the True BASIC language random number generator RND. The True BASIC RND statement supplies a uniformly distributed random number, on each call, between 0 and 1. MATLAB® and many extensions of the FORTRAN language also have their own uniform random number generators. Soon we will see how random numbers with different probability density functions can be constructed from random numbers following the uniform distribution.

From our earlier definitions, and using Fig. 1.5, we can see that the mean value of a uniform distribution is

$$m = E(x) = \int_{-\infty}^{\infty} x p(x)\, dx = \frac{1}{b-a} \int_a^b x\, dx = \frac{b+a}{2}$$

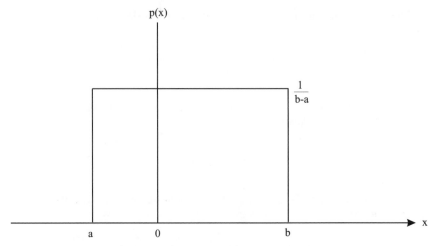

Fig. 1.5　Uniform probability distribution.

This makes sense, because the expected or mean value is halfway between a and b. The variance of a uniform distribution also can be found from our earlier definitions and can be shown to be

$$\sigma^2 = E(x^2) - E^2(x) = \frac{b^3 - a^3}{3(b-a)} - \left(\frac{b+a}{2}\right)^2 = \frac{(b-a)^2}{12}$$

This means that if the random numbers from a uniform distribution vary from 0 to 1, then the mean of the resultant set of numbers should be $\frac{1}{2}$ and the variance should be $\frac{1}{12}$. We will use this property of a uniform distribution in the next section for constructing random numbers with different probability density functions.

Another important probability density function is the Gaussian or normal distribution. In this text zero mean noise following a Gaussian distribution usually will be used to corrupt sensor measurements. The probability density function for Gaussian distribution is shown in Fig. 1.6 and is given by the formula

$$p(x) = \frac{\exp[-(x-m)^2/2\sigma^2]}{\sigma\sqrt{2\pi}}$$

where m and σ are parameters. By using our basic definitions, it is easy to show that the expected or mean value of a Gaussian distribution is given by

$$E(x) = m$$

and its variance is

$$E(x^2) - m^2 = \sigma^2$$

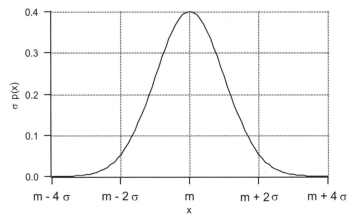

Fig. 1.6   Gaussian or normal probability density function.

Therefore, m and σ in the expression for the Gaussian probability density function correspond to the mean and standard deviation, respectively.

We can see from Fig. 1.6 that this bell-shaped distribution is virtually zero after three standard deviations ($\pm 3\sigma$). Integration of the probability density function, to find the distribution function, shows that there is a 68% probability that the Gaussian random variable is within one standard deviation ($\pm \sigma$) of the mean, 95% probability it is within two standard deviations of the mean, and 99% probability that it is within three standard deviations of the mean.

It can be shown that the resultant probability density function of a sum of Gaussian-distributed random variables is also Gaussian. In addition, under certain circumstances it also can be shown that the sum of independent random variables, regardless of individual density function, tends toward Gaussian as the number of random variables gets larger. This fact is a consequence of the central limit theorem, and an illustration of this phenomenon is presented in the next section. That is in fact why so many random variables are Gaussian distributed.

Gaussian Noise Example[5]

To simulate noise or random events, we have to know how to generate, via the computer, pseudorandom numbers with the appropriate probability density function. The FORTRAN language does not come with a random number generator. However, many microcomputer implementations of FORTRAN provide extensions from which noise, with the desired probability density function, can be constructed. It can be shown from the central limit theorem that the addition of many uniformly distributed variables produces a Gaussian-distributed variable.

The first step in constructing random numbers with the desired probability density function is to normalize the uniform noise generator so that random numbers between -0.5 and 0.5 are produced. The call to the Absoft FORTRAN random number generator Random() produces a uniformly distributed random number ranging from $-32{,}768$ to $32{,}768$ (i.e., -2^{15} to $+2^{15}$). Dividing the resultant random numbers by 65,536 will yield uniformly distributed random numbers between -0.5 and $+0.5$. Therefore, if we add 12 uniformly distributed random variables, we will obtain a zero-mean Gaussian variable with unity standard deviation. (Because the variance of one uniformly distributed random variable is $\frac{1}{12}$, the variance of 12 of them must be 1.) If we add six uniformly generated random numbers, we will still get a Gaussian variable, but the variance will only be $\frac{1}{2}$. Therefore, in this case we have to multiply the resultant random variable by the square root of 2, as shown in the listing, to get the right standard deviation (i.e., the variance is the square of the square root of 2). The GAUSS subroutine is used as part of Listing 1.8 in order to generate 1000 Gaussian-distributed random numbers with zero mean and unity variance. The GAUSS

Listing 1.8   Simulation of a Gaussian random number generator in FORTRAN

```
C THE FIRST THREE STATEMENTS INVOKE THE ABSOFT RANDOM
  NUMBER GENERATOR ON THE MACINTOSH
    GLOBAL DEFINE
              INCLUDE 'quickdraw.inc'
    END
    IMPLICIT REAL*8 (A-H)
    IMPLICIT REAL*8 (O-Z)
    SIGNOISE=1.
    OPEN(1,STATUS='UNKNOWN',FILE='DATFIL')
    DO 10 I=1,1000
    CALL GAUSS(X,SIGNOISE)
    WRITE(9,*)I,X
    WRITE(1,*)I,X
 10 CONTINUE
    CLOSE(1)
    PAUSE
    END

    SUBROUTINE GAUSS(X,SIG)
    IMPLICIT REAL*8(A-H)
    IMPLICIT REAL*8(O-Z)
    INTEGER SUM
    SUM=0
    DO 14 J=1,6
C THE NEXT STATEMENT PRODUCES A UNIF. DISTRIBUTED NUMBER
  FROM -32768 TO +32768
    IRAN=Random()
    SUM=SUM+IRAN
 14 CONTINUE
    X=SUM/65536
    X=1.414*X*SIG
    RETURN
    END
```

subroutine, which is highlighted in bold, also will be used on all other FORTRAN programs in this text requiring the generation of Gaussian-distributed random numbers. MATLAB® can generate Gaussian-distributed random numbers with a single statement.

Figure 1.7 displays the values of each of the 1000 Gaussian random numbers, generated via the program of Listing 1.8, in graphic form. If a Gaussian distribution has a standard deviation of unity, we expect 99% of the Gaussian random numbers to be between −3 and +3 for a peak-to-peak spread of 6. Figure 1.7 illustrates that the mean of the resultant random numbers is indeed about zero and the approximate value of the standard deviation can be eyeballed as

$$\sigma_{\text{APPROX}} \approx \frac{\text{Peak to Peak}}{6} \approx \frac{6}{6} = 1$$

which is the theoretically correct value.

To get an idea of the resultant probability density function of the computer-generated 1000 random numbers, another FORTRAN program was written and appears in Listing 1.9. The program calculates the histogram of the sample data. Each random number is placed in a bin in order to calculate the frequency of occurrence and hence the probability density function. Also included in the listing, for comparative purposes, is the theoretical formula for the probability density function of a zero-mean, unity variance Gaussian distribution.

Figure 1.8 presents the calculated probability density function, generated from the random numbers of Listing 1.9, in graphic form. Superimposed on the figure is a plot of the theoretical Gaussian distribution. The figure indicates that, with a sample size of 1000 random numbers, the computer-generated probability density function approximately follows the theoretical bell-shaped curve of a Gaussian distribution. Figure 1.9 shows that when 5000 random numbers are used the match between the calculated and theoretical Gaussian distributions is even better.

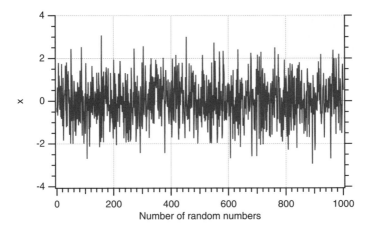

Fig. 1.7   One thousand random numbers with Gaussian distribution.

Listing 1.9   Program used to generate probability density function

```
C THE FIRST THREE STATEMENTS INVOKE THE ABSOFT RANDOM
  NUMBER GENERATOR ON THE MACINTOSH
  GLOBAL DEFINE
            INCLUDE 'quickdraw.inc'
  END
  IMPLICIT REAL*8 (A-H)
  IMPLICIT REAL*8 (O-Z)
  INTEGER BIN
  REAL*8 H(2000),X(2000)
  OPEN(1,STATUS='UNKNOWN',FILE='DATFIL')
  XMAX=6.
  XMIN=-6.
  SIGNOISE=1.
  RANGE=XMAX-XMIN
  TMP=1./SQRT(6.28)
  BIN=50
  N=1000
  DO 10 I=1,N
  CALL GAUSS(Y,SIGNOISE)
  X(I)=Y
10 CONTINUE
  DO 20 I=1,BIN
  H(I)=0
20 CONTINUE
  DO 30 I=1,N
  K=INT((((X(I)-XMIN)/RANGE)*BIN)+.99
  IF(K < 1)K=1
  IF(K>BIN)K=BIN
  H(K)=H(K)+1
30 CONTINUE
  DO 40 K=1,BIN
  PDF=(H(K)/N)*BIN/RANGE
  AB=XMIN+K*RANGE/BIN
  TH=TMP*EXP(-AB*AB/2.)
  WRITE(9,*)AB,PDF,TH
  WRITE(1,*)AB,PDF,TH
40 CONTINUE
  PAUSE
  CLOSE(1)
  END
C SUBROUTINE GAUSS IS SHOWN IN LISTING 1.8
```

Calculating Standard Deviation[5,6]

Often, from simulation outputs, we would like to compute some of the basic random variable properties (i.e., mean, variance, etc.). Stated more mathematically, we wish to compute these basic random variable properties from a finite set of data x_i, when only n samples are available. The discrete equivalent of the

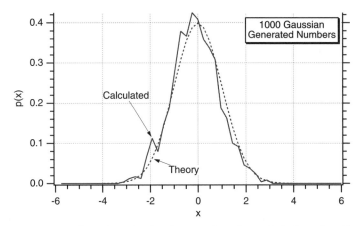

Fig. 1.8   Sampled Gaussian distribution matches theoretical Gaussian distribution for 1000 random numbers.

formulas just presented for basic random variable properties are presented in the following equations:

$$\text{mean} = \frac{\sum\limits_{i=1}^{n} x_i}{n}$$

$$\text{mean square} = \frac{\sum\limits_{i=1}^{n} x_i^2}{n-1}$$

$$\text{standard deviation} = \sqrt{\frac{\sum\limits_{i=1}^{n} x_i - \text{mean})^2}{n-1}}$$

We can see from these equations that integrals from the theoretical or continuous formulas have been replaced with summations in their discrete equivalents. For the theoretical and calculated random variable properties to be equal, the number of samples in the discrete computations must be infinite. Because the sample size is finite, the discrete or calculated formulas are approximations. In fact, the answers generated from these formulas have statistics of their own.

Recognizing that simulation outputs based upon random inputs can vary from run to run, the Monte Carlo approach will sometimes be used in this text to obtain system performance. The Monte Carlo method is approximate and is simply repeated using simulation trials plus postprocessing of the resultant data in order to do ensemble averaging (using the preceding formulas) to get the mean and standard deviation. Usually a large number of simulation trials are required in order to provide confidence in the accuracy of the results.

To demonstrate that our computed statistics are not precise and in fact are random variables with statistics, a simulation of the Gaussian noise was

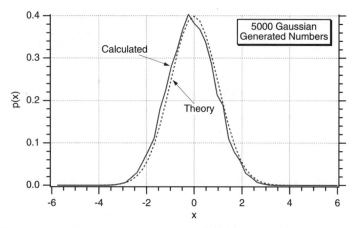

Fig. 1.9   Sampled Gaussian distribution is even better for 5000 random numbers.

generated. Listing 1.10 illustrates the computation of the sampled mean, standard deviation, and rms values. The number of i samples used in the program computation was made a parameter in the study and varied from 1 to 1000. The parts of the simulation that were required for sample statistics calculations are highlighted in bold because these statements can be used to do postprocessing in Monte Carlo simulations.

Figure 1.10 shows that the computed standard deviation (actual standard deviation is unity), mean (actual mean is zero), and rms (actual rms is unity) obtained from the FORTRAN program is a function of the sample size used. Errors in excess of 20% in the standard deviation estimate occur when there are less than 20 samples. The accuracy of the computation improves significantly when many samples are used in computing the standard deviation. In this example, we need more than 100 samples for the computed standard deviation and rms values to be within 5% of the theoretical value of unity. For detailed system performance analysis, when the inputs are random, this information must be taken into account in determining how many simulation trials (Monte Carlo runs) will be required to get reasonably accurate results.

With some Kalman-filter practitioners, filter results are judged by looking at the filter covariance results instead of making many Monte Carlo runs. However, we believe that the covariances (i.e., variances for each filter state variable and cross variances between every pair of state variables) are not as useful as Monte Carlo techniques for detecting unknown errors produced by the filter caused by design deficiencies.

White Noise[2,5]

Up to now we have considered the statistics of a single random variable. However, our equations representing the real world will often change with time, and therefore randomness will be present at each instant of time. Therefore, we now have to consider not one but an infinite number of random variables (one for each instant of time). The mean and standard deviation statistics that we have just

Listing 1.10 Program for computing sampled mean, standard deviation, and rms

```
C THE FIRST THREE STATEMENTS INVOKE THE ABSOFT RANDOM
  NUMBER GENERATOR ON THE MACINTOSH
     GLOBAL DEFINE
          INCLUDE 'quickdraw.inc'
     END
     IMPLICIT REAL*8 (A-H)
     IMPLICIT REAL*8 (O-Z)
     REAL*8 Z(1000)
     OPEN(1,STATUS='UNKNOWN',FILE='DATFIL')
     SIGNOISE=1.
     N=1000
     Z1=0.
     DO 10 I=1,N
     CALL GAUSS(X,SIGNOISE)
     Z(I)=X
     Z1=Z(I)+Z1
     XMEAN=Z1/I
 10  CONTINUE
     SIGMA=0.
     Z1=0.
     Z2=0.
     DO 20 I=1,N
     Z1=(Z(I)-XMEAN)**2+Z1
     Z2=Z(I)**2+Z2
     ZF(I.EQ.1)THEN
          SIGMA=0.
          RMS=0.
     ELSE
          SIGMA=SQRT(Z1/(I-1))
          RMS=SQRT(Z2/(I-1))
     ENDIF
     WRITE(9,*)I,XMEAN,SIGMA,RMS
     WRITE(1,*)I,XMEAN,SIGMA,RMS
 20  CONTINUE
     PAUSE
     CLOSE(1)
     END


C SUBROUTINE GAUSS IS SHOWN IN LISTING 1.8
```

studied summarize the behavior of a single random variable. We need a similar time-varying statistic that will summarize the behavior of random variables that change over time. One such summary statistic is the autocorrelation function, which is defined by

$$\phi_{xx}(t_1, t_2) = E[x(t_1)x(t_2)]$$

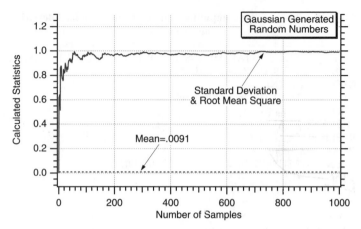

Fig. 1.10 Many samples must be taken before computer-calculated statistics are accurate.

The Fourier transform of the autocorrelation function is called the power spectral density and is defined as

$$\Phi_{xx} = \int_{-\infty}^{\infty} \phi_{xx}(\tau)e^{-j\omega\tau}\, d\tau$$

where the power spectral density, using these definitions, has dimensions of units squared per hertz. In any statistical work presented in this text, the power spectral density will have those units.

One simple and useful form for the power spectral density is that of white noise, in which the power spectral density is constant or

$$\Phi_{xx} = \Phi_0 \quad \text{(white noise)}$$

The autocorrelation function for white noise is a delta function given by

$$\phi_{xx} = \Phi_0 \delta(\tau) \quad \text{(white noise)}$$

Although white noise is not physically realizable, it can serve as an invaluable approximation for situations in which a disturbing noise is wide bandwidth compared to the system bandwidth.

Simulating White Noise[5]

Consider the low-pass filter with time constant T of Fig. 1.11, in which the input is x and the output is y. The transfer function of the low-pass filter is given by

$$\frac{y}{x} = \frac{1}{1 + sT}$$

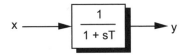

Fig. 1.11   Low-pass filter driven by white noise.

where s is Laplace transform notation. In this example the input x is white noise with spectral density Φ_0. It is important to note that Φ_0 is a number. We would like to find the standard deviation of the output y as a function of time.

One can show that the differential equation describing Fig. 1.11 is given by

$$\dot{y} = \frac{(x-y)}{T}$$

If x is white noise, then the mean square value of the output of the filter can also be shown to be

$$E[y^2(t)] = \frac{\Phi_0(1 - e^{-2t/T})}{2T}$$

where T is the time constant of the low-pass filter and Φ_0 is the spectral density of the white noise input in units squared per hertz. The rms of the output is simply the square root of the preceding expression. Similarly, because the input noise has zero mean, the output will also be of zero mean, and the standard deviation of the output will equal the rms value.

We can simulate the problem of Fig. 1.11 in order to investigate how the theoretical results of the preceding formula agree with simulation. But first, in order to do this, we must be able to simulate Gaussian white noise. We already know how to simulate Gaussian random numbers. Because the Gaussian-distributed random numbers are independent, the resultant Gaussian random numbers will look white to the low-pass filter if the noise bandwidth is much greater than the filter bandwidth. In a simulation of the continuous system of Fig. 1.11, the equivalent Gaussian noise generator is called every integration interval h. Because the integration interval is always chosen to be at least several times smaller than the smallest time constant T ($h \ll T$ in order to get correct answers with numerical integration techniques), the noise will look white to the system.

The standard deviation of the pseudowhite noise (actual white noise over an infinite bandwidth has infinite standard deviation) is related to the desired white noise spectral density Φ_0 and integration interval h according to

$$\sigma = \sqrt{\frac{\Phi_0}{h}}$$

where Φ_0 has dimensions of units squared per hertz. The simulation listing of this white noise-driven low-pass filter is shown in Listing 1.11. We can see from the listing that the differential equation representing the low-pass filter appears twice

Listing 1.11   Simulation of low-pass filter driven by white noise

```
C THE FIRST THREE STATEMENTS INVOKE THE ABSOFT RANDOM
  NUMBER GENERATOR ON THE MACINTOSH
  GLOBAL DEFINE
              INCLUDE 'quickdraw.inc'
  END
  IMPLICIT REAL*8 (A-H)
  IMPLICIT REAL*8 (O-Z)
  OPEN(1,STATUS='UNKNOWN',FILE='DATFIL')
  TAU=.2
  PHI=1.
  T=0.
  H=.01
  SIG=SQRT(PHI/H)
  Y=0.
  WHILE(T < =4.999)
        CALL GAUSS(X,SIG)
        YOLD=Y
        YD=(X-Y)/TAU
        Y=Y+H*YD
        T=T+H
        YD=(X-Y)/TAU
        Y=(YOLD+Y)/2.+.5*H*YD
        SIGPLUS=SQRT(PHI*(1.-EXP(-2.*T/TAU))/(2.*TAU))
        SIGMINUS=-SIGPLUS
        WRITE(9,*)T,Y,SIGPLUS,SIGMINUS
        WRITE(1,*)T,Y,SIGPLUS,SIGMINUS
  END DO
  PAUSE
  CLOSE(1)
  END
C SUBROUTINE GAUSS IS SHOWN IN LISTING 1.8
```

because we are using second-order Runge–Kutta integration. However, we only call the Gaussian random number generator once every integration interval.

We can see from Listing 1.11 that the Gaussian noise with unity standard deviation is modified to get the desired pseudowhite noise spectral density ($\Phi_0 = 1$). The approximate white noise enters the system every integration interval. A sample output for a low-pass filter with a time constant of 0.2 s is shown in Fig. 1.12. The formulas for the plus and minus values of the theoretical standard deviation of the output are displayed with dashed lines. Theoretically, the single flight results should fall within the theoretical $\pm \sigma$ bounds approximately 68% of the time. Therefore, we can visually verify that the experimental and theoretical results appear to be in agreement.

In Kalman-filtering work we shall also attempt to display single run simulation results along with the theoretical bounds in order to determine if the filter is working properly.

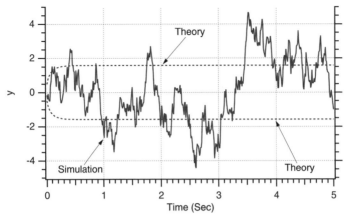

Fig. 1.12   Low-pass filter output agrees with theory.

State-Space Notation[7]

Often in this text we will put linear differential equations into state space form because this will be required for setting up a Kalman-filtering problem. With state-space notation we are saying that any set of linear differential equations can be put in the form of the first-order matrix differential equation

$$\dot{x} = Fx + Gu + w$$

where x is known as the system state vector, F is the systems dynamics matrix, u is a deterministic input sometimes called a control vector, and w is a random forcing function, which is also known as process noise.

In the preceding section we had white noise entering a low-pass filter. The differential equation describing the process was given by

$$\dot{y} = \frac{(x-y)}{T}$$

where y was the output quantity of interest and x was the white noise input. Changing the notation of the preceding equation to avoid confusion, we get

$$\dot{x} = \frac{(n-x)}{T} \quad \text{HOW .? ?}$$

where x represents both the state of the system and the output for this example and n represents the white noise input. If the preceding equation were put in state-space form, the appropriate matrices would all be scalars with values

$$F = -\frac{1}{T}$$

$$G = 0$$

$$w = \frac{n}{T}$$

As another example, consider the second-order differential equation without any random inputs

$$\ddot{y} + 2\dot{y} + 3y = 4$$

First, we rewrite the preceding equation in terms of its highest derivative

$$\ddot{y} = -2\dot{y} - 3y + 4$$

which can be expressed as

$$\begin{bmatrix} \dot{y} \\ \ddot{y} \end{bmatrix} = \begin{bmatrix} 0 & 1 \\ -3 & -2 \end{bmatrix} \begin{bmatrix} y \\ \dot{y} \end{bmatrix} + \begin{bmatrix} 0 \\ 1 \end{bmatrix} 4$$

In this case the appropriate state-space matrices would be

$$x = \begin{bmatrix} y \\ \dot{y} \end{bmatrix}$$

$$F = \begin{bmatrix} 0 & 1 \\ -3 & -2 \end{bmatrix}$$

$$G = \begin{bmatrix} 0 \\ 1 \end{bmatrix}$$

$$u = 4$$

$$w = \begin{bmatrix} 0 \\ 0 \end{bmatrix}$$

Fundamental Matrix[7]

If we have an equation expressed in state-space form as

$$\dot{x} = Fx$$

where the systems dynamics matrix F is time invariant, then we can say that there is a transition or fundamental matrix Φ that can be used to exactly propagate the state forward from any time t_0 to time t according to

$$x(t) = \Phi(t - t_0)x(t_0)$$

Two simple ways of finding the fundamental matrix for time-invariant systems are by Laplace transforms and by Taylor-series expansion. For the Laplace transform technique we use

$$\Phi(t) = \mathcal{L}^{-1}[(sI - F)^{-1}]$$

where $\mathcal{L}^{-1}$ denotes the inverse of the Laplace transform. A Taylor-series expansion for the fundamental matrix can also be used, yielding

$$\Phi(t) = e^{Ft} = I + Ft + \frac{(Ft)^2}{2!} + \cdots + \frac{(Ft)^n}{n!} + \cdots$$

Once we have $\Phi(t)$, we can find $\Phi(t - t_0)$ by simply replacing t by $t - t_0$.

We already showed in the section on numerical integration that the solution to the differential equation

$$\ddot{x} = -\omega^2 x$$

was

$$x = \sin \omega t$$

Using calculus, we can see that the derivative of the solution is given by

$$\dot{x} = \omega \cos \omega t$$

If we rewrite the original second-order differential equation in state-space form, we obtain

$$\begin{bmatrix} \dot{x} \\ \ddot{x} \end{bmatrix} = \begin{bmatrix} 0 & 1 \\ -\omega^2 & 0 \end{bmatrix} \begin{bmatrix} x \\ \dot{x} \end{bmatrix}$$

Therefore, the systems dynamics matrix is given by

$$F = \begin{bmatrix} 0 & 1 \\ -\omega^2 & 0 \end{bmatrix}$$

We already stated that for a time-invariant systems dynamics matrix we can derive the fundamental matrix according to

$$\Phi(t) = \mathcal{L}^{-1}[(sI - F)^{-1}]$$

where I is the identity matrix. Substituting the appropriate matrices, we first
express the inverse of $sI - F$ as

$$(sI - F)^{-1} = \begin{bmatrix} s & -1 \\ \omega^2 & s \end{bmatrix}^{-1}$$

Taking the inverse of a 2×2 matrix can nearly be done by inspection using the
formulas presented in this chapter and can easily be shown to be

$$\Phi(s) = (sI - F)^{-1} = \frac{1}{s^2 + \omega^2} \begin{bmatrix} s & 1 \\ -\omega^2 & 0 \end{bmatrix}$$

We now have the fundamental matrix in the Laplace transform or s domain. We
must take the inverse Laplace transform to express the fundamental matrix as a
function of time. Using inverse Laplace transform tables yields the fundamental
matrix in the time domain as

$$\Phi(t) = \begin{bmatrix} \cos \omega t & \dfrac{\sin \omega t}{\omega} \\ -\omega \sin \omega t & \cos \omega t \end{bmatrix}$$

Before we use the other method for also deriving the fundamental matrix, let us
check the validity of the solution. For our sample problem we know that the initial
conditions on the second-order linear differential equation are given by

$$x(0) = \sin \omega(0) = 0$$
$$\dot{x}(0) = \omega \cos \omega(0) = \omega$$

Because

$$x(t) = \Phi(t - t_0)x(t_0)$$

we can also say that

$$x(t) = \Phi(t)x(0)$$

Let us check the preceding statement by substitution:

$$\begin{bmatrix} x(t) \\ \dot{x}(t) \end{bmatrix} = \begin{bmatrix} \cos \omega t & \dfrac{\sin \omega t}{\omega} \\ \omega \sin \omega t & \cos \omega t \end{bmatrix} \begin{bmatrix} x(0) \\ \dot{x}(0) \end{bmatrix}$$

$$= \begin{bmatrix} \cos \omega t & \dfrac{\sin \omega t}{\omega} \\ \omega \sin \omega t & \cos \omega t \end{bmatrix} \begin{bmatrix} 0 \\ \omega \end{bmatrix} = \begin{bmatrix} \sin \omega t \\ \omega \cos \omega t \end{bmatrix}$$

The preceding statement says that

$$x(t) = \sin \omega t$$
$$\dot{x}(t) = \omega \cos \omega t$$

which is correct. In other words, once we have the fundamental matrix we do not have to integrate differential equations to propagate the states forward. We can propagate states forward by matrix multiplication. In Kalman filtering we have to propagate state estimates ahead by one sampling time. If we have an exact expression for the fundamental matrix, we will be able to perform the propagation by matrix multiplication. When an exact fundamental matrix is not available, we will have to resort to numerical integration to propagate ahead the state estimates. To illustrate that the same answer for the fundamental matrix also could have been obtained by Taylor-series expansion, recall that the systems dynamics matrix was given by

$$F = \begin{bmatrix} 0 & 1 \\ -\omega^2 & 0 \end{bmatrix}$$

Therefore, we can find

$$F^2 = \begin{bmatrix} 0 & 1 \\ -\omega^2 & 0 \end{bmatrix}\begin{bmatrix} 0 & 1 \\ -\omega^2 & 0 \end{bmatrix} = \begin{bmatrix} -\omega^2 & 0 \\ 0 & -\omega^2 \end{bmatrix}$$

$$F^3 = \begin{bmatrix} -\omega^2 & 0 \\ 0 & -\omega^2 \end{bmatrix}\begin{bmatrix} 0 & 1 \\ -\omega^2 & 0 \end{bmatrix} = \begin{bmatrix} 0 & -\omega^2 \\ \omega^4 & 0 \end{bmatrix}$$

$$F^4 = \begin{bmatrix} 0 & -\omega^2 \\ \omega^4 & 0 \end{bmatrix}\begin{bmatrix} 0 & 1 \\ -\omega^2 & 0 \end{bmatrix} = \begin{bmatrix} \omega^4 & 0 \\ 0 & \omega^4 \end{bmatrix}$$

$$F^5 = \begin{bmatrix} \omega^4 & 0 \\ 0 & \omega^4 \end{bmatrix}\begin{bmatrix} 0 & 1 \\ -\omega^2 & 0 \end{bmatrix} = \begin{bmatrix} 0 & \omega^4 \\ -\omega^6 & 0 \end{bmatrix}$$

$$F^6 = \begin{bmatrix} 0 & \omega^4 \\ -\omega^6 & 0 \end{bmatrix}\begin{bmatrix} 0 & 1 \\ -\omega^2 & 0 \end{bmatrix} = \begin{bmatrix} -\omega^6 & 0 \\ 0 & -\omega^6 \end{bmatrix}$$

If we truncate the Taylor-series expansion for the fundamental matrix to six terms, we obtain

$$\Phi(t) = e^{Ft} \approx I + Ft + \frac{(Ft)^2}{2!} + \frac{(Ft)^3}{3!} + \frac{(Ft)^4}{4!} + \frac{(Ft)^5}{5!} + \frac{(Ft)^6}{6!}$$

or

$$\Phi(t) \approx \begin{bmatrix} 1 & 0 \\ 0 & 1 \end{bmatrix} + \begin{bmatrix} 0 & 1 \\ -\omega^2 & 0 \end{bmatrix}t + \begin{bmatrix} -\omega^2 & 0 \\ 0 & -\omega^2 \end{bmatrix}\frac{t^2}{2} + \begin{bmatrix} 0 & -\omega^2 \\ \omega^4 & 0 \end{bmatrix}\frac{t^3}{6}$$
$$+ \begin{bmatrix} \omega^4 & 0 \\ 0 & \omega^4 \end{bmatrix}\frac{t^4}{24} + \begin{bmatrix} 0 & \omega^4 \\ -\omega^6 & 0 \end{bmatrix}\frac{t^5}{120} + \begin{bmatrix} -\omega^6 & 0 \\ 0 & -\omega^6 \end{bmatrix}\frac{t^6}{720}$$

Performing the required additions and subtractions and combining terms yields

$$
\Phi(t) \approx
\begin{bmatrix}
1 - \dfrac{\omega^2 t^2}{2} + \dfrac{\omega^4 t^4}{24} - \dfrac{\omega^6 t^6}{720} & t - \dfrac{\omega^2 t^3}{6} + \dfrac{\omega^4 t^5}{120} \\[2ex]
-\omega^2 t + \dfrac{\omega^4 t^3}{6} - \dfrac{\omega^6 t^5}{120} & 1 - \dfrac{\omega^2 t^2}{2} + \dfrac{\omega^4 t^4}{24} - \dfrac{\omega^6 t^6}{720}
\end{bmatrix}
$$

After recognizing that the trigonometric Taylor-series expansions are given by

$$
\sin \omega t \approx \omega t - \frac{\omega^3 t^3}{3!} + \frac{\omega^5 t^5}{5!} - \cdots
$$

$$
\cos \omega t \approx 1 - \frac{\omega^2 t^2}{2!} + \frac{\omega^4 t^4}{4!} - \frac{\omega^6 t^6}{6!} + \cdots
$$

we obtain a more compact form for the fundamental matrix as

$$
\Phi(t) =
\begin{bmatrix}
\cos \omega t & \dfrac{\sin \omega t}{\omega} \\[2ex]
-\omega \sin \omega t & \cos \omega t
\end{bmatrix}
$$

The preceding expression is the same answer that was obtained with the Laplace transform method.

Strictly speaking, the Taylor-series approach to finding the fundamental matrix only applies when the systems dynamics matrix is time invariant. However, in practical problems where the systems dynamics matrix is time varying, it is often assumed that the systems dynamics matrix is approximately constant over the region of interest (i.e., time between samples) and the Taylor-series approach is still used.

Summary

In this chapter all of the techniques that will be used throughout the text have been introduced. Examples of matrix manipulation have been presented, and subroutines for automating their implementation in computer programs have been provided. The methods for numerically integrating differential equations, which will be used for modeling the real world and for projecting forward state estimates, have been discussed, and sample computer code implementing the various techniques has been presented. A discussion of random phenomenon as applied to understanding Monte Carlo analysis has been briefly introduced. Finally, state-space notation is presented as a technique for representing differential equations.

References

[1] Hohn, F. E., *Elementary Matrix Algebra,* 2nd ed., Macmillian, New York, 1965, pp. 1–29.

[2] Gelb, A., *Applied Optimal Estimation,* Massachusetts Inst. of Technology Press, Cambridge, MA, 1974, pp. 16–18.

[3] Rosko, J.S., *Digital Simulation of Physical Systems*, Addison Wesley Longman, Reading, MA, 1972, pp. 173–179.

[4] Press, W. H., Flannery, B. P., Teukolsky, S. A., and Vetterling, W. T., *Numerical Recipes: The Art of Scientific Computation*, Cambridge Univ. Press, London, 1986, pp. 547–554.

[5] Zarchan, P., *Tactical and Strategic Missile Guidance*, 3rd ed., Progress in Astronautics and Aeronautics, AIAA, Reston, VA, 1998, pp. 51–65.

[6] Johnson, R. A., and Bhattacharyya, G.K., *Statistics Principles and Methods*, 3rd ed., Wiley, New York, 1996, pp. 82–95.

[7] Schwarz, R. J., and Friedland, B., *Linear Systems*, McGraw–Hill, New York, 1965, pp. 28–52.

Chapter 2

Method of Least Squares

Introduction

FINDING the best polynomial curve fit in the least squares sense to a set of data was first developed by Gauss during the 18th century when he was only 17. Hundreds of years later Gauss's least squares technique is still used as the most significant criteria for fitting data to curves. In fact the method of least squares is still used as the basis for many curve-fitting programs used on personal computers. All of the filtering techniques that will be discussed in this text are, one way or the other, based on Gauss's original method of least squares. In this chapter we will investigate the least squares technique as applied to signals, which can be described by a polynomial. Then we shall demonstrate via numerous examples the various properties of the method of least squares when applied to noisy measurement data.

It is not always obvious from the measurement data which is the correct-order polynomial to use to best fit the measurement data in the least squares sense. This type of knowledge is often based on understanding the dynamics of the problem or from information derived from mathematical techniques, such as systems identification, that have been previously applied to the problem. If common sense and good engineering judgement are ignored in formulating the least squares problem, disaster can result.

Overview

Before we begin to look at actual examples using the method of least squares, it may be helpful to step back and take a broad view of what we are about to do. Our goal is to learn something about the actual signal based on collecting measurements of the signal contaminated by noise. We will normally follow a three-step procedure for extracting information or estimating the characteristics of the signal based upon noisy measurements. First, we will assume a polynomial model to represent the actual signal. Second, we will try to estimate the coefficients of the selected polynomial by choosing a goodness of fit criterion. This is where Gauss comes in. The least squares criterion is the sum of the squares of the individual discrepancies between the estimated polynomial and measurement values of the actual signal contaminated by noise. Third, we will use calculus to minimize the sum of the squares of the individual discrepancies in order to obtain the best coefficients for the selected polynomial.

41

Zeroth-Order or One-State Filter

Suppose we have a set of measurement data that we would like to fit with the best constant (i.e., zeroth-order polynomial) in the least squares sense. How should the measurement data be manipulated in order to find the constant that will best estimate the signal? Following the procedure developed by Gauss, we would like to fit the measurements with the best constant in the least squares sense. The method for obtaining the best constant is also known as least squares filtering or estimation.

Suppose we sample n measurements x_k^* of the data every T_s seconds (i.e., k goes from 1 to n). We would like to find the best estimate of x_k (i.e., $\hat{x}_k$, with a caret over it indicating an estimate). To find the best constant in the least squares sense in this example, we first have to formulate the problem mathematically. If we define the state we would like to estimate as $\hat{x}_k$, then we would like the estimate to be close to the measurement data. We would like to minimize the square of the residual or the difference between the estimate and measurement after each measurement. For the zeroth-order case in which the best estimate is supposed to be a constant a_0, a convenient way of stating this problem mathematically is that we desire to minimize

$$R = \sum_{k=1}^{n} (\hat{x}_k - x_k^*)^2 = \sum_{k=1}^{n} (a_0 - x_k^*)^2$$

where R is the square of the summation of all the residuals or differences between the estimate and the measurement. We could perform other operations on the residuals, but squaring the residuals not only can be justified from a practical point of view but is also very convenient from an analytical point of view. As was mentioned before, in this zeroth-order system the estimate will simply be a constant a_0 or

$$\hat{x}_k = a_0$$

Because we are only estimating one constant, this is also sometimes called a one-state system. From calculus we know that a function can be minimized by taking its derivative and setting the result equal to zero. In our example in which we want to minimize R, we simply take the derivative of R with respect to a_0 and set the result to zero. We must also check to see if the second derivative of R with respect to a_0 is positive in order to guarantee that we have a minimum and not a maximum. The reason for this is that the first derivative can be zero at either a maximum or a minimum.

To apply the minimization technique from calculus[1] to our problem, we must first expand R by eliminating the summation sign used in the definition of R and represent all of the terms as a series. Therefore, R becomes

$$R = \sum_{k=1}^{n} (\hat{x}_k - x_k^*)^2 = (a_0 - x_1^*)^2 + (a_0 - x_2^*)^2 + \cdots + (a_0 - x_n^*)^2$$

By taking the derivative of R with respect to a_0 and setting the result to zero, we obtain

$$\frac{\partial R}{\partial a_0} = 0 = 2(a_0 - x_1^*) + 2(a_0 - x_2^*) + \cdots + 2(a_0 - x_n^*)$$

Also,

$$\frac{\partial^2 R}{\partial a_0^2} = 2 + 2 + \cdots + 2 = 2n$$

which is always positive for any positive integer n, thereby ensuring a minimum. Dividing both sides of the first derivative equation by 2 and recognizing that

$$-x_1^* - x_2^* - \cdots - x_n^* = -\sum_{k=1}^{n} x_k^*$$

and

$$a_0 + a_0 + \cdots + a_0 = na_0$$

we obtain the equation

$$0 = na_0 - \sum_{k=1}^{n} x_k^*$$

which has only one unknown a_0. Solving for a_0 yields

$$a_0 = \frac{\sum_{k=1}^{n} x_k^*}{n}$$

The preceding formula shows us that the best constant fit to a set of measurement data in the least squares sense is simply the average value of the measurements. To illustrate the use of the preceding formula, let us consider the case in which we would like to find the best constant in the least squares sense to fit four measurements. The four measurements, which are taken every second, appear in Table 2.1.

Before we proceed we would like to express the data of Table 2.1 in a more mathematical way. If we define the time between measurements as the sampling time T_s, then we can express time t in terms of the sampling time as

$$t = (k - 1)T_s$$

where k varies from 1 to 4, T_s is 1 s (or time varies from 0 to 3 s in steps of 1 s). Because there are four measurements in this example, n is 4. Therefore, for this

Table 2.1 Measurement data for example

t	x^*
0	1.2
1	0.2
2	2.9
3	2.1

example we can expand Table 2.1 to obtain Table 2.2. Therefore, solving for the best constant, in the least squares sense, to the four data points of Table 2.2 yields

$$\hat{x}_k = a_0 = \frac{\sum_{k=1}^{n} x_k^*}{n} = \frac{1.2 + 0.2 + 2.9 + 2.1}{4} = 1.6$$

The meaning of the preceding formula is that, assuming we would like to fit the measurement data of Table 2.2 with a constant, the best constant in the least squares sense has a value of 1.6. Figure 2.1, which compares the least squares estimate to the four measurements, shows that the constant 1.6 is an unreasonable fit to the measurement data. However, the bad fit cannot be blamed on the least squares criterion. The polynomial model we chose, a constant, must bear the brunt of the blame for being an inadequate representation of the measurement data. The example in the next section attempts to overcome this inadequacy by using a more flexible model (i.e., ramp rather than a constant to fit the data). But first we can do some numerical experiments to test the validity of our calculated optimal constant a_0.

To check if we indeed did minimize R, let us first numerically calculate R when the value of the constant is 1.6. Substituting the value of 1.6 for a_0 into the expression for R yields a value of 4.06 or

$$R = \sum_{k=1}^{4}(a_0 - x_k^*)^2 = (1.6 - 1.2)^2 + (1.6 - 0.2)^2 + (1.6 - 2.9)^2 + (1.6 - 2.1)^2$$
$$= 4.06$$

Table 2.2 Measurement data from Table 2.1 expressed more mathematically

k	$(k-1)T_s$	x_k^*
1	0	1.2
2	1	0.2
3	2	2.9
4	3	2.1

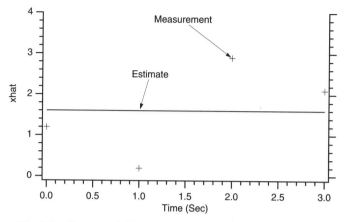

Fig. 2.1 Constant 1.6 is an unreasonable fit to measurement data.

In principal, choosing a value of a_0 other than 1.6 must yield a larger value of R. For example, from Fig. 2.1 we can see that choosing a value for 2 for a_0 also appears to be reasonable. However, substitution of 2 for a_0 in the formula for R yields

$$R = \sum_{k=1}^{4}(a_0 - x_k^*)^2 = (2 - 1.2)^2 + (2 - 0.2)^2 + (2 - 2.9)^2 + (2 - 2.1)^2 = 4.70$$

which is larger than 4.06. Similarly, from Fig. 2.1 we can see that choosing a value for 1 for a_0 also appears to be reasonable. Again, substitution of 1 for a_0 in the formula for R yields

$$R = \sum_{k=1}^{4}(a_0 - x_k^*)^2 = (1 - 1.2)^2 + (1 - 0.2)^2 + (1 - 2.9)^2 + (1 - 2.1)^2 = 5.50$$

which is also larger than 4.06. Thus, we can also see from an experimental point of view that a value of 1.6 for a_0 appears to be the best choice for minimizing R. In the example of this section, finding the best least squares fit to the data used calculations involving all of the measurement data. This implementation of the method of least squares is known as a batch processing technique because all of the measurements must first be collected before an estimate can be made. Generally batch processing techniques are not used in real time because in real time operations estimates are often required as soon as the measurements are being taken.

First-Order or Two-State Filter

Suppose we would like to fit the measurement data with the best straight line (i.e., rather than a constant) in the least squares sense. This means we are seeking estimates to fit the first-order polynomial

$$\hat{x} = a_0 + a_1 t$$

or in discrete form

$$\hat{x}_k = a_0 + a_1(k-1)T_s$$

where k is the number of measurements taken and T_s is the sampling time. After we find a_0 and a_1 we also have enough information to find the derivative of the estimate as

$$\dot{\hat{x}} = a_1$$

or in discrete notation

$$\dot{\hat{x}}_k = a_1$$

Because the straight-line fit gives estimates of x and its derivative (or because we must solve for a_0 and a_1), the resultant least squares fit will also be known as a two-state least squares filter. As before, the quantity to be minimized is still given by the summation of the square of the difference between the estimates and measurements or

$$R = \sum_{k=1}^{n}(\hat{x}_k - x_k^*)^2 = \sum_{k=1}^{n}[a_0 + a_1(k-1)T_s - x_k^*]^2$$

If we expand R, we get

$$R = \sum_{k=1}^{n}[a_0 + a_1(k-1)T_s - x_k^*]^2 = (a_0 - x_1^*)^2 + (a_0 + a_1 T_s - x_2^*)^2$$
$$+ \cdots + [a_0 + a_1(n-1)T_s - x_n^*]^2$$

Again, according to calculus, we can minimize R by setting its derivative with respect to a_0 and a_1 to zero or

$$\frac{\partial R}{\partial a_0} = 0 = 2(a_0 - x_1^*) + 2(a_0 + a_1 T_s - x_2^*) + \cdots + 2[a_0 + a_1(n-1)T_s - x_n^*]$$

$$\frac{\partial R}{\partial a_1} = 0 = 2(a_0 + a_1 T_s - x_2^*)T_s + \cdots + 2(n-1)T_s[a_0 + a_1(n-1)T_s - x_n^*]$$

All of the second derivatives are greater than zero, indicating that we are minimizing R. We now have two equations with two unknowns (i.e., a_0 and a_1). We can rearrange and simplify the two preceding equations as

$$na_0 + a_1 \sum_{k=1}^{n}(k-1)T_s = \sum_{k=1}^{n} x_k^*$$

$$a_0 \sum_{k=1}^{n}(k-1)T_s + a_1 \sum_{k=1}^{n}[(k-1)T_s]^2 = \sum_{k=1}^{n}(k-1)T_s x_k^*$$

To solve the preceding two simultaneous equations for the two unknowns, we must first evaluate the intermediate quantities recognizing that $T_s = 1$. Using the values from Table 2.2, we can obtain for the summations

$$\sum_{k=1}^{n}(k-1)T_s = 0 + 1 + 2 + 3 = 6$$

$$\sum_{k=1}^{n} x_k^* = 1.2 + 0.2 + 2.9 + 2.1 = 6.4$$

$$\sum_{k=1}^{n}[(k-1)T_s]^2 = 0^2 + 1^2 + 2^2 + 3^2 = 14$$

$$\sum_{k=1}^{n}(k-1)T_s x_k^* = 0*1.2 + 1*0.2 + 2*2.9 + 3*2.1 = 12.3$$

Therefore, the two equations with two unknowns simplify to

$$4a_0 + 6a_1 = 6.4$$
$$6a_0 + 14a_1 = 12.3$$

After some algebraic manipulation we can readily see that the solutions for these two constants are $a_0 = 0.79$ and $a_1 = 0.54.$

Although we have shown that these two simultaneous equations can practically be solved by inspection, it is sometimes more beneficial to express them compactly in matrix form. This will especially prove useful when we have more equations and a systematic method for stating their solution will be required. Recall that the we have just shown that the general form of the two simultaneous equations is

$$na_0 + a_1 \sum_{k=1}^{n}(k-1)T_s = \sum_{k=1}^{n} x_k^*$$

$$a_0 \sum_{k=1}^{n}(k-1)T_s + a_1 \sum_{k=1}^{n}[(k-1)T_s]^2 = \sum_{k=1}^{n}(k-1)T_s x_k^*$$

The preceding equations can also be expressed in matrix form as

$$
\begin{bmatrix} n & \sum_{k=1}^{n}(k-1)T_s \\ \sum_{k=1}^{n}(k-1)T_s & \sum_{k=1}^{n}[(k-1)T_s]^2 \end{bmatrix} \begin{bmatrix} a_0 \\ a_1 \end{bmatrix} = \begin{bmatrix} \sum_{k=1}^{n}x_k^* \\ \sum_{k=1}^{n}(k-1)T_s x_k^* \end{bmatrix}
$$

Therefore, we can solve for a_0 and a_1 by matrix inversion or

$$
\begin{bmatrix} a_0 \\ a_1 \end{bmatrix} = \begin{bmatrix} n & \sum_{k=1}^{n}(k-1)T_s \\ \sum_{k=1}^{n}(k-1)T_s & \sum_{k=1}^{n}[(k-1)T_s]^2 \end{bmatrix}^{-1} \begin{bmatrix} \sum_{k=1}^{n}x_k^* \\ \sum_{k=1}^{n}(k-1)T_s x_k^* \end{bmatrix}
$$

A simple program was written to solve the preceding 2×2 matrix equation for the set of four data points of Table 2.2 and appears in Listing 2.1. The listing makes use of the fact established in Chapter 1 that if a matrix A is given by

$$
A = \begin{bmatrix} a & b \\ c & d \end{bmatrix}
$$

then the inverse of this 2×2 matrix is

$$
A^{-1} = \frac{1}{ad - bc} \begin{bmatrix} d & -b \\ -c & a \end{bmatrix}
$$

The coefficients a_0 and a_1 are represented in Listing 2.1 by ANS(1,1) and ANS(2,1), respectively.

Running the program yields coefficients $a_0 = 0.79$ and $a_1 = 0.54$, which is in agreement with our preceding hand calculations for the two simultaneous equations. Therefore, the formula for the best least squares straight-line fit (i.e., first-order fit) to the measurement data of Table 2.2 is given by

$$
\hat{x}_k = 0.79 + 0.54(k-1)T_s
$$

where k takes on integer values from 1 to 4. Because $(k-1)T_s$ is also time, we can say that

$$
\hat{x} = 0.79 + 0.54t
$$

In addition, we can say that the estimate of the derivative (i.e., second state of least squares filter) can be obtained by differentiating the preceding expression, yielding

$$
\dot{\hat{x}}_k = 0.54
$$

**Listing 2.1   Solving for least-squares coefficients in the
first-order filter**

```
IMPLICIT  REAL*8  (A-H)
IMPLICIT  REAL*8  (O-Z)
REAL*8  T(4),X(4),A(2,2),AINV(2,2),B(2,1),ANS(2,1)
T(1)=0
T(2)=1
T(3)=2
T(4)=3
X(1)=1.2
X(2)=0.2
X(3)=2.9
X(4)=2.1
N=4
SUM1=0
SUM2=0
SUM3=0
SUM4=0
DO  10  I=1,4
   SUM1=SUM1+T(I)
   SUM2=SUM2+T(I)*T(I)
   SUM3=SUM3+X(I)
   SUM4=SUM4+T(I)*X(I)
10 CONTINUE
   A(1,1)=N
   A(1,2)=SUM1
   A(2,1)=SUM1
   A(2,2)=SUM2
   DET=A(1,1)*A(2,2)-A(1,2)*A(2,1)
   AINV(1,1)=A(2,2)/DET
   AINV(1,2)=-A(1,2)/DET
   AINV(2,1)=-A(2,1)/DET
   AINV(2,2)=A(1,1)/DET
   B(1,1)=SUM3
   B(2,1)=SUM4
   CALL  MATMUL(AINV,2,2,B,2,1,ANS)
   WRITE(9,*)ANS(1,1),ANS(2,1)
   PAUSE
   END
C SUBROUTINE MATMUL IS SHOWN IN LISTING 1.4
```

The best straight-line fit, in the least squares sense, is plotted alongside the measurement data of Table 2.2 and appears in Fig. 2.2. We can see that the straight-line fit to the data is much better than the constant fit to the data of Fig. 2.1. The fit is better because the straight-line fit captures the slope or upward trend of the measurement data, which by definition must be ignored with the constant fit.

It is also of interest to compute the residual for the best least squares first-order fit to the measurement data and compare it to the residual of the best least-squares

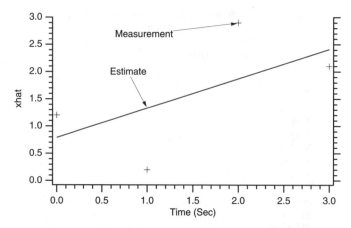

Fig. 2.2   Straight-line fit to data is better than constant fit.

zeroth-order fit. By definition the residual for the straight-line fit to the four sets of measurement data is given by

$$R = \sum_{k=1}^{4} [a_0 + a_1(k-1)T_s - x_k^*]^2$$

For this example $T_s = 1$ s, and we have already shown that the best least-squares fit to the four measurements yields the coefficients $a_0 = 0.79$ and $a_1 = 0.54$. Therefore, expanding the preceding equation using the measurements of Table 2.2, we get

$$R = [0.79 + 0.54(0) - 1.2]^2 + [0.79 + 0.54(1) - 0.2]^2$$
$$+ [0.79 + 0.54(2) - 2.9]^2 + [0.79 + 0.54(3) - 2.1]^2 = 2.61$$

Thus, we can see that for the four measurements the residual obtained by using the best least squares straight-line fit is smaller than the residual obtained using a constant fit to the data (i.e., see last section where we obtained $R = 4.06$).

Second-Order or Three-State Least-Squares Filter

We can also fit the measurement data of Table 1.2 with the best parabola or second-order fit in the least squares sense. This means we are seeking three coefficient estimates to fit the formula for the parabola

$$\hat{x} = a_0 + a_1 t + a_2 t^2$$

After the three coefficients a_0, a_1, and a_2 are found, we have enough information to find estimates of the other two states or derivatives of the first state estimate.

The other state estimates can be found by differentiating the preceding equation, yielding

$$\hat{\dot{x}} = a_1 + 2a_2 t$$

and

$$\hat{\ddot{x}} = 2a_2$$

Recall that for the three-state filter (i.e., filter yields estimates of x and its two derivatives) we are still trying to minimize the sum of the squares of the difference between the estimate of the first state and the measurement. In other words, because the discrete form of the first state estimate is given by

$$\hat{x}_k = a_0 + a_1(k-1)T_s + a_2[(k-1)T_s]^2$$

we are still trying to minimize

$$R = \sum_{k=1}^{n} (\hat{x}_k - x_k^*)^2 = \sum_{k=1}^{n} [a_0 + a_1(k-1)T_s + a_2(k-1)^2 T_s^2 - x_k^*]^2$$

Therefore, if we expand the preceding equation, we obtain

$$R = (a_0 - x_1^*)^2 + [a_0 + a_1 T_s + a_2 T_s^2 - x_2^*)]^2$$
$$+ \cdots + [a_0 + a_1(n-1)T_s + a_2(n-1)^2 T_s^2 - x_n^*]^2$$

Again, we can minimize the preceding equation by setting the derivative of R with respect to a_0, a_1, and a_2 to zero, yielding the following three equations with three unknowns:

$$\frac{\partial R}{\partial a_0} = 0 = 2(a_0 - x_1^*) + 2[a_0 + a_1 T_s + a_2 T_s^2 - x_2^*)]$$
$$+ \cdots + 2[a_0 + a_1(n-1)T_s + a_2(n-1)^2 T_s^2 - x_n^*]$$

$$\frac{\partial R}{\partial a_1} = 0 = 2[a_0 + a_1 T_s + a_2 T_s^2 - x_2^*)]T_s$$
$$+ \cdots + 2[a_0 + a_1(n-1)T_s + a_2(n-1)^2 T_s^2 - x_n^*](n-1)T_s$$

$$\frac{\partial R}{\partial a_2} = 0 = 2[a_0 + a_1 T_s + a_2 T_s^2 - x_2^*)]T_s^2$$
$$+ \cdots + 2[a_0 + a_1(n-1)T_s + a_2(n-1)^2 T_s^2 - x_n^*](n-1)^2 T_s^2$$

We can rearrange and simplify the three preceding equations as

$$na_0 + a_1 \sum_{k=1}^{n}(k-1)T_s + a_2 \sum_{k=1}^{n}[(k-1)T_s]^2 = \sum_{k=1}^{n}x_k^*$$

$$a_0 \sum_{k=1}^{n}(k-1)T_s + a_1 \sum_{k=1}^{n}[(k-1)T_s]^2 + a_2 \sum_{k=1}^{n}[(k-1)T_s]^3 = \sum_{k=1}^{n}(k-1)T_s x_k^*$$

$$a_0 \sum_{k=1}^{n}[(k-1)T_s]^2 + a_1 \sum_{k=1}^{n}[(k-1)T_s]^3 + a_2 \sum_{k=1}^{n}[(k-1)T_s]^4 = \sum_{k=1}^{n}[(k-1)T_s]^2 x_k^*$$

As was the case in the preceding section, these three simultaneous equations can be placed in a more compact matrix form, yielding

$$\begin{bmatrix} n & \sum_{k=1}^{n}(k-1)T_s & \sum_{k=1}^{n}[(k-1)T_s]^2 \\ \sum_{k=1}^{n}(k-1)T_s & \sum_{k=1}^{n}[(k-1)T_s]^2 & \sum_{k=1}^{n}[(k-1)T_s]^3 \\ \sum_{k=1}^{n}[(k-1)T_s]^2 & \sum_{k=1}^{n}[(k-1)T_s]^3 & \sum_{k=1}^{n}[(k-1)T_s]^4 \end{bmatrix} \begin{bmatrix} a_0 \\ a_1 \\ a_2 \end{bmatrix}$$

$$= \begin{bmatrix} \sum_{k=1}^{n}x_k^* \\ \sum_{k=1}^{n}(k-1)T_s x_k^* \\ \sum_{k=1}^{n}[(k-1)T_s]^2 x_k^* \end{bmatrix}$$

Again, the coefficients of the best parabola or second-order fit (i.e., in the least squares sense) can be found by matrix inversion or

$$\begin{bmatrix} a_0 \\ a_1 \\ a_2 \end{bmatrix} = \begin{bmatrix} n & \sum_{k=1}^{n}(k-1)T_s & \sum_{k=1}^{n}[(k-1)T_s]^2 \\ \sum_{k=1}^{n}(k-1)T_s & \sum_{k=1}^{n}[(k-1)T_s]^2 & \sum_{k=1}^{n}[(k-1)T_s]^3 \\ \sum_{k=1}^{n}[(k-1)T_s]^2 & \sum_{k=1}^{n}[(k-1)T_s]^3 & \sum_{k=1}^{n}[(k-1)T_s]^4 \end{bmatrix}^{-1}$$

$$\times \begin{bmatrix} \sum_{k=1}^{n}x_k^* \\ \sum_{k=1}^{n}(k-1)T_s x_k^* \\ \sum_{k=1}^{n}[(k-1)T_s]^2 x_k^* \end{bmatrix}$$

Another simple program was written to solve the preceding 3×3 matrix equation for the three coefficients to best fit the set of four data points of Table 2.2. The three-state least-squares filter appears in Listing 2.2. Listing 2.2 makes use of the fact that was demonstrated in Chapter 1 that if a 3×3 matrix A is given by

$$A = \begin{bmatrix} a & b & c \\ d & e & f \\ g & h & i \end{bmatrix}$$

then the inverse of this matrix is given by

$$A^{-1} = \frac{1}{aei + bfg + cdh - ceg - bdi - afh} \begin{bmatrix} ei - fh & ch - bi & bf - ec \\ gf - di & ai - gc & dc - af \\ dh - ge & gb - ah & ae - bd \end{bmatrix}$$

The program for finding the parabolic polynomial's coefficients is nearly identical to Listing 2.1 except that the dimensions have been increased and more lines of code are required to fill in the larger matrices. The coefficients a_0, a_1, and a_2 are called ANS(1,1), ANS(2,1), and ANS(3,1) respectively in Listing 2.2.

Listing 2.2   Solving for least-squares coefficients with
three-state least-squares filter

```
IMPLICIT REAL*8 (A-H)
IMPLICIT REAL*8 (O-Z)
REAL*8 T(4),X(4),A(3,3),AINV(3,3),B(3,1),ANS(3,1)
T(1)=0
T(2)=1
T(3)=2
T(4)=3
X(1)=1.2
X(2)=0.2
X(3)=2.9
X(4)=2.1
N=4
SUM1=0
SUM2=0
SUM3=0
SUM4=0
SUM5=0
SUM6=0
SUM7=0
DO 10 I=1,4
   SUM1=SUM1+T(I)
   SUM2=SUM2+T(I)*T(I)
   SUM3=SUM3+X(I)
   SUM4=SUM4+T(I)*X(I)
```

(continued)

Listing 2.2 *(Continued)*

```
      SUM5=SUM5+T(I)*T(I)*T(I)
      SUM6=SUM6+T(I)*T(I)*T(I)*T(I)
      SUM7=SUM7+T(I)*T(I)*X(I)
10 CONTINUE
      A(1,1)=N
      A(1,2)=SUM1
      A(1,3)=SUM2
      A(2,1)=SUM1
      A(2,2)=SUM2
      A(2,3)=SUM5
      A(3,1)=SUM2
      A(3,2)=SUM5
      A(3,3)=SUM6
      DET1=A(1,1)*A(2,2)*A(3,3)+A(1,2)*A(2,3)*A(3,1)
      DET2=A(1,3)*A(2,1)*A(3,2)-A(1,3)*A(2,2)*A(3,1)
      DET3=-A(1,2)*A(2,1)*A(3,3)-A(1,1)*A(2,3)*A(3,2)
      DET=DET1+DET2+DET3
      AINV(1,1)=(A(2,2)*A(3,3)-A(2,3)*A(3,2))/DET
      AINV(1,2)=(A(1,3)*A(3,2)-A(1,2)*A(3,3))/DET
      AINV(1,3)=(A(1,2)*A(2,3)-A(2,2)*A(1,3))/DET
      AINV(2,1)=(A(3,1)*A(2,3)-A(2,1)*A(3,3))/DET
      AINV(2,2)=(A(1,1)*A(3,3)-A(3,1)*A(1,3))/DET
      AINV(2,3)=(A(2,1)*A(1,3)-A(1,1)*A(2,3))/DET
      AINV(3,1)=(A(2,1)*A(3,2)-A(3,1)*A(2,2))/DET
      AINV(3,2)=(A(3,1)*A(1,2)-A(1,1)*A(3,2))/DET
      AINV(3,3)=(A(1,1)*A(2,2)-A(1,2)*A(2,1))/DET
      B(1,1)=SUM3
      B(2,1)=SUM4
      B(3,1)=SUM7
      CALL  MATMUL(AINV,3,3,B,3,1,ANS)
      WRITE(9,*)ANS(1,1),ANS(2,1),ANS(3,1)
      PAUSE
      END
C SUBROUTINE MATMUL IS SHOWN IN LISTING 1.4
```

Running the program of Listing 2.2 yields coefficients $a_0 = 0.84$, $a_1 = 0.39$, and $a_2 = 0.05$. Therefore, with this three-state filter the formula for the best least squares estimate of x is given by

$$\hat{x}_k = 0.84 + 0.39(k-1)T_s + 0.05[(k-1)T_s]^2$$

The other two states or estimates of the derivatives of x can be obtained by inspection of the preceding expression and are given by

$$\dot{\hat{x}}_k = 0.39 + 0.1(k-1)T_s$$

and

$$\hat{\hat{x}}_k = 0.1$$

The best least-squares fit (i.e., second-order fit is parabola) to the data for this three-state filter is plotted alongside the four measurements of Table 2.2 in Fig. 2.3. We can see that the best least-squares parabolic fit of this section is of comparable quality to the ramp fit of the preceding section. The reason for this is that both the parabola and ramp captures the slope or upwards trend of the measurement data.

It is also of interest to compute the residual for the best least-squares second-order fit to the measurement data and compare it to the residual of the best least-squares zeroth-order and first-order fits. By definition the residual for the second-order fit to the four sets of measurement data is given by

$$R = \sum_{k=1}^{4}[a_0 + a_1(k-1)T_s + a_2[(k-1)T_s]^2 - x_k^*]^2$$

For this example $T_s = 1$ s, and we have already shown that the best least-squares fit to the four measurements yields the coefficients $a_0 = 0.84$, $a_1 = 0.39$, and $a_2 = 0.05$. Therefore, expanding the preceding equation using the measurements of Table 1.2, we get

$$R = [0.84 + 0.39(0) + 0.05(0) - 1.2]^2 + [0.84 + 0.39(1) + 0.05(1) - 0.2]^2$$
$$+ [0.84 + 0.39(2) + 0.05(4) - 2.9]^2$$
$$+ [0.84 + 0.39(3) + 0.05(9) - 2.1]^2 = 2.60$$

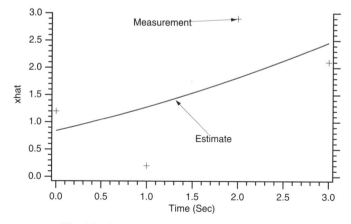

Fig. 2.3   Parabolic fit to data is pretty good, too.

Thus, we can see that for the four measurements the residual obtained by using the best least squares second-order fit is significantly smaller than the residual obtained using a zeroth-order fit to the data (i.e., $R = 4.06$) and slightly smaller than the residual obtained using a first-order fit to the data (i.e., $R = 2.61$). Indeed there was a 35.7% improvement [i.e., $(4.06 - 2.61)/4.06$] in the residual in going from zeroth to first order, but less than a 1% improvement [i.e., $(2.61 - 2.60)/2.61$] in going from first to second order.

Third-Order System

We can also fit the measurement data with the best cubic or third-order curve in the least squares sense. This means we are seeking estimates to fit the formula for the cubic

$$\hat{x} = a_0 + a_1 t + a_2 t^2 + a_3 t^3$$

or

$$\hat{x}_k = a_0 + a_1(k-1)T_s + a_2[(k-1)T_s]^2 + a_3[(k-1)T_s]^3$$

As before, the residual to be minimized is still given by

$$R = \sum_{k=1}^{n} (\hat{x}_k - x_k^*)^2$$

Using the same minimization techniques as in the preceding section, we can obtain the formula for the coefficients to the cubic polynomial according to

$$
\begin{bmatrix} a_0 \\ a_1 \\ a_2 \\ a_3 \end{bmatrix} =
\begin{bmatrix}
n & \sum_{k=1}^{n}(k-1)T_s & \sum_{k=1}^{n}[(k-1)T_s]^2 & \sum_{k=1}^{n}[(k-1)T_s]^3 \\
\sum_{k=1}^{n}(k-1)T_s & \sum_{k=1}^{n}[(k-1)T_s]^2 & \sum_{k=1}^{n}[(k-1)T_s]^3 & \sum_{k=1}^{n}[(k-1)T_s]^4 \\
\sum_{k=1}^{n}[(k-1)T_s]^2 & \sum_{k=1}^{n}[(k-1)T_s]^3 & \sum_{k=1}^{n}[(k-1)T_s]^4 & \sum_{k=1}^{n}[(k-1)T_s]^5 \\
\sum_{k=1}^{n}[(k-1)T_s]^3 & \sum_{k=1}^{n}[(k-1)T_s]^4 & \sum_{k=1}^{n}[(k-1)T_s]^5 & \sum_{k=1}^{n}[(k-1)T_s]^6
\end{bmatrix}^{-1}
$$

$$
\times
\begin{bmatrix}
\sum_{k=1}^{n} x_k^* \\
\sum_{k=1}^{n}(k-1)T_s x_k^* \\
\sum_{k=1}^{n}[(k-1)T_s]^2 x_k^* \\
\sum_{k=1}^{n}[(k-1)T_s]^3 x_k^*
\end{bmatrix}
$$

The solution to the preceding equation involves taking the inverse of a 4×4 matrix. Because no formulas for the matrix inverse were presented in Chapter 1 for this case, numerical techniques for finding the matrix inverse were used. Matrix inversion is part of the MATLAB® and True BASIC languages. For this example the Gauss–Jordan elimination technique[2] was used, but a listing for the technique will not be presented because its application is beyond the scope of the text. Solving the preceding equation yields coefficients $a_0 = 1.2$, $a_1 = -5.25$, $a_2 = 5.45$, and $a_3 = -1.2$. Therefore, the formula for the best least squares cubic fit to the measurement data is given by

$$\hat{x}_k = 1.2 - 5.25(k-1)T_s + 5.45[(k-1)T_s]^2 - 1.2[(k-1)T_s]^3$$

Because $(k-1)T_s$ is also time, we can say that

$$\hat{x}_k = 1.2 - 5.25t + 5.45t^2 - 1.2t^3$$

The best cubic fit is plotted alongside the measurements in Fig. 2.4. We can see that the best least squares cubic fit simply goes through all of the measurements. In other words, no smoothing takes place in this case. The best least squares polynomial will always go through all of the data (i.e., measurements) when the order of the system is one less than the number of measurements. If we tried to increase the order of the fit with the same number of measurements, we would have four equations with five unknowns, which cannot be solved uniquely.

Finally, it is also of interest to compute the residual for the best least squares third-order fit to the measurement data and compare it to the residual of the best least squares zeroth-order and first-order fits. By definition the residual for the second-order fit to the four sets of measurement data is given by

$$R = \sum_{k=1}^{4} [a_0 + a_1(k-1)T_s + a_2[(k-1)T_s]^2 + a_3[(k-1)T_s]^3 - x_k^*]^2$$

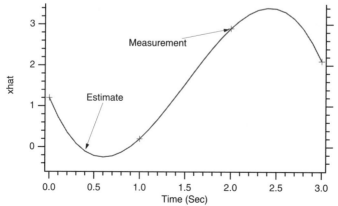

Fig. 2.4   Third-order fit goes through all four measurements.

For this example $T_s = 1$ s, and we have already shown that the best least squares fit to the four measurements yields the coefficients $a_0 = 1.2$, $a_1 = -5.25$, $a_2 = 5.45$, and $a_3 = -1.2$. Therefore, expanding the preceding equation using the measurements of Table 2.2, we get

$$
\begin{aligned}
R = & [1.2 - 5.25(0) + 5.45(0) - 1.2(0) - 1.2]^2 \\
& + [1.2 - 5.25(1) + 5.45(1) - 1.2(1) - 0.2]^2 \\
& + [1.2 - 5.25(2) + 5.45(4) - 1.2(8) - 2.9]^2 \\
& + [1.2 - 5.25(3) + 5.45(9) - 1.2(27) - 2.1]^2 = 0
\end{aligned}
$$

Thus, we can see that for the four measurements the residual obtained by using the best least-squares third-order fit is zero. Obviously the residual is zero because the third-order fit passes through all of the measurement points.

We already showed that for a given order polynomial fit the least squares technique minimized the residual R (i.e., sum of the square of the difference between estimate and measurement). Therefore, once we chose the order of the polynomial, the coefficients we obtained were the best in the sense that any other coefficients would lead to a larger value of R. In the experiment we just concluded, we calculated R for different order polynomials when we had four measurements. Table 2.3, which summarizes the results of the study, shows that as the order of the polynomial increases, R decreases and eventually goes to zero.

We might incorrectly conclude from Table 2.3 that it is best to pick as high an order polynomial as possible. If we do this, all we are doing is trying to ensure that the polynomial passes through each of the measurement points. In fact, this happened in Fig. 2.4. If the measurements are noisy, having a polynomial go through each of the measurements can lead to ridiculous answers. However, it is not always obvious from the measurement data which is the correct order polynomial to use to best fit the measurement data in the least squares sense. This type of knowledge is often based on either understanding the dynamics of the problem or information derived from mathematical techniques (i.e., systems identification) that have been previously applied to the problem. If common sense and good engineering judgement are ignored in formulating the least squares problem, disaster can result.

Table 2.3 Residual decreases as order of least-squares polynomial increases

System order	R
0	4.06
1	2.61
2	2.60
3	0

FUL PAGE DISCUSS **Experiments with Zeroth-Order or One-State Filter**

As noted before, we have seen how the residual error is reduced as we increase the order of the fitting polynomial. This is the error between the noisy data and the estimated coefficient values for the assumed polynomial. *However, what is of most importance to us is not how the estimated polynomial compares with the noisy measurements, but how it compares to the true signal.* We will run experiments to compare our estimates with the true signal. It will then become clear that there can be a real danger in choosing a polynomial whose order is not matched to the signal.

In this section we shall see how the zeroth-order or one-state least squares filter output fits the measurement data in the presence of a zeroth-order and first-order polynomial signal with additive noise. A one-state least squares filter (i.e., zeroth-order fit to measurement data) is programmed to evaluate filter performance, as shown in Listing 2.3. For the nominal case the actual signal is a constant [i.e., $X1(N) = 1$], whereas the measurement [i.e., $X(N)$] is the actual signal plus zero-mean Gaussian noise with a standard deviation of unity. The outputs of the program are time T, the true signal $X1(I)$, the measurement $X(I)$, the estimate or state of the one-state least squares filter XHAT, the error in the estimate of the signal ERRX, and the difference between the estimate and measurement ERRXP. We can also see from Listing 2.3 that we have also calculated the sum of the squares of the difference between the actual signal and estimate SUMPZ1 and the sum of the squares of the difference between the measurement and estimate SUMPZ2.

The nominal case of Listing 2.3 was run for the case in which the actual measurement was a constant buried in noise. The one-state least squares filter was used to estimate the signal after 10 s of data gathering in which measurements were taken every 0.1 s (i.e., $T_s = 0.1$). Figure 2.5 shows that the one-state least squares filter output appears to fit the measurements taken (i.e., constant plus noise) in a reasonable way. Using the least squares method smoothes or filters the noisy measurements.

Figure 2.6 compares the one-state least squares filter estimate after all of the measurements are taken to the actual signal (i.e., measurement minus noise). This is an academic experiment because in the real world the actual signal would not be available to us. Only in simulation do we have the luxury of comparing the estimate to the actual signal. We can see that for the 101 measurements taken, the one-state estimate of the signal is excellent. If more measurements were taken, the match would be even closer because we are averaging noise with zero mean. In this case the zeroth-order or one-state least squares estimator is perfectly matched to the zeroth-order signal (i.e., constant).

Finally, Fig. 2.7 presents the difference between the signal and filter estimate (i.e., error in estimate) and the difference between the measurement and estimate. We can see that the error in the estimate of the signal is a constant and near zero. This indicates that the least squares filter estimate is able to track the actual signal. We can also see from Fig. 2.7 that the difference between the measurement and estimate is noisy and visually appears to have approximately zero mean. This observation simply means that the filter estimate appears to be averaging all of the measurements. This should not be surprising because in the derivation of the least

AIM
How?
NU
NU ?₀
How
Discuss 2.5 and 2.6 above Topics resp.
F16₁ 2.5 F16₁ 2.6 F16₁ 2.7
– Actual & estimated
– measurement & estimated.

Listing 2.3   One-state filter for extracting signal from measurement

```
C THE FIRST THREE STATEMENTS INVOKE THE ABSOFT RANDOM
      NUMBER GENERATOR ON THE MACINTOSH
      GLOBAL DEFINE
              INCLUDE 'quickdraw.inc'
      END
      IMPLICIT REAL*8 (A-H)
      IMPLICIT REAL*8 (O-Z)
      REAL*8 A(1,1),AINV(1,1),B(1,1),ANS(1,1),X(101),X1(101)
      OPEN(1,STATUS='UNKNOWN',FILE='DATFIL')
      SIGNOISE=1.
      N=0
      TS=0.1
      SUM3=0.
      SUMPZ1=0.
      SUMPZ2=0.
      DO 10 T=0.,10.,TS
        N=N+1
        CALL GAUSS(XNOISE,SIGNOISE)
        X1(N)=1
        X(N)=X1(N)+XNOISE
        SUM3=SUM3+X(N)
        NMAX=N
  10  CONTINUE
      A(1,1)=N
      B(1,1)=SUM3
      AINV(1,1)=1./A(1,1)
      ANS(1,1)=AINV(1,1)*B(1,1)
      DO 11 I=1,NMAX
        T=0.1*(I-1)
        XHAT=ANS(1,1)
        ERRX=X1(I)-XHAT
        ERRXP=X(I)-XHAT
        ERRX2=(X1(I)-XHAT)**2
        ERRXP2=(X(I)-XHAT)**2
        SUMPZ1=ERRX2+SUMPZ1
        SUMPZ2=ERRXP2+SUMPZ2
        WRITE(9,*)T,X1(I),X(I),XHAT,ERRX,ERRXP,SUMPZ1,SUMPZ2
        WRITE(1,*)T,X1(I),X(I),XHAT,ERRX,ERRXP,SUMPZ1,SUMPZ2
  11  CONTINUE
      CLOSE(1)
      PAUSE
      END
C SUBROUTINE GAUSS IS SHOWN IN LISTING 1.8
```

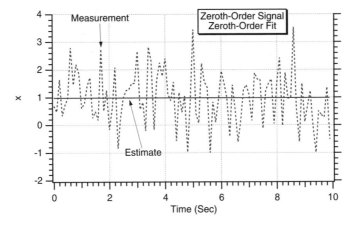

Fig. 2.5 One-state least-squares filter smoothes noisy measurements.

squares filter we were trying to minimize the difference between the estimate (i.e., zeroth-order polynomial or constant in this case) and the measurement.

Using Listing 2.3, the sum of the squares of the difference between the actual signal and estimate SUMPZ1 and the sum of the squares of the difference between the measurement and estimate SUMPZ2 were calculated. For this example the sum of the squares of the difference between the actual signal and estimate was 0.01507, and the sum of the squares of the difference between the measurement and estimate was 91.92.

Another experiment was conducted in which the actual signal was changed from a constant of unity to a straight line with formula

$$x = t + 3$$

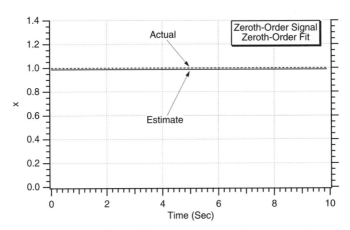

Fig. 2.6 One-state filter yields near perfect estimate of constant signal.

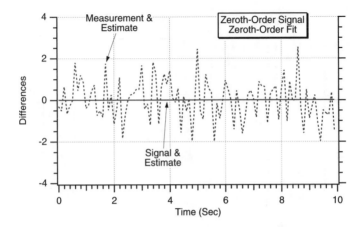

Fig. 2.7   Estimation errors are nearly zero for one-state least-squares filter.

The measurement of the signal was contaminated with zero-mean Gaussian noise with a standard deviation of 5. To reflect this new case, the two lines of code that were modified in Listing 2.3 are

$$\textbf{SIGNOISE} = \textbf{5.}$$
$$\textbf{X1(N)} = \textbf{T} + \textbf{3.}$$

Another case was run with the one-state least squares filter, and the single run simulated results are displayed in Figs. 2.8–2.10. We can see from Fig. 2.8 that the noisy measurements have a slightly upward trend. This is not surprising because the actual signal is a ramp with a positive slope of unity. In this example

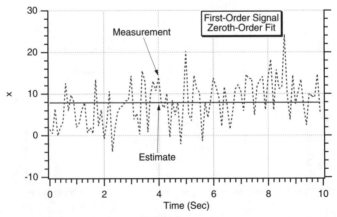

Fig. 2.8   Zeroth-order least-squares filter does not capture upward trend of first-order measurement data.

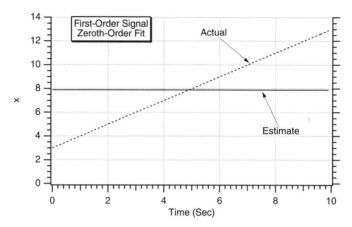

Fig. 2.9 Zeroth-order least-squares filter cannot estimate slope of actual signal.

the zeroth-order least-squares filter still can only provide a constant estimate. Although the estimate is good, it cannot capture the upward trend of the measurements. The zeroth-order filter estimate can only average the upward trend of the data after all of the measurements are taken. Figure 2.9 compares the actual signal (i.e., measurement minus noise) with the one-state (i.e., zeroth-order) least squares filter estimate. Here we can clearly see that the filter order is not high enough to estimate the signal. Finally, Fig. 2.10 presents the difference between the signal and filter estimate (i.e., error in estimate) and the difference between the measurement and estimate. As before, the difference between the measurement and filter estimate is still noisy but no longer has zero mean. Figure 2.10 is telling us that the zeroth-order least-squares filter cannot accurately fit the first-order measurement data. The inability of the filter to track is clarified even

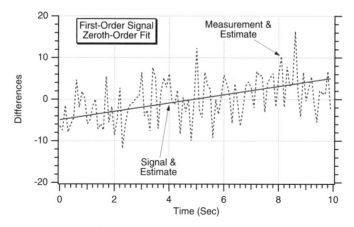

Fig. 2.10 Errors in the estimate of the signal appear to grow with time.

further in this figure because the difference between the signal and filter estimate is growing with time. This is another way of saying that the filter is diverging.

For this example the sum of the squares of the difference between the actual signal and estimate was 834 (up from 0.01507 when the measurement was a constant plus noise), and the sum of the squares of the difference between the measurement and estimate was 2736 (up from 91.92 when the measurement was a constant plus noise). The larger values for these measures of performance also indicate that the filter is diverging.

Experiments with First-Order or Two-State Filter

In this section we will see how the estimates from the first-order or two-state least squares filter output fit the measurement data in the presence of zeroth-order, first-order, and second-order polynomial signals with additive noise. A two-state least squares filter (i.e., first-order fit to measurement data) is programmed to evaluate filter performance as shown in Listing 2.4. The two-state filter also has the ability to estimate the first derivative of the signal. Because the signal in Listing 2.4 is a polynomial, we can also calculate, using calculus, the derivative of the actual signal. In the nominal case of Listing 2.4, where the signal is unity, the derivative is zero [i.e., $XD(N) = 0$.] In this case the actual signal is a constant [i.e., $X1(N) = 1$], whereas the measurement [i.e., $X(N)$] is the signal plus zero-mean Gaussian noise with a standard deviation of unity. The outputs of the program are time T, the true signal $X1(I)$, the measurement $X(I)$, the estimate of the signal of the two-state least squares filter XHAT, the error in the estimate of the signal ERRX, and the difference between the derivative of the actual signal and the filter's estimate of the derivative ERRXD. We can also see from Listing 2.4 that we have calculated and printed out the sum of the squares of the difference between the actual signal and estimate SUMPZ1 and the sum of the squares of the difference between the measurement and estimate SUMPZ2.

Listing 2.4   Two-state least-squares filter for extracting signal from measurement

```
C THE FIRST THREE STATEMENTS INVOKE THE ABSOFT RANDOM
      NUMBER GENERATOR ON THE MACINTOSH
   GLOBAL DEFINE
           INCLUDE 'quickdraw.inc'
   END
   IMPLICIT REAL*8 (A-H)
   IMPLICIT REAL*8 (O-Z)
   REAL*8 A(2,2),AINV(2,2),B(2,1),ANS(2,1),X(101),X1(101)
   REAL*8 XD(101)
   OPEN(1,STATUS='UNKNOWN',FILE='DATFIL')
   SIGNOISE=1.
   N=0
   TS=0.1
   SUM1=0
   SUM2=0
```

(continued)

Listing 2.4 (*Continued*)

```
      SUM3=0
      SUM4=0.
      SUMPZ1=0.
      SUMPZ2=0.
      DO 10 T=0.,10.,TS
        N=N+1
        CALL GAUSS(XNOISE,SIGNOISE)
        X1(N)=1
        XD(N)=0.
        X(N)=X1(N)+XNOISE
        SUM1=SUM1+T
        SUM2=SUM2+T*T
        SUM3=SUM3+X(N)
        SUM4=SUM4+T*X(N)
        NMAX=N
   10 CONTINUE
      A(1,1)=N
      A(1,2)=SUM1
      A(2,1)=SUM1
      A(2,2)=SUM2
      B(1,1)=SUM3
      B(2,1)=SUM4
      DET=A(1,1)*A(2,2)-A(1,2)*A(2,1)
      AINV(1,1)=A(2,2)/DET
      AINV(1,2)=-A(1,2)/DET
      AINV(2,1)=-A(2,1)/DET
      AINV(2,2)=A(1,1)/DET
      CALL MATMUL(AINV,2,2,B,2,1,ANS)
      DO 11 I=1,NMAX
        T=0.1*(I-1)
        XHAT=ANS(1,1)+ANS(2,1)*T
        XDHAT=ANS(2,1)
        ERRX=X1(I)-XHAT
        ERRXD=XD(I)-XDHAT
        ERRXP=X(I)-XHAT
        ERRX2=(X1(I)-XHAT)**2
        ERRXP2=(X(I)-XHAT)**2
        SUMPZ1=ERRX2+SUMPZ1
        SUMPZ2=ERRXP2+SUMPZ2
        WRITE(9,*)T,X1(I),X(I),XHAT,ERRX,ERRXD,SUMPZ1,SUMPZ2
        WRITE(1,*)T,X1(I),X(I),XHAT,ERRX,ERRXD,SUMPZ1,SUMPZ2
   11 CONTINUE
      CLOSE(1)
      PAUSE
      END
      SUBROUTINE MATMUL(A,IROW,ICOL,B,JROW,JCOL,C)
      IMPLICIT REAL*8 (A-H)
      IMPLICIT REAL*8 (O-Z)
      REAL*8 A(IROW,ICOL),B(JROW,JCOL),C(IROW,JCOL)
      DO 110 I=1,IROW
```

(*continued*)

Listing 2.4   (*Continued*)

```
      DO 110 J=1,JCOL
         C(I,J)=0.
         DO 110 K=1,ICOL
         C(I,J)=C(I,J)+A(I,K)*B(K,J)
110 CONTINUE
      RETURN
      END
C SUBROUTINE GAUSS IS SHOWN IN LISTING 1.8
C SUBROUTINE MATMUL IS SHOWN IN LISTING 1.4
```

The nominal case of Listing 2.4 was run for the case in which the actual measurement was a constant buried in noise. The two-state least squares filter was used to estimate the signal after 10 s of data gathering, in which measurements were taken every 0.1 s (i.e., $T_s = 0.1$). Figure 2.11 shows that the filter estimate of the actual signal is on the high side for the first 5 s and on the low side for the last 5 s. Thus, on the average over the 10-s measurement time, our estimate of the actual signal was not as perfect as before because we are now using a first-order filter on zeroth-order signal plus noise. The errors in the estimates of the signal and its derivative are displayed in Fig. 2.12. Here we can see that the error in the estimate of the signal varied between -0.2 and 0.2 over the 10-s period, which is a $\pm 20\%$ variation from the true value, whereas the error in the estimate of the derivative of the signal was approximately 0.05.

For this example the sum of the squares of the difference between the actual signal and estimate was 1.895 (up from 0.01507 when a zeroth-order filter was used), and the sum of the squares of the difference between the measurement and estimate was 90.04 (down from 91.92 when a zeroth-order filter was used). These results indicate that the first-order filter is performing worse than the zeroth-order

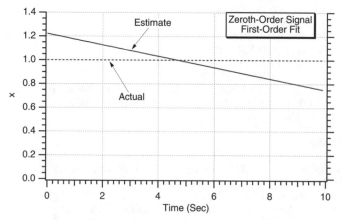

Fig. 2.11   First-order filter has trouble in estimating zeroth-order signal.

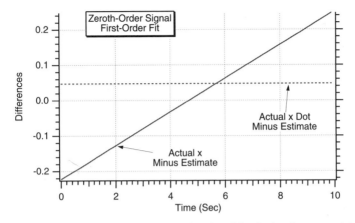

Fig. 2.12 Errors in the estimate of the signal and its derivative are not too large.

filter in estimating the signal but is better in passing through all of the measurements.

Another experiment was conducted in which the actual signal was changed from a constant of unity to a straight line with formula

$$x = t + 3$$

This means that the derivative of the signal is given by

$$\dot{x} = 1$$

The measurement of the signal was contaminated with zero-mean Gaussian noise with a standard deviation of 5. To reflect this new case, the three lines of code, which were modified, are

SIGNOISE = 5.

X1(N) = T + 3.

XD(N) = 1.

Using the new first-order signal plus noise, another case was run with Listing 2.4, and the single flight results are displayed in Figs. 2.13 and 2.14. Figure 2.13 compares the actual signal (i.e., measurement minus noise) with the filter estimate. Here we can clearly see that the filter is doing better in estimating the actual signal. Approximately half of the time the estimate is above the actual signal, whereas the other half of the time the estimate is below the actual signal. These estimates are much closer to the actual signal than those of Fig. 2.9, in which we were only using a zeroth-order least squares filter. Finally, Fig. 2.14 presents the difference between the signal and filter estimate (i.e., error in estimate) and the difference between the first derivative of the signal and the

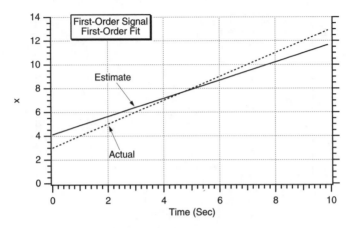

Fig. 2.13 First-order filter does a much better job of estimating first-order signal than does a zeroth-order filter.

filter's estimate of the derivative (i.e., second state of least squares filter). Here we can see that the error in the estimate of the signal varied between −1 and 1 over the 10-s period. This is a great deal better than the zeroth-order filter, which had errors ranging from −5 to 5 for the same problem (see Fig. 2.10). Figure 2.14 also shows that the error in the estimate of the derivative of the signal was approximately 0.2. The zeroth-order filter by definition assumes that the derivative of the signal is zero.

For this example the sum of the squares of the difference between the actual signal and estimate was 47.38 (down from 834 when zeroth-order filter was used), and the sum of the squares of the difference between the measurement and estimate was 2251 (down from 2736 when a zeroth-order filter was used). The

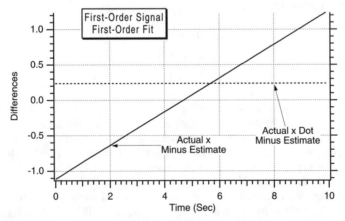

Fig. 2.14 First-order filter is able to estimate derivative of first-order signal accurately.

smaller values for these measures of performance indicate that the first-order filter is doing better than the zeroth-order filter in estimating a first-order signal and in passing through all of the measurements.

Finally, another experiment was conducted in which the actual signal was changed from a straight-line to a parabola with the formula

$$x = 5t^2 - 2t + 2$$

This means that the first derivative of the signal is given by

$$\dot{x} = 10t - 2$$

whereas the second derivative is given by

$$\ddot{x} = 10$$

In this example the measurement of the new polynomial signal was contaminated with zero-mean Gaussian noise with a standard deviation of 50. To reflect this new case, the three lines of code, which were modified in Listing 2.4, are

SIGNOISE = 50.

X1(N) = 5*T*T − 2.*T + 2.

XD(N) = 10.*T − 2.

A new case was run, and the single run simulation results are displayed in Figs. 2.15–2.17. We can see from Fig. 2.15 that the noisy measurements have a parabolic upward trend. This is not surprising because the actual signal is a parabola. We can see that, on the average, the first-order least squares filter estimate is able to follow the upward trend of the measurement data. Figure 2.16

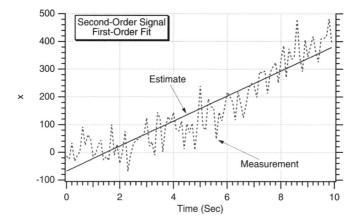

Fig. 2.15 First-order filter attempts to track second-order measurements.

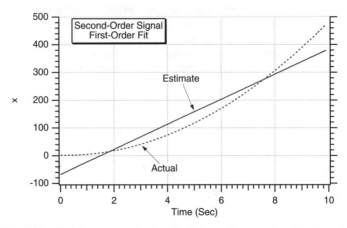

Fig. 2.16   On the average first-order filter estimates second-order signal.

compares the actual parabolic signal with the filter estimate. Here we can clearly see that the filter is doing very well in estimating the actual signal because in this example the ramp fit to the parabola is fairly good. Finally, Fig. 2.17 presents the difference between the signal and filter estimate (i.e., error in the estimate of first state of least squares filter) and the difference between the first derivative of the signal and the filter's estimate of the derivative (i.e., error in the estimate of second state of least squares filter). Here the error in the estimate of the signal varied between −40 and 90 over the 10-s period, whereas the error in the estimate of the derivative of the signal varied between −40 and 40.

For this example the sum of the squares of the difference between the actual signal and estimate was 143,557, and the sum of the squares of the difference

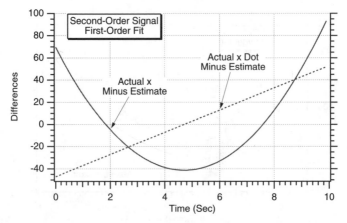

Fig. 2.17   Large estimation errors result when first-order filter attempts to track second-order signal.

between the measurement and estimate was 331,960. The very large values for these measures of performance indicate that the first-order filter is not able to accurately estimate the second-order signal and is also not able to pass through all of the measurements.

Experiments with Second-Order or Three-State Filter

In this section we will see how the second-order or three-state least squares filter output fits the measurement data in the presence of zeroth-order, first-order, and second-order polynomial signals with additive noise. A three-state least squares filter (i.e., second-order fit to measurement data) is programmed to evaluate filter performance, as shown in Listing 2.5. The three-state filter also has the ability to estimate the first and second derivatives of the signal. Because the signal in Listing 2.5 will always be a polynomial, we can also calculate, using calculus, the first and second derivatives of the signal. In the nominal case of Listing 2.5, where the signal is a constant with value unity [i.e., $X1(N) = 1$], the two derivatives are zero [i.e., $XD(N) = 0$ and $XDD(N) = 0$.] In this case the measurement [i.e., $X(N)$] is the signal plus zero-mean Gaussian noise with a standard deviation of unity (i.e., SIGNOISE = 1). The outputs of the program are time T, the true signal $X1(I)$, the measurement $X(I)$, the estimate of the signal of the two-state least squares filter XHAT, the error in the estimate of the signal ERRX, the difference between the derivative of the actual signal and the filter's estimate of the derivative ERRXD, and the difference between the second derivative of the actual signal and the filter's estimate of the second derivative ERRXDD. We can also see from Listing 2.5 that we have calculated and printed out the sum of the squares of the difference between the actual signal and estimate SUMPZ1 and the sum of the squares of the difference between the measurement and estimate SUMPZ2.

Listing 2.5 Three-state least-squares filter for extracting signal from measurement

```
C THE FIRST THREE STATEMENTS INVOKE THE ABSOFT RANDOM
    NUMBER GENERATOR ON THE MACINTOSH
    GLOBAL DEFINE
            INCLUDE 'quickdraw.inc'
    END
    IMPLICIT REAL*8 (A-H)
    IMPLICIT REAL*8 (O-Z)
    REAL*8 A(3,3),AINV(3,3),B(3,1),ANS(3,1),X(101),X1(101)
    REAL*8 XD(101),XDD(101)
    OPEN(1,STATUS='UNKNOWN',FILE='DATFIL')
    SIGNOISE=1.
    TS=0.1
    N=0
    SUM1=0.
    SUM2=0.
    SUM3=0.
```

(continued)

Listing 2.5 *(Continued)*

```
        SUM4=0.
        SUM5=0.
        SUM6=0.
        SUM7=0.
        SUMPZ1=0.
        SUMPZ2=0.
        DO 10 T=0.,10.,TS
          N=N+1
          CALL GAUSS(XNOISE,SIGNOISE)
          X1(N)=1.
          XD(N)=0.
          XDD(N)=0.
          X(N)=X1(N)+XNOISE
          SUM1=SUM1+T
          SUM2=SUM2+T*T
          SUM3=SUM3+X(N)
          SUM4=SUM4+T*X(N)
          SUM5=SUM5+T**3
          SUM6=SUM6+T**4
          SUM7=SUM7+T*T*X(N)
          NMAX=N
10      CONTINUE
        A(1,1)=N
        A(1,2)=SUM1
        A(1,3)=SUM2
        A(2,1)=SUM1
        A(2,2)=SUM2
        A(2,3)=SUM5
        A(3,1)=SUM2
        A(3,2)=SUM5
        A(3,3)=SUM6
        B(1,1)=SUM3
        B(2,1)=SUM4
        B(3,1)=SUM7
        DET1=A(1,1)*A(2,2)*A(3,3)+A(1,2)*A(2,3)*A(3,1)
        DET2=A(1,3)*A(2,1)*A(3,2)-A(1,3)*A(2,2)*A(3,1)
        DET3=-A(1,2)*A(2,1)*A(3,3)-A(1,1)*A(2,3)*A(3,2)
        DET=DET1+DET2+DET3
        AINV(1,1)=(A(2,2)*A(3,3)-A(2,3)*A(3,2))/DET
        AINV(1,2)=(A(1,3)*A(3,2)-A(1,2)*A(3,3))/DET
        AINV(1,3)=(A(1,2)*A(2,3)-A(2,2)*A(1,3))/DET
        AINV(2,1)=(A(3,1)*A(2,3)-A(2,1)*A(3,3))/DET
        AINV(2,2)=(A(1,1)*A(3,3)-A(3,1)*A(1,3))/DET
        AINV(2,3)=(A(2,1)*A(1,3)-A(1,1)*A(2,3))/DET
        AINV(3,1)=(A(2,1)*A(3,2)-A(3,1)*A(2,2))/DET
        AINV(3,2)=(A(3,1)*A(1,2)-A(1,1)*A(3,2))/DET
        AINV(3,3)=(A(1,1)*A(2,2)-A(1,2)*A(2,1))/DET
        CALL MATMUL(AINV,3,3,B,3,1,ANS)
        DO 11 I=1,NMAX
```

(continued)

Listing 2.5 *(Continued)*

```
      T=0.1*(I-1)
      XHAT=ANS(1,1)+ANS(2,1)*T+ANS(3,1)*T*T
      XDHAT=ANS(2,1)+2.*ANS(3,1)*T
      XDDHAT=2.*ANS(3,1)
      ERRX=X1(I)-XHAT
      ERRXD=XD(I)-XDHAT
      ERRXDD=XDD(I)-XDDHAT
      ERRXP=X(I)-XHAT
      ERRX2=(X1(I)-XHAT)**2
      ERRXP2=(X(I)-XHAT)**2
      SUMPZ1=ERRX2+SUMPZ1
      SUMPZ2=ERRXP2+SUMPZ2
      WRITE(9,*)T,X1(I),X(I),XHAT,ERRX,ERRXD,ERRXDD,SUMPZ1,
    1     SUMPZ2
      WRITE(1,*)T,X1(I),X(I),XHAT,ERRX,ERRXD,ERRXDD,SUMPZ1,
    1     SUMPZ2
   11 CONTINUE
      CLOSE(1)
      PAUSE
      END
C SUBROUTINE GAUSS IS SHOWN IN LISTING 1.8
C SUBROUTINE MATMUL IS SHOWN IN LISTING 1.4
```

The nominal case of Listing 2.5 was run for the case in which the actual measurement was a constant buried in noise. Figure 2.18 shows that the three-state least squares filter output is not a constant but a parabola. We can see that, although the estimate of the signal is not too good, the parabola is a rough approximation to the actual signal, which is a constant over the 10-s interval in which the measurements were made. The errors in the estimates of the signal and its two derivatives are shown in Fig. 2.19. Here we can see that the error in the estimate of the signal varied between -0.02 and 0.5 (i.e., 2 to 50% of the true value) over the 10-s period, whereas the error in the estimate of the derivative of the signal was varied between -0.1 and 0.1, and the error in the estimate of the second derivative of the signal was approximately -0.02.

For this example the sum of the squares of the difference between the actual signal and estimate was 2.63 (up from 0.01507 when the zeroth-order filter was used), and the sum of the squares of the difference between the measurement and estimate was 89.3 (down from 91.92 when the zeroth-order filter was used). Thus, we can see that a second-order filter yields worse estimates of the signal than a zeroth-order filter when the measurement is a zeroth-order signal contaminated with noise. As expected, the higher order filter passes through more of the measurements than the zeroth-order filter.

Another experiment was conducted in which the actual signal was changed from a constant of unity to a straight-line with the formula

$$x = t + 3$$

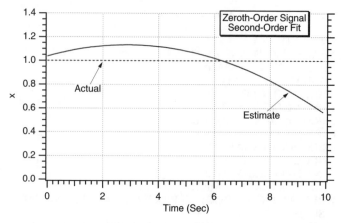

Fig. 2.18   Second-order filter estimates signal is parabola even though it is a constant.

which means that the derivative of the signal is given by

$$\dot{x} = 1$$

The measurement of the signal was contaminated with zero-mean Gaussian noise with a standard deviation of 5. To reflect this new case, the three lines of code, which were modified in Listing 2.5, are

$$\textbf{SIGNOISE} = \textbf{5.}$$

$$\textbf{X1(N)} = \textbf{T} + \textbf{3.}$$

$$\textbf{XD(N)} = \textbf{1.}$$

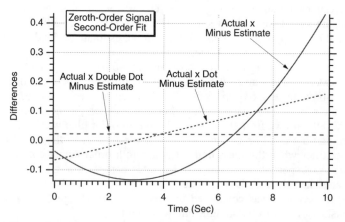

Fig. 2.19   Estimation errors between estimates and states of signal are not terrible when the order of filter is too high.

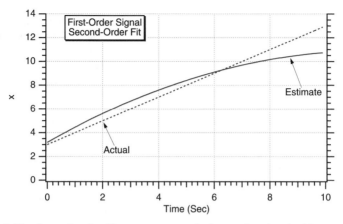

Fig. 2.20 **Second-order filter attempts to fit first-order signal with a parabola.**

Using the new first-order signal plus noise, another case was run with Listing 2.5, and the single run simulation results are displayed in Figs. 2.20 and 2.21. Figure 2.20 compares the actual signal (i.e., measurement minus noise) with the filter estimate. Here we can clearly see that the filter is doing better at trying to estimate the actual signal. However, the second-order filter is still providing a parabolic fit to the measurement data. Approximately 70% of the time the estimate is above the actual signal, whereas the other 30% of the time the estimate is below the actual signal. These estimates are nearly the same as those shown in Fig. 2.13, where the order of the filter was matched to the order of the signal. Finally, Fig. 2.21 presents the difference between the signal and filter estimate (i.e., error in estimate), the difference between the first derivative of the signal and the filter's estimate of the derivative (i.e., second state of least squares

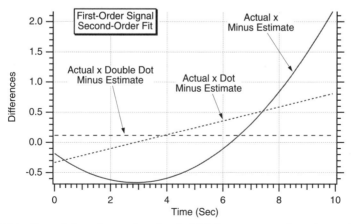

Fig. 2.21 **Second fit to first-order signal yields larger errors than first-order fit.**

filter), and the difference between the second derivative of the signal and the filter's estimate of the second derivative (i.e., third state of least squares filter). Here we can see that that the error in the estimate of the signal varied between -0.2 and 2 over the 10-s period. This is better than the zeroth-order filter, which had errors ranging from -1 to 1 for the same problem (see Fig. 2.14). Figure 2.21 also shows that the error in the estimate of the derivative of the signal varied between -0.4 and 0.7, whereas the error in the second derivative of the signal was approximately 0.1.

For this example the sum of the squares of the difference between the actual signal and estimate was 65.8 (up from 47.38 when a first-order filter was used), and the sum of the squares of the difference between the measurement and estimate was 2232 (down from 2251 when a first-order filter was used). The larger value for the sum of the squares of the difference between the actual signal and estimate indicate that the first-order filter is doing better than the second-order filter in estimating a first-order signal. On the other hand, the second-order filter is closer to passing through all of the measurements.

Finally, another experiment was conducted in which the actual signal was changed from a straight line to a parabola with the formula

$$x = 5t^2 - 2t + 2$$

This means that the first derivative of the signal is given by

$$\dot{x} = 10t - 2$$

whereas the second derivative is given by

$$\ddot{x} = 10$$

In this example the measurement of the new polynomial signal was contaminated with zero-mean Gaussian noise with a standard deviation of 50. To reflect this new case, the three lines of code, which were modified in Listing 2.5, are

SIGNOISE = 50.
X1(N) = 5*T*T − 2.*T + 2.
XD(N) = 10.*T − 2.
XDD(N) = 10.

A new case was run in which the new measurements are a parabolic signal corrupted by noise, and the results are displayed in Figs. 2.22 and 2.23. Figure 2.22 compares the actual signal (i.e., measurement minus noise) with the filter estimate. Here we can clearly see that the filter is doing well in trying to estimate the actual signal because in this example the parabolic fit to the parabola is fairly good. Finally, Fig. 2.23 presents the difference between the signal and filter estimate (i.e., error in estimate), the difference between the first derivative of the signal and the filter's estimate of the derivative (i.e., second state of least squares

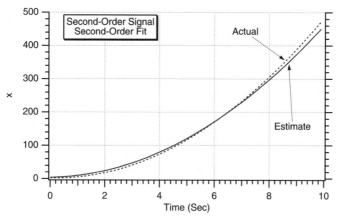

Fig. 2.22   Second-order filter provides near perfect estimates of second-order signal.

filter), and the difference between the second derivative of the signal and the filter's estimate of the second derivative (i.e., third state of least squares filter). Here, the error in the estimate of the signal varied between −2 and 20 over the 10-s period and the error in the estimate of the derivative of the signal varied between −4 and 7, while the error in the estimate of the second derivative of the signal was approximately 1.5.

For this example the sum of the squares of the difference between the actual signal and estimate was 6577 (down from 143,557 when a first-order filter was used), and the sum of the squares of the difference between the measurement and estimate was 223,265 (down from 331,960 when a first-order filter was used). The smaller values for these measures of performance indicate that the second-order

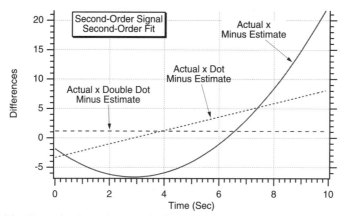

Fig. 2.23   Error in the estimates of all states of second-order filter against second-order signal are better than all other filter fits.

filter is better able to estimate accurately the second-order signal and is also better equipped to pass through all of the measurements than the first-order filter when the measurement is a second-order signal contaminated with noise.

Comparison of Filters

Figure 2.24 compares all of the least squares polynomials to the actual signal. Because the actual signal was a constant, the polynomial with the best fit was zeroth order. Using higher order fits worsened the estimates. Therefore, the method of least squares seems to work best when the polynomial used to fit the measurements matches the order of the polynomial in the measurements. When there is an error in the order of the polynomial fit, the resultant estimates can diverge from the actual signal.

Next, a ramp signal corrupted by noise was considered. The measurement equation was

$$x_k^* = (k - 1)T_s + 3 + 5*\text{Noise}$$

Figure 2.25 compares the estimated polynomials for all of the different order fits to the actual signal. Because the actual signal was a ramp, the closest estimates were obtained by the first-order polynomial fit. Using too low an order least squares fit prevents us from tracking the upward trend of the actual signal. Using too high an order filter enables us to still track the signal, but the estimates are not as good as when the polynomial is matched to the signal.

Another experiment was conducted in which a second-order polynomial measurement corrupted by noise was assumed. The measurement equation was

$$x_k^* = 5[(k - 1)T_s]^2 - 2(k - 1)T_s + 2 + 50*\text{Noise}$$

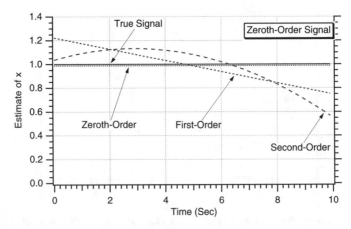

Fig. 2.24   Zeroth-order least-squares filter best tracks zeroth-order measurement.

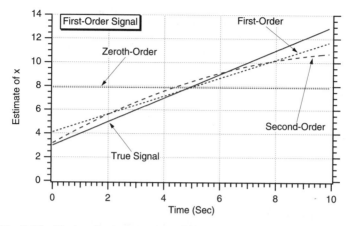

Fig. 2.25   First-order least-squares fit best tracks first-order measurement.

where the noise was considered to be Gaussian with zero mean and a standard deviation of unity. Additional experiments were run with the same measurements applied to second-order least squares polynomial fits. Figure 2.26 compares the estimated polynomials for all of the different order fits to the actual signal. Because the actual signal was a parabola, the best fit was obtained by the second-order polynomial filter. Using too low an order fit in this case prevented the estimate from tracking the signal.

In the experiments conducted in the last section, we also computed the sum of the square of the difference between the actual signal and the estimate and the sum of the square of the difference between the measurement and the estimate. Tables 2.4 and 2.5 summarize the results of those experiments. We can see from Table 2.4 that the sum of the square of the difference between the actual signal

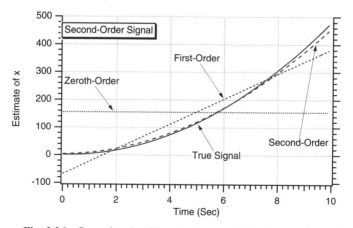

Fig. 2.26   Second-order filter tracks parabolic signal quite well.

Table 2.4   Best estimates of signal are obtained when
filter order matches signal order

| | $\sum(\text{signal} - \text{estimate})^2$ | | |
| | | Signal order | |
Filter order	0	1	2
0	0.01057	834	—
1	1.895	47.38	143,557
2	2.63	65.8	6,577

and the estimate is minimized when the order of the filter matches the order of the signal. There appears to be more of a penalty in underestimating the order of the signal by having too low an order filter. We can see from Table 2.5 that the sum of the square of the difference between the measurement and the estimate gets smaller as the order of the filter gets higher. However, there does not appear to be any practical benefit in making the estimates get closer to the measurements. The main function of the least squares filter is to ensure that the estimates are as close to the actual signal as possible.

Accelerometer Testing Example

In all of the work conducted so far we assumed our estimates were of the form

$$\hat{x}_k = a_0 + a_1(k-1)T_s + a_2[(k-1)T_s]^2 + \cdots + a_0[(k-1)T_s]^n$$

and our goal was to find the best coefficients in the least squares sense for the selected polynomial as a function of time. In other words, we assumed that our measurements were a function of time. In general, least squares filtering does not require that the measurements must be a function of time. In fact, the measure-

Table 2.5   Estimates get closer to measurements when
filter order gets higher

| | $\sum(\text{measurement} - \text{estimate})^2$ | | |
| | | Signal order | |
Filter order	0	1	2
0	91.92	2,736	—
1	90.04	2,251	331,960
2	89.3	2,232	223,265

ments can be a function of any independent variable. For example, we could have assumed that our estimates were of the form

$$\hat{y} = a_0 + a_1 x + a_2 x^2 + \cdots + a_n x^n$$

and our goal would still simply be to find the best coefficients in the least squares sense. Under these circumstances we would obtain for the zeroth-, first-, and second-order polynomial fits to the measurement data polynomial coefficients of the form of Table 2.6.

To demonstrate the partial use of Table 2.6, let us consider an accelerometer testing example in which accelerometer measurements are taken for different accelerometer orientation angles, as shown in Fig. 2.27. When the accelerometer input axis is vertical, the accelerometer reading will be g. As the accelerometer input axis rotates through different angles θ_k, the accelerometer reading will be $g \cos \theta_k$.

Actually, the accelerometer output will also consist of additional terms caused by imperfections in the accelerometer. However, the intrinsic accelerometer noise will be neglected in this example in order to isolate the effects of the angle noise introduced in what follows. The accelerometer output will not only consist of the gravity term but will also contain accelerometer bias, scale factor, and g-sensitive drift errors. The total accelerometer output is given by

$$\text{Accelerometer Output} = g \cos \theta_k + B + SFg \cos \theta_k + K(g \cos \theta_k)^2$$

Table 2.6 General least-squares coefficients for different order polynomial fits

Order	Equations
Zeroth	$a_0 = \dfrac{\sum_{k=1}^{n} y_k^*}{n}$
First	$\begin{bmatrix} a_0 \\ a_1 \end{bmatrix} = \begin{bmatrix} n & \sum_{k=1}^{n} x_k \\ \sum_{k=1}^{n} x_k & \sum_{k=1}^{n} x_k^2 \end{bmatrix}^{-1} \begin{bmatrix} \sum_{k=1}^{n} y_k^* \\ \sum_{k=1}^{n} x_k y_k^* \end{bmatrix}$
Second	$\begin{bmatrix} a_0 \\ a_1 \\ a_2 \end{bmatrix} = \begin{bmatrix} n & \sum_{k=1}^{n} x_k & \sum_{k=1}^{n} x_k^2 \\ \sum_{k=1}^{n} x_k & \sum_{k=1}^{n} x_k^2 & \sum_{k=1}^{n} x_k^3 \\ \sum_{k=1}^{n} x_k^2 & \sum_{k=1}^{n} x_k^3 & \sum_{k=1}^{n} x_k^4 \end{bmatrix}^{-1} \begin{bmatrix} \sum_{k=1}^{n} y_k^* \\ \sum_{k=1}^{n} x_k y_k^* \\ \sum_{k=1}^{n} x_k^2 y_k^* \end{bmatrix}$

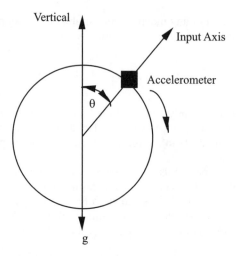

Fig. 2.27   Accelerometer experiment test setup.

where θ_k is the angle the accelerometer makes with the vertical for the kth measurement, g is gravity, B is an accelerometer bias, SF is the accelerometer scale factor error, and K is a gravity-squared or g-squared sensitive drift. Because the theoretical accelerometer output does not have bias, scale factor, or g-sensitive errors, we expect the output to only contain the vertical angle dependent gravity term or

$$\text{Theory} = g \cos \theta_k$$

Therefore, the accelerometer error is simply the difference between the actual output and theory or

$$\text{Error} = \text{Accelerometer Output} - \text{Theory} = B + SFg \cos \theta_k + K(g \cos \theta_k)^2$$

From calculating the preceding error at different angles, we would like to be able to estimate the bias, scale factor, and g-sensitive drift errors. Therefore, using the nomenclature of this section, we can think of the error as y_k (i.e., accelerometer output minus theory) and the gravity component $g \cos \theta_k$ as x_k. However, to make our example slightly more realistic, let us assume that our measurement of the angle θ_k may not be precise but is corrupted by zero-mean, Gaussian noise with standard deviation σ_{Noise}. The error is now given by

$$\text{Error} = \text{Accelerometer Output} - \text{Theory} = g \cos \theta_k^* + B + SFg \cos \theta_k^*$$
$$+ K(g \cos \theta_k^*)^2 - g \cos \theta_k$$

where θ_k is the actual angle and θ_k^* is the measured angle. Under these circumstances we can still think of the error as y_k, but we must now think of

Table 2.7   Nominal values for accelerometer testing example

Term	Scientific value	English units
Bias error	$10\,\mu g$	$10*10^{-6}*32.2 = 0.000322\,\text{ft/s}^2$
Scale factor error	$5\,\text{ppm}$	$5*10^{-6}$
g-squared sensitive drift	$1\,\mu g/g^2$	$1*10^{-6}/32.2 = 3.106*10^{-8}\,\text{s}^2/\text{ft}$

the gravity component $g \cos \theta_k^*$ rather than $g \cos \theta_k$ as x_k. Unfortunately, we no longer get a cancellation in the error equation, which means that

$$g \cos \theta_k^* - g \cos \theta_k \neq 0$$

For our example let us assume that the true values for the bias, scale factor, and g-sensitive drift errors are summarized in Table 2.7.

The simulation of Listing 2.2 was modified for this new experiment and appears in Listing 2.6. The general measurement equation is now given by

$$y_k^* = B + SFg \cos \theta_k^* + K(g \cos \theta_k^*)^2 + g \cos \theta_k^* - g \cos \theta_k$$

We can see from the "do loop" in Listing 2.6 that measurements are taken at angles from 0 to 90 deg in steps of 1 deg. Initially, the simulation is set up so there is no noise on the angle θ_k^*. Under these circumstances

$$\theta_k^* = \theta_k$$

and so the error or measurement equation for the first example will be given by

$$y_k^* = B + SFg \cos \theta_k + K(g \cos \theta_k)^2$$

and the independent variable can be considered to be

$$x_k = g \cos \theta_k$$

Because the measurement equation is second order in our example, we can use the second-order polynomial to fit the data in the least squares sense or

$$\hat{y}_k = a_0 + a_1 x_k + a_2 x_k^2$$

Therefore, we know from Table 1.4 that the coefficients for the best least squares fit to a second-order polynomial are given by

$$
\begin{bmatrix} a_0 \\ a_1 \\ a_2 \end{bmatrix} = \begin{bmatrix} n & \sum_{k=1}^{n} x_k & \sum_{k=1}^{n} x_k^2 \\ \sum_{k=1}^{n} x_k & \sum_{k=1}^{n} x_k^2 & \sum_{k=1}^{n} x_k^3 \\ \sum_{k=1}^{n} x_k^2 & \sum_{k=1}^{n} x_k^3 & \sum_{k=1}^{n} x_k^4 \end{bmatrix}^{-1} \begin{bmatrix} \sum_{k=1}^{n} y_k^* \\ \sum_{k=1}^{n} x_k y_k^* \\ \sum_{k=1}^{n} x_k^2 y_k^* \end{bmatrix}
$$

The simulation of Listing 2.2 was modified for the accelerometer example, and the changes for the input data are highlighted in bold in Listing 2.6. We can see from Listing 2.6 that the nominal values of Table 2.7 are used for the bias, scale factor error, and g-sensitive drift (i.e., BIAS, SF, and XK), whereas the vertical angle measurement noise standard deviation (i.e., SIGTH = 0) is initially assumed to be zero. In other words, initially we are assuming that we have perfect knowledge of the vertical angle. A "do loop" has been added to automatically compute the accelerometer output; otherwise, everything else remains the same. At the end of the program, we have another "do loop" where we recalculate the second-order polynomial, based on the least squares coefficients, and compare the result to the measurement data.

Listing 2.6　Method of least squares applied to accelerometer testing problem

```
C THE FIRST THREE STATEMENTS INVOKE THE ABSOFT RANDOM
  NUMBER GENERATOR ON THE MACINTOSH
      GLOBAL DEFINE
              INCLUDE 'quickdraw.inc'
      END
      IMPLICIT REAL*8 (A-H)
      IMPLICIT REAL*8 (O-Z)
      REAL*8 T(100),X(100),A(3,3),AINV(3,3),B(3,1),ANS(3,1)
      OPEN(1,STATUS='UNKNOWN',FILE='DATFIL')
      BIAS=0.00001*32.2
      SF=0.000005
      XK=0.000001/32.2
      SIGTH=0.
      G=32.2
      JJ=0
      DO 11 THETDEG=0.,180.,2.
      THET=THETDEG/57.3
      CALL GAUSS(THETNOISE,SIGTH)
      THETS=THET+THETNOISE
      JJ=JJ+1
      T(JJ)=32.2*COS(THETS)
      X(JJ)=BIAS+SF*G*COS(THETS)+XK*(G*COS(THETS))**2-G*COS(THET)
    1    +G*COS(THETS)
   11 CONTINUE
```

(continued)

Listing 2.6 (*Continued*)

```
N=JJ
SUM1=0
SUM2=0
SUM3=0
SUM4=0
SUM5=0
SUM6=0
SUM7=0
DO 10 I=1,JJ
   SUM1=SUM1+T(I)
   SUM2=SUM2+T(I)*T(I)
   SUM3=SUM3+X(I)
   SUM4=SUM4+T(I)*X(I)
   SUM5=SUM5+T(I)*T(I)*T(I)
   SUM6=SUM6+T(I)*T(I)*T(I)*T(I)
   SUM7=SUM7+T(I)*T(I)*X(I)
10 CONTINUE
   A(1,1)=N
   A(1,2)=SUM1
   A(1,3)=SUM2
   A(2,1)=SUM1
   A(2,2)=SUM2
   A(2,3)=SUM5
   A(3,1)=SUM2
   A(3,2)=SUM5
   A(3,3)=SUM6
   DET1=A(1,1)*A(2,2)*A(3,3)+A(1,2)*A(2,3)*A(3,1)
   DET2=A(1,3)*A(2,1)*A(3,2)-A(1,3)*A(2,2)*A(3,1)
   DET3=-A(1,2)*A(2,1)*A(3,3)-A(1,1)*A(2,3)*A(3,2)
   DET=DET1+DET2+DET3
   AINV(1,1)=(A(2,2)*A(3,3)-A(2,3)*A(3,2))/DET
   AINV(1,2)=(A(1,3)*A(3,2)-A(1,2)*A(3,3))/DET
   AINV(1,3)=(A(1,2)*A(2,3)-A(2,2)*A(1,3))/DET
   AINV(2,1)=(A(3,1)*A(2,3)-A(2,1)*A(3,3))/DET
   AINV(2,2)=(A(1,1)*A(3,3)-A(3,1)*A(1,3))/DET
   AINV(2,3)=(A(2,1)*A(1,3)-A(1,1)*A(2,3))/DET
   AINV(3,1)=(A(2,1)*A(3,2)-A(3,1)*A(2,2))/DET
   AINV(3,2)=(A(3,1)*A(1,2)-A(1,1)*A(3,2))/DET
   AINV(3,3)=(A(1,1)*A(2,2)-A(1,2)*A(2,1))/DET
   B(1,1)=SUM3
   B(2,1)=SUM4
   B(3,1)=SUM7
   CALL MATMUL(AINV,3,3,B,3,1,ANS)
   WRITE(9,*)ANS(1,1),ANS(2,1),ANS(3,1)
   PAUSE
   DO 12 JJ=1,N
   PZ=S(1,1)+ANS(2,1)*T(JJ)+NS(3,1)*T(JJ)*T(JJ)
   WRITE(9,*)T(JJ),X(JJ),PZ
   WRITE(1,*)T(JJ),X(JJ),PZ
```

(*continued*)

Listing 2.6 (*Continued*)

12 CONTINUE
    PAUSE
    END
C SUBROUTINE GAUSS IS SHOWN IN LISTING 1.8
C SUBROUTINE MATMUL IS SHOWN IN LISTING 1.4

The nominal case of Listing 2.6 was run (i.e., no noise on angle or SIGTH = 0), and the answers for the three coefficients came out to be

$$\text{ANS}(1,1) = a_0 = 0.000322$$
$$\text{ANS}(2,1) = a_1 = 0.000005$$
$$\text{ANS}(3,1) = a_2 = 0.00000003106$$

which are precisely the correct values for the accelerometer bias, scale factor, and g-sensitive drift errors. Therefore, by using the method of least squares, we have estimated the accelerometer bias, scale factor, and drift errors based on accelerometer measurements taken by rotating the input axis of the accelerometer. We can see from Fig. 2.28 that in this case the second-order polynomial with the best least squares coefficients actually passes through all of the measurements. However, accelerometer random noise is being neglected in this example.

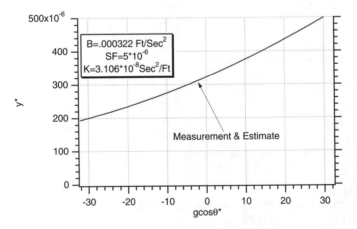

Fig. 2.28   Without measurement noise we can estimate accelerometer errors perfectly.

Another case was run with Listing 2.6 in which there was 1 μr of angle noise on the vertical angle (i.e., $1*10^{-6}$ rad or SIGTH $= 0.000001$). In this case the three coefficients came out to be

$$ANS(1,1) = a_0 = 0.0003206$$
$$ANS(2,1) = a_1 = 0.00000495$$
$$ANS(3,1) = a_2 = 0.00000003203$$

which are approximately the correct values for the accelerometer bias, scale factor, and g-sensitive drift errors. We can see from Fig. 2.29 that in this case the second-order polynomial with the best least squares coefficients passes through the average of the measurements quite well.

Increasing the angle noise by an order of magnitude to 10 μr (i.e., $10*10^{-6}$ rad or SIGTH $= 0.00001$) and running Listing 2.6 again yields coefficients

$$ANS(1,1) = a_0 = 0.0003083$$
$$ANS(2,1) = a_1 = 0.00000451$$
$$ANS(3,1) = a_2 = 0.00000004082$$

which are still to within approximately 10% of the correct values for the accelerometer bias and scale factor error. However, the g-sensitive drift estimate is in error of the true value by approximately 25%. We can see from Fig. 2.30 that in this case the measurements appear to be very noisy. The second-order polynomial with the best least squares coefficients appears to pass through the average of the measurements.

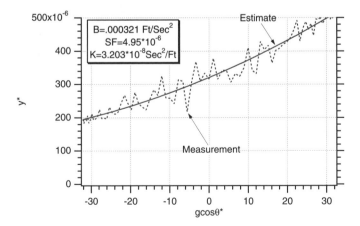

Fig. 2.29 With 1 μr of measurement noise we can nearly estimate accelerometer errors perfectly.

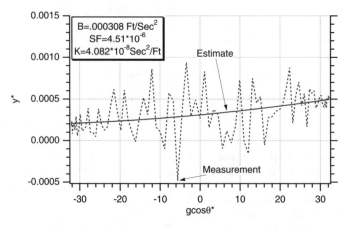

Fig. 2.30 Estimation with 10 μr of measurement noise.

Finally, increasing the angle noise by an order of magnitude to 100 μr (i.e., $100*10^{-6}$ rad or SIGTH = 0.0001) and running Listing 2.6 again yields coefficients

$$ANS(1,1) = a_0 = 0.000185$$
$$ANS(2,1) = a_1 = 0.0000000882$$
$$ANS(3,1) = a_2 = 0.0000001287$$

Our estimates for the accelerometer bias, scale factor error, and g-sensitive drift differ considerably from the actual values. We can see from Fig. 2.31 that,

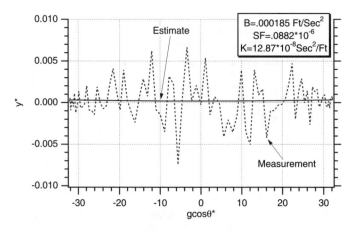

Fig. 2.31 With 100 μr of measurement noise we cannot estimate bias and scale factor errors.

because the measurements are extremely noisy, the second-order polynomial with the best least squares coefficients is hardly a parabola. In other words, for the bias, scale factor, and g-sensitive drift errors of Table 2.5 we require a test set upon which we can very accurately measure the input axis orientation angle.

Thus, we can see in this example that we have to be cautious in applying the method of least squares to measurement data. In this example the noise was not additive on the actual signal but simply corrupted the angle. In effect, we have a pseudonoise in the system. We can see from the results of this section that, despite the pseudonoise, the least squares technique worked reasonably well (i.e., 10 and 25% errors) for small values of measurement noise.

Summary

We have seen that in order to best fit polynomial measurement data corrupted by noise it is best to use a least-squares filter-based on a polynomial order that is matched to the actual signal. If the filter-based polynomial is of lower order than the signal polynomial, the filter will not be able to track the signal over long periods of time because the filter estimate will diverge from the actual signal. If the filter-based polynomial is of higher order than the signal polynomial, bad estimates may be obtained because the filter polynomial is attempting to fit the noise rather than the signal. Table 2.8 summarizes the formulas for the coefficients of the least squares filter used to fit the measurement data for different order systems. We can see from Table 2.8 that this form of the method of least squares is a batch processing technique because all of the measurement data must first be collected before the technique can be applied. It is also apparent from Table 2.8 that the taking of a matrix inverse is required in the

Table 2.8   Least-squares coefficients for different order polynomial fits to measurement data

Order	Equations
Zeroth	$$a_0 = \frac{\sum\limits_{k=1}^{n} x_k^*}{n}$$
First	$$\begin{bmatrix} a_0 \\ a_1 \end{bmatrix} = \begin{bmatrix} n & \sum\limits_{k=1}^{n}(k-1)T_s \\ \sum\limits_{k=1}^{n}(k-1)T_s & \sum\limits_{k=1}^{n}[(k-1)T_s]^2 \end{bmatrix}^{-1} \begin{bmatrix} \sum\limits_{k=1}^{n} x_k^* \\ \sum\limits_{k=1}^{n}(k-1)T_s x_k^* \end{bmatrix}$$
Second	$$\begin{bmatrix} a_0 \\ a_1 \\ a_2 \end{bmatrix} = \begin{bmatrix} n & \sum\limits_{k=1}^{n}(k-1)T_s & \sum\limits_{k=1}^{n}[(k-1)T_s]^2 \\ \sum\limits_{k=1}^{n}(k-1)T_s & \sum\limits_{k=1}^{n}[(k-1)T_s]^2 & \sum\limits_{k=1}^{n}[(k-1)T_s]^3 \\ \sum\limits_{k=1}^{n}[(k-1)T_s]^2 & \sum\limits_{k=1}^{n}[(k-1)T_s]^3 & \sum\limits_{k=1}^{n}[(k-1)T_s]^4 \end{bmatrix}^{-1} \begin{bmatrix} \sum\limits_{k=1}^{n} x_k^* \\ \sum\limits_{k=1}^{n}(k-1)T_s x_k^* \\ \sum\limits_{k=1}^{n}[(k-1)T_s]^2 x_k^* \end{bmatrix}$$

method of least squares. The order of the required inverse is one more than the order of the least squares polynomial fit (i.e., first-order filter requires inverse of 2×2 matrix, second-order filter requires inverse of 3×3 matrix, etc.).

References

[1]Sokolnikoff, I. S., and Redheffer, R. M., *Mathematics of Physics and Modern Engineering*, McGraw–Hill, New York, 1958, pp. 702–715.
[2]Press, W. H., Flannery, B. P., Teukolsky, S. A., and Vetterling, W. T., *Numerical Recipes: The Art of Scientific Computation*, Cambridge Univ. Press, London, 1986, pp. 29–31.

Chapter 3

Recursive Least-Squares Filtering

Introduction

$\mathbf{I}$N THE preceding chapter we saw how the method of least squares could be applied to estimating a signal based upon noisy measurements. This estimation process is also sometimes called filtering. We also observed that the least-squares technique was a batch-processing method because all of the measurements had to be taken before any estimates of the best polynomial coefficients could be made. In addition, a matrix inverse had to be evaluated as part of the required computation. The dimension of the matrix inverse was proportional to the order of the polynomial used to best fit the measurements in the least-squares sense. In this chapter we will see how the batch-processing method of least squares of Chapter 2 can be made recursive. The resulting recursive least-squares filter does not involve taking matrix inverses. Because the new least-squares filter is recursive, estimates are available as soon as measurements are taken. The simple nature of the calculations involved make recursive least-squares filtering ideal for digital computer implementation.

Making Zeroth-Order Least-Squares Filter Recursive

We have shown in the preceding chapter that if we were trying to fit a zeroth-order polynomial or constant to a set of measurement data the best estimate (i.e., minimize sum of squares of the difference between the measurement and estimate) can be expressed as

$$\hat{x}_k = a_0 = \frac{\sum\limits_{i=1}^{k} x_i^*}{k}$$

where x_i^* is the ith measurement and k is the number of measurements taken. In other words, we simply add up the measurements and divide by the number of measurements taken to find the best estimate. By changing subscripts we can rewrite the preceding expression as

$$\hat{x}_{k+1} = \frac{\sum_{i=1}^{k+1} x_i^*}{k+1}$$

Expanding the numerator yields

$$\hat{x}_{k+1} = \frac{\sum_{i=1}^{k} x_i^* + x_{k+1}^*}{k+1}$$

Because

$$\sum_{i=1}^{k} x_i^* = k\hat{x}_k$$

by substitution we can say that

$$\hat{x}_{k+1} = \frac{k\hat{x}_k + x_{k+1}^*}{k+1}$$

Without changing anything we can add and subtract the preceding state estimate to the numerator of the preceding equation, yielding

$$\hat{x}_{k+1} = \frac{k\hat{x}_k + \hat{x}_k + x_{k+1}^* - \hat{x}_k}{k+1} = \frac{(k+1)\hat{x}_k + x_{k+1}^* - \hat{x}_k}{k+1}$$

Therefore, we can rewrite the preceding expression as

$$\hat{x}_{k+1} = \hat{x}_k + \frac{1}{k+1}(x_{k+1}^* - \hat{x}_k)$$

Because we would like k to start from one rather than zero, we can also rewrite the preceding expression, by changing subscripts, to

$$\hat{x}_k = \hat{x}_{k-1} + \frac{1}{k}(x_k^* - \hat{x}_{k-1})$$

The preceding expression is now in the recursive form we desire because the new estimate simply depends on the old estimate plus a gain (i.e., $1/k$ for the zeroth-order filter) times a residual (i.e., current measurement minus preceding estimate).

Properties of Zeroth-Order or One-State Filter

We have just seen in the preceding section that the form of the zeroth-order or one-state recursive least-squares filter is given by

$$\hat{x}_k = \hat{x}_{k-1} + K_{1_k}\text{Res}_k$$

where the gain of the filter is

$$K_{1_k} = \frac{1}{k} \qquad k = 1, 2, \ldots, n$$

and the residual (i.e., difference between the present measurement and the preceding estimate) is given by

$$\text{Res}_k = x_k^* - \hat{x}_{k-1}$$

We can see that the filter gain for the zeroth-order recursive least-squares filter is unity for the first measurement (i.e., $k = 1$) and eventually goes to zero as more measurements are taken.

To better understand how the preceding formulas apply, let us reconsider the four measurement examples of the preceding chapter. For convenience Table 3.1 repeats the sample measurement data of Chapter 2.

Because initially $k = 1$, the first gain of the recursive zeroth-order least-squares filter is computed from

$$K_{1_1} = \frac{1}{k} = \frac{1}{1} = 1$$

To apply the other zeroth-order recursive least-squares filter formulas, we need to have an initial condition for our first estimate in order to get started. For now let us assume that we have no idea how to initialize the filter, and so we simply set the initial estimate to zero or

$$\hat{x}_0 = 0$$

We now calculate the residual as

$$\text{Res}_1 = x_1^* - \hat{x}_0 = 1.2 - 0 = 1.2$$

Table 3.1 Sample measurement data

k	$(k-1)T_s$	x_k^*
1	0	1.2
2	1	0.2
3	2	2.9
4	3	2.1

and the new estimate becomes

$$\hat{x}_1 = \hat{x}_0 + K_{1_1}\mathrm{Res}_1 = 0 + 1*1.2 = 1.2$$

We have now completed the first cycle of the recursive equations. For the next cycle with $k = 2$, the next gain is computed as

$$K_{1_2} = \frac{1}{k} = \frac{1}{2} = 0.5$$

whereas the next residual becomes

$$\mathrm{Res}_2 = x_2^* - \hat{x}_1 = 0.2 - 1.2 = -1$$

Therefore, the second estimate of the recursive least-squares filter is calculated as

$$\hat{x}_2 = \hat{x}_1 + K_{1_2}\mathrm{Res}_2 = 1.2 + 0.5*(-1) = 0.7$$

The calculations for processing the third measurement (i.e., $k = 3$) can be computed as

$$K_{1_3} = \frac{1}{k} = \frac{1}{3} = 0.333$$
$$\mathrm{Res}_3 = x_3^* - \hat{x}_2 = 2.9 - 0.7 = 2.2$$
$$\hat{x}_3 = \hat{x}_2 + K_{1_3}\mathrm{Res}_3 = 0.7 + 0.333*2.2 = 1.43$$

while the calculations for the fourth measurement (i.e., $k = 4$) turn out to be

$$K_{1_4} = \frac{1}{k} = \frac{1}{4} = 0.25$$
$$\mathrm{Res}_4 = x_4^* - \hat{x}_3 = 2.1 - 1.43 = 0.67$$
$$\hat{x}_4 = \hat{x}_3 + K_{1_4}\mathrm{Res}_4 = 1.43 + 0.25*0.67 = 1.6$$

Note that the last estimate of the zeroth-order least-squares recursive filter is identical to the last estimate obtained from the zeroth-order least-squares fit of Chapter 2.

To demonstrate that the recursive least-squares filter estimates are independent of initial conditions, suppose the initial estimate of the zeroth-order least-squares recursive filter was 100 rather than zero or

$$\hat{x}_0 = 100$$

Then the remaining calculations for the first estimate are

$$\mathrm{Res}_1 = \hat{x}_1^* - \hat{x}_0 = 1.2 - 100 = -98.8$$
$$\hat{x}_1 = \hat{x}_0 + K_{1_1}\mathrm{Res}_1 = 100 + 1*(-98.8) = 1.2$$

which is exactly the same answer we obtained when the initial estimate was zero. In other words, *the zeroth-order recursive least-squares filter will yield the same answers regardless of initial conditions.*

Figure 3.1 compares the estimates from the zeroth-order batch-processing method of least squares from Chapter 2 with the zeroth-order recursive method of least squares of this chapter. As was mentioned before, the batch-processing least-squares estimates are only available after all of the measurements are taken (i.e., we solve for polynomial coefficients after all of the measurements are taken and then evaluate the polynomial at different points in time), whereas the recursive least-squares estimates are available as the measurements are being taken. We can see from Fig. 3.1 that after all of the measurements are taken both the batch-processing and recursive least-squares methods yield the same answers.

If the actual measurement data are simply a constant plus noise (i.e., zeroth-order signal plus noise) where the measurement noise is a zero-mean Gaussian process with variance σ_n^2, then a formula can also be derived that describes the variance of the error in the filter's estimate (i.e., variance of actual signal minus filter estimate). Recall that the recursive form of the zeroth-order least-squares filter is given by

$$\hat{x}_k = \hat{x}_{k-1} + \frac{1}{k}(x_k^* - \hat{x}_{k-1})$$

Therefore, the error in the estimate must be the actual signal minus the estimate or

$$x_k - \hat{x}_k = x_k - \hat{x}_{k-1} - \frac{1}{k}(x_k^* - \hat{x}_{k-1})$$

Recognizing that the measurement is simply the signal plus noise or

$$x_k^* = x_k + v_k$$

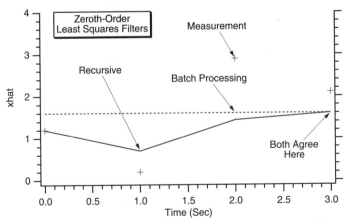

Fig. 3.1 **Batch-processing and recursive least-squares methods yield the same answers after all measurements are taken.**

we obtain for the error in the estimate as

$$x_k - \hat{x}_k = x_k - \hat{x}_{k-1} - \frac{1}{k}(x_k + v_k - \hat{x}_{k-1})$$

However, because the signal is a constant for the zeroth-order system, we can also say that

$$x_k = x_{k-1}$$

Therefore, substitution yields

$$x_k - \hat{x}_k = (x_{k-1} - \hat{x}_{k-1})\left(1 - \frac{1}{k}\right) - \frac{1}{k}v_k$$

After squaring both sides of the preceding equation, we obtain

$$(x_k - \hat{x}_k)^2 = (x_{k-1} - \hat{x}_{k-1})^2\left(1 - \frac{1}{k}\right)^2 - 2\left(1 - \frac{1}{k}\right)(x_{k-1} - \hat{x}_{k-1})\frac{v_k}{k} + \left(\frac{1}{k}v_k\right)^2$$

Taking expectations of both sides of the equation yields

$$E[(x_k - \hat{x}_k)^2] = E[(x_{k-1} - \hat{x}_{k-1})^2]\left(1 - \frac{1}{k}\right)^2 - 2\left(1 - \frac{1}{k}\right)E[(x_{k-1} - \hat{x}_{k-1})v_k]\frac{1}{k}$$
$$+ E\left[\left(\frac{1}{k}v_k\right)^2\right]$$

If we define the variance of the error in the estimate to be P_k and the variance of the measurement noise to be σ_n^2 and also assume that the noise is uncorrelated with the state or the estimate of the state, we can say that

$$E[(x_k - \hat{x}_k)^2] = P_k$$
$$E(v_k^2) = \sigma_n^2$$
$$E[(x_{k-1} - \hat{x}_{k-1})v_k] = 0$$

Substitution of the preceding three equations into the expectation equation yields the difference equation for the variance of the error in the estimate or

$$P_k = P_{k-1}\left(1 - \frac{1}{k}\right)^2 + \frac{\sigma_n^2}{k^2}$$

The preceding difference equation can be solved using Z-transform techniques, but it is far easier in this example to use engineering induction by substituting different values of k. Substituting $k = 1, 2, 3$, and 4 into the preceding equation yields

$$P_1 = P_0\left(1 - \frac{1}{1}\right)^2 + \frac{\sigma_n^2}{1^2} = \sigma_n^2$$

$$P_2 = P_1\left(1 - \frac{1}{2}\right)^2 + \frac{\sigma_n^2}{2^2} = \sigma_n^2\frac{1}{4} + \frac{\sigma_n^2}{4} = \frac{\sigma_n^2}{2}$$

$$P_3 = P_2\left(1 - \frac{1}{3}\right)^2 + \frac{\sigma_n^2}{3^2} = \frac{\sigma_n^2}{2}\frac{4}{9} + \frac{\sigma_n^2}{9} = \frac{\sigma_n^2}{3}$$

$$P_4 = P_3\left(1 - \frac{1}{4}\right)^2 + \frac{\sigma_n^2}{4^2} = \frac{\sigma_n^2}{3}\frac{9}{16} + \frac{\sigma_n^2}{16} = \frac{\sigma_n^2}{4}$$

The trend in the preceding four equations is now obvious, and we can summarize these results by saying in general for the zeroth-order recursive least-squares filter that the variance of the error in the estimate is simply

$$P_k = \frac{\sigma_n^2}{k}$$

where σ_n^2 is the variance of the measurement noise and k is the number of measurements taken (i.e., $k = 1, 2, \ldots, n$).

If, on the other hand, the real signal is a first-order polynomial (one degree higher than the zeroth-order filter) or

$$x_k = a_0 + a_1 t = a_0 + a_1(k - 1)T_s$$

the filter will not be able to track the signal, as we saw in Chapter 2. The resulting error in the estimate is known as truncation error ε_k and by definition is the difference between the true signal and the estimate or

$$\varepsilon_k = x_k - \hat{x}_k$$

Recall that we showed in Chapter 2 that the estimate of the zeroth-order least-squares filter was simply

$$\hat{x}_k = \frac{\sum_{i=1}^{k} x_i^*}{k}$$

In the noise-free case the measurement is the signal, and so the preceding equation can be rewritten as

$$\hat{x}_k = \frac{\sum\limits_{i=1}^{k} x_i}{k} = \frac{\sum\limits_{i=1}^{k}[a_0 + a_1(i-1)T_s]}{k} = \frac{a_0\sum\limits_{i=1}^{k} + a_1 T_s \sum\limits_{i=1}^{k} i - a_1 T_s \sum\limits_{i=1}^{k}}{k}$$

Because mathematical handbooks[1] tell us that

$$\sum_{i=1}^{k} = k$$

$$\sum_{i=1}^{k} i = \frac{k(k+1)}{2}$$

we can eliminate summation signs in the expression for the estimate and obtain

$$\hat{x}_k = \frac{a_0 k + a_1 T_s[k(k+1)/2] - a_1 T_s k}{k} = a_0 + \frac{a_1 T_s}{2}(k-1)$$

Substituting the estimate into the equation for the error in the estimate yields

$$\varepsilon_k = x_k - \hat{x}_k = a_0 + a_1 T_s(k-1) - a_0 - \frac{a_1 T_s}{2}(k-1) = \frac{a_1 T_s}{2}(k-1)$$

which is the equation for the truncation error of the zeroth-order recursive least-squares filter.

Listing 3.1 shows how the zeroth-order recursive least-squares filter, along with the theoretical formulas, were programmed. In this example the listing indicates that the measurements of a constant signal ACT plus noise XS are taken every 0.1 s for 10 s. We can see from Listing 3.1 that the actual error in the estimate XHERR is computed and compared to the theoretical error in the estimate or plus and minus the square root of P_k (i.e., SP11 in Listing 3.1). As was the case in Chapter 2 for the zeroth-order measurement, nominally the simulation is set up so that the true signal is also a zeroth-order polynomial or constant with value unity (A0 = 1, A1 = 0), which is corrupted by noise with unity standard deviation (SIGNOISE = 1). We can also see that the preceding theoretical truncation error formula also appears in Listing 3.1.

A run was made with the nominal case of Listing 3.1, and the true signal, measurement, and filter estimate are compared in Fig. 3.2. We can see that the actual measurement is quite noisy because in this example the amplitude of the signal is equal to the standard deviation of the noise. However, we can also see that the filter estimate is quite good (i.e., solid curve in Fig. 3.2) and appears to be getting closer to the true signal as more measurements are taken.

For more exact filtering work the error in the estimate (i.e., difference between the true signal and estimate) is often used as a measure of performance. The error in the estimate can only be computed in a simulation where the actual signal is known. In the real world, where the filter must eventually work, the real signal or

**Listing 3.1   Simulation for testing zeroth-order recursive
least-squares filter**

```
C THE FIRST THREE STATEMENTS INVOKE THE ABSOFT RANDOM
  NUMBER GENERATOR ON THE MACINTOSH
    GLOBAL DEFINE
            INCLUDE 'quickdraw.inc'
    END
    IMPLICIT REAL*8(A-H,O-Z)
    OPEN(1,STATUS='UNKNOWN',FILE='DATFIL')
    TS=0.1
    SIGNOISE=1.
    A0=1.
    A1=0.
    XH=0.
    XN=0.
    DO 10 T=0.,10.,TS
    XN=XN+1.
    CALL GAUSS(XNOISE,SIGNOISE)
    ACT=A0+A1*T
    XS=ACT+XNOISE
    XK=1./XN
    RES=XS-XH
    XH=XH+XK*RES
    SP11=SIGNOISE/SQRT(XN)
    XHERR=ACT-XH
    EPS=0.5*A1*TS*(XN-1)
    WRITE(9,*)T,ACT,XS,XH,XHERR,SP11,-SP11,EPS
    WRITE(1,*)T,ACT,XS,XH,XHERR,SP11,-SP11,EPS
 10 CONTINUE
    CLOSE(1)
    PAUSE
    END
C SUBROUTINE GAUSS IS SHOWN IN LISTING 1.8
```

truth is never available. A more complete discussion of this topic can be found in Chapter 14. Figure 3.3 compares the error in the estimate of x to the theoretical predictions of this section (i.e., square root of P_k or SP11 in Listing 3.1). We can see that the single-run results are within the theoretical bounds most of the time, indicating that theory and simulation appear to agree. Also, from the formula for the variance of the estimate (i.e., σ_n^2/k), we see that the error in the estimate will tend toward zero as the number of measurements k gets large. For an infinite number of measurements, the variance of the error in the estimate will be zero.

A noise-free case (SIGNOISE $= 0$) was also run to examine the effect of truncation error. In this example the actual signal is first-order (A0 $= 1$, A1 $= 2$), whereas the filter is zeroth-order. We can see from Fig. 3.4 that it is clear that the estimate from the zeroth-order recursive least-squares filter estimate is diverging from the true signal. Therefore, we can say that the zeroth-order recursive least-squares filter cannot track the first-order signal.

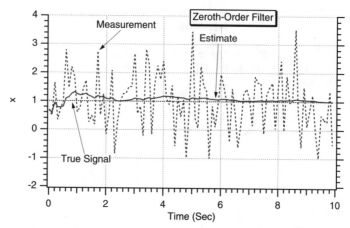

Fig. 3.2 **Zeroth-order recursive least-squares filter is able to track zero-order polynomial plus noise.**

Figure 3.5 displays the error in the estimate from simulation results and the theoretical estimate of the truncation error ε_k. For this particular numerical example the error caused by truncation is given by

$$\varepsilon_k = \frac{a_1 T_s}{2}(k-1) = 0.5*2*0.1(k-1) = 0.1(k-1)$$

where k goes from 1 to 100. We can see from Fig. 3.5 that the error in the estimate of x (i.e., XHERR in Listing 3.1) as a result of running Listing 3.1 and the truncation error formula (i.e., EPS in Listing 3.1) are identical, thus empirically validating the formula.

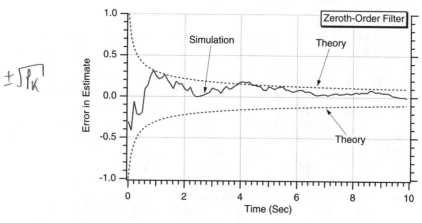

Fig. 3.3 **Single-run simulation results agree with theoretical formula.**

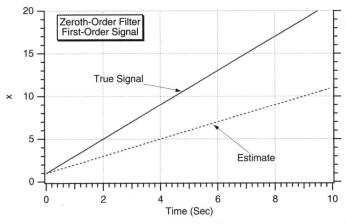

Fig. 3.4   Zeroth-order recursive least-squares filter is unable to track first-order polynomial.

Because the formula for the standard deviation of the error in the estimate caused by noise is given by

$$\sqrt{P_{11_k}} = \frac{\sigma_n}{\sqrt{k}}$$

and the error in the estimate caused by truncation error is

$$\varepsilon_k = 0.5 a_1 T_s (k - 1)$$

we can see that both errors behave in different ways. As already noted, as more measurements are taken k gets larger, and the error in the estimate caused by the

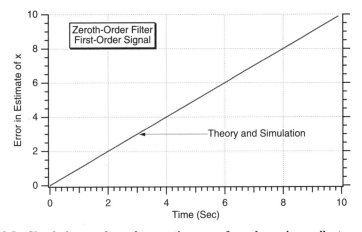

Fig. 3.5   Simulation results and truncation error formula are in excellent agreement.

measurement noise decreases while the error in the estimate caused by truncation error increases. In principle, for any particular numerical example there is an optimal value of k that will minimize the errors in the estimates caused by both measurement noise and truncation error.

If more runs were made in the case in which there was no truncation error, we would expect to find the computed error in the estimate of x to lie between the theoretical abounds approximately 68% of the time. To ensure that the simulated and theoretical errors in the estimates agree in this way, Listing 3.1 was slightly modified so that we could run repeated simulation trials, also known as the Monte Carlo method. The Monte Carlo version of Listing 3.1 appears in Listing 3.2. We

Listing 3.2   Monte Carlo simulation for testing zeroth-order recursive least-squares filter

```
C THE FIRST THREE STATEMENTS INVOKE THE ABSOFT RANDOM
  NUMBER GENERATOR ON THE MACINTOSH
    GLOBAL DEFINE
            INCLUDE 'quickdraw.inc'
    END
    IMPLICIT REAL*8(A-H,O-Z)
    OPEN(1,STATUS='UNKNOWN',FILE='DATFIL1')
    OPEN(2,STATUS='UNKNOWN',FILE='DATFIL2')
    OPEN(3,STATUS='UNKNOWN',FILE='DATFIL3')
    OPEN(4,STATUS='UNKNOWN',FILE='DATFIL4')
    OPEN(5,STATUS='UNKNOWN',FILE='DATFIL5')
    DO 11 K=1,5
    TS=0.1
    SIGNOISE=1.
    A0=1.
    A1=0.
    XH=0.
    XN=0.
    DO 10 T=0.,10.,TS
    XN=XN+1.
    CALL GAUSS(XNOISE,SIGNOISE)
    ACT=A0+A1*T
    XS=ACT+XNOISE
    XK=1./XN
    RES=XS-XH
    XH=XH+XK*RES
    SP11=SIGNOISE/SQRT(XN)
    XHERR=ACT-XH
    EPS=0.5*A1*TS*(XN-1)
    WRITE(9,*)T,XHERR,SP11,-SP11
    WRITE(K,*)T,XHERR,SP11,-SP11
 10 CONTINUE
    CLOSE(K)
 11 CONTINUE
    PAUSE
    END
C SUBROUTINE GAUSS IS SHOWN IN LISTING 1.8
```

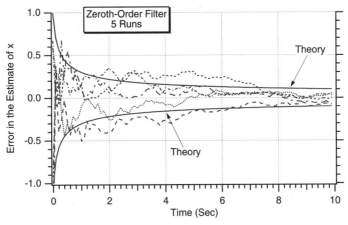

Fig. 3.6 Monte Carlo results lie within the theoretical bounds approximately 68% of the time.

can see that a 5 run "do loop" has been created, and the results of each run are written to files DATFIL1 through DATFIL5. The statements that were changed from Listing 3.1, in order to get a Monte Carlo version of the program, are highlighted in boldface in Listing 3.2.

A five-run Monte Carlo set was run with Listing 3.2. We can see from Fig. 3.6 that the errors in the estimate of x from each of the simulation trials appears, on average, to lie within the theoretical bounds approximately 68% of the time. These results indicate that the recursive zeroth-order least-squares filter appears to be behaving according to the theoretical predictions (i.e., formula for standard deviation of error in the estimate).

Properties of First-Order or Two-State Filter

Using techniques similar to the ones used in the first section of this chapter, the two gains of a first-order or two-state recursive least-squares filter can be shown to be[2]

$$K_{1_k} = \frac{2(2k - 1)}{k(k + 1)} \qquad k = 1, 2, \ldots, n$$

$$K_{2_k} = \frac{6}{k(k + 1)T_s}$$

We can see that, as was the case before, both filter gains eventually go to zero as more measurements are taken (i.e., k gets larger). If we think of the measurement as a position, then the estimates from this two-state filter will be position and velocity (i.e., derivative of position). In other words, position is the first state, whereas the derivative of position or velocity is the second state. For this filter a residual is first formed, which is the difference between the present measurement and a projection of the preceding estimate to the current time or

$$\text{Res}_k = x_k^* - \hat{x}_{k-1} - \hat{\dot{x}}_{k-1} T_s$$

The new filter estimates are a combination of the preceding state estimates projected forward to the current time plus a gain multiplied by the residual or

$$\hat{x}_k = \hat{x}_{k-1} + \hat{\dot{x}}_{k-1} T_s + K_{1_k} \text{Res}_k$$
$$\hat{\dot{x}}_k = \hat{\dot{x}}_{k-1} + K_{2_k} \text{Res}_k$$

To understand better how these formulas apply, let us again reconsider the four measurement examples of the preceding chapter and the preceding section, where the measurements are

$$x_1^* = 1.2$$
$$x_2^* = 0.2$$
$$x_3^* = 2.9$$
$$x_4^* = 2.1$$

The calculations for the gains, residual, and states for each value of k are summarized, in detail, in Table 3.2. For each value of k, we compute gains, a residual, and the two-state estimates.

As was the case in the preceding section, *the last estimate of the first-order recursive least-squares filter (i.e., $\hat{x}_4 = 2.41$) is identical to the estimate obtained from the first-order least-squares fit of Chapter 2.*

Figure 3.7 compares the estimates from both the batch-processing method of least squares for the first-order system of Chapter 2 and the first-order recursive least-squares filter of this section. Recall that the first-order recursive least-squares filter state estimates are based on the calculations of Table 3.2. Again we can see from Fig. 3.7 that, after all of the measurements are taken, both the batch-processing and recursive least-squares filters yield the same answers. The recursive first-order filter estimate passes through the first two measurements.

If the measurement data are also a first-order polynomial and the measurement noise statistics are known, then formulas can also be derived, using techniques similar to those used on the zeroth-order filter of the preceding section, describing the variance of the error in the estimates in both states (i.e., P_{11} and P_{22}) of the first-order filter. The variance of the error in the estimate for the first and second states are given by[2]

$$P_{11_k} = \frac{2(2k-1)\sigma_n^2}{k(k+1)}$$

$$P_{22_k} = \frac{12\sigma_n^2}{k(k^2-1)T_s^2}$$

Table 3.2   First-order recursive least-squares-filter calculations

Value of k	Calculations

$k = 1$

$$K_{1_1} = \frac{2(2k-1)}{k(k+1)} = \frac{2(2*1-1)}{1*1+1)} = 1$$

$$K_{2_k} = \frac{6}{k(k+1)T_s} = \frac{6}{1*(1+1)*1} = 3$$

$$\text{Res}_1 = x_1^* - \hat{x}_0 - \dot{\hat{x}}_0 T_s = 1.2 - 0 - 0*1 = 1.2$$

$$\hat{x}_1 = \hat{x}_0 + \dot{\hat{x}}_0 T_s + K_{1_1}\text{Res}_1 = 0 + 0*1 + 1*1.2 = 1.2$$

$$\dot{\hat{x}}_1 = \dot{\hat{x}}_0 + K_{2_1}\text{Res}_1 = 0 + 3*1.2 = 3.6$$

$k = 2$

$$K_{1_2} = \frac{2(2k-1)}{k(k+1)} = \frac{2(2*2-1)}{2*2+1)} = 1$$

$$K_{2_k} = \frac{6}{k(k+1)T_s} = \frac{6}{2*(2+1)*1} = 1$$

$$\text{Res}_2 = x_2^* - \hat{x}_1 - \dot{\hat{x}}_1 T_s = 0.2 - 1.2 - 3.6*1 = -4.6$$

$$\hat{x}_2 = \hat{x}_1 + \dot{\hat{x}}_1 T_s + K_{1_2}\text{Res}_2 = 1.2 + 3.6*1 + 1*(-4.6) = 0.2$$

$$\dot{\hat{x}}_2 = \dot{\hat{x}}_1 + K_{2_2}\text{Res}_2 = 3.6 + 1*(-4.6) = -1$$

$k = 3$

$$K_{1_3} = \frac{2(2k-1)}{k(k+1)} = \frac{2(2*3-1)}{3*(3+1)} = \frac{5}{6}$$

$$K_{2_3} = \frac{6}{k(k+1)T_s} = \frac{6}{3*(3+1)*1} = 0.5$$

$$\text{Res}_3 = x_3^* - \hat{x}_2 - \dot{\hat{x}}_2 T_s = 2.9 - 0.2 - (-1)*1 = 3.7$$

$$\hat{x}_3 = \hat{x}_2 + \dot{\hat{x}}_2 T_s + K_{1_3}\text{Res}_3 = 0.2 + (-1)*1 + \frac{5*(3.7)}{6} = 2.28$$

$$\dot{\hat{x}}_3 = \dot{\hat{x}}_2 + K_{2_3}\text{Res}_3 = -1 + 0.5*3.7 = 0.85$$

$k = 4$

$$K_{1_4} = \frac{2(2k-1)}{k(k+1)} = \frac{2(2*4-1)}{4*(4+1)} = 0.7$$

$$K_{2_4} = \frac{6}{k(k+1)T_s} = \frac{6}{4*(4+1)*1} = 0.3$$

$$\text{Res}_4 = x_4^* - \hat{x}_3 - \dot{\hat{x}}_3 T_s = 2.1 - 2.28 - 0.85*1 = -1.03$$

$$\hat{x}_4 = \hat{x}_3 + \dot{\hat{x}}_3 T_s + K_{1_4}\text{Res}_4 = 2.28 + 0.85*1 + 0.7*(-1.03) = 2.41$$

$$\dot{\hat{x}}_4 = \dot{\hat{x}}_3 + K_{2_4}\text{Res}_4 = 0.85 + 0.3*(-1.03) = 0.54$$

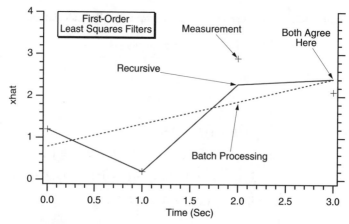

Fig. 3.7   First-order least-squares recursive and batch-processing least-squares filters yield the same answers after all measurements are taken.

where σ_n^2 is the variance of the measurement noise, k is the number of measurements taken (i.e., $k = 1, 2, \ldots, n$), and T_s is the sample time or time between measurements. As before, the variance of the errors in the estimates decrease as the number of measurements increases.

If, on the other hand, the real signal is a second-order polynomial (one degree higher than the first-order filter) or

$$x_k^* = a_0 + a_1 t + a_2 t^2 = a_0 + a_1(k-1)T_s + a_2(k-1)^2 T_s^2$$

the first-order filter will not be able to track the second-order signal as we saw in the preceding section. As was mentioned in the preceding section, the resulting error buildup caused by this lack of tracking ability is known as truncation error ε_k. The truncation errors for the states of the first-order recursive least-squares filter can be derived, using techniques similar to those of the preceding section on the zeroth-order filter, and are given by[2]

$$\varepsilon_k = \frac{1}{6} a_2 T_s^2 (k-1)(k-2)$$
$$\dot{\varepsilon}_k = a_2 T_s (k-1)$$

Listing 3.3 shows how the first-order recursive least-squares filter and second-order measurement signal, along with the theoretical formulas, were programmed. We can see that measurements of a first-order signal X plus noise XS is taken every 0.1 s for 10 s. We can see that the actual error in the estimate XHERR of the first state is compared to the theoretical error in the estimate of the first state (i.e., plus and minus the square root of P_{11}), whereas the actual error in the estimate of the second state XDHERR is compared to the theoretical error in the estimate of the second state (i.e., plus and minus the square root of P_{22}). Nominally the simulation is set so that the measurement is also a

Listing 3.3   Simulation for testing first-order recursive least-squares filter

```
C THE FIRST THREE STATEMENTS INVOKE THE ABSOFT RANDOM
   NUMBER GENERATOR ON THE MACINTOSH
   GLOBAL DEFINE
           INCLUDE 'quickdraw.inc'
   END
   IMPLICIT REAL*8(A-H,O-Z)
   TS=0.1
   SIGNOISE=5.
   OPEN(1,STATUS='UNKNOWN',FILE='DATFIL')
   OPEN(2,STATUS='UNKNOWN',FILE='COVFIL')
   A0=3.
   A1=1.
   A2=0.
   XH=0.
   XDH=0.
   XN=0
   DO 10 T=0.,10.,TS
   XN=XN+1.
   CALL GAUSS(XNOISE,SIGNOISE)
   X=A0+A1*T+A2*T*T
   XD=A1+2*A2*T
   XS=X+XNOISE
   XK1=2*(2*XN-1)/(XN*(XN+1))
   XK2=6/(XN*(XN+1)*TS)
   RES=XS-XH-TS*XDH
   XH=XH+XDH*TS+XK1*RES
   XDH=XDH+XK2*RES
   IF(XN.EQ.1)THEN
      LET SP11=0
      LET SP22=0
   ELSE
      SP11=SIGNOISE*SQRT(2.*(2*XN-1)/(XN*(XN+1)))
      SP22=SIGNOISE*SQRT(12/(XN*(XN*XN-1)*TS*TS))
   ENDIF
   XHERR=X-XH
   XDHERR=XD-XDH
   EPS=A2*TS*TS*(XN-1)*(XN-2)/6
   EPSD=A2*TS*(XN-1)
   WRITE(9,*)T,X,XS,XH,XD,XDH
   WRITE(1,*)T,X,XS,XH,XD,XDH
   WRITE(2,*)T,XHERR,SP11,-SP11,EPS,XDHERR,SP22,-SP22,EPSD
10 CONTINUE
   CLOSE(1)
   CLOSE(2)
   PAUSE
   END
C SUBROUTINE GAUSS IS SHOWN IN LISTING 1.8
```

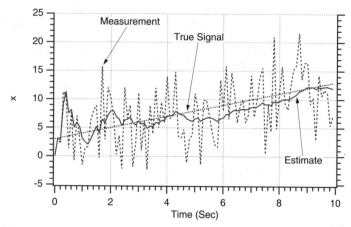

Fig. 3.8   First-order recursive least-squares filter is able to track first-order signal plus noise.

first-order polynomial with values $A0 = 3$ and $A1 = 1$ corrupted by noise with a standard deviation of five (SIGNOISE = 5). These values are identical to the values chosen for the sample first-order example in Chapter 2.

The nominal case of Listing 3.3 was run, and the true signal, measurement, and filter estimate of the first state are compared in Fig. 3.8. We can see that the actual measurement is quite noisy because the amplitude of the signal is comparable to the standard deviation of the noise. We can also see that the filter estimate of the true signal is quite good and is getting better as more measurements are taken. There is no problem in the state estimate tracking the actual signal because the order of the recursive filter is matched to the order of the true signal.

The true second state (i.e., derivative of first state x) for this example is simply

$$\dot{x}_k = 1$$

We can see from Fig. 3.9 that after a brief transient period the recursive first-order least-squares filter has no problem in estimating this state either.

Again, for more exact work the error in the estimate (i.e., difference between true signal and estimate) is used as a measure of performance. Figures 3.10 and 3.11 compare the error in the estimate of the first and second states of the first-order recursive least-squares filter to the theoretical predictions of this section (i.e., square root of P_{11} and P_{22}). We can see that the single-run simulation results are within the theoretical bounds most of the time, indicating that theory and simulation appear to agree. Even though the noise on the first state was of value five, the error in the estimate of x goes down by a factor of 5 after 100 measurements are taken.

We can see from Fig. 3.11 that the single-run simulation results for the error in the estimate of the second state (i.e., derivative of true signal minus second state) also lie within the theoretical error bounds, indicating that the filter is working properly. Figures 3.10 and 3.11 both indicate that the errors in the estimates of the

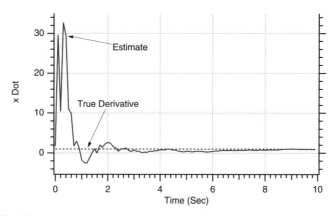

Fig. 3.9 **First-order recursive least-squares filter is able to estimate derivative of signal.**

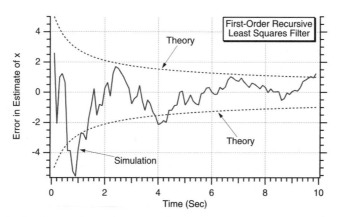

Fig. 3.10 **Single-run simulation results for first state agree with theoretical formula.**

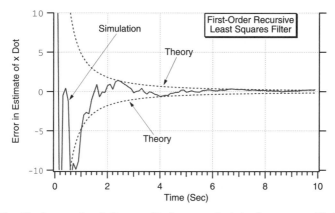

Fig. 3.11 **Single-run simulation results for second state also agree with theoretical formula.**

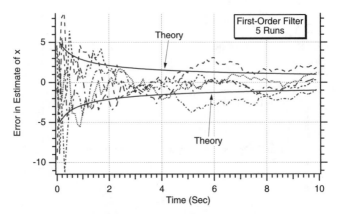

Fig. 3.12   Multiple runs indicate that simulated error in the estimates of first state appear to lie within theoretical error bounds 68% of the time.

first and second states will eventually approach zero as more measurements are taken.

If more runs were made in the case in which there was no truncation error, we would expect to find the computed error in the estimate for both states to lie between the theoretical bounds approximately 68% of the time. To ensure that the simulated and theoretical errors in the estimates agree in this way, Listing 3.3 was slightly modified so that we could run it in the Monte Carlo mode (i.e., repeated simulation trials). A five-run Monte Carlo set was run, and we can see from Figs. 3.12 and 3.13 that the simulation results on average appear to lie within the theoretical bounds approximately 68% of the time. This simple experiment confirms that the theoretical formulas presented in this section for the variance of the errors in the estimates on each of the two states are correct.

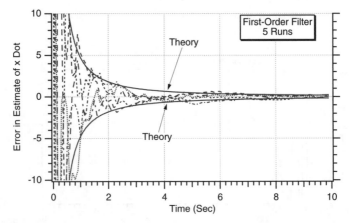

Fig. 3.13   Multiple runs indicate that simulated error in the estimates of second state appear to lie within theoretical error bounds 68% of the time.

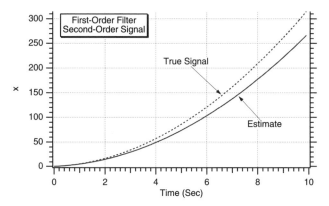

Fig. 3.14 **First-order recursive least-squares filter is unable to track the first state of a second-order polynomial.**

Another case was run with Listing 3.3 to examine the effect of truncation error. In this example the actual signal is now second order without noise (i.e., A0 = 1, A1 = 2, A2 = 3, SIGNOISE = 0), whereas the recursive least-squares filter remains first order. We can see from Figs. 3.14 and 3.15 that it is clear from the two filter estimates that the first-order filter cannot track the first or second state of the signal.

Figures 3.16 and 3.17 display the actual errors in the estimates of the first and second states, respectively (i.e., XHERR and XDHERR in Listing 3.3), obtained from running Listing 3.3. Also superimposed on the figures are the theoretical truncation error formulas already presented in this section (i.e., EPS and EPSD in Listing 3.3). We can see from both figures that the theoretical truncation error formulas and simulation results are in excellent agreement. Therefore, the truncation error formulas for the first-order filter are empirically validated.

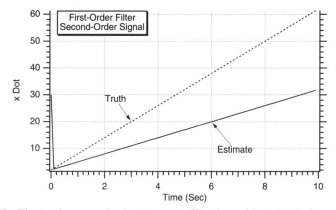

Fig. 3.15 **First-order recursive least-squares filter is unable to track the second state of a second-order polynomial.**

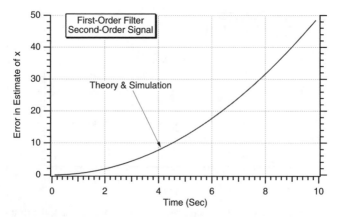

Fig. 3.16   Simulation results and truncation error for first state are in excellent agreement.

Properties of Second-Order or Three-State Filter

Using the same techniques used in the section on the recursive zeroth-order least-squares filter, the three gains of a second-order or three-state recursive least-squares filter can be shown to be[2]

$$K_{1_k} = \frac{3(3k^2 - 3k + 2)}{k(k + 1)(k + 2)} \qquad k = 1, 2, \ldots, n$$

$$K_{2_k} = \frac{18(2k - 1)}{k(k + 1)(k + 2)T_s}$$

$$K_{3_k} = \frac{60}{k(k + 1)(k + 2)T_s^2}$$

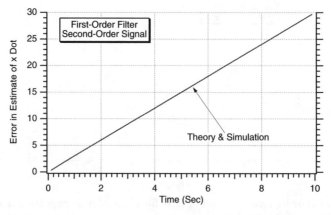

Fig. 3.17   Simulation results and truncation error for second state are in excellent agreement.

Again, we can see that all of the gains eventually go to zero as more measurements are taken (i.e., k gets larger). If we think of the measurement as a position, then the estimates from this three-state filter will be position, velocity, and acceleration (i.e., derivative of position and second derivative of position). For this filter a residual is also formed, which is the difference between the present measurement and a projection of the preceding estimate to the current time or

$$\text{Res}_k = x_k^* - \hat{x}_{k-1} - \hat{\dot{x}}_{k-1}T_s - 0.5\hat{\ddot{x}}_{k-1}T_s^2$$

The new filter estimates are a combination of the preceding state estimates projected forward to the current time plus a gain multiplied by the residual or

$$\hat{x}_k = \hat{x}_{k-1} + \hat{\dot{x}}_{k-1}T_s + 0.5\hat{\ddot{x}}_{k-1}T_s^2 + K_{1_k}\text{Res}_k$$
$$\hat{\dot{x}}_k = \hat{\dot{x}}_{k-1} + \hat{\ddot{x}}_{k-1}T_s^2 + K_{2_k}\text{Res}_k$$
$$\hat{\ddot{x}}_k = \hat{\ddot{x}}_{k-1} + K_{3_k}\text{Res}_k$$

To understand better how these formulas apply, let us again reconsider the four measurement examples of the preceding chapter and the preceding sections where the four measurements are

$$x_1^* = 1.2$$
$$x_2^* = 0.2$$
$$x_3^* = 2.9$$
$$x_4^* = 2.1$$

Using the gain formulas from this section, the first set of gains for $k = 1$ can be calculated as

$$K_{1_1} = \frac{3(3k^2 - 3k + 2)}{k(k+1)(k+2)} = \frac{3(3*1 - 3*1 + 2)}{1(2)(3)} = 1$$

$$K_{2_1} = \frac{18(2k-1)}{k(k+1)(k+2)T_s} = \frac{18(2-1)}{1(2)(3)(1)} = 3$$

$$K_{3_1} = \frac{60}{k(k+1)(k+2)T_s^2} = \frac{60}{1(2)(3)(1)} = 10$$

Let us assume that the initial state estimates of the filter are zero or

$$\hat{x}_0 = 0$$
$$\hat{\dot{x}}_0 = 0$$
$$\hat{\ddot{x}}_0 = 0$$

We can now calculate the residual as

$$\text{Res}_1 = x_1^* - \hat{x}_0 - \dot{\hat{x}}_0 T_s - 0.5\ddot{\hat{x}}_0 T_s^2 = 1.2 - 0 - 0 - 0 = 1.2$$

Therefore, the new state estimates become

$$\hat{x}_1 = \hat{x}_0 + \dot{\hat{x}}_0 T_s + 0.5\ddot{\hat{x}}_0 T_s^2 + K_{1_1}\text{Res}_1 = 0 + 0 + 0 + 1*1.2 = 1.2$$
$$\dot{\hat{x}}_1 = \dot{\hat{x}}_0 + \ddot{\hat{x}}_0 T_s + K_{2_1}\text{Res}_1 = 0 + 0 + 3*1.2 = 3.6$$
$$\ddot{\hat{x}}_1 = \ddot{\hat{x}}_0 + K_{3_1}\text{Res}_1 = 0 + 10*1.2 = 12$$

Repeating the calculations for $k = 2$ yields the next set of gains:

$$K_{1_2} = \frac{3(3k^2 - 3k + 2)}{k(k+1)(k+2)} = \frac{3(3*4 - 3*2 + 2)}{2(3)(4)} = 1$$

$$K_{2_2} = \frac{18(2k - 1)}{k(k+1)(k+2)T_s} = \frac{18(2*2 - 1)}{2(3)(4)(1)} = 2.25$$

$$K_{3_2} = \frac{60}{k(k+1)(k+2)T_s^2} = \frac{60}{2(3)(4)(1)} = 2.5$$

The next residual is based on the preceding state estimates and can be calculated as

$$\text{Res}_2 = x_2^* - \hat{x}_1 - \dot{\hat{x}}_1 T_s - 0.5\ddot{\hat{x}}_1 T_s^2 = 0.2 - 1.2 - 3.6 - 0.5*12 = -10.6$$

whereas the new state estimates become

$$\hat{x}_2 = \hat{x}_1 + \dot{\hat{x}}_1 T_s + 0.5\ddot{\hat{x}}_1 T_s^2 + K_{1_2}\text{Res}_2 = 1.2 + 3.6 + 0.5*12 + 1*(-10.6) = 0.2$$
$$\dot{\hat{x}}_2 = \dot{\hat{x}}_1 + \ddot{\hat{x}}_1 T_s + K_{2_2}\text{Res}_2 = 3.6 + 12 + 2.25*(-10.6) = -8.25$$
$$\ddot{\hat{x}}_2 = \ddot{\hat{x}}_1 + K_{3_2}\text{Res}_2 = 12 + 2.5*(-10.6) = -14.5$$

The remaining calculations for processing the third measurement are

$$K_{1_3} = \frac{3(3k^2 - 3k + 2)}{k(k + 1)(k + 2)} = \frac{3(3*9 - 3*3 + 2)}{3(4)(5)} = 1$$

$$K_{2_3} = \frac{18(2k - 1)}{k(k + 1)(k + 2)T_s} = \frac{18(2*3 - 1)}{3(4)(5)(1)} = 1.5$$

$$K_{3_3} = \frac{60}{k(k + 1)(k + 2)T_s^2} = \frac{60}{3(4)(5)(1)} = 1$$

$$\text{Res}_3 = x_3^* - \hat{x}_2 - \hat{\dot{x}}_2 T_s - 0.5\hat{\ddot{x}}_2 T_s^2$$
$$= 2.9 - 0.2 - (-8.25) - 0.5*(-14.5) = 18.2$$

$$\hat{x}_3 = \hat{x}_2 + \hat{\dot{x}}_2 T_s + 0.5\hat{\ddot{x}}_2 T_s^2 + K_{1_3}\text{Res}_3$$
$$= 0.2 - 8.25 + 0.5*(-14.5) + 1*18.2 = 2.9$$

$$\hat{\dot{x}}_3 = \hat{\dot{x}}_2 + \hat{\ddot{x}}_2 T_s + K_{2_3}\text{Res}_3 = -8.25 - 14.5 + 1.5*18.2 = 4.55$$

$$\hat{\ddot{x}}_3 = \hat{\ddot{x}}_2 + K_{3_3}\text{Res}_3 = -14.5 + 1*18.2 = 3.7$$

and, finally, the calculations for the fourth measurement are given by

$$K_{1_4} = \frac{3(3k^2 - 3k + 2)}{k(k + 1)(k + 2)} = \frac{3(3*16 - 3*4 + 2)}{4(5)(6)} = \frac{19}{20}$$

$$K_{2_4} = \frac{18(2k - 1)}{k(k + 1)(k + 2)T_s} = \frac{18(2*4 - 1)}{4(5)(6)(1)} = \frac{21}{20}$$

$$K_{3_4} = \frac{60}{k(k + 1)(k + 2)T_s^2} = \frac{60}{4(5)(6)(1)} = 0.5$$

$$\text{Res}_4 = x_4^* - \hat{x}_3 - \hat{\dot{x}}_3 T_s - 0.5\hat{\ddot{x}}_3 T_s^2 = 2.1 - 2.9 - 4.55 - 0.5*3.7 = -7.2$$

$$\hat{x}_4 = \hat{x}_3 + \hat{\dot{x}}_3 T_s + 0.5\hat{\ddot{x}}_3 T_s^2 + K_{1_4}\text{Res}_4$$
$$= 2.9 + 4.55 + 0.5*3.7 + \frac{19}{20}*(-7.2) = 2.46$$

$$\hat{\dot{x}}_4 = \hat{\dot{x}}_3 + \hat{\ddot{x}}_3 T_s + K_{2_4}\text{Res}_4 = 4.55 + 3.7*1 + \frac{21}{20}*(-7.2) = 0.69$$

$$\hat{\ddot{x}}_4 = \hat{\ddot{x}}_3 + K_{3_4}\text{Res}_4 = 3.7 + 0.5*(-7.2) = 0.1$$

Notice that the last estimate of the second-order recursive least-squares filter (i.e., $\hat{x}_4 = 2.46$) is again identical to the estimate obtained from the second-order least squares fit of Chapter 2.

Figure 3.18 compares the estimates from the batch-processing method of least squares for the second-order system of Chapter 2 and the second-order recursive least-squares filter of this section. Again we see from Fig. 3.18 that after all of the measurements are taken both the batch-processing and recursive least-squares filters yield identical answers. Of course, the major difference between the two techniques is that the batch-processing method requires that all of the measurements must first be taken before any estimates can be made, whereas the recursive

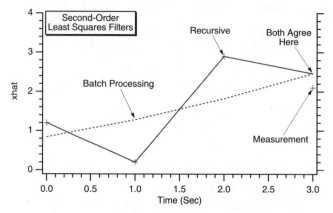

Fig. 3.18 Recursive and batch-processing second-order least-squares filters yield the same answers after all of the measurements are taken.

filter provides estimates after each measurement. In this example the recursive filter estimates pass through the first three measurements.

If the measurement data are a second-order polynomial corrupted by noise and the measurement noise statistics are known, then formulas can be derived, using techniques similar to those used on the zeroth-order filter, describing the variance of the error in the estimates of all three states (i.e., position, velocity, and acceleration) of the second-order filter. The variance of the error in the estimate for the first, second, and third states can be shown to be[2]

$$P_{11_k} = \frac{3(3k^2 - 3k + 2)\sigma_n^2}{k(k+1)(k+2)}$$

$$P_{22_k} = \frac{12(16k^2 - 30k + 11)\sigma_n^2}{k(k^2 - 1)(k^2 - 4)T_s^2}$$

$$P_{33_k} = \frac{720\sigma_n^2}{k(k^2 - 1)(k^2 - 4)T_s^4}$$

where σ_n^2 is the variance of the measurement noise and k is the number of measurements taken (i.e., $k = 1, 2, \ldots, n$). Again the variances of the errors in the estimates of all of the states decrease with increasing k. If the real signal is a third-order polynomial (one degree higher than the second-order filter) or

$$x_k^* = a_0 + a_1 t + a_2 t^2 + a_3 t^3 = a_0 + a_1(k - 1)T_s + a_2(k - 1)^2 T_s^2 + a_3(k - 1)^3 T_s^3$$

the filter will not be able to track the signal as we saw in the preceding section. As was already mentioned, the resulting error buildup caused by this lack of tracking ability is known as truncation error ε. The truncation errors for the errors in the estimates of the first, second, and third states of the second-order recursive least-squares filter are given by[2]

$$\varepsilon_k = \frac{1}{20}a_3 T_s^3(k-1)(k-2)(k-3)$$

$$\dot{\varepsilon}_k = \frac{1}{10}a_3 T_s^2(6k^2 - 15k + 11)$$

$$\ddot{\varepsilon}_k = 3a_3 T_s(k-1)$$

Listing 3.4 shows how the second-order recursive least-squares filter, along with the theoretical formulas, was programmed. We can see that measurements of a second-order signal X plus noise XS is taken every 0.1 s for 10 s. From Listing 3.4 we can tell that the actual error in the estimate XHERR in the first state is compared to plus and minus the square root of P_{11}, the actual error in the estimate of the second state XDHERR is compared to plus and minus the square root of P_{22}, and the actual error in the estimate of the third state XDDHERR is compared to plus and minus the square root of P_{33}. Nominally the simulation is set so that the true signal is also a second-order polynomial with values identical to those of the example in Chapter 2 (i.e., A0 = 2, A1 = − 2, and A2 = 5). The measurement noise has a standard deviation of 50 (SIGNOISE = 50).

**Listing 3.4   Simulation for testing second-order recursive
least-squares filter**

```
C THE FIRST THREE STATEMENTS INVOKE THE ABSOFT RANDOM
  NUMBER GENERATOR ON THE MACINTOSH
    GLOBAL DEFINE
              INCLUDE 'quickdraw.inc'
    END
    IMPLICIT REAL*8(A-H,O-Z)
    OPEN(1,STATUS='UNKNOWN',FILE='DATFIL')
    OPEN(2,STATUS='UNKNOWN',FILE='COVFIL')
    TS=0.1
    SIGNOISE=50.
    A0=2.
    A1=-2.
    A2=5.
    A3=0.
    XH=0.
    XDH=0.
    XDDH=0.
    XN=0.
    DO 10 T=0,10.,TS
    XN=XN+1.
    CALL GAUSS(XNOISE,SIGNOISE)
    X=A0+A1*T+A2*T*T+A3*T*T*T
    XD=A1+2*A2*T+3.*A3*T*T
```

(continued)

Listing 3.4 *(Continued)*

```
      XDD=2*A2+6*A3*T
      XS=X+XNOISE
      XK1=3*(3*XN*XN-3*XN+2)/(XN*(XN+1)*(XN+2))
      XK2=18*(2*XN-1)/(XN*(XN+1)*(XN+2)*TS)
      XK3=60/(XN*(XN+1)*(XN+2)*TS*TS)
      RES=XS-XH-TS*XDH-0.5*TS*TS*XDDH
      XH=XH+XDH*TS+0.5*XDDH*TS*TS+XK1*RES
      XDH=XDH+XDDH*TS+XK2*RES
      XDDH=XDDH+XK3*RES
      IF(XN.EQ.1.OR.XN.EQ.2)THEN
         SP11=0
         SP22=0
         SP33=0
      ELSE
         SP11=SIGNOISE*SQRT(3*(3*XN*XN-3*XN+2)/(XN*(XN+1)*
     1      (XN+2)))
         SP22=SIGNOISE*SQRT(12*(16*XN*XN-30*XN+11)/
     1      (XN*(XN*XN-1)*(XN*XN-4)*TS*TS))
         SP33=SIGNOISE*SQRT(720/(XN*(XN*XN-1)*(XN*XN-4)
     1      *TS*TS*TS*TS))
      ENDIF
      XHERR=X-XH
      XDHERR=XD-XDH
      XDDHERR=XDD-XDDH
      EPS=A3*TS*TS*TS*(XN-1)*(XN-2)*(XN-3)/20
      EPSD=A3*TS*TS*(6*XN*XN-15*XN+11)/10
      EPSDD=3*A3*TS*(XN-1)
      WRITE(9,*)T,X,XS,XH,XD,XDH,XDD,XDDH
      WRITE(1,*)T,X,XS,XH,XD,XDH,XDD,XDDH
      WRITE(2,*)T,XHERR,SP11,-SP11,EPS,XDHERR,SP22,-SP22,EPSD,
     1      XDDHERR,SP33,-SP33,EPSDD
   10 CONTINUE
      CLOSE(1)
      CLOSE(2)
      PAUSE
      END
C SUBROUTINE GAUSS IS SHOWN IN LISTING 1.8
```

The nominal case of Listing 3.4 was run, and the true signal, measurement, and filter estimate are compared in Fig. 3.19. We can see that the actual measurement is not too noisy in this example because the amplitude of the signal is much larger than the standard deviation of the noise. We can also see that the filter estimate of the first state is excellent. Figures 3.20 and 3.21 show that after an initial transient period the second-order recursive least-squares filter is able to estimate the first and second derivatives of x quite accurately.

As was already mentioned, for more exact work the error in the estimate (i.e., difference between true signal and estimate) is used as a measure of performance. Figures 3.22–3.24 compare the error in the estimate of the first, second, and third

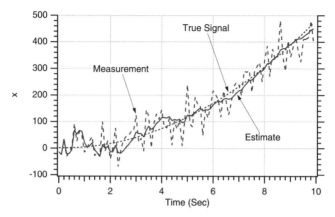

Fig. 3.19   Second-order growing memory filter is able to track second-order signal plus noise.

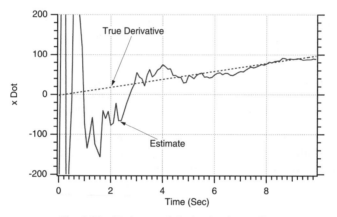

Fig. 3.20   Estimate of derivative is excellent.

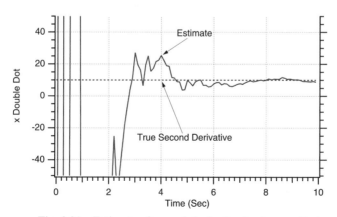

Fig. 3.21   Estimate of second derivative is also excellent.

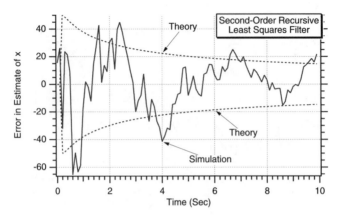

Fig. 3.22   Error in estimate of first state appears to be within theoretical error bounds.

states, respectively, to the theoretical predictions of this section (i.e., square root of P_{11}, P_{22}, and P_{33}). We can see that because the single-run simulation results are within the theoretical bounds most of the time theory and simulation appear to agree.

If more runs were made for this case, we would expect to find the computed error in the estimate for all three states to lie between the theoretical error bounds approximately 68% of the time. To ensure that the simulated and theoretical errors in the estimates agree in this way, Listing 3.4 was slightly modified so that we could run it in the Monte Carlo mode (i.e., repeated simulation trials). A five-run Monte Carlo set was run, and we can see from Figs. 3.25–3.27 that the five runs, on average, appear to lie within the theoretical error bounds approximately

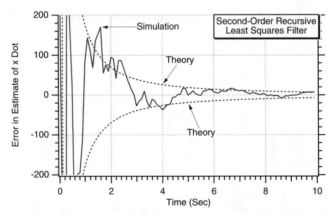

Fig. 3.23   Error in estimate of second state appears to be within theoretical error bounds.

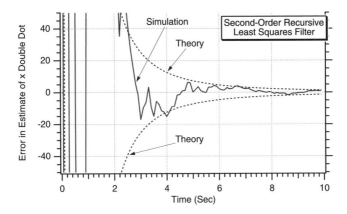

Fig. 3.24   Error in estimate of third state appears to be within theoretical error bounds.

68% of the time. This simple experiment confirms that the theoretical formulas presented in this section for the variance of the errors in the estimates on each of the three states appear to be correct.

Another case was run with Listing 3.4 to examine the effect of truncation error. In this example the actual signal is now a noise-free third-order polynomial (i.e., $A0 = 1$, $A1 = 2$, $A2 = 3$, $A3 = 4$, SIGNOISE $= 0$), whereas the recursive least-squares filter remains second order. We can see from Figs. 3.28–3.30 that it is clear that the second-order filter cannot track the first, second, or third state of the third-order signal (i.e., signal or its first and second derivatives). In all cases the filter estimate is diverging from either the true signal or the true state.

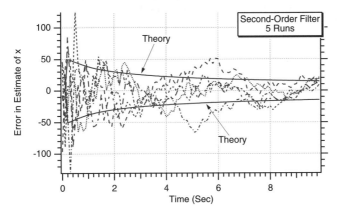

Fig. 3.25   Multiple runs indicate that on average the error in the estimate of first state appears to be within error bounds 68% of the time.

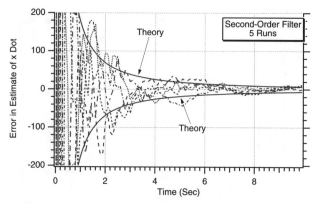

Fig. 3.26   Multiple runs indicate that on average the error in the estimate of second state appears to be within error bounds 68% of the time.

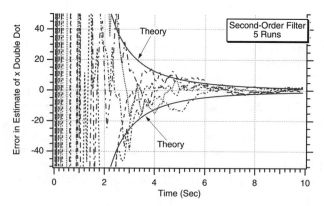

Fig. 3.27   Multiple runs indicate that on average the error in the estimate of third state appears to be within error bounds 68% of the time.

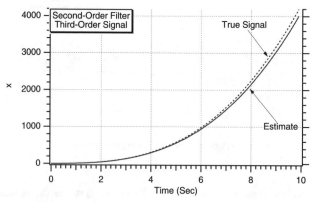

Fig. 3.28   Second-order recursive least-squares filter is unable to track the first state of a third-order polynomial.

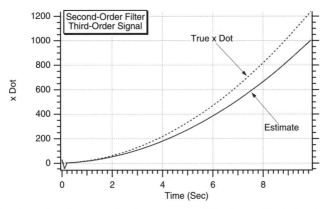

Fig. 3.29   Second-order recursive least-squares filter is unable to track the second state of a third-order polynomial.

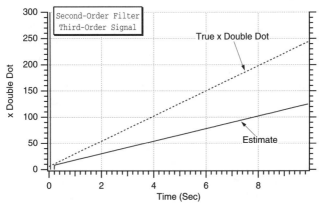

Fig. 3.30   Second-order recursive least-squares filter is unable to track the third state of a third-order polynomial.

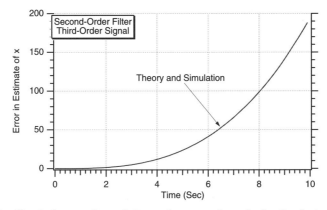

Fig. 3.31   Simulation results and truncation error formula for the first state are in excellent agreement.

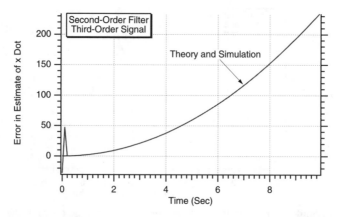

Fig. 3.32   Simulation results and truncation error formula for the second state are in excellent agreement.

Figures 3.31–3.33 display the actual errors in the estimates of the first, second, and third states, respectively, obtained from running Listing 3.4. Also superimposed on the figures are the theoretical truncation error formulas already presented in this section. We can see from the three figures that the theoretical truncation error formulas and simulation results are in excellent agreement. Therefore, the truncation error formulas for the second-order recursive least-squares filter can also be considered empirically validated.

Summary

In this chapter we have first shown that the batch-processing method of least squares could be made recursive. Table 3.3 presents the equations for the one-,

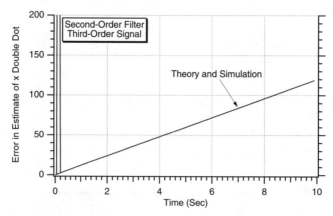

Fig. 3.33   Simulation results and truncation error formula for the third state are in excellent agreement.

two-, and three-state (i.e., also known as zeroth-, first-, and second-order) recursive least-squares filters. We can see from the second column of Table 3.3 that all three filters have the same structure in that the new state estimate is always the old state estimate projected forward to the current time plus a gain times a residual. The gain formulas for all three filters are summarized in the third column of Table 3.3.

Table 3.4 presents important theoretical performance formulas for the one-, two-, and three-state (i.e., also known as zeroth-, first-, and second-order) recursive least-squares filters. The second column represents the standard deviation of the error in the estimate for any state for each of the three filters caused by measurement noise. The third column represents the error in the estimate for any state because the measurement signal may be of higher order than the filter. All of the formulas presented in Table 3.4 were empirically validated by simulation experiment.

The formulas of Table 3.4 for the standard deviation of the errors in the estimates of the three different recursive least-squares filters were programmed so that they could be compared on a normalized basis. Figures 3.34–3.36 present normalized plots showing how the errors in the state estimates caused by measurement noise vary with the number of measurements taken and with the order of the filter used. We can see from Figs. 3.34 and 3.35 that, from a measurement noise reduction point of view, it is best to use as low an order filter as possible because that will tend to reduce the error in the estimate caused by measurement noise. However, we also know from Table 3.4 that if we make the

Table 3.3   Recursive least-square filter comparison

State	Filter	Gains
1	$\mathrm{Res}_k = x_k^* - \hat{x}_{k-1}$ $\hat{x}_k = \hat{x}_{k-1} + K_{1_k}\mathrm{Res}_k$	$K_{1_k} = \dfrac{1}{k}$
2	$\mathrm{Res}_k = x_k^* - \hat{x}_{k-1} - \hat{\dot{x}}_{k-1}T_s$ $\hat{x}_k = \hat{x}_{k-1} + \hat{\dot{x}}_{k-1}T_s + K_{1_k}\mathrm{Res}_k$ $\hat{\dot{x}}_k = \hat{\dot{x}}_{k-1} + K_{2_k}\mathrm{Res}_k$	$K_{1_k} = \dfrac{2(2k-1)}{k(k+1)}$ $K_{2_k} = \dfrac{6}{k(k+1)T_s}$
3	$\mathrm{Res}_k = x_k^* - \hat{x}_{k-1} - \hat{\dot{x}}_{k-1}T_s - 0.5\hat{\ddot{x}}_{k-1}T_s^2$ $\hat{x}_k = \hat{x}_{k-1} + \hat{\dot{x}}_{k-1}T_s + 0.5\hat{\ddot{x}}_{k-1}T_s^2 + K_{1_k}\mathrm{Res}_k$ $\hat{\dot{x}}_k = \hat{\dot{x}}_{k-1} + \hat{\ddot{x}}_{k-1}T_s + K_{2_k}\mathrm{Res}_k$ $\hat{\ddot{x}}_k = \hat{\ddot{x}}_{k-1} + K_{3_k}\mathrm{Res}_k$	$K_{1_k} = \dfrac{3(3k^2 - 3k + 2)}{k(k+1)(k+2)}$ $K_{2_k} = \dfrac{18(2k-1)}{k(k+1)(k+2)T_s}$ $K_{3_k} = \dfrac{60}{k(k+1)(k+2)T_s^2}$

Table 3.4 Standard deviation of errors in estimates and truncation error formulas for various order recursive least-squares filters

State	Standard deviation	Truncation error
1	$\sqrt{P_k} = \dfrac{\sigma_n}{\sqrt{k}}$	$\varepsilon_k = \dfrac{a_1 T_s}{2}(k-1)$
2	$\sqrt{P_{11_k}} = \sigma_n \sqrt{\dfrac{2(2k-1)}{k(k+1)}}$ $\sqrt{P_{22_k}} = \dfrac{\sigma_n}{T_s} \sqrt{\dfrac{12}{k(k^2-1)}}$	$\varepsilon_k = \dfrac{1}{6} a_2 T_s^2 (k-1)(k-2)$ $\dot{\varepsilon}_k = a_2 T_s (k-1)$
3	$\sqrt{P_{11_k}} = \sigma_n \sqrt{\dfrac{3(3k^2-3k+2)}{k(k+1)(k+2)}}$ $\sqrt{P_{22_k}} = \dfrac{\sigma_n}{T_s} \sqrt{\dfrac{12(16k^2-30k+11)}{k(k^2-1)(k^2-4)}}$ $\sqrt{P_{33_k}} = \dfrac{\sigma_n}{T_s} \sqrt{\dfrac{720}{k(k^2-1)(k^2-4)}}$	$\varepsilon_k = \dfrac{1}{20} a_3 T_s^3 (k-1)(k-2)(k-3)$ $\dot{\varepsilon}_k = \dfrac{1}{10} a_3 T_s^2 (6k^2-15k+11)$ $\ddot{\varepsilon}_k = 3a_3 T_s (k-1)$

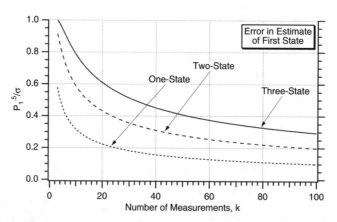

Fig. 3.34 Error in the estimate of the first state decreases with decreasing filter order and increasing number of measurements taken.

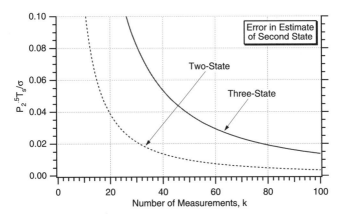

Fig. 3.35   Error in the estimate of the second state decreases with decreasing filter order and increasing number of measurements taken.

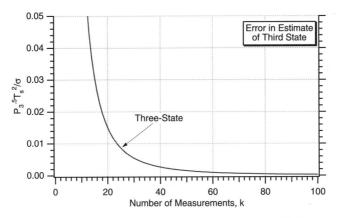

Fig. 3.36   Error in the estimate of the third state decreases with increasing number of measurements taken.

order of the filter too low there might be excessive truncation error. In each case the best-order filter to use will depend on the number of measurements to be taken and the actual order of the signal.

References

[1] Selby, S. M., *Standard Mathematical Tables*, 20th ed., Chemical Rubber Co., Cleveland, OH, 1972, p. 37.

[2] Morrison, N., *Introduction to Sequential Smoothing and Prediction*, McGraw–Hill, New York, 1969, pp. 339–376.

Chapter 4

Polynomial Kalman Filters

Introduction

IN THE previous chapter we saw how the batch-processing method of least-squares could be made recursive when the measurement was a polynomial signal corrupted by noise. We saw that the recursive least-squares filter provided the exact same estimates, for a given number of measurements, as the batch-processing least-squares filter. In this chapter we will first provide the general equations for the discrete Kalman filter and then show under what conditions it is completely equivalent to the recursive least-squares filter.

We also saw in the preceding chapter that if the order of the recursive least-squares filter was less than the order of the polynomial measurement signal then the filter's estimates would diverge from the true signal and its derivatives. The engineering fix was to use a higher-order filter. We will also demonstrate the divergence problem with the Kalman filter, but we will show that there are a variety of engineering fixes.

General Equations[1,2]

To apply Kalman-filtering theory, our model of the real world must be described by a set of differential equations. These equations must be cast in matrix or state-space form as

$$\dot{x} = Fx + Gu + w$$

where x is a column vector with the states of the system, F the system dynamics matrix, u is a known vector, which is sometimes called the control vector, and w is a white-noise process, which is also expressed as a vector. There is a process-noise matrix Q that is related to the process-noise vector according to

$$Q = E[ww^T]$$

We will also see that, although process noise might not always have physical meaning, it is sometimes used as a device for telling the filter that we know the filter's model of the real world is not precise. The Kalman-filter formulation requires that the measurements be linearly related to the states according to

$$z = Hx + v$$

where z is the measurement vector, H is the measurement matrix, and v is white measurement noise, which is also expressed as a vector. The measurement noise matrix R is related to the measurement noise vector v according to

$$R = E[vv^T]$$

The preceding relationships must be discretized before a discrete Kalman filter can be built. If we take measurements every T_s seconds, we first need to find a fundamental matrix Φ. Recall from Chapter 1 that for a time-invariant system the fundamental matrix can be found from the system dynamics matrix according to

$$\Phi(t) = \mathscr{L}^{-1}[(sI - F)^{-1}]$$

where I is the identity matrix, $\mathscr{L}^{-1}$ is the inverse Laplace transform, and F is the systems dynamics matrix. Typically, inverse Laplace transforms can be found from tables in engineering handbooks.[3] As was also mentioned in Chapter 1, another way of finding the fundamental matrix is to evaluate the Taylor-series expansion

$$\Phi(t) = e^{Ft} = I + Ft + \frac{(Ft)^2}{2!} + \cdots + \frac{(Ft)^n}{n!} + \cdots$$

The discrete fundamental or transition matrix can easily be found by evaluating the fundamental matrix at the sampling time T_s or

$$\Phi_k = \Phi(T_s)$$

The discrete form of the Kalman-filtering measurement equation becomes

$$z_k = Hx_k + v_k$$

and

$$R_k = E(v_k v_k^T)$$

where R_k is a matrix consisting of the variances of each of the measurement noise sources. In the case of polynomial Kalman filters, R_k is a scalar. The resultant Kalman-filtering equation is given by

$$\hat{x}_k = \Phi_k \hat{x}_{k-1} + G_k u_{k-1} + K_k(z_k - H\Phi_k \hat{x}_{k-1} - HG_k u_{k-1})$$

where K_k represents the Kalman gain matrix and G_k is obtained from

$$G_k = \int_0^{T_s} \Phi(\tau)G \, d\tau$$

if u_{k-1} is assumed to be constant between sampling instants. The Kalman gains are computed, while the filter is operating, from the matrix Riccati equations. The Riccati equations are a set of recursive matrix equations given by

$$M_k = \Phi_k P_{k-1} \Phi_k^T + Q_k$$
$$K_k = M_k H^T (HM_k H^T + R_k)^{-1}$$
$$P_k = (I - K_k H)M_k$$

where P_k is a covariance matrix representing errors in the state estimates (i.e., variance of truth minus estimate) after an update and M_k is the covariance matrix representing errors in the state estimates before an update. The discrete process-noise matrix Q_k can be found from the continuous process-noise matrix Q and the fundamental matrix according to

$$Q_k = \int_0^{T_s} \Phi(\tau) Q \Phi^T(\tau)\, dt$$

To start the Riccati equations, we need an initial covariance matrix P_0.

Derivation of Scalar Riccati Equations

A special form of the Kalman filter occurs when there is no deterministic disturbance or control vector. In this special case the discrete Kalman-filtering equation simplifies to

$$\hat{x}_k = \Phi_k \hat{x}_{k-1} + K_k(z_k - H\Phi\hat{x}_{k-1})$$

while the matrix Riccati equations remain as

$$M_k = \Phi_k P_{k-1} \Phi_k^T + Q_k$$
$$K_k = M_k H^T (HM_k H^T + R_k)^{-1}$$
$$P_k = (I - K_k H)M_k$$

Although the derivation of the matrix Riccati equations is beyond the scope of this text, the derivation process to illustrate when the matrices are scalars is quite simple. If there is no deterministic disturbance or control scalar, our discrete model of the real world is given by

$$x_k = \Phi_k x_{k-1} + w_k$$

where Φ_k is the scalar that propagates the states from one sampling instant to the next and w_k is white process noise. From the first equation of this section, we can see the scalar Kalman-filtering equation is

$$\hat{x}_k = \Phi_k \hat{x}_{k-1} + K_k(z_k - H\Phi_k \hat{x}_{k-1})$$

Our zeroth-order recursive least-squares filter of Chapter 3 was of the same form. In the preceding scalar equation z_k is the measurement, which is given by

$$z_k = Hx_k + v_k$$

where H is the measurement scalar that relates the state to the measurement and v_k is the measurement noise. From the preceding equations we can say that the error in the estimate is

$$\tilde{x}_k = x_k - \hat{x}_k = x_k - \Phi_k \hat{x}_{k-1} - K_k(z_k - H\Phi_k \hat{x}_{k-1})$$

Recognizing that we can express the measurement in terms of the state, we obtain

$$\tilde{x}_k = x_k - \Phi_k \hat{x}_{k-1} - K_k(Hx_k + v_k - H\Phi_k \hat{x}_{k-1})$$

If we also note that the state at time k can be replaced by an alternate expression at time $k - 1$, we get

$$\tilde{x}_k = \Phi_k x_{k-1} + w_k - \Phi_k \hat{x}_{k-1} - K_k(H\Phi_k x_{k-1} + Hw_k + v_k - H\Phi_k \hat{x}_{k-1})$$

Because

$$\tilde{x}_k = x_k - \hat{x}_k$$

we can also say that

$$\tilde{x}_{k-1} = x_{k-1} - \hat{x}_{k-1}$$

Therefore, combining similar terms in the error in the estimate equation yields

$$\tilde{x}_k = (1 - K_k H)\tilde{x}_{k-1}\Phi_k + (1 - K_k H)w_k - K_k v_k$$

If we define the covariance P_k to be

$$P_k = E(\tilde{x}_k^2)$$

and recognize that

$$Q_k = E(w_k^2)$$
$$R_k = E(v_k^2)$$

we can square and take expectations of both sides of the error in the estimate equation to obtain

$$P_k = (1 - K_k H)^2(P_{k-1}\Phi_k^2 + Q_k) + K_k^2 R_k$$

To simplify the preceding equation, let us define

$$M_k = P_{k-1}\Phi_k^2 + Q_k$$

which is analogous to the first Riccati equation. The covariance equation now simplifies to

$$P_k = (1 - K_k H)^2 M_k + K_k^2 R_k$$

If we want to find the gain that will minimize the variance of the error in the estimate, we can simply take the derivative of the preceding expression with respect to the gain and set the result equal to zero or

$$\frac{\partial P_k}{\partial K_k} = 0 = 2(1 - K_k H)M_k(-H) + 2K_k R_k$$

Solving the preceding equation for the gain yields

$$K_k = \frac{M_k H}{H^2 M_k + R_k} = M_k H (H^2 M_k + R_k)^{-1}$$

which is analogous to the second Riccati equation. Substitution of the optimal gain into the covariance equation yields

$$P_k = \left(1 - \frac{M_k H^2}{H^2 M_k + R_k}\right) M_k + \left(\frac{M_k H}{H^2 M_k + R_k}\right)^2 R_k$$

which simplifies to

$$P_k = \frac{R_k M_k}{H^2 M_k + R_k} = \frac{R_k K_k}{H}$$

By inverting the optimal gain equation, we obtain

$$K_k R_k = M_k H - H^2 M_k K_k$$

and substituting the preceding equation back into the variance equation yields

$$P_k = \frac{R_k K_k}{H} = \frac{M_k H - H^2 M_k K_k}{H} = M_k - H M_k K_k$$

or, more simply,

$$P_k = (1 - K_k H)M_k$$

which is analogous to the third Riccati equation.

From the preceding derivation we can see that the gain of the Kalman filter is chosen to minimize the variance of the error in the estimate. The Riccati equations are simply an iterative way of finding the optimal gain at each time step. However, the derivation of the matrix Riccati equations is far more complex than the scalar derivation of this section because matrices do not always obey the same rules as scalars. For example, with matrices the order of their place in multiplication is important, whereas this is not true with scalars (i.e., $P_1 P_2 \neq P_2 P_1$, whereas $P_1 P_2 = P_2 P_1$). In addition, the matrix inverse cannot be computed in the same way in which we computed the scalar inverse.

Polynomial Kalman Filter (Zero Process Noise)

We have already mentioned that a special form of the Kalman filter occurs when there is no control or deterministic vector. If the differential equations describing system behavior yield polynomial signals, the resultant polynomial Kalman filter will have a special type of fundamental matrix. The simplest possible polynomial filter occurs when there is no process noise (i.e., $Q_k = 0$). In this special case the filtering equation simplifies to

$$\hat{x}_k = \Phi_k \hat{x}_{k-1} + K_k(z_k - H\Phi_k \hat{x}_{k-1})$$

while the Riccati equations simplify to

$$M_k = \Phi_k P_{k-1} \Phi_k^T$$
$$K_k = M_k H^T (H M_k H^T + R_k)^{-1}$$
$$P_k = (I - K_k H) M_k$$

If the actual signal or polynomial is a constant, then we can say that

$$x = a_0$$

which means that the derivative of the signal is zero or

$$\dot{x} = 0$$

If the state-space representation of the preceding equation is given by

$$\dot{x} = Fx$$

then we can say for this zeroth-order system that

$$F = 0$$

If the polynomial under consideration was a ramp or

$$x = a_0 + a_1 t$$

then its derivative is given by

$$\dot{x} = a_1$$

whereas its second derivative is given by

$$\ddot{x} = 0$$

We can express the preceding two equations in state-space form as

$$\begin{bmatrix} \dot{x} \\ \ddot{x} \end{bmatrix} = \begin{bmatrix} 0 & 1 \\ 0 & 0 \end{bmatrix} \begin{bmatrix} x \\ \dot{x} \end{bmatrix}$$

which means the system dynamics matrix for a ramp or first-order polynomial is given by

$$F = \begin{bmatrix} 0 & 1 \\ 0 & 0 \end{bmatrix}$$

We can proceed using similar logic assuming the signal is a parabola. That work has been done and Table 4.1 tabulates the system dynamics matrix, the resultant fundamental matrix, the measurement matrix, and the noise matrix (scalar) for different-order systems, where it is assumed that the measurement is a polynomial signal plus noise.

The matrices of Table 4.1 form the basis of the different-order polynomial Kalman filters. For the polynomial Kalman filters considered, the fundamental matrices can be derived exactly from the system dynamics matrix according to

$$\Phi(t) = \mathscr{L}^{-1}[(sI - F)^{-1}]$$

Table 4.1   Important matrices for different-order polynomial Kalman filters

Order	Systems dynamics	Fundamental	Measurement	Noise
0	$F = 1$	$\Phi_k = 1$	$H = 1$	$R_k = \sigma_n^2$
1	$F = \begin{bmatrix} 0 & 1 \\ 0 & 0 \end{bmatrix}$	$\Phi_k = \begin{bmatrix} 1 & T_s \\ 0 & 1 \end{bmatrix}$	$H = \begin{bmatrix} 1 & 0 \end{bmatrix}$	$R_k = \sigma_n^2$
2	$F = \begin{bmatrix} 0 & 1 & 0 \\ 0 & 0 & 1 \\ 0 & 0 & 0 \end{bmatrix}$	$\Phi_k = \begin{bmatrix} 1 & T_s & 0.5T_s^2 \\ 0 & 1 & T_s \\ 0 & 0 & 1 \end{bmatrix}$	$H = \begin{bmatrix} 1 & 0 & 0 \end{bmatrix}$	$R_k = \sigma_n^2$

or exactly from the Taylor-series expansion

$$\Phi(t) = e^{Ft} = I + Ft + \frac{(Ft)^2}{2!} + \cdots + \frac{(Ft)^n}{n!} + \cdots$$

For example, in the first-order or two-state system the systems dynamics matrix is given by

$$F = \begin{bmatrix} 0 & 1 \\ 0 & 0 \end{bmatrix}$$

Therefore, we can find the square of the systems dynamics matrix to be

$$F^2 = \begin{bmatrix} 0 & 1 \\ 0 & 0 \end{bmatrix} \begin{bmatrix} 0 & 1 \\ 0 & 0 \end{bmatrix} = \begin{bmatrix} 0 & 0 \\ 0 & 0 \end{bmatrix}$$

which means that the fundamental matrix as a function of time is exactly a three-term Taylor-series expansion or

$$\Phi(t) = e^{Ft} = \begin{bmatrix} 1 & 0 \\ 0 & 1 \end{bmatrix} + \begin{bmatrix} 0 & 1 \\ 0 & 0 \end{bmatrix} t + \begin{bmatrix} 0 & 0 \\ 0 & 0 \end{bmatrix} \frac{t^2}{2} = \begin{bmatrix} 1 & t \\ 0 & 1 \end{bmatrix}$$

Substituting T_s for t in the preceding equation yields the discrete form of the fundamental matrix to be

$$\Phi_k = \begin{bmatrix} 1 & T_s \\ 0 & 1 \end{bmatrix}$$

which is precisely what appears in Table 4.1. The H and R_k matrices shown in Table 4.1 indicate that for the polynomial Kalman filters considered the measurement is always the first state corrupted by noise with variance σ_n^2. Therefore R_k is a scalar, which means it will be easy to take the matrix inverse as required by the second Riccati equation. In this case the matrix inverse and scalar inverse are identical.

Comparing Zeroth-Order Recursive Least-Squares and Kalman Filters

Table 4.1 provides enough information so that we can examine the structure of the polynomial Kalman filter. First we will consider the zeroth-order filter. Substitution of the fundamental and measurement matrices of Table 4.1 into the Kalman-filtering equation

$$\hat{x}_k = \Phi_k \hat{x}_{k-1} + K_k(z_k - H\Phi_k \hat{x}_{k-1})$$

yields the scalar equation

$$\hat{x}_k = \hat{x}_{k-1} + K_{1_k}(x_k^* - \hat{x}_{k-1})$$

If we define the residual Res_k as

$$\text{Res}_k = x_k^* - \hat{x}_{k-1}$$

then the preceding zeroth-order or one-state polynomial Kalman-filter equation simplifies to

$$\hat{x}_k = \hat{x}_{k-1} + K_{1_k}\text{Res}_k$$

which is the same as the equation for the zeroth-order recursive least-squares filter of the preceding chapter. The only possible difference between the zeroth-order recursive least-squares filter and polynomial Kalman filter is in the gain.

Table 4.1 also provides enough information so that we can solve the Riccati equations for the Kalman gains. If we assume that the initial covariance matrix is infinite (i.e., this means that we have no a priori information on how to initialize the filter state), then for the zeroth-order system we have

$$P_0 = \infty$$

From the first Riccati equation we can say that

$$M_1 = \Phi_1 P_0 \Phi_1^T = 1 * \infty * 1 = \infty$$

Therefore, the second Riccati equation tells us that

$$K_1 = M_1 H^T (HM_1 H^T + R_1)^{-1} = \frac{M_1}{M_1 + \sigma_n^2} = \frac{\infty}{\infty + \sigma_n^2} = 1$$

The zeroth-order polynomial Kalman-filter gain is the same value of the gain with $k = 1$ as the zeroth-order recursive least-squares filter of Chapter 3. Solving the third Riccati equation yields

$$P_1 = (I - K_1 H)M_1 = \left(1 - \frac{M_1}{M_1 + \sigma_n^2}\right)M_1 = \frac{\sigma_n^2 M_1}{M_1 + \sigma_n^2} = \sigma_n^2$$

which is also the same value for the variance of the error in the estimate obtained for $k = 1$ with the zeroth-order recursive least-squares filter of Chapter 3. Repeating the first Riccati equation for $k = 2$ yields

$$M_2 = \Phi_2 P_1 \Phi_2^T = 1 * \sigma_n^2 * 1 = \sigma_n^2$$

Substitution into the second Riccati equation yields the Kalman gain

$$K_2 = \frac{M_2}{M_2 + \sigma_n^2} = \frac{\sigma_n^2}{\sigma_n^2 + \sigma_n^2} = 0.5$$

Again, the second value of the zeroth-order polynomial Kalman-filter gain is the same value of the gain with $k = 2$ in the zeroth-order recursive least-squares filter of Chapter 3. Solving the third Riccati equation yields the covariance of the error in the estimate

$$P_2 = \frac{\sigma_n^2 M_2}{M_2 + \sigma_n^2} = \frac{\sigma_n^2 \sigma_n^2}{\sigma_n^2 + \sigma_n^2} = \frac{\sigma_n^2}{2}$$

which is also the same value for the variance of the error in the estimate obtained for $k = 2$ with the zeroth-order recursive least-squares filter of Chapter 3. Repeating the first Riccati equation for $k = 3$ yields

$$M_3 = P_2 = \frac{\sigma_n^2}{2}$$

Solving the second Riccati equation for the Kalman gain yields

$$K_3 = \frac{M_3}{M_3 + \sigma_n^2} = \frac{0.5\sigma_n^2}{0.5\sigma_n^2 + \sigma_n^2} = \frac{1}{3}$$

Again, this is the same value of the gain with $k = 3$ in the zeroth-order recursive least-squares filter. Solving the third Riccati equation

$$P_3 = \frac{\sigma_n^2 M_3}{M_3 + \sigma_n^2} = \frac{\sigma_n^2 * 0.5\sigma_n^2}{0.5\sigma_n^2 + \sigma_n^2} = \frac{\sigma_n^2}{3}$$

which is also the same value for the variance of the error in the estimate obtained for $k = 3$ with the zeroth-order recursive least-squares filter of Chapter 3. Now we solve the Riccati equations the last time for $k = 4$. Solving the first equation yields

$$M_4 = P_3 = \frac{\sigma_n^2}{3}$$

Solving the second equation yields the Kalman gain

$$K_4 = \frac{M_4}{M_4 + \sigma_n^2} = \frac{0.333\sigma_n^2}{0.333\sigma_n^2 + \sigma_n^2} = \frac{1}{4}$$

which again is the same as the zeroth-order recursive least-squares filter gain with $k = 4$. Solving the last Riccati equation yields

$$P_4 = \frac{\sigma_n^2 M_4}{M_4 + \sigma_n^2} = \frac{\sigma_n^2 * 0.333\sigma_n^2}{0.333\sigma_n^2 + \sigma_n^2} = \frac{\sigma_n^2}{4}$$

which is also the same value for the variance of the error in the estimate obtained for $k = 3$ with the zeroth-order recursive least-squares filter of the preceding chapter. *Thus, when the zeroth-order polynomial Kalman filter has zero process noise and infinite initial covariance matrix, it has the same gains and variance predictions as the zeroth-order recursive least-squares filter. We will soon see that this is not a coincidence.* Our particular recursive least-squares filter was developed without any assumption concerning the statistical information about the states being estimated.

Comparing First-Order Recursive Least-Squares and Kalman Filters

Table 4.1 also provides enough information so that we can examine the structure of the first-order polynomial Kalman filter. Substitution of the funda-mental and measurement matrices of Table 4.1 into

$$\hat{x}_k = \Phi_k \hat{x}_{k-1} + K_k(z_k - H\Phi_k \hat{x}_{k-1})$$

yields

$$\begin{bmatrix} \hat{x}_k \\ \hat{\dot{x}}_k \end{bmatrix} = \begin{bmatrix} 1 & T_s \\ 0 & 1 \end{bmatrix} \begin{bmatrix} \hat{x}_{k-1} \\ \hat{\dot{x}}_{k-1} \end{bmatrix} + \begin{bmatrix} K_{1_k} \\ K_{2_k} \end{bmatrix} \left(x_k^* - \begin{bmatrix} 1 & 0 \end{bmatrix} \begin{bmatrix} 1 & T_s \\ 0 & 1 \end{bmatrix} \begin{bmatrix} \hat{x}_{k-1} \\ \hat{\dot{x}}_{k-1} \end{bmatrix} \right)$$

Multiplying out the terms of the matrix equation yields the two scalar equations

$$\hat{x}_k = \hat{x}_{k-1} + T_s \hat{\dot{x}}_{k-1} + K_{1_k}(x_k^* - \hat{x}_{k-1} - T_s \hat{\dot{x}}_{k-1})$$
$$\hat{\dot{x}}_k = \hat{\dot{x}}_{k-1} + K_{2_k}(x_k^* - \hat{x}_{k-1} - T_s \hat{\dot{x}}_{k-1})$$

If we define the residual Res_k as

$$\mathrm{Res}_k = x_k^* - \hat{x}_{k-1} - T_s \hat{\dot{x}}_{k-1}$$

then the two preceding equations simplify to

$$\hat{x}_k = \hat{x}_{k-1} + T_s \hat{\dot{x}}_{k-1} + K_{1_k} \mathrm{Res}_k$$
$$\hat{\dot{x}}_k = \hat{\dot{x}}_{k-1} + K_{2_k} \mathrm{Res}_k$$

which is precisely the same as the equations for the first-order recursive least-squares filter of Chapter 3. Again, the only possible difference between the first-order recursive least-squares filter and polynomial Kalman filter are in the gains.

Next, we would like to compare the gains and variance predictions of both the first-order recursive least-squares filter and the first-order Kalman filter with zero process-noise and infinite initial covariance matrix. Recall that the formula presented in Chapter 3 for the first-order recursive least-square filter gains is given by

$$K_{1_k} = \frac{2(2k-1)}{k(k+1)} \qquad k = 1, 2, \ldots, n$$

$$K_{2_k} = \frac{6}{k(k+1)T_s}$$

whereas the formula for the variance of the errors in the estimates (i.e., variance of truth minus estimate) of the first and second states is given by

$$P_{11_k} = \frac{2(2k-1)\sigma_n^2}{k(k+1)}$$

$$P_{22_k} = \frac{12\sigma_n^2}{k(k^2-1)T_s^2}$$

A program was written in which the Riccati equations were solved for the first-order polynomial Kalman filter and is presented in Listing 4.1. The Riccati equations not only yield values for the Kalman gains but also values for the entire covariance matrix $\boldsymbol{P}_k$. Logic was added to the program so that the gains and square root of the covariance matrix diagonal elements computed from Listing 4.1 could be compared to the preceding set of formulas for the first-order recursive least-squares filter.

Figure 4.1 compares the square root of the first diagonal element of the covariance matrix of the first-order polynomial Kalman filter to the square root of the formula for variance of the error in the estimate of the first state in the first-order recursive least-squares filter. We can see that the comparison is exact, meaning that the filters are identical when the Kalman filter has zero process noise and infinite initial covariance matrix (i.e., infinite value for diagonal terms and zero value for off-diagonal terms). In addition, this plot also represents how the standard deviation of the error in the estimate of the first state (i.e., truth minus estimate) improves as more measurements are taken. In this example the input noise had a standard deviation of unity. After approximately 15 measurements the standard deviation of the error in the estimate decreases to approximately 0.5, meaning that filtering has effectively reduced the effect of noise by a factor of two.

Figure 4.2 shows that both the first-order polynomial Kalman and recursive least-squares filters also agree exactly in their predictions of the standard deviation in the error in the estimate of the second state (i.e., derivative of first state). We can also see that the standard deviation in the error of the estimate of the second state diminishes quite rapidly as more measurements are taken.

We can also see from Figures 4.3 and 4.4 that both gains of the first-order polynomial Kalman and recursive least-squares filters agree exactly. *In other words, we have confirmed by simulation that the first-order polynomial Kalman filter with zero process-noise and infinite initial covariance matrix and the*

Listing 4.1   Comparing first-order polynomial Kalman and recursive least-squares filter gains and covariance matrices

```
      IMPLICIT  REAL*8(A-H,O-Z)
      REAL*8  M(2,2),P(2,2),K(2,1),PHI(2,2),H(1,2),R(1,1),PHIT(2,2)
      REAL*8  PHIP(2,2),HT(2,1),KH(2,2),IKH(2,2)
      REAL*8  MHT(2,1),HMHT(1,1),HMHTR(1,1),HMHTRINV(1,1),IDN(2,2)
      REAL*8  K1GM,K2GM
      INTEGER  ORDER
      OPEN(1,STATUS='UNKNOWN',FILE='DATFIL')
      ORDER=2
      TS=1.
      SIGNOISE=1.
      DO  1000  I=1,ORDER
      DO  1000  J=1,ORDER
            PHI(I,J)=0.
            P(I,J)=0.
            IDN(I,J)=0.
 1000 CONTINUE
      IDN(1,1)=1.
      IDN(2,2)=1.
      P(1,1)=99999999999999.
      P(2,2)=99999999999999.
      PHI(1,1)=1
      PHI(1,2)=TS
      PHI(2,2)=1
      DO  1100  I=1,ORDER
            H(1,I)=0.
 1100 CONTINUE
      H(1,1)=1
      CALL  MATTRN(H,1,ORDER,HT)
      R(1,1)=SIGNOISE**2
      CALL  MATTRN(PHI,ORDER,ORDER,PHIT)
      DO  10  XN=1.,100.
            CALL  MATMUL(PHI,ORDER,ORDER,P,ORDER,ORDER,PHIP)
            CALL  MATMUL(PHIP,ORDER,ORDER,PHIT,ORDER,ORDER,M)
            CALL  MATMUL(M,ORDER,ORDER,HT,ORDER,1,MHT)
            CALL  MATMUL(H,1,ORDER,MHT,ORDER,1,HMHT)
            HMHTR(1,1)=HMHT(1,1)+R(1,1)
            HMHTRINV(1,1)=1./HMHTR(1,1)
            CALL  MATMUL(MHT,ORDER,1,HMHTRINV,1,1,K)
            CALL  MATMUL(K,ORDER,1,H,1,ORDER,KH)
            CALL  MATSUB(IDN,ORDER,ORDER,KH,IKH)
            CALL  MATMUL(IKH,ORDER,ORDER,M,ORDER,ORDER,P)
            IF(XN<2)THEN
                  P11GM=9999999999.
                  P22GM=9999999999.
            ELSE
                  P11GM=2.*(2.*XN-1)*SIGNOISE*SIGNOISE/(XN*(XN+1.))
                  P22GM=12.*SIGNOISE*SIGNOISE/(XN*(XN*XN-1.)
     1                  *TS*TS)
            ENDIF
            SP11=SQRT(P(1,1))
```

(continued)

Listing 4.1   (*Continued*)

```
      SP22=SQRT(P(2,2))
      SP11GM=SQRT(P11GM)
      SP22GM=SQRT(P22GM)
      K1GM=2.*(2.*XN-1.)/(XN*(XN+1.))
      K2GM=6./(XN*(XN+1.)*TS)
      WRITE(9,*)XN,K(1,1),K1GM,K(2,1),K2GM,SP11,SP11GM,SP22,
1            SP22GM
      WRITE(1,*)XN,K(1,1),K1GM,K(2,1),K2GM,SP11,SP11GM,SP22,
1            SP22GM
10    CONTINUE
      CLOSE(1)
      PAUSE
      END


C SUBROUTINE MATTRN IS SHOWN IN LISTING 1.3
C SUBROUTINE MATMUL IS SHOWN IN LISTING 1.4
C SUBROUTINE MATADD IS SHOWN IN LISTING 1.1
C SUBROUTINE MATSUB IS SHOWN IN LISTING 1.2
```

recursive polynomial least-squares filters not only have identical structure but also have identical gains.

Comparing Second-Order Recursive Least-Squares and Kalman Filters

Table 4.1 also provides enough information so that we can examine the structure of the second-order polynomial Kalman filter. Substitution of the fundamental and measurement matrices of Table 4.1 into the fundamental Kalman-filtering equation

$$\hat{x}_k = \Phi_k \hat{x}_{k-1} + K_k(z_k - H\Phi_k \hat{x}_{k-1})$$

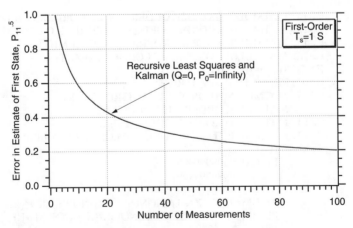

Fig. 4.1   First-order polynomial Kalman and recursive least-squares filters have identical standard deviations for errors in the estimate of the first state.

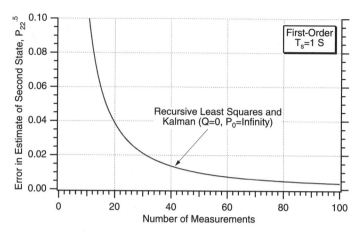

Fig. 4.2   First-order polynomial Kalman and recursive least-squares filters have identical standard deviations for errors in the estimate of the second state.

yields

$$
\begin{bmatrix} \hat{x}_k \\ \hat{\dot{x}}_k \\ \hat{\ddot{x}}_k \end{bmatrix} = \begin{bmatrix} 1 & T_s & 0.5T_s^2 \\ 0 & 1 & T_s \\ 0 & 0 & 1 \end{bmatrix} \begin{bmatrix} \hat{x}_{k-1} \\ \hat{\dot{x}}_{k-1} \\ \hat{\ddot{x}}_{k-1} \end{bmatrix}
$$
$$
+ \begin{bmatrix} K_{1_k} \\ K_{2_k} \\ K_{3_k} \end{bmatrix} \begin{bmatrix} x_k^* - [1 \ 0 \ 0] \begin{bmatrix} 1 & T_s & 0.5T_s^2 \\ 0 & 1 & T_s \\ 0 & 0 & 1 \end{bmatrix} \begin{bmatrix} \hat{x}_{k-1} \\ \hat{\dot{x}}_{k-1} \\ \hat{\ddot{x}}_{k-1} \end{bmatrix} \end{bmatrix}
$$

Multiplying out the terms of the preceding matrix equation yields the three scalar equations for the second-order or three-state polynomial Kalman filter

$$
\hat{x}_k = \hat{x}_{k-1} + T_s\hat{\dot{x}}_{k-1} + 0.5T_s^2\hat{\ddot{x}}_{k-1} + K_{1_k}(x_k^* - \hat{x}_{k-1} - T_s\hat{\dot{x}}_{k-1} - 0.5T_s^2\hat{\ddot{x}}_{k-1})
$$
$$
\hat{\dot{x}}_k = \hat{\dot{x}}_{k-1} + T_s\hat{\ddot{x}}_{k-1} + K_{2_k}(x_k^* - \hat{x}_{k-1} - T_s\hat{\dot{x}}_{k-1} - .5T_s^2\hat{\ddot{x}}_{k-1})
$$
$$
\hat{\ddot{x}}_k = \hat{\ddot{x}}_{k-1} + K_{3_k}(x_k^* - \hat{x}_{k-1} - T_s\hat{\dot{x}}_{k-1} - .5T_s^2\hat{\ddot{x}}_{k-1})
$$

If we define the residual Res_k as

$$
\text{Res}_k = x_k^* - \hat{x}_{k-1} - T_s\hat{\dot{x}}_{k-1} - .5T_s^2\hat{\ddot{x}}_{k-1}
$$

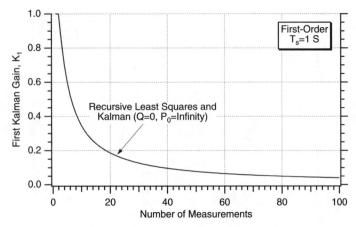

Fig. 4.3   First gain of first-order polynomial Kalman and recursive least-squares filters are identical.

then the three preceding equations simplify to

$$\hat{x}_k = \hat{x}_{k-1} + T_s\dot{\hat{x}}_{k-1} + 0.5T_s^2\ddot{\hat{x}}_{k-1} + K_{1_k}\text{Res}_k$$
$$\dot{\hat{x}}_k = \dot{\hat{x}}_{k-1} + T_s\ddot{\hat{x}}_{k-1} + K_{2_k}\text{Res}_k$$
$$\ddot{\hat{x}}_k = \ddot{\hat{x}}_{k-1} + K_{3_k}\text{Res}_k$$

which is precisely the same form as the equations used for the second-order recursive least-squares filter of the preceding chapter. Therefore, the only possible difference between the second-order recursive least-squares filter and polynomial Kalman filter would be in the gains.

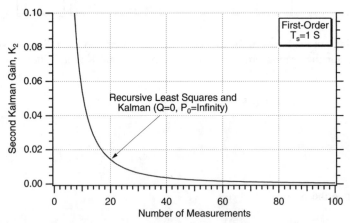

Fig. 4.4   Second gain of first-order polynomial Kalman and recursive least-squares filters are identical.

Next, we would like to compare the gains and variance predictions of both the second-order recursive least-squares filter and the second-order polynomial Kalman filter with zero process-noise and infinite initial covariance matrix. Recall from the preceding chapter that the formulas for the three second-order recursive least-squares filter gains are given by

$$K_{1_k} = \frac{3(3k^2 - 3k + 2)}{k(k+1)(k+2)} \qquad k = 1, 2, \ldots, n$$

$$K_{2_k} = \frac{18(2k - 1)}{k(k+1)(k+2)T_s}$$

$$K_{3_k} = \frac{60}{k(k+1)(k+2)T_s^2}$$

whereas the formulas for the variance of the errors in the estimates of the first, second, and third states are given by

$$P_{11_k} = \frac{3(3k^2 - 3k + 2)\sigma_n^2}{k(k+1)(k+2)}$$

$$P_{22_k} = \frac{12(16K^2 - 30k + 11)\sigma_n^2}{k(k^2 - 1)(k^2 - 4)T_s^2}$$

$$P_{33_k} = \frac{720\sigma_n^2}{k(k^2 - 1)(k^2 - 4)T_s^4}$$

A program was written and is displayed in Listing 4.2 in which the second-order polynomial Kalman-filter Riccati equations, assuming zero process noise and infinite diagonal elements of the initial covariance matrix, were solved for the gains and covariance matrix. The program included the preceding formulas for the second-order recursive least-squares filter so that both the covariance matrix projections and gain calculations could be compared.

Figure 4.5 compares the square root of the first diagonal element of the second-order polynomial Kalman-filter covariance matrix to the second-order recursive least-squares filter formula for the error in the estimate of the first state. We can see from Fig. 4.5 that the comparison is exact. In addition, this plot also represents how the standard deviation of the error in the estimate of the first state improves as more measurements are taken. In this example the input noise had a standard deviation of unity. After approximately 30 measurements (only 15 measurements for first-order filter) the standard deviation of the error in the estimate is approximately 0.5, meaning that filtering has effectively reduced the effect of noise by a factor of two.

Figure 4.6 compares the square root of the second diagonal element of the second-order polynomial Kalman-filter covariance matrix to the P_{22} formula of the second-order recursive least-squares filter for the error in the estimate of the second state. Again we can see that the comparison is exact. The plot also shows how rapidly the error in the estimate of the second state (i.e., derivative of first state) diminishes after only a few measurements are taken.

Listing 4.2   Comparing second-order Kalman and recursive least-squares filter gains and covariance matrices

```
      IMPLICIT  REAL*8(A-H,O-Z)
      REAL*8  M(3,3),P(3,3),K(3,1),PHI(3,3),H(1,3),R(1,1),PHIT(3,3)
      REAL*8  PHIP(3,3),HT(3,1),KH(3,3),IKH(3,3)
      REAL*8  MHT(3,1),HMHT(1,1),HMHTR(1,1),HMHTRINV(1,1),IDN(3,3)
      REAL*8  K1GM,K2GM,K3GM
      INTEGER ORDER
      OPEN(1,STATUS='UNKNOWN',FILE='DATFIL')
      ORDER =3
      TS=1.
      SIGNOISE=1.
      DO 1000 I=1,ORDER
      DO 1000 J=1,ORDER
              PHI(I,J)=0.
              P(I,J)=0.
              IDN(I,J)=0.
1000  CONTINUE
      IDN(1,1)=1.
      IDN(2,2)=1.
      IDN(3,3)=1.
      P(1,1)=99999999999999.
      P(2,2)=99999999999999.
      P(3,3)=99999999999999.
      PHI(1,1)=1
      PHI(1,2)=TS
      PHI(1,3)=.5*TS*TS
      PHI(2,2)=1
      PHI(2,3)=TS
      PHI(3,3)=1
      DO 1100 I=1,ORDER
              H(1,I)=0.
1100  CONTINUE
      H(1,1)=1
      CALL  MATTRN(H,1,ORDER,HT)
      R(1,1)=SIGNOISE**2
      CALL  MATTRN(PHI,ORDER,ORDER,PHIT)
      DO 10 XN=1.,100
              CALL  MATMUL(PHI,ORDER,ORDER,P,ORDER,ORDER,PHIP)
              CALL  MATMUL(PHIP,ORDER,ORDER,PHIT,ORDER,ORDER,M)
              CALL  MATMUL(M,ORDER,ORDER,HT,ORDER,1,MHT)
              CALL  MATMUL(H,1,ORDER,MHT,ORDER,1,HMHT)
              HMHTR(1,1)=HMHT(1,1)+R(1,1)
              HMHTRINV(1,1)=1./HMHTR(1,1)
              CALL  MATMUL(MHT,ORDER,1,HMHTRINV,1,1,K)
              CALL  MATMUL(K,ORDER,1,H,1,ORDER,KH)
              CALL  MATSUB(IDN,ORDER,ORDER,KH,IKH)
              CALL  MATMUL(IKH,ORDER,ORDER,M,ORDER,ORDER,P)
              IF(XN<3)THEN
                      P11GM=9999999999.
                      P22GM=9999999999.
```

(*continued*)

Listing 4.2 *(Continued)*

```
                       P33GM=9999999999.
             ELSE
                       P11GM=(3*(3*XN*XN-3*XN+2)/(XN*(XN+1)*(XN+2)))
1                        *SIGNOISE**2
                       P22GM=(12*(16*XN*XN-30*XN+11)/(XN*(XN*XN-1)
1                        *(XN*XN-4)*TS*TS))*SIGNOISE**2
                       P33GM=(720/(XN*(XN*XN-1)*(XN*XN-4)*TS*TS*TS*TS
1                        ))*SIGNOISE**2
             ENDIF
             SP11=SQRT(P(1,1))
             SP22=SQRT(P(2,2))
             SP33=SQRT(P(3,3))
             SP11GM=SQRT(P11GM)
             SP22GM=SQRT(P22GM)
             SP33GM=SQRT(P33GM)
             K1GM=3*(3*XN*XN-3*XN+2)/(XN*(XN+1)*(XN+2))
             K2GM=18*(2*XN-1)/(XN*(XN+1)*(XN+2)*TS)
             K3GM=60/(XN*(XN+1)*(XN+2)*TS*TS)
             IF(XN>=3)THEN
             WRITE(9,*)XN,K(1,1),K1GM,K(2,1),K2GM,K(3,1),K3GM
             WRITE(1,*)XN,K(1,1),K1GM,K(2,1),K2GM,K(3,1),K3GM,
1                        SP11,SP11GM,SP22,SP22GM,SP33,SP33GM

             ENDIF
10   CONTINUE
             PAUSE
             CLOSE(1)
             END

C SUBROUTINE MATTRN IS SHOWN IN LISTING 1.3
C SUBROUTINE MATMUL IS SHOWN IN LISTING 1.4
C SUBROUTINE MATADD IS SHOWN IN LISTING 1.1
C SUBROUTINE MATSUB IS SHOWN IN LISTING 1.2
```

Figure 4.7 compares the square root of the third diagonal element of the second-order polynomial Kalman-filter covariance matrix to the P_{33} formula for the second-order recursive least-squares filter. Again, we can see from Fig. 4.7 that the comparison is exact. The plot also shows how rapidly the error in the estimate of the third state (i.e., second derivative of first state) diminishes after only a few measurements are taken.

Figures 4.8–4.10 compare the gains of both the second-order polynomial Kalman and recursive least-squares filters. We can see that in all cases the comparison is exact. Because the structure and gains of both second-order filters are exactly the same, the filters must be identical. Therefore, we can say once again that simulation experiments have confirmed *that a polynomial Kalman filter with zero process-noise and infinite initial covariance matrix is identical to a recursive least-squares filter of the same order.*

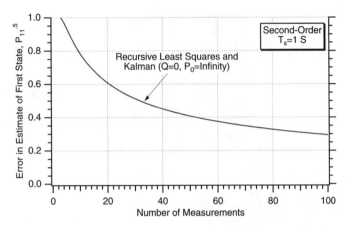

Fig. 4.5 Covariance matrix projections of first state are identical for both second-order polynomial Kalman and recursive least-squares filters.

Comparing Different-Order Filters

We have already demonstrated that the zeroth-order, first-order, and second-order polynomial Kalman and recursive least-squares filters are identical in every way when the Kalman filter has zero process noise and infinite initial covariance matrix. It is now of interest to see how the different-order filters compare in terms of their performance projections. Figure 4.11 compares the standard deviation of the error in the estimate of the first state (i.e., true first state minus estimated first state) of a zeroth, first- and second-order, zero process-noise polynomial Kalman filter as a function of the number of measurements taken. As expected, the errors in the estimates decrease as the number of measurements taken increase. We can

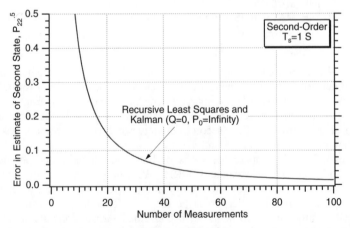

Fig. 4.6 Covariance matrix projections of second state are identical for both second-order polynomial Kalman and recursive least-squares filters.

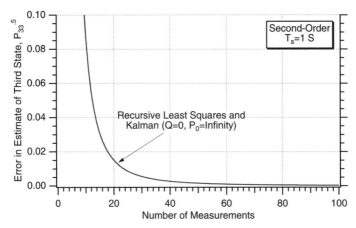

Fig. 4.7 **Covariance matrix projections of third state are identical for both second-order polynomial Kalman and recursive least-squares filters.**

see that the errors in the estimate are reduced more rapidly for the lower-order filters. This means that maximum noise reduction will be obtained by using the lowest-order filter possible. In practice, truncation error or the ability to track higher-order signals will place the limit on how low an order filter can be used.

Figure 4.12 displays the standard deviation of the error in the estimate of the second state as a function of the number of measurements taken for the first- and second-order, zero process-noise polynomial Kalman filters. The zeroth-order filter was not included because it does not have the ability to estimate the derivative of the first state (i.e., it assumes signal is a constant or of zeroth order and therefore, by definition, has a zero derivative). Again, we can see that the

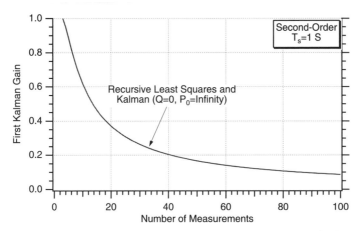

Fig. 4.8 **First gain of second-order polynomial Kalman and recursive least-squares filters are identical.**

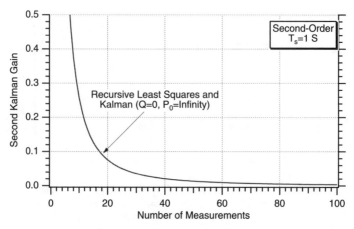

Fig. 4.9 Second gain of second-order polynomial Kalman and recursive least-squares filters are identical.

errors in the estimates decrease as the number of measurements taken increase. As was the case before, the lower-order filter reduces the errors in the estimate of the second state more rapidly.

Finally, Fig. 4.13 displays the standard deviation of the error in the estimate of the third state as a function of the number of measurements taken for the second-order, zero process-noise polynomial Kalman filter. The zeroth-order and first-order filters were not included because they do not have the ability to estimate the derivative of the third state (i.e., second derivative of first state) because they both implicitly assume that the derivative is zero by definition. Again, we can see that the errors in the estimates decrease as the number of measurements taken increase.

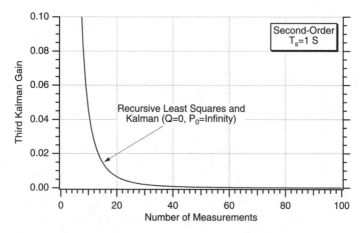

Fig. 4.10 Third gain of second-order polynomial Kalman and recursive least-squares filters are identical.

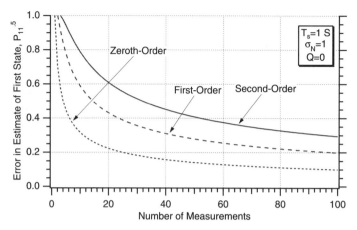

Fig. 4.11   Errors in estimate of first state are smaller with lower-order filters.

Initial Covariance Matrix

So far we have assumed that the polynomial Kalman filter had an initial covariance matrix whose diagonal elements were infinite. This really meant that we had no a priori information on how to initialize the states of the Kalman filter. In practice, information from other sources may be available to help in the initialization process. Under these circumstances the diagonal elements of the initial covariance matrix can be made smaller to reflect more favorable circumstances (i.e., we are not as ignorant).

An experiment was conducted in which the initial covariance matrix (i.e., a number in this example because the matrix is a scalar) for the zeroth-order polynomial Kalman filter was made a parameter. Figure 4.14 shows that the error

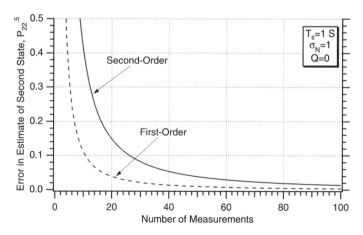

Fig. 4.12   Errors in estimate of second state are smaller with lower-order filter.

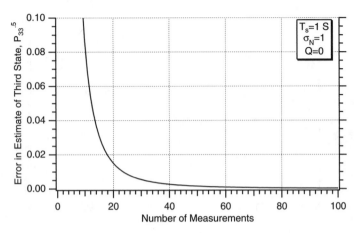

Fig. 4.13 Errors in estimate of third state decrease as number of measurements taken increase.

in the estimate of the first state for the zeroth-order filter does not change as the initial covariance matrix is reduced from infinity to 100. Reducing the initial covariance matrix further by another two orders of magnitude to unity only slightly changes the errors in the estimates when only a few measurements are taken (i.e., less than 20). When many measurements are taken, the answers are the same, regardless of the value for the initial covariance. Reducing the initial covariance matrix by another order of magnitude reduces the errors in the estimates for the first 30 measurements. If the initial covariance matrix is made zero, there will not be any errors in the estimates. Under these circumstances it is assumed that the filter has been perfectly initialized. Because there is no process

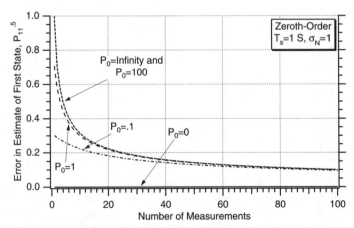

Fig. 4.14 Errors in estimates of first state of a zeroth-order polynomial Kalman filter are fairly insensitive to the initial covariance matrix.

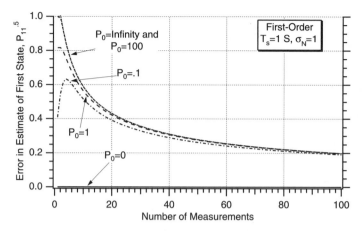

Fig. 4.15   Errors in estimates of first state of a first-order polynomial Kalman filter are fairly insensitive to the initial covariance matrix.

noise, we are assuming that the filter has a perfect model of the real world. Under these circumstances future estimates will be perfect, and there will not be any errors in the estimates.

For the first-order polynomial Kalman filter the initial covariance matrix has two diagonal terms and two off-diagonal terms. The off-diagonal terms were set to zero, whereas the diagonal terms were set equal to each other and reduced from infinity to one tenth. Figures 4.15 and 4.16 show that under these circumstances the error in the estimate of the first and second states of the first-order filter are also fairly insensitive to the initial value of the diagonal terms. Only when the diagonal terms are set to zero do we get significantly different answers. Therefore,

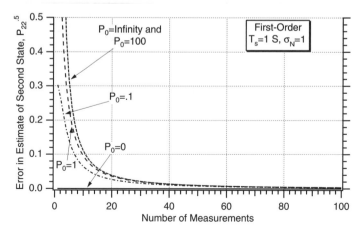

Fig. 4.16   Errors in estimates of second state of a first-order polynomial Kalman filter are fairly insensitive to the initial covariance matrix.

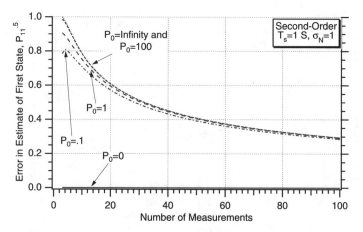

Fig. 4.17 Errors in estimates of first state of a second-order polynomial Kalman filter are fairly insensitive to the initial covariance matrix.

we can conclude that the theoretical performance (i.e., standard deviation of the error in the estimate of the first and second states) of the first-order polynomial Kalman filter is also insensitive to the initial value of the covariance matrix.

For the second-order polynomial Kalman filter the initial covariance matrix has three diagonal terms and six off-diagonal terms (i.e., nine terms in a 3×3 matrix). For this experiment the off-diagonal terms were set to zero while the diagonal terms were set equal to each other and reduced from infinity to one tenth. Figures 4.17–4.19 show that under these circumstances the error in the estimate of the first, second, and third states of the second-order filter is also fairly insensitive to the initial value of the diagonal terms. Again, only when the

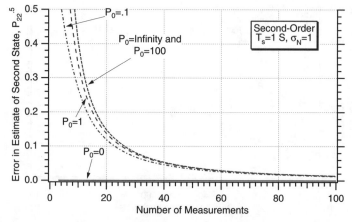

Fig. 4.18 Errors in estimates of second state of a second-order polynomial Kalman filter are fairly insensitive to the initial covariance matrix.

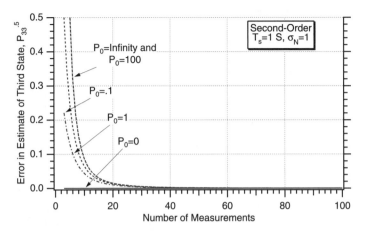

Fig. 4.19 **Errors in estimates of third state of a second-order polynomial Kalman filter are fairly insensitive to the initial covariance matrix.**

diagonal terms are set to zero do we get significantly different answers. Therefore, we can conclude that the theoretical performance (i.e., standard deviation of the error in the estimate of the first, second, and third states) of the second-order polynomial Kalman filter is also insensitive to the initial value of the covariance matrix.

Riccati Equations with Process Noise

Recall that the Riccati equations are a set of recursive matrix equations given by

$$M_k = \Phi_k P_{k-1} \Phi_k^T + Q_k$$
$$K_k = M_k H^T (H M_k H^T + R_k)^{-1}$$
$$P_k = (I - K_k H) M_k$$

where P_k is a covariance matrix representing errors in the state estimates (i.e., diagonal elements represent variance of true state minus estimated state) after an update and M_k is the covariance matrix representing errors in the state estimates before an update. The discrete process-noise matrix Q_k, which until now in this chapter was assumed to be zero, can be found from the continuous process-noise matrix Q and the fundamental matrix according to

$$Q_k = \int_0^{T_s} \Phi(\tau) Q \Phi^T(\tau)\, dt$$

Table 4.2 lists particularly useful Q matrices for different-order polynomial Kalman filters, along with the appropriate fundamental matrices (i.e., derived at the beginning of this chapter) and the derived discrete Q_k matrices. The

Table 4.2   Discrete process-noise matrix varies with system order

Order	Continuous Q	Fundamental	Discrete Q
0	$Q = \Phi_s$	$\Phi_k = 1$	$Q_k = \Phi_s T_s$
1	$Q = \Phi_s \begin{bmatrix} 0 & 0 \\ 0 & 1 \end{bmatrix}$	$\Phi_k = \begin{bmatrix} 1 & T_s \\ 0 & 1 \end{bmatrix}$	$Q_k = \Phi_s \begin{bmatrix} \dfrac{T_s^3}{3} & \dfrac{T_s^2}{2} \\ \dfrac{T_s^2}{2} & T_s \end{bmatrix}$
2	$Q = \Phi_s \begin{bmatrix} 0 & 0 & 0 \\ 0 & 0 & 0 \\ 0 & 0 & 1 \end{bmatrix}$	$\Phi_k = \begin{bmatrix} 1 & T_s & 0.5T_s^2 \\ 0 & 1 & T_s \\ 0 & 0 & 1 \end{bmatrix}$	$Q_k = \Phi_s \begin{bmatrix} \dfrac{T_s^5}{20} & \dfrac{T_s^4}{8} & \dfrac{T_s^3}{6} \\ \dfrac{T_s^4}{8} & \dfrac{T_s^3}{3} & \dfrac{T_s^2}{2} \\ \dfrac{T_s^3}{6} & \dfrac{T_s^2}{2} & T_s \end{bmatrix}$

continuous process-noise matrix assumes that the process noise always enters the system model on the derivative of the last state (i.e., highest derivative). We can see that the discrete process-noise matrix depends on the continuous process-noise spectral density Φ_s and the sampling time T_s. When the continuous process-noise spectral density Φ_s does not represent actual model noise uncertainty, it simply becomes a fudge factor that accounts for our lack of knowledge of the real world. If Φ_s is zero or small, it means that we believe our Kalman filter model of the real world is excellent. Large values of Φ_s mean that we really do not believe that the Kalman filter has a good model of the real world and the filter must be prepared for a great deal of uncertainty. Usually Φ_s is determined experimentally by performing many Monte Carlo experiments under a variety of scenarios.

Let us first examine how the zeroth-order Kalman filter's ability to estimate is influenced by process noise. Figure 4.20 shows how the error in the estimate of the state (i.e., true state minus estimated state) worsens as more process noise is added to the zeroth-order filter for the case in which the standard deviation of the measurement noise is unity and the sampling time is 1 s. We can see from Fig. 4.20 that when the process noise is zero the error in the estimate will eventually approach zero as more measurements are taken. However, when process noise is present, the error in the estimate will approach a steady-state value after only a few measurements, and the quality of the estimates will not improve beyond a certain point no matter how many more measurements are taken. Increasing the value of the process noise increases the error in the estimate because the filter is assuming that process noise is continually contributing uncertainty to the states even when the actual process noise does not exist.

Figures 4.21 and 4.22 show how the error in the estimate of the two states (i.e., true state minus estimated state) of the first-order polynomial Kalman filter increases as more process noise is added for the case in which the standard

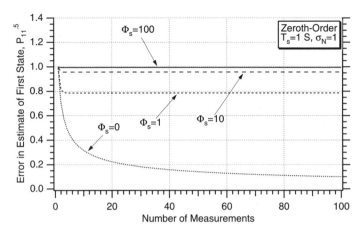

Fig. 4.20   Error in estimate of the state degrades as process noise increases for zeroth-order filter.

deviation of the measurement noise is unity and the sampling time is 1 s. As was the case earlier, we can see from Figs. 4.21 and 4.22 that when the process noise is zero the error in the state estimate will eventually approach zero as more measurements are taken. However, when process noise is present, the error in the state estimates will approach a steady-state value after only a few measurements are taken. As was the case earlier, increasing the value of the process noise increases the error in the estimate.

Figures 4.23–4.25 show how the error in the estimate of the three states (i.e., true state minus estimated state) of the second-order polynomial Kalman filter increase as more process noise is added for the case in which the standard

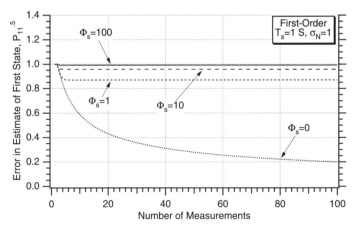

Fig. 4.21   Increasing the process noise increases the error in the estimate of the first state of a first-order polynomial Kalman filter.

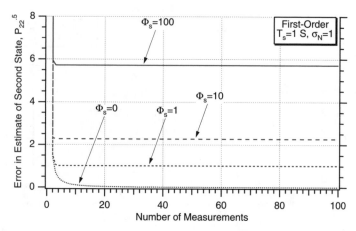

Fig. 4.22 Increasing the process noise increases the error in the estimate of the second state of a first-order polynomial Kalman filter.

deviation of the measurement noise is unity and the sampling time is 1 s. As was the case for both the zeroth- and first-order polynomial Kalman filters, we can see from Figs. 4.23–4.25 that when the process noise is zero the error in the state estimate will eventually approach zero as more measurements are taken. Again, when process noise is present, the error in the state estimates will approach a steady-state value after only a few measurements are taken. As was the case for the zeroth- and first-order polynomial Kalman filters, increasing the value of the process noise also increases the error in the estimate for the second-order polynomial Kalman filter.

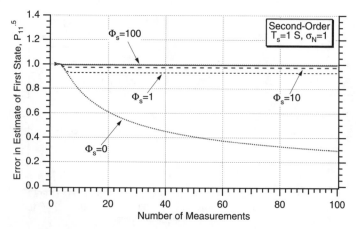

Fig. 4.23 Increasing the process noise increases the error in the estimate of the first state of a second-order polynomial Kalman filter.

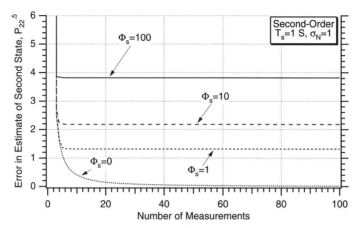

Fig. 4.24   Increasing the process noise increases the error in the estimate of the second state of a second-order polynomial Kalman filter.

Example of Kalman Filter Tracking a Falling Object

To illustrate the utility of a polynomial Kalman filter with process noise, let us consider the one-dimensional example of an object falling rather quickly on a tracking radar, as shown in Fig. 4.26. The object is initially 400,000 ft above the radar and has a velocity of 6000 ft/s toward the radar, which is located on the surface of a flat Earth. In this example we are neglecting drag or air resistance so that only gravity g (i.e., $g = 32.2$ ft/s^2) acts on the object. Let us pretend that the radar measures the range from the radar to the target (i.e., altitude of the target) with a 1000-ft standard deviation measurement accuracy. The radar takes measurement 10 times a second for 30 s. We would like to build a filter to

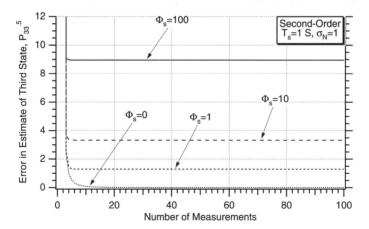

Fig. 4.25   Increasing the process noise increases the error in the estimate of the third state of a second-order polynomial Kalman filter.

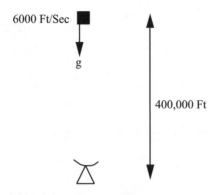

6000 Ft/Sec

g

400,000 Ft

Fig. 4.26 Radar tracking falling object.

estimate the altitude and velocity of the object without any a priori information (i.e., knowing initial altitude and velocity of the object).

We know from basic physics that if x is the distance from the radar to the object, then the value of x at any time t is given by

$$x = 400{,}000 - 6000t - \frac{gt^2}{2}$$

As a check, we set time to zero in the preceding expression and get the correct initial value for x of 400,000 ft. The velocity of the object at any time can be found by taking the derivative of the preceding expression, yielding

$$\dot{x} = -6000 - gt$$

Again, as a check we set time to zero in the preceding expression and get the correct initial velocity of -6000 ft/s. We can see that the expression for altitude x is a second-order polynomial in time. Because, in this example, the real world is actually a second-order polynomial, a second-order polynomial Kalman filter with zero process noise can be used to track the object.

Listing 4.3 presents the simulation that will be used to illustrate how the second-order polynomial Kalman filter's processing of the radar measurements can be used to estimate the altitude and velocity of the target. We can see from Listing 4.3 that the filter states are initialized to zero (i.e., XH = XDH = XDDH = 0) and that the diagonal elements of the initial covariance matrix are infinite [i.e., $P(1,1) = P(2,2) = P(3,3) = 9999999999999$] because we do not have any a priori information concerning the initial target altitude or speed. The initial state estimates are in considerable error because the initial altitude is 400,000 ft. The initial velocity is -6000 ft/s, and the acceleration is always -32.2 ft/s^2. Other parameters have been set in accordance with the problem statement. We print out the filter state estimates along with the true states. In addition, we compute the actual error in the state estimates along with the theoretical predictions from the diagonal elements of the covariance matrix.

Figure 4.27 displays the actual altitude of the object along with the filter's estimate of altitude as a function of time. We can see that the filter is doing an

Listing 4.3   Simulation of radar tracking falling object

```
C THE FIRST THREE STATEMENTS INVOKE THE ABSOFT RANDOM
  NUMBER GENERATOR ON THE MACINTOSH
      GLOBAL DEFINE
              INCLUDE 'quickdraw.inc'
      END
      IMPLICIT REAL*8(A-H,O-Z)
      REAL*8 M(3,3),P(3,3),K(3,1),PHI(3,3),H(1,3),R(1,1),PHIT(3,3)
      REAL*8 PHIP(3,3),HT(3,1),KH(3,3),IKH(3,3)
      REAL*8 MHT(3,1),HMHT(1,1),HMHTR(1,1),HMHTRINV(1,1),IDN(3,3)
      REAL*8 Q(3,3),PHIPPHIT(3,3)
      INTEGER ORDER
      OPEN(1,STATUS='UNKNOWN',FILE='DATFIL')
      OPEN(2,STATUS='UNKNOWN',FILE='COVFIL')
      ORDER =3
      PHIS=0.
      TS=.1
      A0=400000
      A1=-6000.
      A2=-16.1
      XH=0
      XDH=0
      XDDH=0
      SIGNOISE=1000.
      DO 1000 I=1,ORDER
      DO 1000 J=1,ORDER
              PHI(I,J)=0.
              P(I,J)=0.
              IDN(I,J)=0.
              Q(I,J)=0.
1000  CONTINUE
      IDN(1,1)=1.
      IDN(2,2)=1.
      IDN(3,3)=1.
      P(1,1)=99999999999999.
      P(2,2)=99999999999999.
      P(3,3)=99999999999999.
      PHI(1,1)=1
      PHI(1,2)=TS
      PHI(1,3)=.5*TS*TS
      PHI(2,2)=1
      PHI(2,3)=TS
      PHI(3,3)=1
      DO 1100 I=1,ORDER
              H(1,I)=0.
1100  CONTINUE
      H(1,1)=1
      CALL MATTRN(H,1,ORDER,HT)
      R(1,1)=SIGNOISE**2
      CALL MATTRN(PHI,ORDER,ORDER,PHIT)
      Q(1,1)=PHIS*TS**5/20
```

(*continued*)

Listing 4.3 (*Continued*)

```
      Q(1,2)=PHIS*TS**4/8
      Q(1,3)=PHIS*TS**3/6
      Q(2,1)=Q(1,2)
      Q(2,2)=PHIS*TS**3/3
      Q(2,3)=PHIS*TS*TS/2
      Q(3,1)=Q(1,3)
      Q(3,2)=Q(2,3)
      Q(3,3)=PHIS*TS
      DO 10 T=0.,30.,TS
            CALL MATMUL(PHI,ORDER,ORDER,P,ORDER,ORDER,PHIP)
            CALL MATMUL(PHIP,ORDER,ORDER,PHIT,ORDER,ORDER,
                  PHIPPHIT)
            CALL MATADD(PHIPPHIT,ORDER,ORDER,Q,M)
            CALL MATMUL(M,ORDER,ORDER,HT,ORDER,1,MHT)
            CALL MATMUL(H,1,ORDER,MHT,ORDER,1,HMHT)
            HMHTR(1,1)=HMHT(1,1)+R(1,1)
            HMHTRINV(1,1)=1./HMHTR(1,1)
            CALL MATMUL(MHT,ORDER,1,HMHTRINV,1,1,K)
            CALL MATMUL(K,ORDER,1,H,1,ORDER,KH)
            CALL MATSUB(IDN,ORDER,ORDER,KH,IKH)
            CALL MATMUL(IKH,ORDER,ORDER,M,ORDER,ORDER,P)
            CALL GAUSS(XNOISE,SIGNOISE)
            X=A0+A1*T+A2*T*T
            XD=A1+2*A2*T
            XDD=2*A2
            XS=X+XNOISE
            RES=XS-XH-TS*XDH-.5*TS*TS*XDDH
            XH=XH+XDH*TS+.5*TS*TS*XDDH+K(1,1)*RES
            XDH=XDH+XDDH*TS+K(2,1)*RES
            XDDH=XDDH+K(3,1)*RES
            SP11=SQRT(P(1,1))
            SP22=SQRT(P(2,2))
            SP33=SQRT(P(3,3))
            XHERR=X-XH
            XDHERR=XD-XDH
            XDDHERR=XDD-XDDH
            WRITE(9,*)T,X,XH,XD,XDH,XDD,XDDH
            WRITE(1,*)T,X,XH,XD,XDH,XDD,XDDH
            WRITE(2,*)T,XHERR,SP11,-SP11,XDHERR,SP22,-SP22,XDDHERR,
     1      SP33,-SP33
10    CONTINUE
      CLOSE(1)
      CLOSE(2)
      PAUSE
      END

C SUBROUTINE GAUSS IS SHOWN IN LISTING 1.8
C SUBROUTINE MATTRN IS SHOWN IN LISTING 1.3
C SUBROUTINE MATMUL IS SHOWN IN LISTING 1.4
C SUBROUTINE MATADD IS SHOWN IN LISTING 1.1
C SUBROUTINE MATSUB IS SHOWN IN LISTING 1.2
```

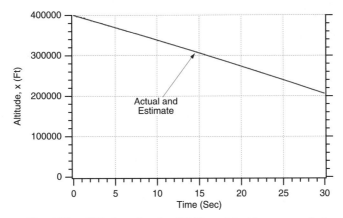

Fig. 4.27   Altitude estimate of falling object is near perfect.

excellent job because the estimate of the object's altitude appears to be virtually exact.

Figure 4.28 displays the actual velocity of the object along with the filter's estimate of velocity. We can see that it takes approximately 10 s for the filter to obtain a highly accurate estimate of the target velocity. The large excursions in the velocity estimate at the beginning of the measurement and estimation process is because the initial estimate of target velocity was 0 ft/s, whereas the actual target velocity was −6000 ft/s. This large initialization error was the main reason it took the filter 10 s to establish an accurate velocity track.

Although we really do not have to estimate the object's acceleration because we know it must be 32.2 ft/s² (i.e., only gravity acts on the object), the acceleration estimate is free with a second-order polynomial Kalman filter. Figure 4.29 shows that for nearly 10 s the filter's acceleration estimate is terrible

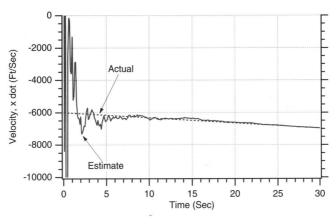

Fig. 4.28   It takes approximately 10 s to accurately estimate velocity of falling object with second-order filter.

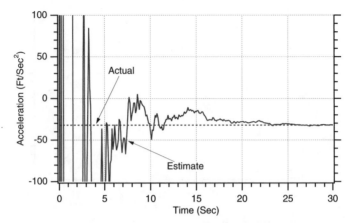

Fig. 4.29 It takes nearly 20 s to accurately estimate acceleration of falling object with second-order filter.

(i.e., because the filter had terrible initial conditions). Only after 20 s do we receive accurate acceleration estimates. Again, the filter had a bad initial estimate of acceleration because it assumed zero, whereas the actual acceleration of the object was -32.2 ft/s^2. Later in this chapter we will see if total filter performance could be improved if we made use of the fact that we know in advance the exact acceleration acting on the object.

As we saw in Chapter 3, just comparing the state estimates to the actual states is not sufficient for establishing that the filter is working according to theory. To be sure that the filter is working properly, we must look at the actual errors in the estimates and compare them to the theoretical answers obtained from the covariance matrix (i.e., square root of first diagonal element for first state, square root of second diagonal element for second state, etc.). Figures 4.30

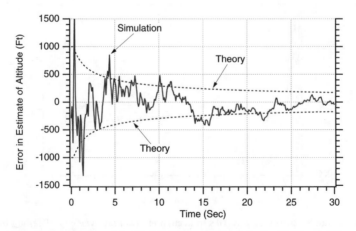

Fig. 4.30 Second-order polynomial Kalman filter single flight results appear to match theory for errors in estimate of altitude.

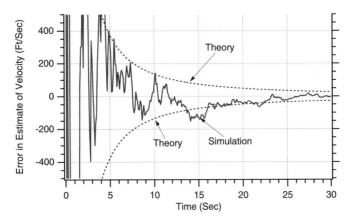

Fig. 4.31   Second-order polynomial Kalman filter single flight results appear to match theory for errors in estimate of velocity.

and 4.31 show that the simulated errors in the estimate of the first and second states lie within the theoretical bounds approximately 68% of the time. In other words, the second-order polynomial Kalman filter appears to be working correctly.

If we were not satisfied with the estimates from the second-order polynomial Kalman filter, we could try a first-order filter. Based on work in this chapter and the preceding chapter, we know that a lower-order filter will have superior noise-reduction properties. However, if we use a first-order filter with zero process noise, Fig. 4.32 shows that our position estimate diverges from the truth because of the gravity term. Filter divergence is expected because we have a first-order filter operating in a second-order world due to gravity.

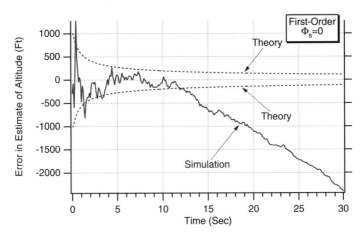

Fig. 4.32   First-order polynomial Kalman filter without process noise can not track second-order signal.

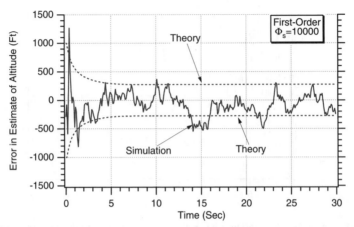

Fig. 4.33   Adding process noise prevents altitude errors of first-order filter from diverging.

One engineering fix to the divergence problem is the use of process noise. Figures 4.33 and 4.34 show that the errors in the estimates of the first-order polynomial Kalman filter can be kept from diverging if we add process noise (i.e., $\Phi_s = 10,000$). However, although the addition of process noise to the first-order filter prevents divergence, it also increases the errors in the estimate to the point where we would have been better off using the second-order polynomial Kalman filter without process noise. For example, Fig. 4.33 shows the error in the estimate of altitude for the first-order polynomial Kalman filter with process noise approaches 300 ft in the steady state, whereas Fig. 4.30 shows that the error in the estimate of altitude for the second-order polynomial Kalman filter without process noise approaches 200 ft in the steady state. Figure 4.34 shows that the

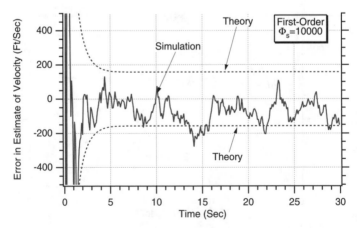

Fig. 4.34   Adding process noise prevents velocity errors of first-order filter from diverging.

error in the estimate of velocity for the first-order polynomial Kalman filter with process noise approaches $150\,\text{ft/s}$ in the steady state, whereas Fig. 4.31 shows that the error in the estimate of velocity for the second-order polynomial Kalman filter without process noise approaches only $25\,\text{ft/s}$ in the steady state.

We can reduce the estimation errors with a first-order filter even further and without the use of process noise by making use of a priori information. We know that only gravity (i.e., $g = 32.2\,\text{ft/s}^2$) acts on the falling body. Gravitational information can be incorporated into the Kalman filter by first recalling that if we have a priori deterministic information the real world can be described in state-space form by

$$\dot{x} = Fx + Gu + w$$

With the preceding equation representing the real world the resultant Kalman-filter equation will be

$$\hat{x}_k = \Phi_k \hat{x}_{k-1} + G_k u_{k-1} + K_k(z_k - H\Phi_k \hat{x}_{k-1} - HG_k u_{k-1})$$

If u_{k-1} is assumed to be constant between sampling instants, G_k is obtained from

$$G_k = \int_0^{T_s} \Phi(\tau)G\,d\tau$$

For our particular problem we have already pointed out that gravity is known and does not have to be estimated. Therefore, for our particular problem the only differential equation describing the real world is

$$\ddot{x} = -g$$

The preceding second-order differential equation can be recast in state-space form as

$$\begin{bmatrix} \dot{x} \\ \ddot{x} \end{bmatrix} = \begin{bmatrix} 0 & 1 \\ 0 & 0 \end{bmatrix}\begin{bmatrix} x \\ \dot{x} \end{bmatrix} + \begin{bmatrix} 0 \\ -1 \end{bmatrix}g$$

From the preceding matrix differential equation we recognize that the systems dynamics matrix is given by

$$F = \begin{bmatrix} 0 & 1 \\ 0 & 0 \end{bmatrix}$$

For the preceding system dynamics matrix we have already shown that the fundamental matrix turns out to be

$$\Phi_k = \begin{bmatrix} 1 & T_s \\ 0 & 1 \end{bmatrix}$$

From the state-space equation we can also see that

$$G = \begin{bmatrix} 0 \\ -1 \end{bmatrix}$$

and that

$$u_{k-1} = g$$

Therefore, the discrete matrix G_k can be found to be

$$G_k = \int_0^{T_s} \Phi(\tau)G \, d\tau = \int_0^{T_s} \begin{bmatrix} 1 & \tau \\ 0 & 1 \end{bmatrix} \begin{bmatrix} 0 \\ -1 \end{bmatrix} d\tau = \begin{bmatrix} \dfrac{-T_s^2}{2} \\ -T_s \end{bmatrix}$$

Because the formula for the Kalman filter is

$$\hat{x}_k = \Phi_k \hat{x}_{k-1} + G_k u_{k-1} + K_k(z_k - H\Phi_k \hat{x}_{k-1} - HG_k u_{k-1})$$

substitution yields

$$\begin{bmatrix} \hat{x}_k \\ \hat{\dot{x}}_k \end{bmatrix} = \begin{bmatrix} 1 & T_s \\ 0 & 1 \end{bmatrix} \begin{bmatrix} \hat{x}_{k-1} \\ \hat{\dot{x}}_{k-1} \end{bmatrix} + \begin{bmatrix} -0.5T_s^2 \\ -T_s \end{bmatrix} g$$

$$+ \begin{bmatrix} K_{1_k} \\ K_{2_k} \end{bmatrix} \left[x_k^* - [1 \quad 0] \begin{bmatrix} 1 & T_s \\ 0 & 1 \end{bmatrix} \begin{bmatrix} \hat{x}_{k-1} \\ \hat{\dot{x}}_{k-1} \end{bmatrix} - [1 \quad 0] \begin{bmatrix} -0.5T_s^2 \\ -T_s \end{bmatrix} g \right]$$

Multiplying out the terms of the preceding equation yields two scalar equations:

$$\hat{x}_k = \hat{x}_{k-1} + \hat{\dot{x}}_{k-1}T_s - 0.5gT_s^2 + K_{1_k}(x_k^* - \hat{x}_{k-1} - \hat{\dot{x}}_{k-1}T_s + 0.5gT_s^2)$$
$$\hat{\dot{x}}_k = \hat{\dot{x}}_{k-1} - gT_s + K_{2_k}(x_k^* - \hat{x}_{k-1} - \hat{\dot{x}}_{k-1}T_s + 0.5gT_s^2)$$

If we define the residual as

$$\text{Res}_k = x_k^* - \hat{x}_{k-1} - \hat{\dot{x}}_{k-1}T_s + 0.5gT_s^2$$

then the two equations for the Kalman filter simplify to

$$\hat{x}_k = \hat{x}_{k-1} + \hat{\dot{x}}_{k-1}T_s - 0.5gT_s^2 + K_{1_k}\text{Res}_k$$
$$\hat{\dot{x}}_k = \hat{\dot{x}}_{k-1} - gT_s + K_{2_k}\text{Res}_k$$

The a priori information on gravity only affects the structure of the Kalman filter. Gravity does not influence the Riccati equations.

Listing 4.4 is a simulation of the falling object being tracked and estimated with the preceding first-order polynomial Kalman filter that compensates for gravity without using process noise. The errors in the estimate of altitude and

Listing 4.4   First-order polynomial Kalman filter with gravity compensation

```
C THE FIRST THREE STATEMENTS INVOKE THE ABSOFT RANDOM
  NUMBER GENERATOR ON THE MACINTOSH
      GLOBAL DEFINE
            INCLUDE 'quickdraw.inc'
      END
      IMPLICIT REAL*8(A-H,O-Z)
      REAL*8 P(2,2),Q(2,2),M(2,2),PHI(2,2),HMAT(1,2),HT(2,1),PHIT(2,2)
      REAL*8 RMAT(1,1),IDN(2,2),PHIP(2,2),PHIPPHIT(2,2),HM(1,2)
      REAL*8 HMHT(1,1),HMHTR(1,1),HMHTRINV(1,1),MHT(2,1),K(2,1)
      REAL*8 KH(2,2),IKH(2,2)
      INTEGER STEP,ORDER
      TS=.1
      PHIS=0.
      A0=400000.
      A1=-6000.
      A2=-16.1
      XH=0.
      XDH=0.
      SIGNOISE=1000.
      ORDER=2
      OPEN(1,STATUS='UNKNOWN',FILE='DATFIL')
      OPEN(2,STATUS='UNKNOWN',FILE='COVFIL')
      T=0.
      S=0.
      H=.001
      DO 14 I=1,ORDER
      DO 14 J=1,ORDER
      PHI(I,J)=0.
      P(I,J)=0.
      Q(I,J)=0.
      IDN(I,J)=0.
14    CONTINUE
      RMAT(1,1)=SIGNOISE**2
      IDN(1,1)=1.
      IDN(2,2)=1.
      P(1,1)=99999999999.
      P(22)=99999999999.
      PHI(1,1)=1.
      PHI(1,2)=TS
      PHI(2,2)=1.
      Q(1,1)=TS*TS*TS*PHIS/3.
      Q(1,2)=.5*TS*TS*PHIS
      Q(2,1)=Q(1,2)
      Q(2,2)=PHIS*TS
      HMAT(1,1)=1.
      HMAT(1,2)=0.
      DO 10 T=0.,30.,TS
      CALL MATTRN(PHI,ORDER,ORDER,PHIT)
      CALL MATTRN(HMAT,1,ORDER,HT)
```

(*continued*)

Listing 4.4   (*Continued*)

```
      CALL  MATMUL(PHI,ORDER,ORDER,P,ORDER,ORDER,PHIP)
      CALL  MATMUL(PHIP,ORDER,ORDER,PHIT,ORDER,ORDER,PHIPPHIT)
      CALL  MATADD(PHIPPHIT,ORDER,ORDER,Q,M)
      CALL  MATMUL(HMAT,1,ORDER,M,ORDER,ORDER,HM)
      CALL  MATMUL(HM,1,ORDER,HT,ORDER,1,HMHT)
      CALL  MATADD(HMHT,ORDER,ORDER,RMAT,HMHTR)
         HMHTRINV(1,1)=1./HMHTR(1,1)
      CALL  MATMUL(M,ORDER,ORDER,HT,ORDER,1,MHT)
      CALL  MATMUL(MHT,ORDER,1,HMHTRINV,1,1,K)
      CALL  MATMUL(K,ORDER,1,HMAT,1,ORDER,KH)
      CALL  MATSUB(IDN,ORDER,ORDER,KH,IKH)
      CALL  MATMUL(IKH,ORDER,ORDER,M,ORDER,ORDER,P)
      CALL  GAUSS(XNOISE,SIGNOISE)
      X=A0+A1*T+A2*T*T
      XD=A1+2*A2*T
      XS=X+XNOISE
      RES=XS-XH-TS*XDH+16.1*TS*TS
      XH=XH+XDH*TS-16.1*TS*TS+K(1,1)*RES
      XDH=XDH-32.2*TS+K(2,1)*RES
      SP11=SQRT(P(1,1))
      SP22=SQRT(P(2,2))
      XHERR=X-XH
      XDHERR=XD-XDH
      WRITE(9,*)T,XD,XDH,K(1,1),K(2,1)
      WRITE(1,*)T,X,XH,XD,XDH
      WRITE(2,*)T,XHERR,SP11,-SP11,XDHERR,SP22,-SP22
10    CONTINUE
      PAUSE
      CLOSE(1)
      END


C SBROUTINE GAUSS IS SHOWN IN LISTING 1.8
C SUBROUTINE MATTRN IS SHOWN IN LISTING 1.3
C SUBROUTINE MATMUL IS SHOWN IN LISTING 1.4
C SUBROUTINE MATADD IS SHOWN IN LISTING 1.1
C SUBROUTINE MATSUB IS SHOWN IN LISTING 1.2
```

velocity, along with the theoretical covariance matrix projections, are printed out every tenth of a second.

Figures 4.35 and 4.36 show that the first-order polynomial Kalman filter with gravity compensation and without process noise is superior to the second-order polynomial Kalman filter without process noise. Figure 4.35 shows that the error in the estimate of altitude has been reduced from 200 ft (i.e., see Fig. 4.30) to 100 ft. Figure 4.36 shows that the error in the estimate of velocity has been reduced from 25 ft/s (i.e., see Fig. 4.31) to less than 5 ft/s.

Thus, we can see that if a priori information is available it pays to use it. A priori information may allow us to use lower-order filters that will in turn reduce the errors in all state estimates.

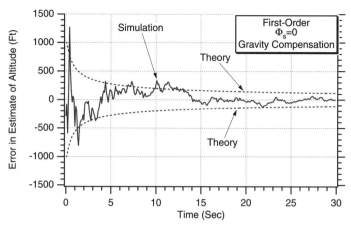

Fig. 4.35 Adding gravity compensation to first-order filter without process noise reduces altitude errors from that of a second-order filter without process noise.

Revisiting Accelerometer Testing Example

To demonstrate another example of the application of polynomial Kalman filters, let us revisit the accelerometer testing example of Chapter 2. Recall that in this example accelerometer measurements are taken for different accelerometer orientation angles, as shown in Fig. 4.37. When the accelerometer input axis is vertical, the accelerometer reading will be g. As the accelerometer input axis rotates through different angles θ_k the accelerometer reading will be $g \cos \theta_k$.

We showed in Chapter 2 that the accelerometer output will not only consist of the gravity term but also the accelerometer bias, scale-factor, and g-sensitive

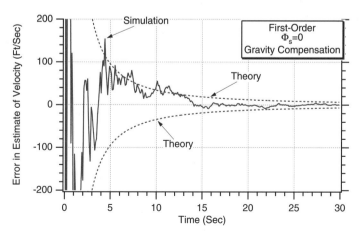

Fig. 4.36 Adding gravity compensation to first-order filter without process noise reduces velocity errors from that of a second-order filter without process noise.

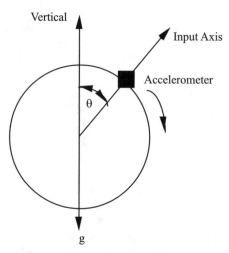

Fig. 4.37   Accelerometer experiment test setup.

drift errors. Therefore, the total accelerometer output is given by

$$\text{Accelerometer Output} = g \cos \theta_k + B + SFg \cos \theta_k + K(g \cos \theta_k)^2$$

where θ_k is the angle the accelerometer makes with the vertical for the kth measurement, g is gravity, B is an accelerometer bias, SF is the accelerometer scale-factor error, and K is a gravity squared or g-squared sensitive drift. Because the theoretical accelerometer output does not have bias, scale-factor, or g-sensitive errors, we expect the output to only contain the vertical angle dependent gravity term or

$$\text{Theory} = g \cos \theta_k$$

Therefore, the accelerometer error is simply the difference between the actual output and theory or

$$\text{Error} = \text{Accelerometer Output} - \text{Theory} = B + SFg \cos \theta_k + K(g \cos \theta_k)^2$$

From a Kalman-filtering point of view, we would like to treat the preceding error equation as measurements at different angles. In addition, we would like to be able to estimate the bias, scale-factor, and g-sensitive drift errors. We also assumed that our measurement of the angle θ_k may not be precise but is corrupted by zero mean Gaussian noise with standard deviation σ_{Noise}. Then the total error will actually be given by

$$\text{Error} = \text{Accelerometer Output} - \text{Theory} = g \cos \theta_K^* + B + SFg \cos \theta_k^*$$
$$+ K(g \cos \theta_K^*)^2 - g \cos \theta_K$$

where θ_k is the actual angle and θ_k^* is the measured angle. The true values for the bias, scale-factor, and g-sensitive drift errors are repeated from Chapter 2 and are summarized in Table 4.3.

Table 4.3   Nominal values for accelerometer testing example

Term	Scientific value	English units
Bias error	$10 \, \mu g$	$10*10^{-6}*32.2 = 0.000322 \, \text{ft/s}^2$
Scale-factor error	5 ppm	$5*10^{-6}$
g-squared sensitive drift	$1 \, \mu g/g^2$	$1*10^{-6}/32.2 = 3.106*10^{-8} \text{s}^2/\text{ft}$

To put the information of this section into a format suitable for Kalman filtering, we first have to choose a set of states for the filter to estimate. Because we would like to estimate the bias, scale-factor error, and g-squared sensitive drift, these would be obvious choices for states. If these states are constants, their derivatives must be zero and the state-space equation, upon which our Kalman filter will be designed, is given by

$$
\begin{bmatrix} \dot{B} \\ \dot{SF} \\ \dot{K} \end{bmatrix} = \begin{bmatrix} 0 & 0 & 0 \\ 0 & 0 & 0 \\ 0 & 0 & 0 \end{bmatrix} \begin{bmatrix} B \\ SF \\ K \end{bmatrix}
$$

We have neglected process noise in this formulation. Because the systems dynamics matrix can be seen from the preceding equation to be zero or

$$
F = \begin{bmatrix} 0 & 0 & 0 \\ 0 & 0 & 0 \\ 0 & 0 & 0 \end{bmatrix}
$$

the discrete fundamental matrix must be the identity matrix because

$$
\Phi_k = I + F T_s + \cdots = \begin{bmatrix} 1 & 0 & 0 \\ 0 & 1 & 0 \\ 0 & 0 & 1 \end{bmatrix} + \begin{bmatrix} 0 & 0 & 0 \\ 0 & 0 & 0 \\ 0 & 0 & 0 \end{bmatrix} T_s = \begin{bmatrix} 1 & 0 & 0 \\ 0 & 1 & 0 \\ 0 & 0 & 1 \end{bmatrix}
$$

We can assume the measurement is simply the error already defined:

$$
z_k = g \cos \theta_K^* + B + SFg \cos \theta_K^* + K(g \cos \theta_K^*)^2 - g \cos \theta_K
$$

Therefore, the measurement is linearly related to the states if we have

$$
z_k = [1 \quad g \cos \theta_K^* \quad (g \cos \theta_K^*)^2] \begin{bmatrix} B \\ SF \\ K \end{bmatrix} + v_k
$$

where the measurement matrix can be seen from the preceding equation to be

$$
H = [1 \quad g \cos \theta_K^* \quad (g \cos \theta_K^*)^2]
$$

and v_k is the measurement noise. The measurement noise is considered to be the difference between the actual accelerometer reading and the theoretical accelerometer output or

$$v_k = g \cos \theta_K^* - g \cos \theta_K = g(\cos \theta_K^* - \cos \theta_K)$$

The actual noise is on the angle (i.e., θ_k^*), and our job is to find an equivalent noise v_k. The method for finding the equivalent noise is to first assume that the noisy angle is simply the true angle plus a small term. Therefore, the noise term v_k becomes

$$v_k = g[\cos(\theta_K + \Delta\theta_k) - \cos \theta_K]$$

Using the trigonometric expansion

$$\cos(\theta_K + \Delta\theta_k) = \cos \theta_k \cos \Delta\theta_k - \sin \theta_k \sin \Delta\theta_k$$

and making the small angle approximation, we obtain

$$v_k = g(\cos \theta_k \cos \Delta\theta_k - \sin \theta_k \sin \Delta\theta_k - \cos \theta_K) \approx -g\Delta\theta_k \sin \theta_k$$

Squaring and taking expectations of both sides of the preceding equation yields an expression for the variance of the effective measurement noise as

$$R_k = E(v_k^2) = (g \sin \theta_k)^2 E(\Delta\theta_k^2) = g^2 \sin^2 \theta_k \sigma_\theta^2$$

In other words, we have just developed an expression for the variance of the equivalent noise or pseudonoise R_k in terms of the variance of the actual noise σ_θ^2. Assuming zero process noise, we now have enough information to build the Kalman filter and solve the Riccati equations for the Kalman gains. Because there is no deterministic input in this formulation of the problem, the Kalman-filtering equation simplifies to

$$\hat{x}_k = \Phi_k \hat{x}_{k-1} + K_k(z_k - H\Phi_k \hat{x}_{k-1})$$

Substitution of the appropriate matrices yields

$$
\begin{bmatrix} \hat{B}_k \\ \widehat{SF}_k \\ \hat{K}_k \end{bmatrix} = \begin{bmatrix} 1 & 0 & 0 \\ 0 & 1 & 0 \\ 0 & 0 & 1 \end{bmatrix} \begin{bmatrix} \hat{B}_{k-1} \\ \widehat{SF}_{k-1} \\ \hat{K}_{k-1} \end{bmatrix}
$$
$$
+ \begin{bmatrix} K_{1_k} \\ K_{2_k} \\ K_{3_k} \end{bmatrix} \left[z_k - \begin{bmatrix} 1 & g \cos \theta_K^* & (g \cos \theta_K^*)^2 \end{bmatrix} \begin{bmatrix} 1 & 0 & 0 \\ 0 & 1 & 0 \\ 0 & 0 & 1 \end{bmatrix} \begin{bmatrix} \hat{B}_{k-1} \\ SF_{k-1} \\ \hat{K}_{k-1} \end{bmatrix} \right]
$$

where the measurement z_k has already been defined. Multiplying out the preceding matrix difference equation and defining a residual yields the scalar equations

$$\text{Res}_k = z_k - \widehat{BIAS}_{k-1} - \widehat{SF}_{k-1}g\cos\theta_k^* - \hat{K}_{k-1}(g\cos\theta_k^*)^2$$
$$\hat{B}_k = \hat{B}_{k-1} + K_{1_k}\text{Res}_k$$
$$\widehat{SF}_k = \widehat{SF}_{k-1} + K_{2_k}\text{Res}_k$$
$$\hat{K}_k = \hat{K}_{k-1} + K_{3_k}\text{Res}_k$$

Listing 4.5 presents the simulation that uses the Kalman filter for estimating the accelerometer bias, scale-factor, and g-sensitive drift. We can see from Listing 4.5 that the filter states are initialized to zero (i.e., $BIASH = SFH = XKH = 0$) because we do not have any a priori information. The angle measurement noise is initially set to 1 μr. Other parameters have been set in accordance with Table 4.3. We print out the filter state estimates along with the true states. In addition, we compute the actual error in the state estimates along with the covariance matrix predictions. The simulation also checks to see if the actual single-run measurement noise

$$v_k = g\cos\theta_K^* - g\cos\theta_K$$

falls between the theoretically calculated plus and minus value of the standard deviation of the measurement noise

$$\sigma_{v_k} = g\sin\theta_k\sigma_\theta$$

The nominal case of Listing 4.5 was run, and the first check was to see if the single-run measurement noise was within the theoretical bounds of the measurement noise standard deviation which was derived in this section. We can see from Fig. 4.38 that there appears to be an agreement between the single-run simulation results and the derived formula (denoted "Theory" in the figure), which indicates that we derived the pseudonoise formula correctly.

The first test to see if the filter is working properly is to see if the actual errors in the estimates lie within the theoretical bounds as determined by the square root of the appropriate diagonal element of the covariance matrix. The nominal case of Listing 4.5 was run, and Figs. 4.39–4.41 present the errors in the estimates for all three states. We can see that because in all three cases the single flight results are within the theoretical bounds the filter appears to be working properly.

We would also like to see how well the actual state estimates compared to the true error terms. Because there is noise on the measurement angle, the state estimates will vary from run to run. Listing 4.5 was modified slightly so that it could be run in the Monte Carlo mode. Cases were run in which the angle measurement noise was 1 and 10 μr. We can see from Figs. 4.42–4.44 that when the angle measurement noise is 1 μr the filter is able to estimate each of the error terms quite well. Of course, for this case there was 1 μr of angle noise. We

Listing 4.5 Simulation of Kalman filter for estimating accelerometer bias, scale-factor, and *g*-sensitive drift

```
C THE FIRST THREE STATEMENTS INVOKE THE ABSOFT RANDOM
  NUMBER GENERATOR ON THE MACINTOSH
      GLOBAL DEFINE
              INCLUDE 'quickdraw.inc'
      END
      IMPLICIT REAL*8 (A-H)
      IMPLICIT REAL*8 (O-Z)
      REAL*8 PHI(3,3),P(3,3),M(3,3),PHIP(3,3),PHIPPHIT(3,3),K(3,1)
      REAL*8 Q(3,3),HMAT(1,3),HM(1,3),MHT(3,1)
      REAL*8 PHIT(3,3),R(1,1)
      REAL*8 HMHT(1,1),HT(3,1),KH(3,3),IDN(3,3),IKH(3,3)
      INTEGER ORDER
      OPEN(1,STATUS='UNKNOWN',FILE='DATFIL')
      OPEN(2,STATUS='UNKNOWN',FILE='COVFIL')
      ORDER=3
      BIAS=.00001*32.2
      SF=.000005
      XK=.000001/32.2
      SIGTH=.000001
      G=32.2
      BIASH=0.
      SFH=0.
      XKH=0
      SIGNOISE=.000001
      S=0.
      DO 1000 I=1,ORDER
      DO 1000 J=1,ORDER
              PHI(I,J)=0.
              P(I,J)=0.
              Q(I,J)=0.
              IDN(I,J)=0.
 1000 CONTINUE
      IDN(1,1)=1.
      IDN(2,2)=1.
      IDN(3,3)=1.
      PHI(1,1)=1.
      PHI(2,2)=1.
      PHI(3,3)=1.
      CALL MATTRN(PHI,ORDER,ORDER,PHIT)
      P(1,1)=9999999999.
      P(2,2)=9999999999.
      P(3,3)=9999999999.
      DO 1100 I=1,ORDER
              HMAT(1,I)=0.
              HT(I,1)=0.
 1100 CONTINUE
      DO 10 THETDEG=0.,180.,2.
              THET=THETDEG/57.3
              CALL GAUSS(THETNOISE,SIGTH)
```

(continued)

Listing 4.5   (*Continued*)

```
                  THETS=THET+THETNOISE
                  HMAT(1,1)=1
                  HMAT(1,2)=G*COS(THETS)
                  HMAT(1,3)=(G*COS(THETS))**2
                  CALL  MATTRN(HMAT,1,ORDER,HT)
                  R(1,1)=(G*SIN(THETS)*SIGTH)**2
                  CALL  MATMUL(PHI,ORDER,ORDER,P,ORDER,ORDER,PHIP)
                  CALL  MATMUL(PHIP,ORDER,ORDER,PHIT,ORDER,ORDER,
                       PHIPPHIT)
                  CALL  MATADD(PHIPPHIT,ORDER,ORDER,Q,M)
                  CALL  MATMUL(HMAT,1,ORDER,M,ORDER,ORDER,HM)
                  CALL  MATMUL(HM,1,ORDER,HT,ORDER,1,HMHT)
                  HMHTR=HMHT(1,1)+R(1,1)
                  HMHTRINV=1./HMHTR
                  CALL  MATMUL(M,ORDER,ORDER,HT,ORDER,1,MHT)
                  DO  150  I=1,ORDER
                       K(I,1)=MHT(I,1)*HMHTRINV
150               CONTINUE
                  CALL  MATMUL(K,ORDER,1,HMAT,1,ORDER,KH)
                  CALL  MATSUB(IDN,ORDER,ORDER,KH,IKH)
                  CALL  MATMUL(IKH,ORDER,ORDER,M,ORDER,ORDER,P)
                  Z=BIAS+SF*G*COS(THETS)+XK*(G*COS(THETS))**2-
                  G*COS(THET)+G*COS(THETS)
                  RES=Z-BIASH-SFH*G*COS(THETS)-XKH*(G*COS(THETS))**2
                  BIASH=BIASH+K(1,1)*RES
                  SFH=SFH+K(2,1)*RES
                  XKH=XKH+K(3,1)*RES
                  SP11=SQRT(P(1,1))
                  SP22=SQRT(P(2,2))
                  SP33=SQRT(P(3,3))
                  BIASERR=BIAS-BIASH
                  SFERR=SF-SFH
                  XKERR=XK-XKH
                  ACTNOISE=G*COS(THETS)-G*COS(THET)
                  SIGR=SQRT(R(1,1))
                  WRITE(9,*)THETDEG,BIAS,BIASH,SF,SFH,XK,XKH
                  WRITE(1,*)THETDEG,BIAS,BIASH,SF,SFH,XK,XKH
                  WRITE(2,*)THETDEG,BIASERR,SP11,-SP11,SFERR,SP22,-SP22,
1                      XKERR,SP33,-SP33,ACTNOISE,SIGR,-SIGR
10     CONTINUE
       CLOSE(1)
       PAUSE
       END

C SUBROUTINE GAUSS IS SHOWN IN LISTING 1.8
C SUBROUTINE MATTRN IS SHOWN IN LISTING 1.3
C SUBROUTINE MATMUL IS SHOWN IN LISTING 1.4
C SUBROUTINE MATADD IS SHOWN IN LISTING 1.1
C SUBROUTINE MATSUB IS SHOWN IN LISTING 1.2
```

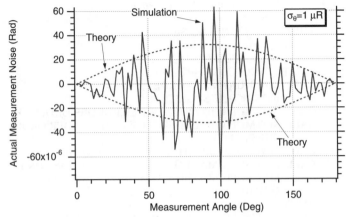

Fig. 4.38 Derived formula for standard deviation of pseudonoise appears to be correct.

showed in Chapter 2 that the method of least squares also worked quite well. However, we can also see from the three figures that when the measurement angle noise is increased to 10 μr all the estimates deteriorate. In fact for 10 μr of angle measurement noise, it is now impossible to determine the g-sensitive drift.

In Chapter 2 we saw that when using the method of least squares we were not able to estimate bias, scale-factor, and g-sensitive drift if there were 100 μr of measurement noise. To see if things get better when a Kalman filter is used, we now rerun Listing 4.5 with 100 μr of measurement noise. We can see from Figs. 4.45–4.47 that the Kalman filter also cannot estimate bias, scale-factor, and g-sensitive drift when there are 100 μr of measurement noise. Thus, nothing magical happens when a Kalman filter is used.

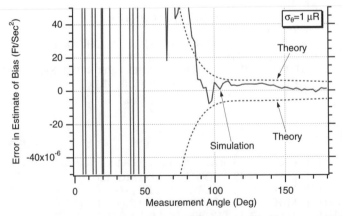

Fig. 4.39 Kalman filter appears to be working correctly because actual error in estimate of accelerometer bias is within theoretical bounds.

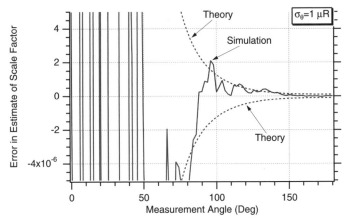

Fig. 4.40   Kalman filter appears to be working correctly because actual error in estimate of accelerometer scale-factor error is within theoretical bounds.

Summary

In this chapter we presented the theoretical equations for making a discrete Kalman filter. We showed, via several simulation examples, that a polynomial Kalman filter with zero process noise and infinite covariance matrix (i.e., diagonal elements are infinite and off-diagonal elements are zero) was equivalent to the recursive least-squares filters of the preceding chapter. We also showed that the performance of the polynomial Kalman filter was approximately independent of the initial covariance matrix for covariances above relatively small values. A numerical example was presented illustrating various Kalman-filtering options for the problem of tracking a falling object. We showed that when the Kalman filter

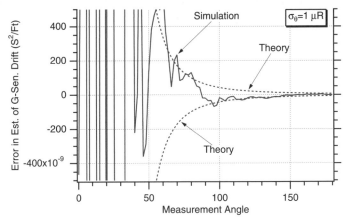

Fig. 4.41   Kalman filter appears to be working correctly because actual error in estimate of accelerometer *g*-sensitive drift is within theoretical bounds.

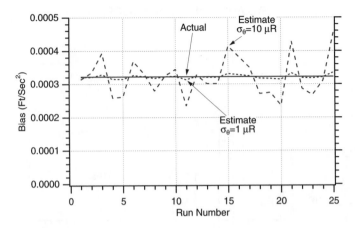

Fig. 4.42 Kalman filter estimates accelerometer bias accurately.

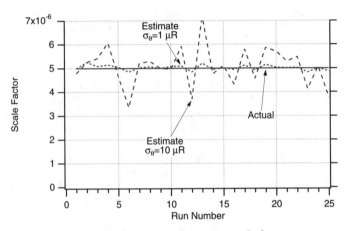

Fig. 4.43 Kalman filter estimates accelerometer scale-factor error accurately.

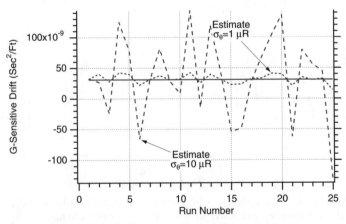

Fig. 4.44 Kalman filter estimates accelerometer g-sensitive drift accurately.

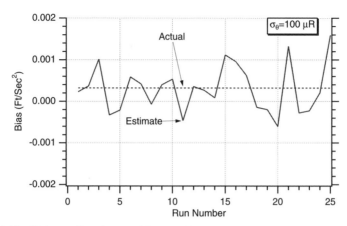

Fig. 4.45 **Kalman filter is not able to estimate accelerometer bias when there are 100 μr of measurement noise.**

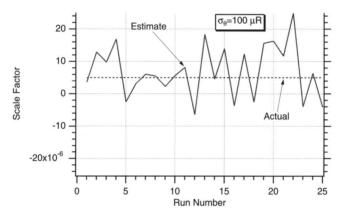

Fig. 4.46 **Kalman filter is not able to estimates accelerometer scale-factor error when there are 100 μr of measurement noise.**

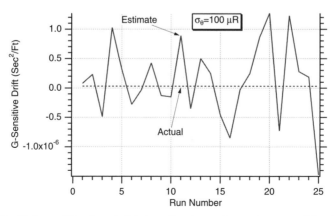

Fig. 4.47 **Kalman filter is not able to estimate accelerometer g-sensitive drift when there are 100 μr of measurement noise.**

was able to make use of a priori information superior performance could be obtained. We also showed that the addition of process noise could sometimes be used as a tool for preventing filter divergence.

References

[1] Kalman, R. E., "A New Approach to Linear Filtering and Prediction Problems," *Journal of Basic Engineering*, Vol. 82, No. 1, 1960, pp. 35–46.

[2] Gelb., A., *Applied Optimal Estimation*, Massachusetts Inst. of Technolgy, Cambridge, MA 1974, pp. 102–155.

[3] Selby, S. M., *Standard Mathematical Tables*, 20th ed. Chemical Rubber Co., Cleveland, OH, 1972, pp. 491–499.

Chapter 5

Kalman Filters in a Nonpolynomial World

Introduction

S O FAR we have seen how polynomial Kalman filters perform when the measurement is a polynomial signal plus noise. In the real world most measurements cannot be described exactly by simple polynomials, and it is therefore of considerable practical interest to see how well the polynomial Kalman filter performs under these circumstances and if anything can be done to improve performance. Two examples will be picked in which the real world cannot be described by a polynomial signal corrupted by noise. Several possible Kalman-filter designs will be presented to illustrate key concepts.

Polynomial Kalman Filter and Sinusoidal Measurement

Suppose that the actual measurement is a pure sine wave of unity amplitude corrupted by noise or

$$x^* = \sin \omega t + \text{noise}$$

where ω is the frequency of the sinusoidal signal and the noise is zero-mean Gaussian with a standard deviation of unity. We would like to build a Kalman filter that will be able to track the sinusoid and estimate its states based on the noisy sinusoidal measurement.

Because the true signal is the sinusoid

$$x = \sin \omega t$$

its derivative is given by

$$\dot{x} = \omega \cos \omega t$$

For the sinusoidal measurement signal we can first attempt to use the first-order polynomial Kalman filter. The polynomial Kalman filter is a reasonable choice because at this point we have not yet studied any other kind of Kalman filter.

183

Recall that the general formula for the Kalman filter, assuming no known deterministic inputs, is given by

$$\hat{x}_k = \Phi_k \hat{x}_{k-1} + K_k(z_k - H\Phi_k \hat{x}_{k-1})$$

For the first-order polynomial Kalman filter we have already shown that the fundamental and measurement matrices are given by

$$\Phi_k = \begin{bmatrix} 1 & T_s \\ 0 & 1 \end{bmatrix}$$

$$H = \begin{bmatrix} 1 & 0 \end{bmatrix}$$

Therefore, after some algebra the scalar equations representing the first-order polynomial Kalman filter can be shown to be

$$\hat{x}_k = \hat{x}_{k-1} + T_s\dot{\hat{x}}_{k-1} + K_{1_k} \text{Res}_k$$
$$\dot{\hat{x}}_k = \dot{\hat{x}}_{k-1} + K_{2_k} \text{Res}_k$$

where the residual is defined as

$$\text{Res}_k = x_k^* - \hat{x}_{k-1} - T_s\dot{\hat{x}}_{k-1}$$

We have shown already that the Kalman gains K_k, required by the preceding set of discrete filtering equations, are obtained from the following recursive set of discrete matrix Riccati equations:

$$M_k = \Phi_k P_{k-1} \Phi_k^T + Q_k$$
$$K_k = M_k H^T (H M_k H^T + R_k)^{-1}$$
$$P_k = (I - K_k H)M_k$$

In this case the matrix describing the variance of the measurement noise turns out to be a scalar given by

$$R_K = \sigma_n^2$$

whereas the matrix describing the process noise has already been shown in Chapter 4 to be

$$Q_k = \Phi_s \begin{bmatrix} \dfrac{T_s^3}{3} & \dfrac{T_s^2}{2} \\ \dfrac{T_s^2}{2} & T_s \end{bmatrix}$$

Listing 5.1 presents a first-order polynomial Kalman filter measuring the noisy sinusoidal signal and attempting to estimate the first two states of the sinusoid (i.e., X and XD). In the nominal case shown in Listing 5.1, the frequency of the sinusoid is 1 rad/s while its amplitude is unity. As was already mentioned, for the nominal example the measurement noise is Gaussian with zero mean and unity standard deviation. A measurement is being taken every 0.1 s (i.e $TS = 0.1$). Initially the Kalman filter is set up without process noise (PHIS = 0), and the initial state estimates are set to zero. As before, the diagonal terms of the initial covariance matrix are set to infinity to indicate that we have no idea how to initialize the filter. Printed out every sampling time are the true signal X and its estimate XH and the true derivative XD and its estimate XDH.

Listing 5.1 First-order polynomial Kalman filter and sinusoidal measurement

```
C THE FIRST THREE STATEMENTS INVOKE THE ABSOFT RANDOM
  NUMBER GENERATOR ON THE MACINTOSH
      GLOBAL  DEFINE
              INCLUDE 'quickdraw.inc'
      END
      IMPLICIT  REAL*8(A-H,O-Z)
      REAL*8  P(2,2),Q(2,2),M(2,2),PHI(2,2),HMAT(1,2),HT(2,1),PHIT(2,2)
      REAL*8  RMAT(1,1),IDN(2,2),PHIP(2,2),PHIPPHIT(2,2),HM(1,2)
      REAL*8  HMHT(1,1),HMHTR(1,1),HMHTRINV(1,1),MHT(2,1),K(2,1)
      REAL*8  KH(2,2),IKH(2,2)
      INTEGER ORDER
      OPEN(1,STATUS='UNKNOWN',FILE='DATFIL')
      ORDER=2
      PHIS=0.
      TS=.1
      XH=0.
      XDH=0.
      SIGNOISE=1.
      DO 14 I=1,ORDER
      DO 14 J=1,ORDER
      PHI(I,J)=0.
      P(I,J)=0.
      Q(I,J)=0.
      IDN(I,J)=0.
14    CONTINUE
      RMAT(1,1)=SIGNOISE**2
      IDN(1,1)=1.
      IDN(2,2)=1.
      P(1,1)=99999999999.
      P(2,2)=99999999999.
      PHI(1,1)=1
      PHI(1,2)=TS
      PHI(2,2)=1
      HMAT(1,1)=1.
      HMAT(1,2)=0.
      CALL  MATTRN(PHI,ORDER,ORDER,PHIT)
      CALL  MATTRN(HMAT,1,ORDER,HT)
```

(continued)

Listing 5.1 *(Continued)*

```
Q(1,1)=PHIS*TS**3/3
Q(1,2)=PHIS*TS*TS/2
Q(2,1)=Q(1,2)
Q(2,2)=PHIS*TS
DO 10 T=0.,20.,TS
        CALL  MATMUL(PHI,ORDER,ORDER,P,ORDER,ORDER,PHIP)
        CALL  MATMUL(PHIP,ORDER,ORDER,PHIT,ORDER,ORDER,
           PHIPPHIT)
        CALL  MATADD(PHIPPHIT,ORDER,ORDER,Q,M)
        CALL  MATMUL(HMAT,1,ORDER,M,ORDER,ORDER,HM)
        CALL  MATMUL(HM,1,ORDER,HT,ORDER,1,HMHT)
        CALL  MATADD(HMHT,ORDER,ORDER,RMAT,HMHTR)
        HMHTRINV(1,1)=1./HMHTR(1,1)
        CALL  MATMUL(M,ORDER,ORDER,HT,ORDER,1,MHT)
        CALL  MATMUL(MHT,ORDER,1,HMHTRINV,1,1,K)
        CALL  MATMUL(K,ORDER,1,HMAT,1,ORDER,KH)
        CALL  MATSUB(IDN,ORDER,ORDER,KH,IKH)
        CALL  MATMUL(IKH,ORDER,ORDER,M,ORDER,ORDER,P)
        CALL  GAUSS(XNOISE,SIGNOISE)
        X=SIN(T)
        XD=COS(T)
        XS=X+XNOISE
        RES=XS-XH-TS*XDH
        XH=XH+XDH*TS+K(1,1)*RES
        XDH=XDH+K(2,1)*RES
        WRITE(9,*)T,X,XS,XH,XD,XDH
        WRITE(1,*)T,X,XS,XH,XD,XDH
10   CONTINUE
     PAUSE
     CLOSE(1)
     END


C    SUBROUTINE GAUSS IS SHOWN IN LISTING 1.8
C    SUBROUTINE MATTRN IS SHOWN IN LISTING 1.3
C    SUBROUTINE MATMUL IS SHOWN IN LISTING 1.4
C    SUBROUTINE MATADD IS SHOWN IN LISTING 1.1
C    SUBROUTINE MATSUB IS SHOWN IN LISTING 1.2
```

Figure 5.1 shows the true signal along with the measurement as a function of time. We can see that the measurement appears to be very noisy for this example. From this noisy measurement the first-order polynomial Kalman filter will attempt to estimate the true signal and its derivative.

Figure 5.2 shows that even though the true signal is not a first-order polynomial the polynomial Kalman filter is not doing a bad job in extracting the true signal from the noisy measurement. However, in all honesty it would be difficult to use the filter's estimate of the first state to determine with certainty that the true signal was a sinusoid.

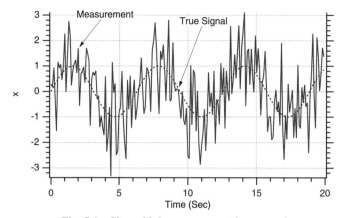

Fig. 5.1   Sinusoidal measurement is very noisy.

Figure 5.3 shows that matters get worse when the filter attempts to estimate the derivative to the actual signal, x dot. There is a large transient at the beginning because the filter has been initialized incorrectly (i.e., second filter state initialized to zero rather than unity). Although the filter transient settles out quickly, there is no sinusoidal motion in the estimate. Therefore, we can conclude that the first-order polynomial Kalman filter is doing a poor job of estimating the derivative of the true signal.

In a sense we have a modeling error when we are attempting to track a sinusoidal signal with a polynomial Kalman filter. We saw in Chapter 4 that one way of handling modeling errors was to increase the process noise. Figures 5.4 and 5.5 show that if we increase the process noise from zero to 10 (i.e., $\Phi_s = 0$ to $\Phi_s = 10$) matters improve. For example, Fig. 5.4 shows that the filter estimate of

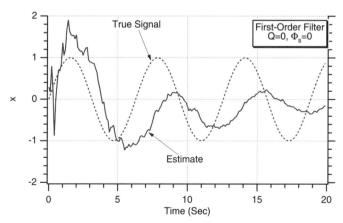

Fig. 5.2   First-order polynomial Kalman-filter has difficulty in tracking the sinusoidal signal.

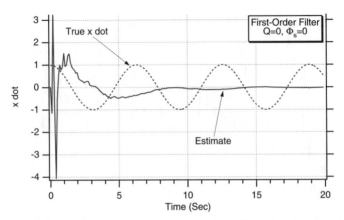

Fig. 5.3   First-order polynomial Kalman-filter does poorly in estimating the derivative of the true signal.

the first state now looks more sinusoidal. The price paid for being better able to track the signal is that the estimate is also noisier. Increasing the process noise effectively increases the bandwidth of the filter, which improves its tracking capabilities at the expense of more noise transmission. Figure 5.5 shows that the filter is now able to provide a noisy track of the derivative of the true signal. However, it is still difficult to discern the sinusoidal nature of this state from the noisy estimate.

It may be possible to improve tracking performance by using a higher-order polynomial Kalman filter. The higher-order polynomial should be a better match to the true sinusoidal signal. The reason for this is that one can show via a Taylor-series expansion that a sinusoid is really an infinite-order polynomial (see Chapter 1). Therefore, although we cannot use an infinite-order Kalman filter, we can try

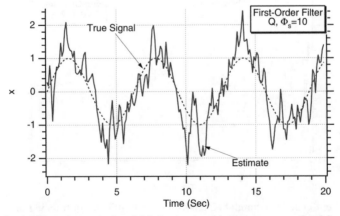

Fig. 5.4   Adding process noise yields better tracking at expense of noisier estimate.

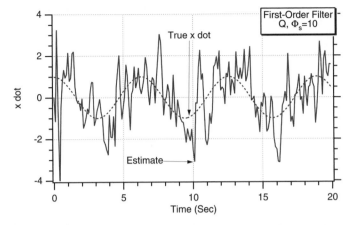

Fig. 5.5   First-order filter with process noise is now able to provide noisy estimate of derivative of true signal.

to use the second-order polynomial Kalman filter, which should be an improvement over the first-order Kalman filter. Again, recall that the general formula for the Kalman filter without any deterministic disturbances is given by

$$\hat{x}_k = \Phi_k \hat{x}_{k-1} + K_k(z_k - H\Phi_k \hat{x}_{k-1})$$

For the second-order polynomial Kalman filter we have already shown in Chapter 4 that the fundamental and measurement matrices are given by

$$\Phi_k = \begin{bmatrix} 1 & T_s & 0.5T_s^2 \\ 0 & 1 & T_s \\ 0 & 0 & 1 \end{bmatrix}$$

$$H = [1 \quad 0 \quad 0]$$

Therefore, after substitution and some algebra, the scalar equations representing the second-order filter can be shown to be

$$\hat{x}_k = \hat{x}_{k-1} + T_s \hat{\dot{x}}_{k-1} + 0.5T_s^2 \hat{\ddot{x}}_{k-1} + K_{1_k} \text{Res}_k$$

$$\hat{\dot{x}}_k = \hat{\dot{x}}_{k-1} + T_s \hat{\ddot{x}}_{k-1} + K_{2_k} \text{Res}_k$$

$$\hat{\ddot{x}}_k = \hat{\ddot{x}}_{k-1} + K_{3_k} \text{Res}_k$$

where the residual is defined as

$$\text{Res}_k = x_k^* - \hat{x}_{k-1} - T_s \hat{\dot{x}}_{k-1} - 0.5T_s^2 \hat{\ddot{x}}_{k-1}$$

The Kalman gains K_k, required by the filter, are still obtained from the recursive matrix Riccati equations

$$M_k = \Phi_k P_{k-1} \Phi_k^T + Q_k$$
$$K_k = M_k H^T (H M_k H^T + R_k)^{-1}$$
$$P_k = (I - K_k H) M_k$$

As was also true for the first-order filter, the matrix describing the variance of the measurement noise for the second-order filter turns out to be a scalar and is given by

$$R_k = \sigma_n^2$$

whereas the matrix describing the process noise has already been shown in Chapter 4 to be

$$Q_k = \Phi_s \begin{bmatrix} \dfrac{T_s^5}{20} & \dfrac{T_s^4}{8} & \dfrac{T_s^3}{6} \\[2mm] \dfrac{T_s^4}{8} & \dfrac{T_s^3}{3} & \dfrac{T_s^2}{2} \\[2mm] \dfrac{T_s^3}{6} & \dfrac{T_s^2}{2} & T_s \end{bmatrix}$$

Listing 5.2 presents the simulation of the second-order polynomial Kalman filter measuring the noisy sinusoidal signal. As was also the case for the first-order filter, the frequency of the sinusoid is again 1 rad/s while the amplitude is unity. The Gaussian measurement noise has zero mean and unity standard deviation. Again, initially the Kalman filter is set up without process noise, and the initial state estimates are set to zero because of our lack of knowledge of

Listing 5.2   Second-order polynomial Kalman filter and sinusoidal measurement

```
C   THE FIRST THREE STATEMENTS INVOKE THE ABSOFT RANDOM
    NUMBER GENERATOR ON THE MACINTOSH
       GLOBAL DEFINE
             INCLUDE 'quickdraw.inc'
    END
    IMPLICIT REAL*8(A-H,O-Z)
    REAL*8  P(3,3),Q(3,3),M(3,3),PHI(3,3),HMAT(1,3),HT(3,1),PHIT(3,3)
    REAL*8  RMAT(1,1),IDN(3,3),PHIP(3,3),PHIPPHIT(3,3),HM(1,3)
    REAL*8  HMHT(1,1),HMHTR(1,1),HMHTRINV(1,1),MHT(3,1),K(3,1)
    REAL*8  KH(23,3),IKH(3,3)
    INTEGER ORDER
```

(continued)

Listing 5.2   (*Continued*)

```
      OPEN(1,STATUS='UNKNOWN',FILE='DATFIL')
      ORDER =3
      PHIS=0.
      TS=.1
      XH=0.
      XDH=0.
      XDDH=0
      SIGNOISE=1.
      DO 14 I=1,ORDER
      DO 14 J=1,ORDER
      PHI(I,J)=0.
      P(I,J)=0.
      Q(I,J)=0.
      IDN(I,J)=0.
14    CONTINUE
      RMAT(1,1)=SIGNOISE**2
      IDN(1,1)=1.
      IDN(2,2)=1.
      IDN(3,3)=1.
      P(1,1)=99999999999.
      P(2,2)=99999999999.
      P(3,3)=99999999999.
      PHI(1,1)=1
      PHI(1,2)=TS
      PHI(1,3)=.5*TS*TS
      PHI(2,2)=1
      PHI(2,3)=TS
      PHI(3,3)=1
      HMAT(1,1)=1.
      HMAT(1,2)=0.
      HMAT(1,3)=0.
      CALL  MATTRN(PHI,ORDER,ORDER,PHIT)
      CALL  MATTRN(HMAT,1,ORDER,HT)
      Q(1,1)=PHIS*TS**5/20
      Q(1,2)=PHIS*TS**4/8
      Q(1,3)=PHIS*TS**3/6
      Q(2,1)=Q(1,2)
      Q(2,2)=PHIS*TS**3/3
      Q(2,3)=PHIS*TS*TS/2
      Q(3,1)=Q(1,3)
      Q(3,2)=Q(2,3)
      Q(3,3)=PHIS*TS
      DO 10 T=0.,20.,TS
              CALL  MATMUL(PHI,ORDER,ORDER,P,ORDER,ORDER,PHIP)
              CALL  MATMUL(PHIP,ORDER,ORDER,PHIT,ORDER,ORDER,
                 PHIPPHIT)
              CALL  MATADD(PHIPPHIT,ORDER,ORDER,Q,M)
              CALL  MATMUL(HMAT,1,ORDER,M,ORDER,ORDER,HM)
              CALL  MATMUL(HM,1,ORDER,HT,ORDER,1,HMHT)
```

(*continued*)

Listing 5.2 *(Continued)*

```
      CALL  MATADD(HMHT,ORDER,ORDER,RMAT,HMHTR)
      HMHTRINV(1,1)=1./HMHTR(1,1)
      CALL  MATMUL(M,ORDER,ORDER,HT,ORDER,1,MHT)
      CALL  MATMUL(MHT,ORDER,1,HMHTRINV,1,1,K)
      CALL  MATMUL(K,ORDER,1,HMAT,1,ORDER,KH)
      CALL  MATSUB(IDN,ORDER,ORDER,KH,IKH)
      CALL  MATMUL(IKH,ORDER,ORDER,M,ORDER,ORDER,P)
      CALL  GAUSS(XNOISE,SIGNOISE)
      X=SIN(T)
      XD=COS(T)
      XDD=-SIN(T)
      XS=X+XNOISE
      RES=XS-XH-TS*XDH-.5*TS*TS*XDDH
      XH=XH+XDH*TS+.5*TS*TS*XDDH+K(1,1)*RES
      XDH=XDH+XDDH*TS+K(2,1)*RES
      XDDH=XDDH+K(3,1)*RES
      WRITE(9,*)T,X,XH,XD,XDH,XDD,XDDH
      WRITE(1,*)T,X,XH,XD,XDH,XDD,XDDH
10    CONTINUE
      PAUSE
      CLOSE(1)
      END


C   SUBROUTINE GAUSS IS SHOWN IN LISTING 1.8
C   SUBROUTINE MATTRN IS SHOWN IN LISTING 1.3
C   SUBROUTINE MATMUL IS SHOWN IN LISTING 1.4
C   SUBROUTINE MATADD IS SHOWN IN LISTING 1.1
C   SUBROUTINE MATSUB IS SHOWN IN LISTING 1.2
```

the real world. As before, the diagonal terms of the initial covariance matrix are set to infinity to reflect the fact that we have no a priori information.

Figure 5.6 shows that going to a higher-order polynomial Kalman filter with zero process noise definitely improves the filter's ability to track the actual signal. By comparing Fig. 5.6 with Fig. 5.2, we can see that the price paid for the better tracking ability is a somewhat noisier estimate. This is to be expected because we have already shown that higher-order filters have more noise transmission than lower-order filters. The fact that the process noise is zero makes this filter sluggish (i.e., lower bandwidth), which causes the filter estimate to lag the actual signal.

We can see from Fig. 5.7 that the second-order filter still has trouble in estimating the derivative of the true signal. Although the estimate is better than that of Fig. 5.3, it is still difficult to discern the sinusoidal nature of the derivative.

We saw in the case of the first-order polynomial Kalman filter that the estimates could be improved somewhat by adding process noise. Process noise was also added to the second-order polynomial Kalman filter. Figure 5.8 shows that the lag in the case in which there was no process noise (i.e., $\Phi_s = 0$) can be

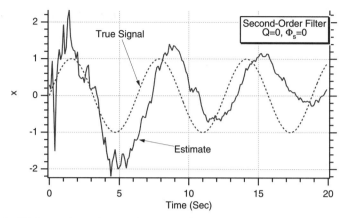

Fig. 5.6   Higher-order polynomial Kalman filter with zero process noise yields better but noisier estimates.

removed by adding process noise (i.e., $\Phi_s = 10$). We can see that the estimate of Fig. 5.8 is much better than that of Fig. 5.6. Figure 5.9 shows that adding process noise also improves the estimate of the derivative of the signal. The improvement in the estimate can be seen by comparing Fig. 5.9, in which there was process noise (i.e., $\Phi_s = 10$), with Fig. 5.7, in which there was no process noise (i.e., $\Phi_s = 0$). For the first time we are able to discern the sinusoidal nature of the derivative of the signal. In all of the preceding examples, we could have also handled the measurement $A \sin \omega t$, where A is the amplitude of the sinusoid and is different from unity. In this case the polynomial filters would not have any difficulty, and the estimates would simply be scaled by A.

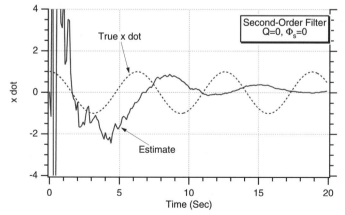

Fig. 5.7   Higher-order polynomial Kalman filter does better job of tracking derivative of true signal.

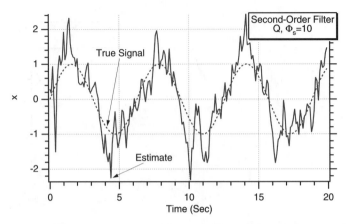

Fig. 5.8   Filter lag has been removed by the addition of process noise.

Sinusoidal Kalman Filter and Sinusoidal Measurement

So far we have seen how a second-order polynomial Kalman filter with process noise could track a sinusoidal signal. Its estimate of the derivative of the signal was not terrific but possibly acceptable for certain applications. The virtue of the polynomial Kalman filter was that no a priori information concerning the true signal was required. Could we have done better with another type of Kalman filter had we known that the true signal was sinusoidal? Let us see if we can improve filter performance if a priori information is available. Recall that the actual signal is

$$x = A \sin \omega t$$

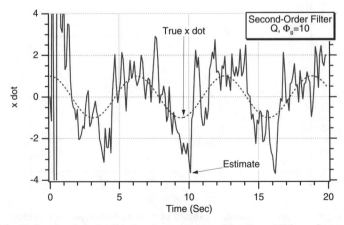

Fig. 5.9   Estimate of derivative has been improved by the addition of process noise.

If we take the derivative of this signal, we also get a sinusoid or

$$\dot{x} = A\omega \cos \omega t$$

Taking the derivative again yields

$$\ddot{x} = -A\omega^2 \sin \omega t$$

The preceding equation can be expressed in terms of the first equation. Therefore, we can rewrite the preceding equation as

$$\ddot{x} = -\omega^2 x$$

The sinusoidal term has been eliminated, and the second derivative of the signal or state has been expressed in terms of a state. The preceding differential equation does not depend on the amplitude of the sinusoid but only its frequency. We can rewrite the preceding equation as a matrix differential equation in state-space form or

$$\begin{bmatrix} \dot{x} \\ \ddot{x} \end{bmatrix} = \begin{bmatrix} 0 & 1 \\ -\omega^2 & 0 \end{bmatrix} \begin{bmatrix} x \\ \dot{x} \end{bmatrix}$$

This equation is our new model of the real world. Remember from Chapter 1 that the state-space form is just another way of expressing the scalar second-order differential equation. Although the state-space form adds no additional information, it is required for the work that follows. From the preceding equation we can see that the system dynamics matrix is

$$F = \begin{bmatrix} 0 & 1 \\ -\omega^2 & 0 \end{bmatrix}$$

For a time-invariant systems dynamics matrix we can derive the fundamental matrix according to[1]

$$\Phi(t) = \mathcal{L}^{-1}[(sI - F)^{-1}]$$

where I is the identity matrix. First, we express the inverse of $sI - F$ as

$$(sI - F)^{-1} = \begin{bmatrix} s & -1 \\ \omega^2 & s \end{bmatrix}^{-1}$$

Taking the inverse of a 2×2 matrix can nearly be done by inspection or following the formula from Chapter 1 and can easily be shown to be

$$\Phi(s) = (sI - F)^{-1} = \frac{1}{s^2 + \omega^2} \begin{bmatrix} s & 1 \\ -\omega^2 & 0 \end{bmatrix}$$

We now have the fundamental matrix in the Laplace transform or s domain. We must take the inverse Laplace transform to express the fundamental matrix as a function of time. Using inverse Laplace transform tables[2] yields

$$\Phi(t) = \begin{bmatrix} \cos \omega t & \dfrac{\sin \omega t}{\omega} \\ -\omega \sin \omega t & \cos \omega t \end{bmatrix}$$

Thus, we can see that the fundamental matrix is now sinusoidal. To derive the discrete fundamental matrix, required for the Kalman filter and Riccati equations, we simply substitute the sampling time T_s for time or

$$\Phi_k = \begin{bmatrix} \cos \omega T_s & \dfrac{\sin \omega T_s}{\omega} \\ -\omega \sin \omega T_s & \cos \omega T_s \end{bmatrix}$$

Again, it is important to note that we can propagate the states forward exactly with the fundamental matrix exactly as long as we know the frequency of the sinusoid. We do not have to know the amplitude of the sinusoid. Substituting the new fundamental matrix into the Kalman-filtering equation

$$\hat{x}_k = \Phi_k \hat{x}_{k-1} + K_k(z_k - H\Phi_k \hat{x}_{k-1})$$

yields a matrix difference equation, which can easily be converted to several scalar difference equations or

$$\hat{x}_k = \cos \omega T_s \hat{x}_{k-1} + \frac{\sin \omega T_s}{\omega} \hat{\dot{x}}_{k-1} + K_{1_k} \mathrm{Res}_k$$

$$\hat{\dot{x}}_k = -\omega \sin \omega T_s \hat{x}_{k-1} + \cos \omega T_s \hat{\dot{x}}_{k-1} + K_{2_k} \mathrm{Res}_k$$

where the residual is defined as

$$RES_k = x_k^* - \cos \omega T_s \hat{x}_{k-1} - \frac{\sin \omega T_s}{\omega} \hat{\dot{x}}_{k-1}$$

In a similar way the Kalman gains can also be obtained using the new fundamental matrix. Remember that this Kalman filter assumes that we know the frequency of the sinusoid. If the frequency of the sinusoid is unknown and must also be estimated, nonlinear filtering techniques must be used (see Chapter 10, which uses an extended Kalman filter to estimate the same states plus the frequency of the sinusoid).

Listing 5.3 shows the new two-state or first-order Kalman filter that makes use of the fact that the signal is sinusoidal. In the nominal case the frequency of the sinusoid is again 1 rad/s, while the amplitude is unity (i.e., running with different amplitudes will not change the results qualitatively). The Gaussian measurement noise again has zero mean and unity standard deviation. Initially the Kalman filter

Listing 5.3   New first-order Kalman filter and sinusoidal measurement

```
C THE FIRST THREE STATEMENTS INVOKE THE ABSOFT RANDOM
  NUMBER GENERATOR ON THE MACINTOSH
  GLOBAL DEFINE
          INCLUDE 'quickdraw.inc'
  END
  IMPLICIT REAL*8(A-H,O-Z)
  REAL*8 P(2,2),Q(2,2),M(2,2),PHI(2,2),HMAT(1,2),HT(2,1),PHIT(2,2)
  REAL*8 RMAT(1,1),IDN(2,2),PHIP(2,2),PHIPPHIT(2,2),HM(1,2)
  REAL*8 HMHT(1,1),HMHTR(1,1),HMHTRINV(1,1),MHT(2,1),K(2,1)
  REAL*8 KH(2,2),IKH(2,2)
  INTEGER ORDER
  OPEN(1,STATUS='UNKNOWN',FILE='DATFIL')
  ORDER=2
  PHIS=0.
  W=1
  A=1
  TS=.1
  XH=0.
  XDH=0.
  SIGNOISE=1.
  DO 14 I=1,ORDER
  DO 14 J=1,ORDER
  PHI(I,J)=0.
  P(I,J)=0.
  Q(I,J)=0.
  IDN(I,J)=0.
14    CONTINUE
  RMAT(1,1)=SIGNOISE**2
  IDN(1,1)=1.
  IDN(2,2)=1.
  P(1,1)=99999999999.
  P(2,2)=99999999999.
  PHI(1,1)=COS(W*TS)
  PHI(1,2)=SIN(W*TS)/W
  PHI(2,1)=-W*SIN(W*TS)
  PHI(2,2)=COS(W*TS)
  HMAT(1,1)=1.
  HMAT(1,2)=0.
  CALL MATTRN(PHI,ORDER,ORDER,PHIT)
  CALL MATTRN(HMAT,1,ORDER,HT)
  Q(1,1)=PHIS*TS**3/3
  Q(1,2)=PHIS*TS*TS/2
  Q(2,1)=Q(1,2)
  Q(2,2)=PHIS*TS
  DO 10 T=0.,20.,TS
          CALL MATMUL(PHI,ORDER,ORDER,P,ORDER,ORDER,PHIP)
          CALL MATMUL(PHIP,ORDER,ORDER,PHIT,ORDER,ORDER,
          PHIPPHIT)
          CALL MATADD(PHIPPHIT,ORDER,ORDER,Q,M)
```

(*continued*)

Listing 5.3 (*Continued*)

```
      CALL  MATMUL(HMAT,1,ORDER,M,ORDER,ORDER,HM)
      CALL  MATMUL(HM,1,ORDER,HT,ORDER,1,HMHT)
      CALL  MATADD(HMHT,ORDER,ORDER,RMAT,HMHTR)
         HMHTRINV(1,1)=1./HMHTR(1,1)
      CALL  MATMUL(M,ORDER,ORDER,HT,ORDER,1,MHT)
      CALL  MATMUL(MHT,ORDER,1,HMHTRINV,1,1,K)
      CALL  MATMUL(K,ORDER,1,HMAT,1,ORDER,KH)
      CALL  MATSUB(IDN,ORDER,ORDER,KH,IKH)
      CALL  MATMUL(IKH,ORDER,ORDER,M,ORDER,ORDER,P)
      CALL  GAUSS(XNOISE,SIGNOISE)
      X=A*SIN(W*T)
      XD=A*W*COS(W*T)
      XS=X+XNOISE
      XHOLD=XH
      RES=XS-XH*COS(W*TS)-SIN(W*TS)*XDH/W
      XH=COS(W*TS)*XH+XDH*SIN(W*TS)/W+K(1,1)*RES
      XDH=-W*SIN(W*TS)*XHOLD+XDH*COS(W*TS)+K(2,1)*RES
      WRITE(9,*)T,X,XS,XH,XD,XDH
      WRITE(1,*)T,X,XS,XH,XD,XDH
10 CONTINUE
      PAUSE
      CLOSE(1)
      END


C   SUBROUTINE GAUSS IS SHOWN IN LISTING 1.8
C   SUBROUTINE MATTRN IS SHOWN IN LISTING 1.3
C   SUBROUTINE MATMUL IS SHOWN IN LISTING 1.4
C   SUBROUTINE MATADD IS SHOWN IN LISTING 1.1
C   SUBROUTINE MATSUB IS SHOWN IN LISTING 1.2
```

is set up without process noise, and the initial state estimates are set to zero. As before, the diagonal terms of the initial covariance matrix are set to infinity.

The nominal case of Listing 5.3 was run, and Fig. 5.10 shows that the new Kalman filter, which makes use of the fact that the signal is sinusoidal, does a spectacular job in estimating the signal when there is no process noise. The process noise was set to zero because we knew there were no modeling errors. We can see that not having process noise reduced the noise transmission. Note also from Fig. 5.10 that there is virtually no lag between the actual signal and the estimate. We can also see from Fig. 5.11 that the new Kalman filter's estimate of the derivative of the signal is near perfect. Thus, we can see that if a priori information is really available (i.e., actual signal is a sinusoid) it pays to use it. Of course if the real signal were really a polynomial and we thought it was supposed to be a sinusoid, a serious error could result.

In practice it is often too difficult to find the fundamental matrix exactly if we are using the Taylor-series approach. We have already derived the exact fundamental matrix using the Taylor-series approach for a sinusoidal signal in Chapter 1. We shall now investigate the influence on system performance when an

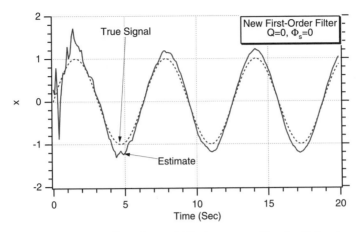

Fig. 5.10   New filter dramatically improves estimate of signal.

approximate fundamental matrix is used instead. The fundamental matrix can be expressed in terms of the systems dynamics matrix as

$$\Phi_k = e^{FT_s} \approx I + FT_s + \frac{(FT_s)^2}{2} + \cdots$$

If we approximate the preceding expression with a two-term Taylor series for the fundamental matrix, we obtain

$$\Phi_k \approx I + FT_s = \begin{bmatrix} 1 & 0 \\ 0 & 1 \end{bmatrix} + \begin{bmatrix} 0 & 1 \\ -\omega^2 & 0 \end{bmatrix} T_s$$

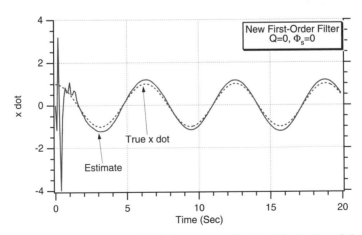

Fig. 5.11   New filter dramatically improves estimate of derivative of signal.

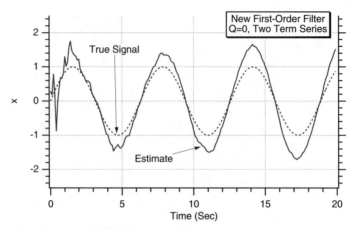

Fig. 5.12   Estimate of signal is worse when fundamental matrix is approximate.

or, more compactly,

$$\Phi_k \approx \begin{bmatrix} 1 & T_s \\ -\omega^2 T_s & 1 \end{bmatrix}$$

The new Kalman filter was rerun, but using the approximate fundamental matrix rather than the exact one. We can see from Figs. 5.12 and 5.13 that although the estimates of the signal and its derivative still appear to be sinusoidal in shape they are not as good as the estimates based on the exact fundamental matrix in Figs. 5.10 and 5.11.

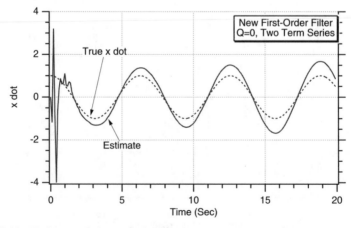

Fig. 5.13   Estimate of signal derivative is also worse when fundamental matrix is approximate.

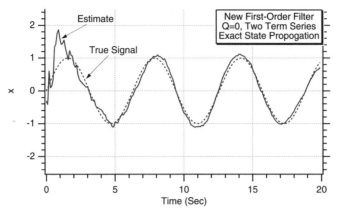

Fig. 5.14   Estimate of signal is excellent when two-term Taylor-series approximation for fundamental matrix is only used in Riccati equation.

Let us now assume that the approximate fundamental matrix, based on the two-term Taylor-series expansion, was used only in the Riccati equations to compute the Kalman gains, and the exact fundamental matrix was used to propagate the states in the filter. This example represents an important case, which we shall examine in more detail when we discuss extended Kalman filters. With extended Kalman filters the filtering equations are propagated forward by actual numerical integration of the state equations (i.e., an exact fundamental matrix in this case represents a perfect integration), and the fundamental matrix (i.e., often approximate because it is based on a few terms from a Taylor-series expansion) is only used in the Riccati equations. We can see from Figs. 5.14 and 5.15 that now the performance of the Kalman filter is excellent when the states

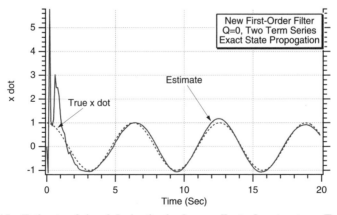

Fig. 5.15   Estimate of signal derivative is also excellent when two-term Taylor-series approximation for fundamental matrix is only used in Riccati equation.

are propagated correctly even though we are using suboptimal Kalman gains. This leads us to believe that the filter structure is more important than the gain computation.

Let us now assume that the fundamental matrix was known precisely. However, our knowledge of the signal frequency was in error by 100%. The real signal frequency was 1 rad/s, but in the fundamental matrix and in the filter we thought the weave frequency was actually 2 rad/s. Figure 5.16 shows that the estimate of the signal deteriorates substantially. This shows the sensitivity of this type of Kalman filter to modeling errors. The polynomial Kalman filter would have been insensitive to this type of error because it made no use of knowledge of the frequency ω of the sinusoidal signal.

To further demonstrate that the deterioration in performance caused by the incorrect fundamental matrix was because of an error in filter structure, rather than an error in filter gains, another experiment was conducted based on the results of Fig. 5.16. The preceding example was repeated, except this time the frequency mismatch only applied to the Riccati equation fundamental matrix. The actual fundamental matrix was used to propagate the states in the Kalman filter. We can see from Fig. 5.17 that now the filter estimates are excellent. This experiment again demonstrates that the filter structure is more important than the gain computation.

Often there really is a mismatch between the real world and the filtering world—both in the filter structure and in the gain computation. Under these circumstances the engineering fix to the mismatched frequency situation of Fig. 5.16 is to add process noise to the Kalman filter. The addition of process noise tells the filter that our model of the real world may be in error. The more process noise we add, the less reliable our model of the real world. Figure 5.18 shows that if we added process noise by making $\Phi_s = 10$ rather than $\Phi_s = 0$, then the estimate of the true signal improves substantially. However, the quality of the estimate is approximately similar to that provided by an ordinary polynomial Kalman filter (see Fig. 5.8). We will revisit this problem in Chapter 10.

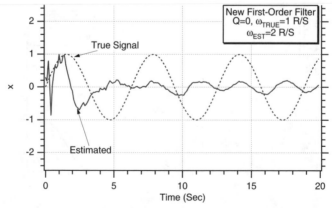

Fig. 5.16 **Kalman filter that depends on knowing signal frequency is sensitive to modeling errors.**

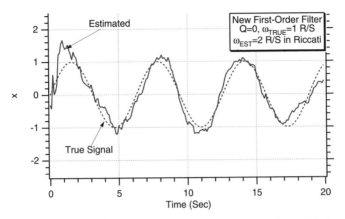

Fig. 5.17   If frequency mismatch is in Riccati equations only, filter still gives excellent estimates.

Suspension System Example

Consider the case of a car riding over a bumpy road. A suspension system that is attached to the wheel axle and car body frame protects the car from the bumps in the road. Figure 5.19 simplifies the example considerably by assuming that the car has only one wheel. For analytical convenience the bumpy road is represented by the sinusoid

$$x_2 = A_2 \sin \omega t$$

The wheel has radius b_1, and the length of the suspension is denoted $x_1 + b_2$. As one would expect, the purpose of the suspension is to provide the car with a

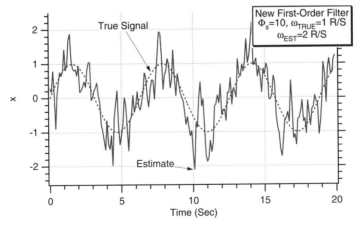

Fig. 5.18   Adding process noise enables Kalman filter with bad a priori information to provide good estimates.

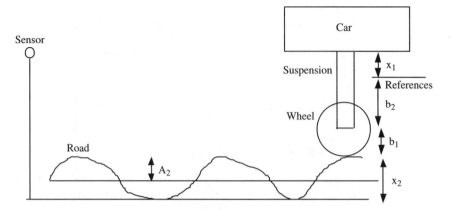

Fig. 5.19   Car riding over a bumpy road.

smooth ride. Therefore, if the suspension system is working properly, the variations in the height of the car $x_1 + b_2 + b_1 + x_2$ will be much smaller than that of the road amplitude A_2. A sensor measures the height of the car at all times so we can see if the suspension is working properly.

The suspension system of Fig. 5.19 is drawn in more detail in Fig. 5.20. For analytical convenience we have represented the suspension system by a spring and a dashpot.[3,4] The stiffness of the spring is denoted by its spring constant K, which by definition is the number of pounds tension necessary to extend the spring 1 in. In this example the car is considered to have mass M. The dashpot of

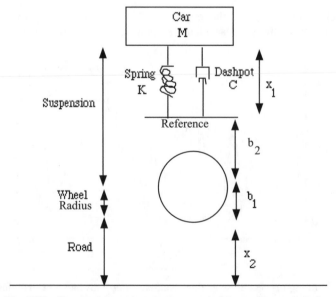

Fig. 5.20   Suspension system is represented by spring and dashpot.

Fig. 5.20 is not supposed to transmit any force to the mass as long as it is at rest, but, as soon as the mass moves, the damping force of the dashpot is proportional to the velocity and directed opposite to it. The quantity C is known as the damping constant or as the coefficient of viscous damping.

From Newton's second law, where force is mass times acceleration, we obtain from Fig. 5.20

$$M(\ddot{x}_1 + \ddot{b}_2 + \ddot{b}_1 + \ddot{x}_2) = -C\dot{x}_1 - Kx_1$$

Because the radius of the wheel b_1 is a constant, its first and second derivative must be zero. In addition, because b_2 is a constant, its second derivative must also be zero. By dividing both sides of the preceding equation by the mass of the car we obtain

$$\ddot{x}_1 + \frac{C}{M}\dot{x}_1 + \frac{K}{M}x_1 = -\ddot{x}_2$$

For convenience we can express the damping ζ and natural frequency ω_n of the suspension mechanism in terms of the spring constant and coefficient of viscous damping. Therefore, we can define

$$2\zeta\omega_n = \frac{C}{M}$$

$$\omega_n^2 = \frac{K}{M}$$

Using the preceding definitions, the second-order differential equation can now be rewritten in more convenient form as

$$\ddot{x}_1 = -2\zeta\omega_n\dot{x}_1 - \omega_n^2 x_1 - \ddot{x}_2$$

We have already mentioned that the bumpy road is represented by

$$x_2 = A_2 \sin \omega t$$

Therefore, the first and second derivatives of x_2 can be obtained by inspection of the preceding equation as

$$\dot{x}_2 = A_2\omega \cos \omega t$$

$$\ddot{x}_2 = -A_2\omega^2 \sin \omega t$$

The last equation is all we need for the second-order differential equation of the effective suspension length x_1.

Let us now consider a numerical example in which the amplitude and frequency of the sinusoidal road are 0.1 ft and 6.28 rad/s, respectively, while the initial suspension length is 1.5 ft. Initially we assume that $x_1 = 0.25$ ft and that $b_2 = 1.25$ ft. The wheel radius is considered to be 1 ft. We now have enough information to integrate the second-order differential equation. Listing 5.4 presents the program that numerically integrates the suspension differential equation using the second-order Runge–Kutta integration technique described in Chapter 1. The structure of this program is identical to that of Listing 1.7 in Chapter 1. Printed out are the length of the suspension X1, the height of the road X2, and the height of the car from the reference DIST.

Listing 5.4   Simulation that integrates the suspension differential equation

```
OPEN(1,STATUS='UNKNOWN',FILE='DATFIL')
WN=6.28*.1
W=6.28*1.
Z=.7
A2=.1
X1=.25
B2=1.25
X1D=0.
B1=1.
T=0.
S=0.
H=.001
WHILE(T<=20.)
        S=S+H
        X1OLD=X1
        X1DOLD=X1D
        X2=A2*SIN(W*T)
        X2D=A2*W*COS(W*T)
        X2DD=-A2*W*W*SIN(W*T)
        X1DD=-2.*Z*WN*X1D-WN*WN*X1-X2DD
        X1=X1+H*X1D
        X1D=X1D+H*X1DD
        T=T+H
        X2=A2*SIN(W*T)
        X2D=A2*W*COS(W*T)
        X2DD=-A2*W*W*SIN(W*T)
        X1DD=-2.*Z*WN*X1D-WN*WN*X1-X2DD
        X1=.5*(X1OLD+X1+H*X1D)
        X1D=.5*(X1DOLD+X1D+H*X1DD)
        IF(S>=.09999)THEN
                S=0.
                DIST=X1+X2+B1+B2
                SUSP=X1+B2
                WRITE(9,*)T,SUSP,X2,DIST
                WRITE(1,*)T,SUSP,X2,DIST
        ENDIF
END DO
PAUSE
CLOSE(1)
END
```

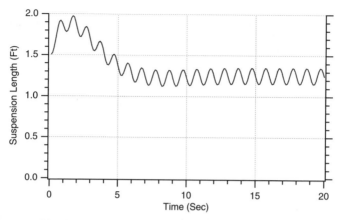

Fig. 5.21 Suspension oscillates at the road frequency.

The nominal case of Listing 5.4 was run. We can see from Fig. 5.21 that the suspension length oscillates at the frequency of the road. The amplitude of the suspension oscillations are approximately the amplitude of the road bumps. However, we can see from Fig. 5.22 that the suspension oscillations cancel out the road oscillations because the car height $x_1 + b_2 + b_1 + x_2$ above the reference appears to be approximately a constant after an initial transient period. In other words, it appears that everything is working properly because the suspension is protecting the car from the bumpy ride.

Kalman Filter for Suspension System

Recall from Fig. 5.19 that we have a sensor that is taking measurements of the distance from the car body to a reference. We desire to estimate x_1 based on noisy

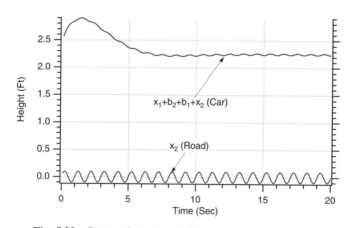

Fig. 5.22 Suspension enables the car to have a smooth ride.

measurements of the distance from the car body to the reference. Before we can build a Kalman filter for the estimation process we must first express the second-order differential equation for the suspension as two first-order equations in state-space form as

$$\begin{bmatrix} \dot{x}_1 \\ \ddot{x}_1 \end{bmatrix} = \begin{bmatrix} 0 & 1 \\ -\omega_n^2 & -2\zeta\omega_n \end{bmatrix} \begin{bmatrix} x_1 \\ \dot{x}_1 \end{bmatrix} + \begin{bmatrix} 0 \\ -1 \end{bmatrix} \ddot{x}_2$$

The actual measurement is a function of the suspension length, road height, and wheel radius. The true measurement equation can be written as

$$\text{Meas} = x_1 + x_2 + b_1 + b_2 + v$$

where v is the measurement noise. To fit the Kalman-filtering equations where the measurement must be a linear function of the states, we can define a pseudo-measurement as

$$x_1^* = \text{Meas} - x_2 - b_1 - b_2 = x_1 + v = \begin{bmatrix} 1 & 0 \end{bmatrix} \begin{bmatrix} x_1 \\ \dot{x}_1 \end{bmatrix} + v$$

where we are assuming prior knowledge of x_2, b_1, and b_2. From the preceding equation we can see that the measurement matrix is given by

$$H = \begin{bmatrix} 1 & 0 \end{bmatrix}$$

The systems dynamics matrix can be obtained directly from the state-space equation as

$$F = \begin{bmatrix} 0 & 1 \\ -\omega_n^2 & -2\zeta\omega_n \end{bmatrix}$$

Therefore, the fundamental matrix in Laplace transform notation is given by

$$\Phi(s) = (sI - F)^{-1} = \begin{bmatrix} s & -1 \\ \omega_n^2 & s + 2\zeta\omega_n \end{bmatrix}^{-1}$$

Taking the inverse of a two-dimensional square matrix yields

$$\Phi(s) = \frac{1}{s^2 + 2\zeta\omega_n s + \omega_n^2} \begin{bmatrix} s + 2\zeta\omega_n & 1 \\ -\omega_n^2 & s \end{bmatrix}$$

If we define

$$a = -\zeta\omega_n$$

$$b = \omega_n\sqrt{1 - \zeta^2}$$

then the denominator in the expression for the fundamental matrix becomes

$$s^2 + 2\zeta\omega_n s + \omega_n^2 = (s-a)^2 + b^2$$

From inverse Laplace transform tables[2] we know that the Laplace transforms of interest can be converted to the time domain according to

$$\frac{1}{(s-a)^2 + b^2} = \frac{e^{at}\sin bt}{b}$$

$$\frac{s}{(s-a)^2 + b^2} = \frac{e^{at}(a\sin bt + b\cos bt)}{b}$$

Therefore, we can find the fundamental matrix in the time domain to be

$$\Phi(t) = \begin{bmatrix} \dfrac{e^{at}(-a\sin bt + b\cos bt)}{b} & \dfrac{e^{at}\sin bt}{b} \\ \dfrac{-\omega_n^2 e^{at}\sin bt}{b} & \dfrac{e^{at}(a\sin bt + b\cos bt)}{b} \end{bmatrix}$$

The discrete fundamental matrix can be found from the preceding expression by simply replacing time with the sampling time or

$$\Phi_k = \begin{bmatrix} \dfrac{e^{aT_s}(-a\sin bT_s + b\cos bT_s)}{b} & \dfrac{e^{aT_s}\sin bT_s}{b} \\ \dfrac{-\omega_n^2 e^{aT_s}\sin bT_s}{b} & \dfrac{e^{aT_s}(a\sin bT_s + b\cos bT_s)}{b} \end{bmatrix} = \begin{bmatrix} \Phi_{11} & \Phi_{12} \\ \Phi_{21} & \Phi_{22} \end{bmatrix}$$

From the state-space equation we can also see that

$$G = \begin{bmatrix} 0 \\ -1 \end{bmatrix}$$

The discrete G matrix can be found from the continuous G matrix according to

$$G_k = \int_0^{T_s} \Phi(\tau)G \, d\tau$$

Strictly speaking, the preceding equation is only valid if the deterministic input is constant between the sampling instants. Because the input is a high-frequency sinusoid, we know that this approximation is not good. However, we will proceed nonetheless and correct any resultant errors by using process noise. Substitution of the appropriate matrices into the preceding integral equation yields

$$G_k = \int_0^{T_s} \begin{bmatrix} \dfrac{e^{a\tau}(-a\sin b\tau + b\cos b\tau)}{b} & \dfrac{e^{a\tau}\sin b\tau}{b} \\ -\dfrac{\omega_n^2 e^{a\tau}\sin b\tau}{b} & \dfrac{e^{a\tau}(a\sin b\tau + b\cos b\tau)}{b} \end{bmatrix} \begin{bmatrix} 0 \\ -1 \end{bmatrix} d\tau$$

Matrix multiplication simplifies the preceding expression to

$$G_k = \int_0^{T_s} \begin{bmatrix} \dfrac{e^{a\tau}\sin b\tau}{b} \\ -\dfrac{e^{a\tau}(a\sin b\tau + b\cos b\tau)}{b} \end{bmatrix} d\tau$$

From integration tables[2] we know that

$$\int e^{ax}\sin bx \, dx = \frac{e^{ax}(a\sin bx - b\cos bx)}{a^2 + b^2}$$

$$\int e^{ax}\cos bx \, dx = \frac{e^{ax}(a\cos bx + b\sin bx)}{a^2 + b^2}$$

Therefore, the discrete G matrix becomes

$$G_k = \begin{bmatrix} -\dfrac{e^{aT_s}(a\sin bT_s - b\cos bT_s) + b}{b(a^2 + b^2)} \\ -\dfrac{e^{aT_s}\sin bT_s}{b} \end{bmatrix} = \begin{bmatrix} G_1 \\ G_2 \end{bmatrix}$$

The discrete linear Kalman-filtering equation is given by

$$\hat{x}_k = \Phi_k \hat{x}_{k-1} + G_k u_{k-1} + K_k(z_k - H\Phi_k \hat{x}_{k-1} - HG_k U_{k-1})$$

Substitution of the appropriate matrices into the preceding equation yields

$$\begin{bmatrix} \hat{x}_{1_k} \\ \hat{\dot{x}}_{1_k} \end{bmatrix} = \begin{bmatrix} \Phi_{11} & \Phi_{12} \\ \Phi_{21} & \Phi_{22} \end{bmatrix} \begin{bmatrix} \hat{x}_{1_{k-1}} \\ \hat{\dot{x}}_{1_{k-1}} \end{bmatrix} + \begin{bmatrix} G_1 \\ G_2 \end{bmatrix} \ddot{x}_2$$

$$+ \begin{bmatrix} K_{1_k} \\ K_{2_k} \end{bmatrix} \left(x_{1_k}^* - \begin{bmatrix} 1 & 0 \end{bmatrix} \begin{bmatrix} \Phi_{11} & \Phi_{12} \\ \Phi_{21} & \Phi_{22} \end{bmatrix} \begin{bmatrix} \hat{x}_{1_{k-1}} \\ \hat{\dot{x}}_{1_{k-1}} \end{bmatrix} - \begin{bmatrix} 1 & 0 \end{bmatrix} \begin{bmatrix} G_1 \\ G_2 \end{bmatrix} \ddot{x}_2 \right)$$

After multiplying out the terms of the preceding matrix difference equation, we obtain the following scalar equations for the Kalman filter

$$\text{Res}_k = x^*_{1_k} - \Phi_{11}\hat{x}_{1_{k-1}} - \Phi_{12}\hat{\dot{x}}_{1_{k-1}} - G_1\ddot{x}_2$$

$$\hat{x}_{1_k} = \Phi_{11}\hat{x}_{1_{k-1}} + \Phi_{12}\hat{\dot{x}}_{1_{k-1}} + G_1\ddot{x}_2 + K_{1_k}\text{Res}_k$$

$$\hat{\dot{x}}_{1_k} = \Phi_{21}\hat{x}_{1_{k-1}} + \Phi_{22}\hat{\dot{x}}_{1_{k-1}} + G_2\ddot{x}_2 + K_{2_k}\text{Res}_k$$

Listing 5.5 presents the code for the Kalman filter that is attempting to measure the suspension parameters based on measurements of the car's distance from the reference. Both the single-run errors in the estimates of the filter states plus the covariance matrix predictions are printed out every sampling interval.

Listing 5.5   Using a Kalman filter for measuring suspension length

```
C THE FIRST THREE STATEMENTS INVOKE THE ABSOFT RANDOM
  NUMBER GENERATOR ON THE MACINTOSH
  GLOBAL DEFINE
          INCLUDE 'quickdraw.inc'
  END
  IMPLICIT REAL*8(A-H,O-Z)
  REAL*8 P(2,2),Q(2,2),M(2,2),PHI(2,2),HMAT(1,2),HT(2,1),PHIT(2,2)
  REAL*8 RMAT(1,1),IDN(2,2),PHIP(2,2),PHIPPHIT(2,2),HM(1,2)
  REAL*8 HMHT(1,1),HMHTR(1,1),HMHTRINV(1,1),MHT(2,1),K(2,1)
  REAL*8 KH(2,2),IKH(2,2),G(2,1)
  INTEGER ORDER
  OPEN(1,STATUS='UNKNOWN',FILE='DATFIL')
  OPEN(2,STATUS='UNKNOWN',FILE='COVFIL')
  ORDER=2
  TF=20.
  SIGX=.1
  TS=.01
  WN=6.28*.1
  W=6.28*1.
  Z=.7
  A=-Z*WN
  B=WN*SQRT(1.-Z*Z)
  A2=.1
  X1=.25
  B2=1.25
  X1D=0.
  B1=1.
  T=0.
  S=0.
  H=.001
  X2DDOLD=0.
  DO 14 I=1,ORDER
  DO 14 J=1,ORDER
```

(continued)

Listing 5.5 *(Continued)*

```
      PHI(I,J)=0.
      P(I,J)=0.
      Q(I,J)=0.
      IDN(I,J)=0.
14    CONTINUE
      IDN(1,1)=1.
      IDN(2,2)=1.
      PHI(1,1)=EXP(A*TS)*(-A*SIN(B*TS)+B*COS(B*TS))/B
      PHI(1,2)=EXP(A*TS)*SIN(B*TS)/B
      PHI(2,1)=-WN*WN*EXP(A*TS)*SIN(B*TS)/B
      PHI(2,2)=EXP(A*TS)*(A*SIN(B*TS)+B*COS(B*TS))/B
      HMAT(1,1)=1.
      HMAT(1,2)=0.
      G(1,1)=-(EXP(A*TS)*(A*SIN(B*TS)-B*COS(B*TS))+B)/(B*(A*A+B*B))
      G(2,1)=-EXP(A*TS)*SIN(B*TS)/B
      CALL MATTRN(PHI,ORDER,ORDER,PHIT)
      CALL MATTRN(HMAT,1,ORDER,HT)
      P(1,1)=9999999.
      P(2,2)=9999999.
      Q(2,2)=0.
      X1H=X1
      X1DH=X1D
      RMAT(1,1)=SIGX**2
      WHILE(T < =20.)
            S=S+H
            X1OLD=X1
            X1DOLD=X1D
            X2=A2*SIN(W*T)
            X2D=A2*W*COS(W*T)
            X2DD=-A2*W*W*SIN(W*T)
            X1DD=-2.*Z*WN*X1D-WN*WN*X1-X2DD
            X1=X1+H*X1D
            X1D=X1D+H*X1DD
            T=T+H
            X2=A2*SIN(W*T)
            X2D=A2*W*COS(W*T)
            X2DD=-A2*W*W*SIN(W*T)
            X1DD=-2.*Z*WN*X1D-WN*WN*X1-X2DD
            X1=.5*(X1OLD+X1+H*X1D)
            X1D=.5*(X1DOLD+X1D+H*X1DD)
            IF(S>=(TS-.00001))THEN
                        S=0.
                        CALL MATMUL(PHI,ORDER,ORDER,P,ORDER,
                           ORDER,PHIP)
                        CALL MATMUL(PHIP,ORDER,ORDER,PHIT,
                           ORDER,ORDER,PHIPPHIT)
                        CALL MATADD(PHIPPHIT,ORDER,ORDER,Q,M)
                        CALL MATMUL(HMAT,1,ORDER,M,ORDER,
                           ORDER,HM)
```

(continued)

Listing 5.5 *(Continued)*

```
                    CALL  MATMUL(HM,1,ORDER,HT,ORDER,1,HMHT)
                    CALL  MATADD(HMHT,1,1,RMAT,HMHTR)
                    HMHTRINV(1,1)=1./HMHTR(1,1)
                    CALL  MATMUL(M,ORDER,ORDER,HT,ORDER,1,MHT)
                    CALL  MATMUL(MHT,ORDER,1,HMHTRINV,1,1,K)
                    CALL  MATMUL(K,ORDER,1,HMAT,1,ORDER,KH)
                    CALL  MATSUB(IDN,ORDER,ORDER,KH,IKH)
                    CALL  MATMUL(IKH,ORDER,ORDER,M,
                        ORDER,ORDER,P)
                    CALL  GAUSS(XNOISE,SIGX)
                    XMEAS=X1+B1+X2+B2+XNOISE
                    XS=XMEAS-X2-B1-B2
                    RES=XS-PHI(1,1)*X1H-PHI(1,2)*X1DH-G(1,1)*
                        X2DDOLD
                    X1HOLD=X1H
                    X1H=PHI(1,1)*X1H+PHI(1,2)*X1DH+G(1,1)*
                        X2DDOLD+K(1,1)*RES
                    X1DH=PHI(2,1)*X1HOLD+PHI(2,2)*X1DH+G(2,1)*
                        X2DDOLD+K(2,1)*RES
                    ERRX1=X1-X1H
                    SP11=SQRT(P(1,1))
                    ERRX1D=X1D-X1DH
                    SP22=SQRT(P(2,2))
                    X2DDOLD=X2DD
                    WRITE(9,*)T,X1,X1H,X1D,X1DH
                    WRITE(1,*)T,X1,X1H,X1D,X1DH
                    WRITE(2,*)T,ERRX1,SP11,-SP11,ERRX1D,SP22,-SP22
          ENDIF
     END  DO
     PAUSE
     CLOSE(1)
     END

C SUBROUTINE GAUSS IS SHOWN IN LISTING 1.8
C SUBROUTINE MATTRN IS SHOWN IN LISTING 1.3
C SUBROUTINE MATMUL IS SHOWN IN LISTING 1.4
C SUBROUTINE MATADD IS SHOWN IN LISTING 1.1
C SUBROUTINE MATSUB IS SHOWN IN LISTING 1.2
```

The nominal case of Listing 5.5 was run in which there was no process noise. We can see from Figs. 5.23 and 5.24 that it appears that the filter is working well because the estimates of the suspension parameter x_1 and its derivative appear to be very close to the actual values.

To see if the Kalman filter is performing as expected we must examine the errors in the estimates of both filter states. We expect that the errors in the estimates should go to zero because there is no process noise. However, Figs. 5.25 and 5.26 show that the single-run simulation results of the errors in the

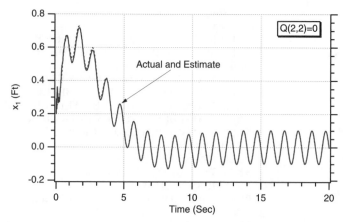

Fig. 5.23   Kalman filter provides excellent estimates of the first state of the suspension.

estimates oscillate, rather than go to zero, as more measurements are made. These results do not match the theoretical predictions of the covariance matrix, indicating that something is wrong.

To see if a mistake was made in deriving Φ_k and G_k, another experiment was conducted. The filter propagation was tested with the filter states initialized perfectly and the Kalman gains set to zero. In other words, the filter simply coasts and the modified filtering equations of Listing 5.5 become

X1H=PHI(1,1)*X1H+PHI(1,2)*X1DH+G(1,1)*X2DDOLD
X1DH=PHI(2,1)*X1HOLD+PHI(2,1)*X1DH+G(2,1)*X2DDOLD

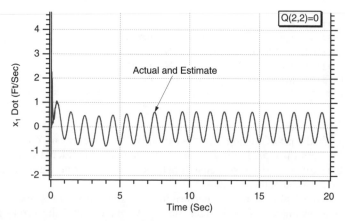

Fig. 5.24   Kalman filter provides excellent estimates of the derivative of the first state of the suspension.

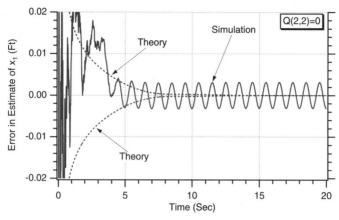

Fig. 5.25   Error in the estimate of first state does not agree with covariance matrix predictions.

We can see from Fig 5.27 that the error in the estimate of the first state still oscillates even though the filter was perfectly initialized. Under these circumstances the error in the estimate of the first state should have been close to zero. These results indicate that there is a mistake in either or both Φ_k and G_k.

To isolate the cause of the problem, the simulation was further simplified by setting X2DD to zero. By doing this any errors in G_k will disappear, and we will only be checking the fundamental matrix Φ_k. We can see from Fig. 5.28 that when X2DD is set to zero the errors in the estimate of x_1 stop oscillating and become many orders of magnitude smaller than they were before indicating that the fundamental matrix is correct.

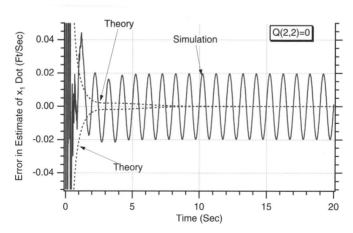

Fig. 5.26   Error in the estimate of second state does not agree with covariance matrix predictions.

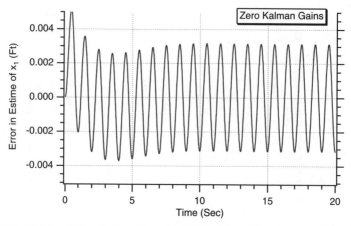

Fig. 5.27 Kalman filter does not coast properly when filter is perfectly initialized.

Strictly speaking G_k is only valid if the deterministic input X2DD is constant between sampling instants. To see if a mistake was made in deriving G_k, another experiment was conducted in which the deterministic input was set to unity. We can see from Fig. 5.29 that now, because the deterministic input is indeed constant between sampling instants, the error in the estimate of the first state stops oscillating. These results prove that G_k is correct for the assumptions made. However G_k is not correct for this problem.

At this juncture we have two possible ways to go. We could derive a new discrete G matrix based on the fact that the deterministic input is sinusoidal between sampling instants, or we could simply add process noise to the filter to account for the fact that we do not have a perfect model of the real world. Because the first method requires a great deal of mathematics while the second method

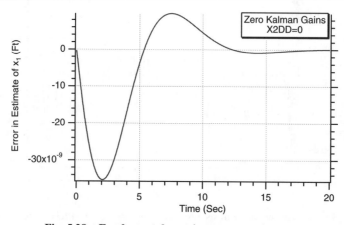

Fig. 5.28 Fundamental matrix appears to be correct.

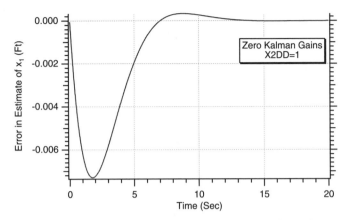

Fig. 5.29 Discrete G matrix is correct if deterministic input is constant.

only requires a one-line program change, the second method was chosen. When process noise is added, we can now see from Figs. 5.30 and 5.31 that the errors in the estimates of the first and second states are within the theoretical error bounds, indicating that the Kalman filter is now working properly.

The suspension length is $x_1 + b_2$, with b_2 being fixed at 1.25 ft. We know from Fig. 5.23 that x_1 varies between ± 0.1 ft and that the standard deviation of the measurement noise is 0.1 ft. Therefore, the suspension length varies between 1.24 and 1.26 ft. We can see from Fig. 5.30 that we have effectively reduced the effect of the measurement noise by nearly an order of magnitude because the error in the estimate of x_1 is approximately ± 0.015 ft. We know from Fig. 5.24 that the actual derivative of x_1 varies between ± 0.75 ft/s. From Fig. 5.31 we can see that

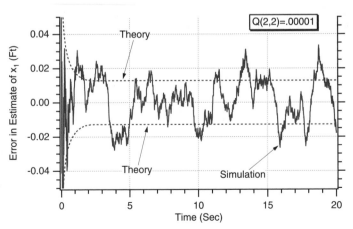

Fig. 5.30 Addition of process noise ensures that errors in the estimate of the first state are within the theoretical error bounds.

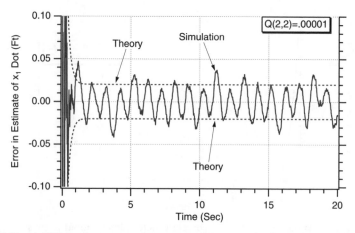

Fig. 5.31 Addition of process noise ensures that errors in the estimate of the second state are within the theoretical error bounds.

the Kalman filter is able to estimate the derivative of x_1 to within $\pm.025$ ft/s. Thus, we can see that even with process noise the Kalman filter is very effective.

Summary

In this chapter we have provided examples of various types of Kalman filters that could be used when the measurement signal was not a polynomial. If possible, we saw that it was important to have an accurate fundamental matrix. However, we also saw that the main importance of the fundamental matrix was in the propagation of the state estimates for the Kalman filter. Using an exact fundamental matrix was not as important for the computation of the Kalman gains. We also demonstrated that adding process noise to the filter was often the engineering fix for accounting for errors in the fundamental matrix or any other errors that could cause the filter to be mismatched from the real world.

References

[1] Schwarz, R. J., and Friedland, B., *Linear Systems*, McGraw–Hill, New York, 1965, pp. 106–128.

[2] Selby, S. M., *Standard Mathematical Tables*, 20th ed., Chemical Rubber Co., Cleveland, OH, 1972, pp. 491–499.

[3] Den Hartog, J. P., *Mechanical Vibrations*, Dover, New York, 1984, pp. 24–29.

[4] Maymon, G., *Some Engineering Applications in Random Vibrations and Random Structures*, Progress in Astronautics and Aeronautics, AIAA, Reston, VA, 1998, pp. 1–12.

Chapter 6

Continuous Polynomial Kalman Filter

Introduction

S O FAR we have only studied the discrete Kalman filter because that is the filter that is usually implemented in real-world applications. However, there is also a continuous version of the Kalman filter. Although the continuous Kalman filter is usually not implemented, it can be used as an aid in better understanding the properties of the discrete Kalman filter. For small sampling times the discrete Kalman filter in fact becomes a continuous Kalman filter. Because the continuous Kalman filter does not require the derivation of a fundamental matrix for either the Riccati equations or filtering equations, the continuous filter can also be used as a check of the discrete filter. In addition, we shall see that with the continuous polynomial Kalman filter it is possible, under steady-state conditions, to derive transfer functions that exactly represent the filter. These transfer functions cannot only be used to better understand the operation of the Kalman filter (i.e., both continuous and discrete) but can also aid with such mundane issues as helping to choose the correct amount of process noise to use.

First, we will present the theoretical differential equations for the continuous Riccati equations and Kalman filter. Next, we will derive different-order continuous polynomial Kalman filters. We will demonstrate for the case of zero process noise that the Kalman gains and covariance matrix predictions of the continuous polynomial Kalman filter exactly match the formulas we have already derived for the discrete polynomial Kalman filter. We will then show that when process noise is introduced we can derive closed-form solutions for both the gains and covariance matrix predictions of the continuous polynomial Kalman filters. These closed-form solutions can then be used to derive transfer functions for the continuous polynomial Kalman filter. Finally, formulas will be derived showing how the bandwidth of the continuous polynomial Kalman filter transfer function is related to the amount of process and measurement noise we tell the filter.

Theoretical Equations

As was the case with the discrete filter, continuous Kalman filtering theory[1] first requires that our model of the real world be described by a set of differential equations. These equations must be cast in matrix or state-space form as

$$\dot{x} = Fx + Gu + w$$

where x is a column vector with the states of the system, F the system dynamics matrix, u a known deterministic or control vector, and w a white noise process represented as a column vector. There is a process noise matrix Q that is related to the process noise column vector according to

$$Q = E[ww^T]$$

We have seen in preceding chapters that although the process noise sometimes has no physical meaning it is often used as a device for telling the filter that we understand that the filter's model of the real world may not be precise (i.e., larger values of process noise indicate less confidence in our model of the real world). The continuous Kalman filter also requires that the measurements be linearly related to the states according to

$$z = Hx + v$$

where z is the measurement vector, H the measurement matrix, and v white measurement noise. The measurement noise matrix R is related to the measurement noise vector v according to

$$R = E[vv^T]$$

where the continuous measurement noise matrix consists of spectral densities describing the measurement noise. In preceding chapters we saw that the discrete measurement noise matrix consisted of variances describing the measurement noise.

For the model of the real world just described, the resultant continuous Kalman filter is described by the matrix differential equation

$$\dot{\hat{x}} = F\hat{x} + Gu + K(z - H\hat{x})$$

The continuous Kalman filter no longer requires the fundamental matrix for the propagation of the states. State propagation is accomplished by numerically integrating the matrix differential equation involving the system dynamics matrix. The Kalman gains K, required by the preceding Kalman-filtering differential equation, are now obtained by first integrating the nonlinear matrix differential Riccati equation for the covariance matrix

$$\dot{P} = -PH^T R^{-1} HP + PF^T + FP + Q$$

and then solving the matrix equation for the gain in terms of the covariance matrix or

$$K = PH^T R^{-1}$$

In the next few sections we will derive continuous polynomial Kalman filters for different-order systems.

However, let us first visualize how the filtering equations work in the continuous case. From the continuous Riccati equation we can see that there are two sources of uncertainty involved, represented by the covariance matrices Q and R. If we are trying to quantify the correctness of our model of the real world, we must also include its uncertainty as part of its behavior (i.e., characterized by the process noise Q). This means we want the process noise to directly influence our estimates. The measurement noise R is a hindrance to estimating the states of the real world, and we would like to reduce its influence in the estimation process. We can see from the filtering equation that the filter gain K operates on the difference between a measurement and the prior estimate of what the measurement should be. Consequently, a larger Q increases P, which results in a larger K, whereas a larger R tends to decrease K. A decreased K does not weight the measurement as much because it is noisy, whereas a larger K (caused by more Q) causes more of the measurement to be incorporated in the estimate. It will be seen that this logic leads directly to the concept of how filter bandwidth is determined (i.e., larger Q or smaller R will increase the filter bandwidth, whereas smaller Q or larger R will decrease the filter bandwidth). These considerations apply similarly to the discrete Kalman filter presented earlier.

Zeroth-Order or One-State Continuous Polynomial Kalman Filter

In a zeroth-order system the differential equation representing the real world is given by

$$\dot{x} = u_s$$

where x is the state and u_s is white process noise with spectral density Φ_s. Therefore, for the preceding zeroth-order polynomial case, matrices became scalars, and the continuous process noise matrix is also a scalar given by

$$Q = E(u_s^2) = \Phi_s$$

From the state-space equation or the equation representing the real world the system dynamics matrix can be obtained by inspection and is the scalar

$$F = 0$$

Let us assume that the measurement equation is simply

$$x^* = x + v_n$$

where v_n is white noise with spectral density Φ_n. We can see from the preceding equation that the measurement matrix is a scalar given by

$$H = 1$$

whereas the measurement noise matrix is a scalar spectral density or

$$R = E(v_n^2) = \Phi_n$$

Substitution of the preceding scalar matrices into the nonlinear matrix Riccati differential equation for the covariance yields

$$\dot{P} = -PH^T R^{-1} HP + PF^T + FP + Q = -P\Phi_n^{-1}P + \Phi_s$$

or

$$\dot{P} = \frac{-P^2}{\Phi_n} + \Phi_s$$

whereas the Kalman gain can be obtained from

$$K = PH^T R^{-1} = P\Phi_n^{-1}$$

or

$$K = \frac{P}{\Phi_n}$$

We will often call the continuous Kalman gain K_c.

We would like to see if the formulas we just derived for the gain and covariance of the zeroth-order continuous polynomial Kalman filter are equivalent in some way to their discrete counter parts. We have already shown in Chapter 4 that a discrete polynomial Kalman filter without process noise and infinite initial covariance matrix is equivalent to a recursive least-squares filter. Therefore, we will use our closed-form solutions for the recursive least-squares filter as if they were solutions for the discrete Kalman filter (i.e., without process noise). For the zeroth-order system we have demonstrated in Chapter 3 that the recursive least-squares filter gain (i.e., discrete Kalman gain) is given by

$$K_k = \frac{1}{k} \qquad k = 1, 2, \ldots, n$$

whereas the variance of the error in the estimate of the state (i.e., covariance matrix) is

$$P_k = \frac{\sigma_n^2}{k}$$

In this chapter we will sometimes call the discrete or recursive Kalman gain K_d and the discrete error in the estimate of the state P_d.

A simulation was written to integrate the continuous differential equation derived from the zeroth-order covariance differential equation

$$\dot{P} = \frac{-P^2}{\Phi_n} + \Phi_s$$

and the resultant program is shown in Listing 6.1. However, to make the comparison with the discrete filter we will initially set the process noise to zero (i.e., PHIS = 0). Also shown in the program are the relationships between the discrete noise variance and continuous noise spectral density. If the sampling interval T_s in the discrete world is small, we can relate the continuous noise spectral density to the discrete noise variance σ_n^2 according to

$$\Phi_n = \sigma_n^2 T_s$$

Similarly, Listing 6.1 also describes the theoretical relationship between the continuous Kalman gain and discrete recursive least-squares gain. One can show that as the sampling interval gets small the continuous Kalman gain K_c and the discrete recursive least-squares gain K_d are related according to[2]

$$K_c = \frac{K_d}{T_s}$$

As also can be seen from Listing 6.1, the simulation uses second-order Runge–Kutta numerical integration and double-precision arithmetic to solve the nonlinear Riccati differential equation for the covariance of the error in the estimate. The program had numerical difficulties when the initial covariance was set to infinity, and so the initial covariance was set to 100 to avoid those difficulties. We have already shown in Chapter 4 that the solution for the steady-state covariance matrix was virtually independent of initial condition. Therefore, it was hypothesized that choosing an initial covariance value to avoid numerical problems should not influence the final result. The integration step size used to integrate the Riccati equation was set to 0.001 s. Using smaller values of the integration step size did not appear to change the answers significantly but did dramatically increase the computer running time.

The nominal case of Listing 6.1 was run. From Fig. 6.1 we can conclude that, for the case considered, the match between the continuous Kalman gain and the discrete recursive least-squares gain (i.e., or discrete Kalman gain) is excellent. From Fig. 6.2 we can conclude that the continuous and discrete covariances are also nearly identical. It therefore appears that we are integrating the nonlinear covariance differential equation correctly because the theoretical relationships have been verified. In other words, for small sampling times we have verified that for the one-state system the continuous Kalman gain evaluated at the sampling

Listing 6.1 Integrating one-state covariance nonlinear Riccati differential equation

```
IMPLICIT  REAL*8(A-H)
IMPLICIT  REAL*8(O-Z)
REAL*8  F(1,1),P(1,1),Q(1,1),POLD(1,1),HP(1,1)
REAL*8  PD(1,1)
REAL*8  HMAT(1,1),HT(1,1),FP(1,1),PFT(1,1),PHT(1,1),K(1,1)
REAL*8  PHTHP(1,1),PHTHPR(1,1),PFTFP(1,1),PFTFPQ(1,1)
INTEGER ORDER
OPEN(1,STATUS='UNKNOWN',FILE='DATFIL')
ORDER=1
T=0.
S=0.
H=.001
TS=.1
TF=10.
PHIS=0.
XJ=1.
DO 14 I=1,ORDER
DO 14 J=1,ORDER
F(I,J)=0.
P(I,J)=0.
Q(I,J)=0.
14 CONTINUE
DO 11 I=1,ORDER
HMAT(1,I)=0.
HT(I,1)=0.
11 CONTINUE
Q(1,1)=PHIS
HMAT(1,1)=1.
HT(1,1)=1.
SIGN2=1.**2
PHIN=SIGN2*TS
P(1,1)=100.
WHILE(T<=TF)
        DO 20 I=1,ORDER
        DO 20 J=1,ORDER
                POLD(I,J)=P(I,J)
20        CONTINUE
        CALL  MATMUL(F,ORDER,ORDER,P,ORDER,ORDER,FP)
        CALL  MATTRN(FP,ORDER,ORDER,PFT)
        CALL  MATMUL(P,ORDER,ORDER,HT,ORDER,1,PHT)
        CALL  MATMUL(HMAT,1,ORDER,P,ORDER,ORDER,HP)
        CALL  MATMUL(PHT,ORDER,1,HP,1,ORDER,PHTHP)
        DO 12 I=1,ORDER
        DO 12 J=1,ORDER
                PHTHPR(I,J)=PHTHP(I,J)/PHIN
12.        CONTINUE
        CALL  MATADD(PFT,ORDER,ORDER,FP,PFTFP)
```
(continued)

Listing 6.1 *(Continued)*

```
              CALL  MATADD(PFTFP,ORDER,ORDER,Q,PFTFPQ)
              CALL  MATSUB(PFTFPQ,ORDER,ORDER,PHTHPR,PD)
              DO  13  I=1,ORDER
                  K(I,1)=PHT(I,1)/PHIN
13            CONTINUE
              DO  50  I=1,ORDER
              DO  50  J=1,ORDER
                  P(I,J)=P(I,J)+H*PD(I,J)
50            CONTINUE
              T=T+H
              CALL  MATMUL(F,ORDER,ORDER,P,ORDER,ORDER,FP)
              CALL  MATTRN(FP,ORDER,ORDER,PFT)
              CALL  MATMUL(P,ORDER,ORDER,HT,ORDER,1,PHT)
              CALL  MATMUL(HMAT,1,ORDER,P,ORDER,ORDER,HP)
              CALL  MATMUL(PHT,ORDER,1,HP,1,ORDER,PHTHP)
              DO  15  I=1,ORDER
              DO  15  J=1,ORDER
                  PHTHPR(I,J)=PHTHP(I,J)/PHIN
15            CONTINUE
              CALL  MATADD(PFT,ORDER,ORDER,FP,PFTFP)
              CALL  MATADD(PFTFP,ORDER,ORDER,Q,PFTFPQ)
              CALL  MATSUB(PFTFPQ,ORDER,ORDER,PHTHPR,PD)
              DO  16  I=1,ORDER
              K(I,1)=PHT(I,1)/PHIN
16            CONTINUE
              DO  60  I=1,ORDER
              DO  60  J=1,ORDER
                  P(I,J)=.5*(POLD(I,J)+P(I,J)+H*PD(I,J))
60            CONTINUE
              S=S+H
              IF(S>=(TS-.00001))THEN
                      S=0.
                      XK1=1./XJ
                      PDISC=SIGN2/XJ;
                      WRITE(9,*)T,K(1,1)*TS,XK1,P,PDISC
                      WRITE(1,*)T,K(1,1)*TS,XK1,P,PDISC
                      XJ=XJ+1.
              ENDIF
      END  DO
      PAUSE
      CLOSE(1)
      END

C SUBROUTINE MATTRN IS SHOWN IN LISTING 1.3
C SUBROUTINE MATMUL IS SHOWN IN LISTING 1.4
C SUBROUTINE MATADD IS SHOWN IN LISTING 1.1
C SUBROUTINE MATSUB IS SHOWN IN LISTING 1.2
```

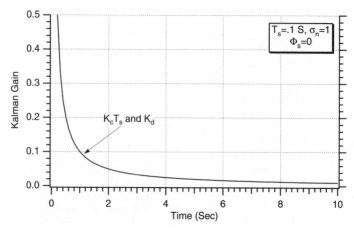

Fig. 6.1 Continuous and discrete Kalman gains are identical for zeroth-order system.

time is simply the discrete recursive least-squares gain (or discrete Kalman gain evaluated at discrete sampling times) divided by the sampling time, or

$$K_c = \frac{K_d}{T_s}$$

In addition we have verified that the continuous and discrete covariances are equal or

$$P_c = P_d$$

We would also get a similarly excellent match between the continuous Riccati equations and discrete Riccati equations when there is process noise. However,

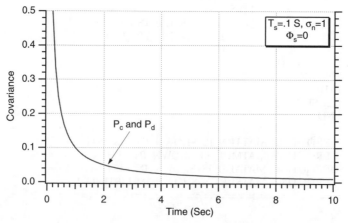

Fig. 6.2 Continuous and discrete covariances are identical for zeroth-order system.

with process noise the gain and covariance will not go to zero as time increases (i.e., more measurements are taken).

First-Order or Two-State Continuous Polynomial Kalman Filter

In a first-order system the differential equation representing the real world is given by

$$\ddot{x} = u_s$$

where x is the state and u_s is white process noise with spectral density Φ_s. We can represent the preceding differential equation in state-space form as

$$\begin{bmatrix} \dot{x} \\ \ddot{x} \end{bmatrix} = \begin{bmatrix} 0 & 1 \\ 0 & 0 \end{bmatrix} \begin{bmatrix} x \\ \dot{x} \end{bmatrix} + \begin{bmatrix} 0 \\ u_s \end{bmatrix}$$

We can see from the preceding equation that the continuous process noise matrix can be described by

$$\boldsymbol{Q} = E\left[\begin{bmatrix} 0 \\ u_s \end{bmatrix} \begin{bmatrix} 0 & u_s \end{bmatrix}\right] = \Phi_s \begin{bmatrix} 0 & 0 \\ 0 & 1 \end{bmatrix}$$

whereas the system dynamics matrix can be obtained by inspection of the state-space equation as

$$\boldsymbol{F} = \begin{bmatrix} 0 & 1 \\ 0 & 0 \end{bmatrix}$$

We assume that the measurement equation is simply

$$x^* = x + v_n$$

where v_n is white noise with spectral density Φ_n. The preceding equation can be put into the form

$$x^* = \begin{bmatrix} 1 & 0 \end{bmatrix} \begin{bmatrix} x \\ \dot{x} \end{bmatrix} + v_n$$

Therefore, the measurement matrix can be obtained by inspection of the preceding equation as

$$\boldsymbol{H} = \begin{bmatrix} 1 & 0 \end{bmatrix}$$

In addition, the measurement noise matrix, which was already described by a scalar variance in the discrete world, now becomes a scalar spectral density in the continuous world or

$$\boldsymbol{R} = \Phi_n$$

Substitution of the preceding matrices into the differential equation for the covariance and numerically integrating

$$\dot{P} = -PH^T R^{-1} HP + PF^T + FP + Q$$

yields the covariance P, which is now a two-dimensional matrix. The Kalman gain can be obtained from the covariance matrix according to

$$K = PH^T R^{-1}$$

where K is a two-dimensional column vector.

We would like see if the formulas we just derived for the gain and covariance matrices for the first-order filter are equivalent in some way to their discrete counterparts. To facilitate the comparison, we will again consider the case of zero process noise. We have already shown that a discrete polynomial Kalman filter without process noise and infinite initial covariance matrix is equivalent to a recursive least-squares filter. For the first-order system we have demonstrated in Chapter 3 that the two recursive least-squares filter gains, which are equivalent to the Kalman gains (i.e., for the case of zero process noise), are given by

$$K_{1_k} = \frac{2(2k-1)}{k(k+1)} \qquad k = 1, 2, \ldots, n$$

$$K_{2_k} = \frac{6}{k(k+1)T_s}$$

We have also demonstrated in Chapter 4 that the two diagonal terms of the Kalman or recursive least-squares covariance matrix for the case of zero process noise are given by

$$P_{11_k} = \frac{2(2k-1)\sigma_n^2}{k(k+1)}$$

$$P_{22_k} = \frac{12\sigma_n^2}{k(k^2-1)T_s^2}$$

A simulation was written to integrate the second-order continuous nonlinear matrix differential equation for the covariance matrix, and the resultant program is shown in Listing 6.2. The simulation is initially set up without process noise (PHIS = 0). Also shown in the program are the relationships between the discrete noise variance and continuous noise spectral density. As was the case in the preceding section, if the sampling interval T_s in the discrete world is small we can relate the continuous noise spectral density to the discrete noise variance according to

$$\Phi_n = \sigma_n^2 T_s$$

In addition the program also describes the relationship between the continuous and discrete Kalman (i.e., recursive least squares) gains. We have already demonstrated that as the sampling interval gets small the continuous Kalman gain K_c and the discrete recursive least-squares gain K_d at the sampling times are related according to

$$K_c = \frac{K_d}{T_s}$$

As was the case with Listing 6.1, Listing 6.2 also uses second-order Runge–Kutta numerical integration and double-precision arithmetic to solve the nonlinear second-order matrix differential equation for the covariance matrix. This program also had numerical difficulties when the diagonal elements of the initial covariance matrix were set to infinity, and so the diagonal terms of the initial covariance matrix were set to 100 and the off-diagonal terms were set to zero. Recall that we have already shown in Chapter 4 that the solution for the steady-state covariance matrix was virtually independent of initial condition. Therefore, we again hypothesized that choosing an initial covariance matrix to avoid numerical problems should not influence the final result. As was the case with Listing 6.1, the integration step size in Listing 6.2 was set to 0.001 s. As was the case in the preceding section, using smaller values did not seem to change the answers significantly but did dramatically increase the run time.

The nominal case of Listing 6.2 was run, and the continuous and discrete Kalman gains are displayed in Figs. 6.3 and 6.4. From these figures we can conclude that for the case considered, the match between the continuous and discrete Kalman gains at the sampling times are excellent. We can feel confident that we are numerically integrating the nonlinear covariance matrix differential equation correctly because the theoretical relationships and simulation results agree. Again, for small sampling times we have verified that for the two-state system the continuous Kalman gain is simply the discrete Kalman gain divided by the sampling time or

$$K_{1c} = \frac{K_{1d}}{T_s}$$

$$K_{2c} = \frac{K_{2d}}{T_s}$$

Similarly Figs. 6.5 and 6.6 show that the covariance matrix diagonal terms obtained by numerically integrating the nonlinear matrix Riccati differential equation of Listing 6.2 matches the formulas from the recursive least-squares filter. Therefore, we have confidence that for the first-order polynomial Kalman filter there is also a good match between the continuous and discrete systems. In addition we have verified that the continuous and discrete covariances are equal or

$$P_{11c} = P_{11d}$$

$$P_{22c} = P_{22d}$$

One would also get a similar excellent match between the continuous Riccati equations and discrete Riccati equations when there is process noise for the first-order system. However, as was already mentioned, the gains and covariances will not go to zero as more measurements are taken when there is process noise.

Listing 6.2   Integrating two-state covariance nonlinear Riccati differential equation

```
      IMPLICIT  REAL*8(A-H)
      IMPLICIT  REAL*8(O-Z)
      REAL*8  F(2,2),P(2,2),Q(2,2),POLD(2,2),HP(1,2)
      REAL*8  PD(2,2)
      REAL*8  HMAT(1,2),HT(2,1),FP(2,2),PFT(2,2),PHT(2,1),K(2,1)
      REAL*8  PHTHP(2,2),PHTHPR(2,2),PFTFP(2,2),PFTFPQ(2,2)
      INTEGER  ORDER
      OPEN(1,STATUS='UNKNOWN',FILE='DATFIL')
      ORDER=2
      T=0.
      S=0.
      H=.001
      TS=.1
      TF=10.
      PHIS=0.
      XJ=1.
      DO  14  I=1,ORDER
      DO  14  J=1,ORDER
      F(I,J)=0.
      P(I,J)=0.
      Q(I,J)=0.
14    CONTINUE
      DO  11  I=1,ORDER
      HMAT(1,I)=0.
      HT(I,1)=0.
11    CONTINUE
      F(1,2)=1.
      Q(2,2)=PHIS
      HMAT(1,1)=1.
      HT(1,1)=1.
      SIGN2=1.**2
      PHIN=SIGN2*TS
      P(1,1)=100.
      P(2,2)=100.
      WHILE(T<=TF)
              DO  20  I=1,ORDER
              DO  20  J=1,ORDER
              POLD(I,J)=P(I,J)
20            CONTINUE
              CALL  MATMUL(F,ORDER,ORDER,P,ORDER,ORDER,FP)
              CALL  MATTRN(FP,ORDER,ORDER,PFT)
              CALL  MATMUL(P,ORDER,ORDER,HT,ORDER,1,PHT)
              CALL  MATMUL(HMAT,1,ORDER,P,ORDER,ORDER,HP)
              CALL  MATMUL(PHT,ORDER,1,HP,1,ORDER,PHTHP)
              DO  12  I=1,ORDER
              DO  12  J=1,ORDER
                    PHTHPR(I,J)=PHTHP(I,J)/PHIN
12            CONTINUE
              CALL  MATADD(PFT,ORDER,ORDER,FP,PFTFP)
```

(continued)

Listing 6.2 *(Continued)*

```
        CALL  MATADD(PFTFP,ORDER,ORDER,Q,PFTFPQ)
        CALL  MATSUB(PFTFPQ,ORDER,ORDER,PHTHPR,PD)
        DO 13 I=1,ORDER
            K(I,1)=PHT(I,1)/PHIN
13      CONTINUE
        DO 50 I=1,ORDER
        DO 50 J=1,ORDER
            P(I,J)=P(I,J)+H*PD(I,J)
50      CONTINUE
        T=T+H
        CALL  MATMUL(F,ORDER,ORDER,P,ORDER,ORDER,FP)
        CALL  MATTRN(FP,ORDER,ORDER,PFT)
        CALL  MATMUL(P,ORDER,ORDER,HT,ORDER,1,PHT)
        CALL  MATMUL(HMAT,1,ORDER,P,ORDER,ORDER,HP)
        CALL  MATMUL(PHT,ORDER,1,HP,1,ORDER,PHTHP)
        DO 15 I=1,ORDER
        DO 15 J=1,ORDER
            PHTHPR(I,J)=PHTHP(I,J)/PHIN
15      CONTINUE
        CALL  MATADD(PFT,ORDER,ORDER,FP,PFTFP)
        CALL  MATADD(PFTFP,ORDER,ORDER,Q,PFTFPQ)
        CALL  MATSUB(PFTFPQ,ORDER,ORDER,PHTHPR,PD)
        DO 16 I=1,ORDER
            K(I,1)=PHT(I,1)/PHIN
16      CONTINUE
        DO 60 I=1,ORDER
        DO 60 J=1,ORDER
            P(I,J)=.5*(POLD(I,J)+P(I,J)+H*PD(I,J))
60      CONTINUE
        S=S+H
        IF(S>=(TS-.00001))THEN
            S=0.
            XK1=2.*(2.*XJ-1.)/(XJ*(XJ+1))
            XK2=6./(XJ*(XJ+1)*TS)
            P11DISC=2.*(2.*XJ-1)*SIGN2/(XJ*(XJ+1.))
            IF(XJ.EQ.1)THEN
                P22DISC=0.
            ELSE
                P22DISC=12*SIGN2/(XJ*(XJ*XJ-1)*TS*TS)
            ENDIF
            WRITE(9,*)T,K(1,1)*TS,XK1,K(2,1)*TS,XK2,P(1,1)
1               ,P11DISC,P(2,2),P22DISC
            WRITE(1,*)T,K(1,1)*TS,XK1,K(2,1)*TS,XK2,P(1,1)
1               ,P11DISC,P(2,2),P22DISC
            XJ=XJ+1.
        ENDIF
    END DO
    PAUSE
    CLOSE(1)
    END
```

(continued)

Listing 6.2 (*Continued*)

```
C SUBROUTINE MATTRN IS SHOWN IN LISTING 1.3
C SUBROUTINE MATMUL IS SHOWN IN LISTING 1.4
C SUBROUTINE MATADD IS SHOWN IN LISTING 1.1
C SUBROUTINE MATSUB IS SHOWN IN LISTING 1.2
```

Second-Order or Three-State Continuous Polynomial Kalman Filter

Using techniques similar to that of the preceding two sections, one can show that for the second-order case the continuous process noise matrix is described by

$$Q = \Phi_s \begin{bmatrix} 0 & 0 & 0 \\ 0 & 0 & 0 \\ 0 & 0 & 1 \end{bmatrix}$$

whereas the system dynamics and measurement matrices are given by

$$F = \begin{bmatrix} 0 & 1 & 0 \\ 0 & 0 & 1 \\ 0 & 0 & 0 \end{bmatrix}$$

$$H = \begin{bmatrix} 1 & 0 & 0 \end{bmatrix}$$

The measurement noise matrix remains a scalar spectral density in the continuous world or

$$R = \Phi_n$$

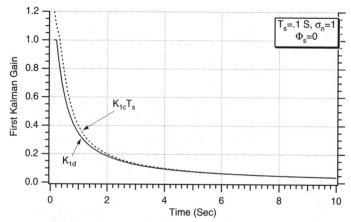

Fig. 6.3 Integrating two-state nonlinear matrix Riccati differential equation yields good match with formula for first gain.

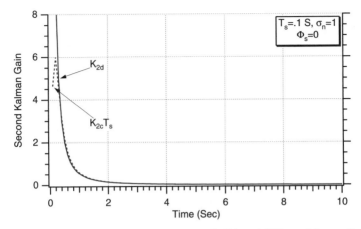

Fig. 6.4   Integrating two-state nonlinear matrix Riccati differential equation yields good match with formula for second gain.

Substitution of the preceding matrices into the differential equation for the covariance yields

$$\dot{P} = -PH^T R^{-1} HP + PF^T + FP + Q$$

where P is now a three-dimensional matrix and the Kalman gain can be obtained from

$$K = PH^T R^{-1}$$

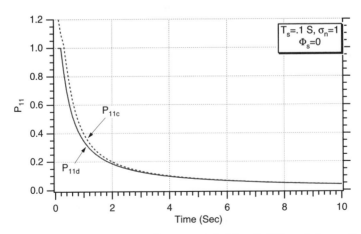

Fig. 6.5   Integrating two-state nonlinear matrix Riccati differential equation yields good match for first diagonal element of covariance matrix.

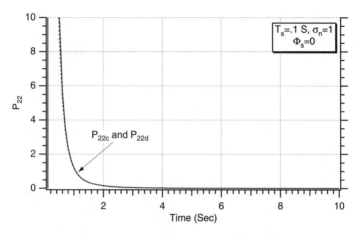

Fig. 6.6 Integrating two-state nonlinear matrix Riccati differential equation yields good match for second diagonal element of covariance matrix.

where K is a three-dimensional column vector.

We have already shown that a discrete polynomial Kalman filter without process noise and infinite initial covariance matrix is equivalent to a recursive least-squares filter. In Chapter 3 we have demonstrated that for the second-order system the three recursive least-squares gains are given by

$$K_{1_k} = \frac{3(3k^2 - 3k + 2)}{k(k+1)(k+2)} \qquad k = 1, 2, \ldots, n$$

$$K_{2_k} = \frac{18(2k-1)}{k(k+1)(k+2)T_s}$$

$$K_{3_k} = \frac{60}{k(k+1)(k+2)T_s^2}$$

and that the three diagonal terms of the covariance matrix are given by

$$P_{11_k} = \frac{3(3k^2 - 3k + 2)\sigma_n^2}{k(k+1)(k+2)}$$

$$P_{22_k} = \frac{12(16k^2 - 30k + 11)\sigma_n^2}{k(k^2 - 1)(k^2 - 4)T_s^2}$$

$$P_{33_k} = \frac{720\sigma_n^2}{k(k^2 - 1)(k^2 - 4)T_s^4}$$

A simulation was written to integrate the preceding second-order nonlinear continuous matrix differential equation, and the resultant program is shown in Listing 6.3. The process noise in the simulation is initially set to zero (i.e., PHIS = 0). The simulation of Listing 6.3 uses the second-order Runge–Kutta numerical integration technique of Chapter 1 and double-precision arithmetic to solve the nonlinear matrix differential equation. This program also had numerical difficulties when the diagonal elements of the initial covariance matrix were set to infinity, and so the diagonal elements of the initial covariance matrix were each set to 100. Again, we have already shown in Chapter 4 that the solution for the steady-state covariance matrix was virtually independent of initial condition. Therefore, we again hypothesize that choosing an initial covariance matrix to avoid numerical problems should not influence the final result. The integration step size of Listing 6.3 was set to 0.001 s. Again, using smaller values of the integration step size did not seem to change the answers significantly but did dramatically increase the computer run time.

The nominal case of Listing 6.3 was run, and the three Kalman gains and their discrete equivalents appear in Figs. 6.7–6.9. From these figures we can see that, for the case considered, the match between the continuous and discrete Kalman gains is close. We can again conclude that we are integrating the nonlinear covariance matrix differential equation correctly because the theoretical relationships have been verified. Again, for small sampling times we have verified that the continuous Kalman gains of the three-state system are simply the discrete Kalman gains divided by the sampling time or

$$K_{1c} = \frac{K_{1d}}{T_s}$$

$$K_{2c} = \frac{K_{2d}}{T_s}$$

$$K_{3c} = \frac{K_{3d}}{T_s}$$

Similarly Figs. 6.10–6.12 show that the covariance matrix diagonal terms obtained by integration of the continuous matrix Riccati differential equation match the formulas from the recursive least-squares filter. Therefore, we have confidence that for the second-order polynomial Kalman filter there is also a good match between the continuous and discrete systems. In addition we have verified that the continuous and discrete covariances are equal or

$$P_{11c} = P_{11d}$$

$$P_{22c} = P_{22d}$$

$$P_{33c} = P_{33d}$$

Again, one would also get a similar excellent match between the continuous Riccati equations and discrete Riccati equations when there is process noise for the second-order system. As was mentioned before, the gains and covariances will not go to zero as more measurements are taken when process noise is present.

Listing 6.3 Integrating three-state covariance nonlinear Riccati differential equation

```
        IMPLICIT  REAL*8(A-H)
        IMPLICIT  REAL*8(O-Z)
        REAL*8  F(3,3),P(3,3),Q(3,3),POLD(3,3),HP(1,3)
        REAL*8  PD(3,3)
        REAL*8  HMAT(1,3),HT(3,1),FP(3,3),PFT(3,3),PHT(3,1),K(3,1)
        REAL*8  PHTHP(3,3),PHTHPR(3,3),PFTFP(3,3),PFTFPQ(3,3)
        INTEGER  ORDER
        OPEN(1,STATUS='UNKNOWN',FILE='DATFIL')
        ORDER=3
        T=0.
        S=0.
        H=.001
        TS=.1
        TF=10.
        PHIS=0.
        XJ=1.
        DO 14 I=1,ORDER
        DO 14 J=1,ORDER
        F(I,J)=0.
        P(I,J)=0.
        Q(I,J)=0.
14      CONTINUE
        DO 11 I=1,ORDER
        HMAT(1,I)=0.
        HT(I,1)=0.
11      CONTINUE
        F(1,2)=1.
        F(2,3)=1.
        Q(3,3)=PHIS
        HMAT(1,1)=1.
        HT(1,1)=1.
        SIGN2=1.**2
        PHIN=SIGN2*TS
        P(1,1)=100.
        P(2,2)=100.
        P(3,3)=100.
        WHILE(T<=TF)
              DO 20 I=1,ORDER
              DO 20 J=1,ORDER
                     POLD(I,J)=P(I,J)
20            CONTINUE
              CALL  MATMUL(F,ORDER,ORDER,P,ORDER,ORDER,FP)
              CALL  MATTRN(FP,ORDER,ORDER,PFT)
              CALL  MATMUL(P,ORDER,ORDER,HT,ORDER,1,PHT)
              CALL  MATMUL(HMAT,1,ORDER,P,ORDER,ORDER,HP)
              CALL  MATMUL(PHT,ORDER,1,HP,1,ORDER,PHTHP)
              DO 12 I=1,ORDER
              DO 12 J=1,ORDER
```

(continued)

Listing 6.3   (*Continued*)

```
                  PHTHPR(I,J)=PHTHP(I,J)/PHIN
12.       CONTINUE
          CALL  MATADD(PFT,ORDER,ORDER,FP,PFTFP)
          CALL  MATADD(PFTFP,ORDER,ORDER,Q,PFTFPQ)
          CALL  MATSUB(PFTFPQ,ORDER,ORDER,PHTHPR,PD)
          DO  13  I=1,ORDER
                  K(I,1)=PHT(I,1)/PHIN
13        CONTINUE
          DO  50  I=1,ORDER
          DO  50  J=1,ORDER
                  P(I,J)=P(I,J)+H*PD(I,J)
50        CONTINUE
          T=T+H
          CALL  MATMUL(F,ORDER,ORDER,P,ORDER,ORDER,FP)
          CALL  MATTRN(FP,ORDER,ORDER,PFT)
          CALL  MATMUL(P,ORDER,ORDER,HT,ORDER,1,PHT)
          CALL  MATMUL(HMAT,1,ORDER,P,ORDER,ORDER,HP)
          CALL  MATMUL(PHT,ORDER,1,HP,1,ORDER,PHTHP)
          DO  15  I=1,ORDER
          DO  15  J=1,ORDER
                  PHTHPR(I,J)=PHTHP(I,J)/PHIN
15        CONTINUE
          CALL  MATADD(PFT,ORDER,ORDER,FP,PFTFP)
          CALL  MATADD(PFTFP,ORDER,ORDER,Q,PFTFPQ)
          CALL  MATSUB(PFTFPQ,ORDER,ORDER,PHTHPR,PD)
          DO  16  I=1,ORDER
                  K(I,1)=PHT(I,1)/PHIN
16        CONTINUE
          DO  60  I=1,ORDER
          DO  60  J=1,ORDER
                  P(I,J)=.5*(POLD(I,J)+P(I,J)+H*PD(I,J))
60        CONTINUE
          S=S+H
          IF(S>=(TS-.00001))THEN
                  S=0.
                  XK1=3.*(3*XJ*XJ-3.*XJ+2.)/(XJ*(XJ+1)*(XJ+2))
                  XK2=18.*(2.*XJ-1.)/(XJ*(XJ+1)*(XJ+2)*TS)
                  XK3=60./(XJ*(XJ+1)*(XJ+2)*TS*TS)
                  P11DISC=3*(3*XJ*XJ-3*XJ+2)*SIGN2/(XJ*(XJ+1)*
1                        (XJ+2))
                  IF(XJ.EQ.1.OR.XJ.EQ.2)THEN
                          P22DISC=0.
                          P33DISC=0.
                  ELSE
                          P22DISC=12*(16*XJ*XJ-30*XJ+11)*SIGN2/
1                                (XJ*(XJ*XJ-1)*(XJ*XJ-2)*TS*TS)
                          P33DISC=720*SIGN2/(XJ*(XJ*XJ-1)*(XJ*XJ
1                                -2)*TS**4)
                  ENDIF
```

(*continued*)

Listing 6.2   (*Continued*)

```
                   WRITE(9,*)T,K(1,1)*TS,XK1,K(2,1)*TS,XK2,
1                       K(3,1)*TS,XK3,P(1,1),P11DISC,P(2,2),
2                       P22DISC,P(3,3),P33DISC
                   WRITE(1,*)T,K(1,1)*TS,XK1,K(2,1)*TS,XK2,
1                       K(3,1)*TS,XK3,P(1,1),P11DISC,P(2,2),
2                       P22DISC,P(3,3),P33DISC

              XJ=XJ+1.
       ENDIF
     END DO
     PAUSE
     CLOSE(1)
     END


C SUBROUTINE MATTRN IS SHOWN IN LISTING 1.3
C SUBROUTINE MATMUL IS SHOWN IN LISTING 1.4
C SUBROUTINE MATADD IS SHOWN IN LISTING 1.1
C SUBROUTINE MATSUB IS SHOWN IN LISTING 1.2
```

Transfer Function for Zeroth-Order Filter

We saw in Chapter 5 that when process noise was present the gain and covariance matrix approached steady state. We can derive this steady-state equation for the continuous Kalman filter by noting that in the covariance nonlinear differential equation steady state is reached when the covariance no longer changes. For the zeroth-order continuous polynomial Kalman filter this means that the steady-state solution can be found by setting the derivative of the covariance to zero or

$$\dot{P} = \frac{-P^2}{\Phi_n} + \Phi_s = 0$$

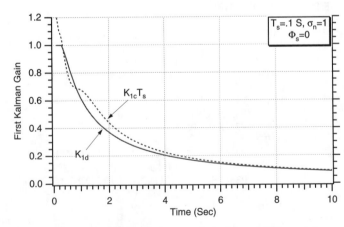

Fig. 6.7   Integrating three-state nonlinear matrix Riccati differential equation yields good match with formula for first gain.

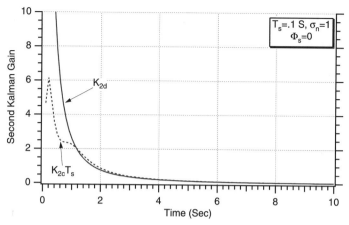

Fig. 6.8 Integrating three-state nonlinear matrix Riccati differential equation yields good match with formula for second gain.

The preceding equation can be solved algebraically to yield the covariance in terms of the measurement and process noise spectral densities or

$$P = (\Phi_s \Phi_n)^{1/2}$$

Because the continuous Kalman gain is computed from the covariance, we can find the closed-form solution for the gain to be given by

$$K = \frac{P}{\Phi_n} = \frac{(\Phi_s \Phi_n)^{1/2}}{\Phi_n}$$

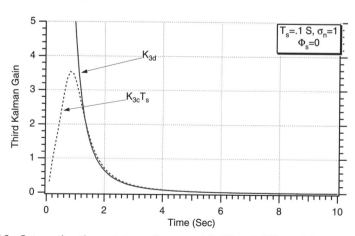

Fig. 6.9 Integrating three-state nonlinear matrix Riccati differential equation yields good match with formula for third gain.

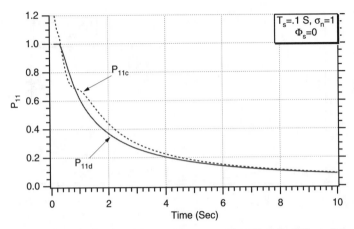

Fig. 6.10 Integrating three-state nonlinear matrix Riccati differential equation yields good match for first diagonal element of covariance matrix.

or

$$K = \left(\frac{\Phi_s}{\Phi_n}\right)^{1/2}$$

Thus, the continuous steady-state Kalman gain only depends on the ratio of the process and measurement noise spectral densities for the zeroth-order continuous polynomial Kalman filter. The interpretation of the preceding formula follows the same logic that was already described. The Kalman filter gain increases with increasing uncertainty of the real world (i.e., Φ_s increases), and new measure-

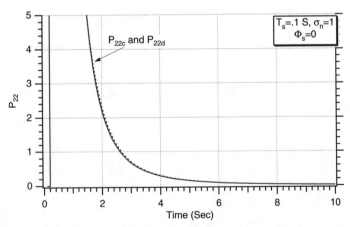

Fig. 6.11 Integrating three-state nonlinear matrix Riccati differential equation yields good match for second diagonal element of covariance matrix.

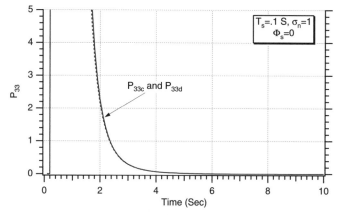

Fig. 6.12 **Integrating three-state nonlinear matrix Riccati differential equation yields good match for third diagonal element of covariance matrix.**

ments are weighted more heavily. The gain decreases with more uncertainty in the measurements (i.e., Φ_n increases) or with the belief that our model of the real world is near perfect (i.e., Φ_s decreases), and therefore we pay less attention to new measurements.

To check the preceding steady-state expressions, Listing 6.1 was rerun with process noise (i.e., $\Phi_s = 10$ or PHIS = 10 in Listing 6.1). We can see from Figs. 6.13 and 6.14 that after an initial transient period numerically integrating the nonlinear covariance differential equation yields exactly the same results as the steady-state formulas for both the covariance and Kalman gain. Thus, we can conclude that the steady-state formulas for both the gain and covariance for the zeroth-order continuous polynomial Kalman filter are correct.

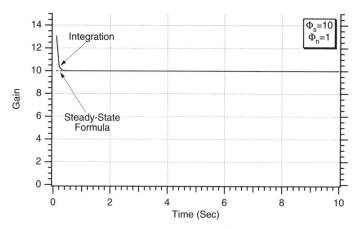

Fig. 6.13 **Steady-state formula accurately predicts Kalman gain for zeroth-order continuous polynomial Kalman filter.**

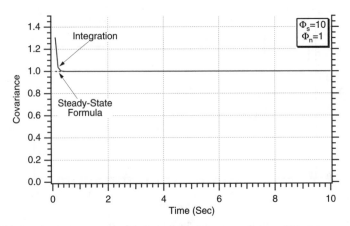

Fig. 6.14 Steady-state formula accurately predicts covariance for zeroth-order continuous polynomial Kalman filter.

Now that we have confidence in the steady-state gain calculation for the zeroth-order continuous polynomial Kalman filter, let us see if we can develop a transfer function for the filter. For our problem there are no deterministic inputs (i.e., no known deterministic or control vector), and so the zeroth-order continuous polynomial Kalman filter can be found from the Kalman filtering differential equation

$$\dot{\hat{x}} = F\hat{x} + K(z - H\hat{x})$$

to yield

$$\dot{\hat{x}} = K(x^* - \hat{x})$$

where x^* is the measurement. We can convert the preceding scalar differential equation to Laplace transform notation and get

$$s\hat{x} = K(x^* - \hat{x})$$

After some algebraic manipulations we get the Kalman filter transfer function relating the estimate to the measurement to be

$$\frac{\hat{x}}{x^*} = \frac{K}{s + K}$$

If we define a natural frequency ω_0 to be the square root of the ratio of the process and measurement noise spectral densities or

$$\omega_0 = \left(\frac{\Phi_s}{\Phi_n}\right)^{1/2}$$

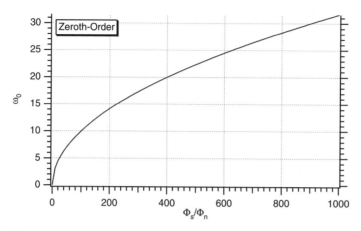

Fig. 6.15 **Zeroth-order continuous polynomial Kalman filter's natural frequency increases as the ratio of process to measurement noise increases.**

then we can redefine the Kalman gain in terms of the natural frequency

$$K = \left(\frac{\Phi_s}{\Phi_n}\right)^{1/2} = \omega_0$$

and substitute the expression for the steady-state Kalman gain into the filter transfer function and obtain

$$\frac{\hat{x}}{x^*} = \frac{1}{1 + \dfrac{s}{\omega_0}}$$

In other words, in the steady state the continuous zeroth-order polynomial Kalman filter is simply a first-order transfer function with a natural frequency that depends on the ratio of the spectral densities of the process noise to the measurement noise. Having more process noise or less measurement noise will tend to increase the bandwidth of the Kalman filter. Figure 6.15 shows more precisely how the bandwidth of the filter varies with the spectral density ratio.

Transfer Function for First-Order Filter

As was the case in the preceding section, we can also derive the steady-state equation for the covariance matrix by noting that in the covariance matrix differential equation steady state is reached when the covariance no longer changes or

$$\begin{bmatrix} \dot{P}_{11} & \dot{P}_{12} \\ \dot{P}_{12} & \dot{P}_{22} \end{bmatrix} = -\begin{bmatrix} P_{11} & P_{12} \\ P_{12} & P_{22} \end{bmatrix}\begin{bmatrix} 1 \\ 0 \end{bmatrix}\Phi_n^{-1}[1 \quad 0]\begin{bmatrix} P_{11} & P_{12} \\ P_{12} & P_{22} \end{bmatrix}$$
$$+ \begin{bmatrix} P_{11} & P_{12} \\ P_{12} & P_{22} \end{bmatrix}\begin{bmatrix} 0 & 0 \\ 1 & 0 \end{bmatrix}$$
$$+ \begin{bmatrix} 0 & 1 \\ 0 & 0 \end{bmatrix}\begin{bmatrix} P_{11} & P_{12} \\ P_{12} & P_{22} \end{bmatrix} + \begin{bmatrix} 0 & 0 \\ 0 & \Phi_s \end{bmatrix} = 0$$

Because the preceding matrix equation is symmetric, it simplifies to the following three scalar equations after multiplying out the terms or

$$0 = 2P_{12} - \frac{P_{11}^2}{\Phi_n}$$

$$0 = P_{22} - \frac{P_{11}P_{12}}{\Phi_n}$$

$$0 = \frac{-P_{12}^2}{\Phi_n} + \Phi_s$$

The preceding equations can be solved algebraically to yield the covariance in terms of the measurement and process noise spectral densities or

$$P_{11} = \sqrt{2}\Phi_s^{1/4}\Phi_n^{3/4}$$
$$P_{22} = \sqrt{2}\Phi_s^{3/4}\Phi_n^{1/4}$$
$$P_{12} = \Phi_s^{1/2}\Phi_n^{1/2}$$

Because the continuous Kalman gains are computed directly from the covariance matrix, they turn out to be

$$K_1 = \frac{P_{11}}{\Phi_n} = \sqrt{2}\left(\frac{\Phi_s}{\Phi_n}\right)^{1/4}$$

$$K_2 = \frac{P_{12}}{\Phi_n} = \left(\frac{\Phi_s}{\Phi_n}\right)^{1/2}$$

To check the preceding steady-state expressions, Listing 6.2 was rerun with process noise (i.e., $\Phi_s = 10$). We can see from Figs. 6.16 and 6.17 that the steady-state gain formulas match the gains obtained by numerical integration after a brief transient period.

We can also see from Figs. 6.18 and 6.19 that the steady-state formulas for the diagonal elements of the covariance matrix match numerical integration experiments. Thus, the formulas for the covariance matrix elements can help us predict the filtering properties of the first-order continuous or discrete polynomial

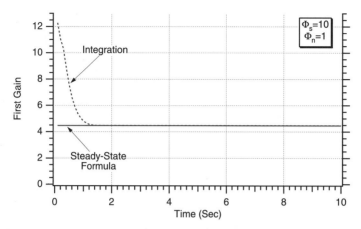

Fig. 6.16 Steady-state gain formula is accurate for first gain in continuous first-order polynomial Kalman filter.

Kalman filters from only knowledge of the process and measurement noise spectral densities.

Now that we have confidence in the steady-state gain calculation for the first-order continuous polynomial Kalman filter let us see if we can develop a transfer function for the filter. For our problem, in which there is no known deterministic input or control vector, we have shown that the polynomial Kalman filter equation in matrix form can be expressed as

$$\dot{\hat{x}} = F\hat{x} + K(z - H\hat{x})$$

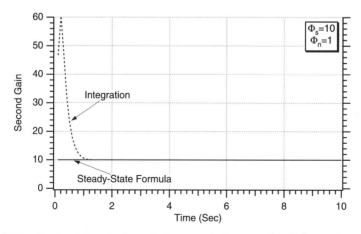

Fig. 6.17 Steady-state gain formula is accurate for second gain in continuous first-order polynomial Kalman filter.

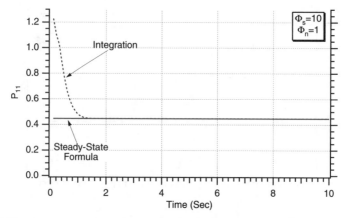

Fig. 6.18 Steady-state formula for first diagonal element of covariance matrix is accurate for continuous first-order polynomial Kalman filter.

Substitution of the appropriate matrices into the preceding equation yields for the first-order system the two scalar differential equations for the Kalman filter state estimates

$$\dot{\hat{x}} = \hat{\hat{x}} + K_1(x^* - \hat{x})$$
$$\dot{\hat{\hat{x}}} = K_2(x^* - \hat{x})$$

where x^* is the measurement. We can convert the preceding two scalar differential equations to Laplace transform notation and get

$$s\hat{x} = \hat{\hat{x}} + K_1(x^* - \hat{x})$$
$$s\hat{\hat{x}} = K_2(x^* - \hat{x})$$

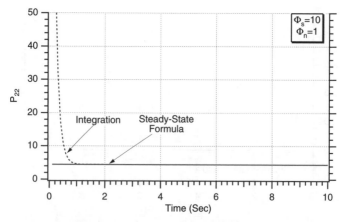

Fig. 6.19 Steady-state formula for second diagonal element of covariance matrix is accurate for continuous first-order polynomial Kalman filter.

After some algebraic manipulation we get the Kalman filter transfer function from the state estimate to the measurement to be

$$\frac{\hat{x}}{x^*} = \frac{K_2 + K_1 s}{s^2 + K_2 + K_1 s}$$

If we define a new natural frequency for the first-order filter (i.e., this natural frequency is different from the natural frequency of zeroth-order filter) as

$$\omega_0 = \left(\frac{\Phi_s}{\Phi_n}\right)^{1/4}$$

then we can express the Kalman gains in terms of the natural frequency as

$$K_1 = \frac{P_{11}}{\Phi_n} = \sqrt{2}\left(\frac{\Phi_s}{\Phi_n}\right)^{1/4} = \sqrt{2}\omega_0$$

$$K_2 = \frac{P_{12}}{\Phi_n} = \left(\frac{\Phi_s}{\Phi_n}\right)^{1/2} = \omega_0^2$$

Substitution of the new gain relationships into the transfer function yields a new transfer function expressed in terms of a natural frequency or

$$\frac{\hat{x}}{x^*} = \frac{1 + \sqrt{2}s/\omega_0}{1 + \sqrt{2}s/\omega_0 + s^2/\omega_0^2}$$

In other words, in the steady state the first-order continuous polynomial Kalman filter is simply a second-order quadratic transfer function with a damping of 0.7 and a natural frequency that depends on the ratio of the spectral densities of the process noise to the measurement noise. Having more process noise or less measurement noise will tend to increase the bandwidth of the

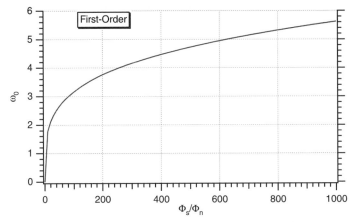

Fig. 6.20 Filter natural frequency increases as the ratio of the process to measurement noise spectral densities increases.

Kalman filter. Figure 6.20 shows more precisely how the bandwidth of the filter varies with the spectral density ratio.

Transfer Function for Second-Order Filter[3]

We can derive the steady-state equation for the second-order filter by noting that in the covariance matrix differential equation steady state is reached when the covariance no longer changes or

$$
\begin{bmatrix} \dot{P}_{11} & \dot{P}_{12} & \dot{P}_{13} \\ \dot{P}_{12} & \dot{P}_{22} & \dot{P}_{23} \\ \dot{P}_{13} & \dot{P}_{23} & \dot{P}_{33} \end{bmatrix} = - \begin{bmatrix} P_{11} & P_{12} & P_{13} \\ P_{12} & P_{22} & P_{23} \\ P_{13} & P_{23} & P_{33} \end{bmatrix} \begin{bmatrix} 1 \\ 0 \\ 0 \end{bmatrix} \Phi_n^{-1} [1 \quad 0 \quad 0]
$$

$$
\times \begin{bmatrix} P_{11} & P_{12} & P_{13} \\ P_{12} & P_{22} & P_{23} \\ P_{13} & P_{23} & P_{33} \end{bmatrix}
$$

$$
+ \begin{bmatrix} P_{11} & P_{12} & P_{13} \\ P_{12} & P_{22} & P_{23} \\ P_{13} & P_{23} & P_{33} \end{bmatrix} \begin{bmatrix} 0 & 0 & 0 \\ 1 & 0 & 0 \\ 0 & 1 & 0 \end{bmatrix} + \begin{bmatrix} 0 & 1 & 0 \\ 0 & 0 & 1 \\ 0 & 0 & 0 \end{bmatrix}
$$

$$
\times \begin{bmatrix} P_{11} & P_{12} & P_{13} \\ P_{12} & P_{22} & P_{23} \\ P_{13} & P_{23} & P_{33} \end{bmatrix} + \begin{bmatrix} 0 & 0 & 0 \\ 0 & 0 & 0 \\ 0 & 0 & \Phi_s \end{bmatrix} = 0
$$

Because the preceding matrix equation is symmetric, the matrix equation simplifies to the following six scalar algebraic equations after multiplying out the terms or

$$
P_{11}^2 = 2P_{12}\Phi_n
$$

$$
P_{12}^2 = 2P_{23}\Phi_n
$$

$$
P_{13}^2 = \Phi_s\Phi_n
$$

$$
P_{11}P_{12} = \Phi_n(P_{22} + P_{13})
$$

$$
P_{11}P_{13} = P_{23}\Phi_n
$$

$$
P_{12}P_{13} = P_{33}\Phi_n
$$

The preceding equations can be solved algebraically to yield the covariance matrix elements in terms of the measurement and process noise spectral densities.

After some algebra we obtain

$$P_{11} = 2\Phi_s^{1/6}\Phi_n^{5/6}$$
$$P_{12} = 2\Phi_s^{1/3}\Phi_n^{2/3}$$
$$P_{13} = \Phi_s^{1/2}\Phi_n^{1/2}$$
$$P_{22} = 3\Phi_s^{1/2}\Phi_n^{1/2}$$
$$P_{23} = 2\Phi_s^{2/3}\Phi_n^{1/3}$$
$$P_{33} = 2\Phi_s^{5/6}\Phi_n^{1/6}$$

The continuous gains are computed from some of the covariance matrix elements. The three Kalman gains turn out to be

$$K_1 = \frac{P_{11}}{\Phi_n} = 2\left(\frac{\Phi_s}{\Phi_n}\right)^{1/6}$$
$$K_2 = \frac{P_{12}}{\Phi_n} = 2\left(\frac{\Phi_s}{\Phi_n}\right)^{1/3}$$
$$K_3 = \frac{P_{13}}{\Phi_n} = \left(\frac{\Phi_s}{\Phi_n}\right)^{1/2}$$

To check the preceding steady-state expressions, Listing 6.3 was rerun with process noise. We can see from Figs. 6.21–6.23 that the steady-state formulas for the Kalman gains are in close agreement with the gains obtained by numerical integration.

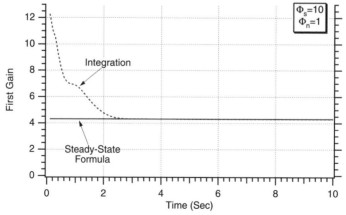

Fig. 6.21 **Steady-state gain formula is accurate for first gain in continuous second-order polynomial Kalman filter.**

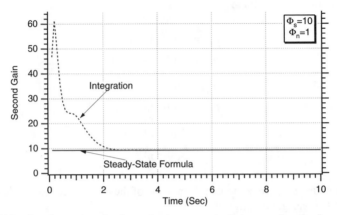

Fig. 6.22 Steady-state gain formula is accurate for second gain in continuous second-order polynomial Kalman filter.

Now that we have confidence in the steady-state gain calculation for the second-order continuous polynomial Kalman filter let us see if we can also develop a transfer function for the filter. For our problem, in which there is no control vector, the polynomial Kalman filter equation in matrix form is given by

$$\dot{\hat{x}} = F\hat{x} + K(z - H\hat{x})$$

Substitution of the appropriate matrices yields the three scalar differential equations

$$\dot{\hat{x}} = \hat{\dot{x}} + K_1(x^* - \hat{x})$$

$$\dot{\hat{\dot{x}}} = \hat{\ddot{x}} + K_2(x^* - \hat{x})$$

$$\dot{\hat{\ddot{x}}} = K_3(x^* - \hat{x})$$

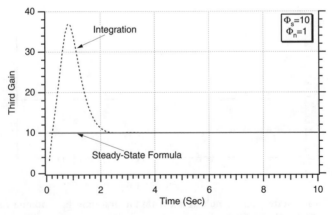

Fig. 6.23 Steady-state gain formula is accurate for third gain in continuous second-order polynomial Kalman filter.

where x^* is the measurement. We can convert the preceding three scalar differential equations to Laplace transform notation and get

$$s\hat{x} = \hat{\dot{x}} + K_1(x^* - \hat{x})$$
$$s\hat{\dot{x}} = \hat{\ddot{x}} + K_2(x^* - \hat{x})$$
$$s\hat{\ddot{x}} = K_3(x^* - \hat{x})$$

After some algebraic manipulation we obtain the Kalman filter transfer function from state estimate to measurement as

$$\frac{\hat{x}}{x^*} = \frac{K_3 + sK_2 + s^2K_1}{K_3 + sK_2 + s^2K_1 + s^3}$$

If we define a natural frequency for this second-order filter (i.e., different from zeroth-order and first-order natural frequencies) to be

$$\omega_0 = \left(\frac{\Phi_s}{\Phi_n}\right)^{1/6}$$

we can express the Kalman gains in terms of the natural frequency as

$$K_1 = \frac{P_{11}}{\Phi_n} = 2\left(\frac{\Phi_s}{\Phi_n}\right)^{1/6} = 2\omega_0$$

$$K_2 = \frac{P_{12}}{\Phi_n} = 2\left(\frac{\Phi_s}{\Phi_n}\right)^{1/3} = 2\omega_0^2$$

$$K_3 = \frac{P_{13}}{\Phi_n} = \left(\frac{\Phi_s}{\Phi_n}\right)^{1/2} = \omega_0^3$$

Substituting the preceding three expressions into the transfer function yields

$$\frac{\hat{x}}{x^*} = \frac{1 + 2s/\omega_0 + 2s^2/\omega_0^2}{1 + 2s/\omega_0 + 2s^2/\omega_0^2 + s^3/\omega_0^3}$$

In other words, in the steady state the continuous second-order polynomial Kalman filter is simply a third-order transfer function whose poles follow a Butterworth[4] distribution with a natural frequency that depends on the ratio of the spectral densities of the process noise to the measurement noise. Having more process noise or less measurement noise will tend to increase the bandwidth of the Kalman filter. Figure 6.24 shows more precisely how the bandwidth of the filter varies with the spectral density ratio.

Filter Comparison

It is sometimes easier to understand the properties of the continuous polynomial Kalman filters from a frequency domain point of view. Table 6.1 presents

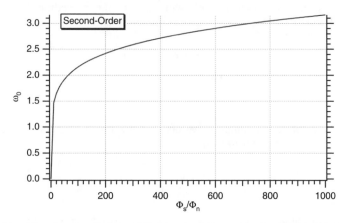

Fig. 6.24 Second-order Kalman filter natural frequency increases with increasing ratio of process to measurement noise spectral density.

the zeroth-, first- and second-order polynomial Kalman filter transfer functions, in the Laplace transform domain, that we just derived in this chapter. The other column of Table 6.1 presents the magnitude of the various filter transfer functions in the complex frequency domain. Recall that the magnitude of a transfer function can be found by replacing s with $j\omega$ and then finding the square root of the real part squared plus the imaginary part squared.

Listing 6.4 presents a program that was written to evaluate the polynomial Kalman filter transfer function magnitudes that are displayed in Table 6.1. We can see that the listing consists of one loop in which the frequency is varied from 1 rad/s to 100 rad/s in steps of 1 rad/s. The magnitudes of each of the filter transfer functions are computed and printed out for each frequency. Nominally the program is set up with a natural frequency of 10 rad/s.

The nominal case of Listing 6.4 was run, and the results for each of the three different filters are displayed in Fig. 6.25. We can see that each of the filters has a magnitude of approximately unity until 10 rad/s. This means that each filter simply passes the measurement, without amplification or attenuation, until the natural frequency (i.e., 10 rad/s in this example). For frequencies higher than the natural frequency, the output is attenuated (i.e., magnitude less than unity). This means that the filter is not passing those frequencies. In other words the filter is behaving as a low-pass filter. Everything gets passed until the natural frequency is reached, and afterwards everything is attenuated. We also can tell from Fig. 6.25 that the higher-order filters have less attenuation after the natural frequency than the lower-order filters. This simply indicates that there is more noise transmission with the higher-order filters (i.e., higher frequencies are less attenuated). We have already seen in the preceding chapters that higher-order filters result in more noise transmission.

Next, the magnitude of the second-order polynomial Kalman filter transfer function was computed for natural frequencies of 5, 10, and 25 rad/s. We can see from Fig. 6.26 that, as expected, the magnitude of the transfer function peaks at

Table 6.1 Transfer functions and magnitudes for different-order polynomial Kalman filters

Name	Laplace transform	Magnitude
Zeroth order	$\dfrac{\hat{x}}{x^*} = \dfrac{1}{1 + s/\omega_0}$	$\left\|\dfrac{\hat{x}}{x^*}\right\| = \dfrac{1}{\sqrt{1 + (\omega/\omega_0)^2}}$
First order	$\dfrac{\hat{x}}{x^*} = \dfrac{1 + \sqrt{2}s/\omega_0}{1 + \sqrt{2}s/\omega_0 + s^2/\omega_0^2}$	$\left\|\dfrac{\hat{x}}{x^*}\right\| = \sqrt{\dfrac{1 + (\sqrt{2}\omega/\omega_0)^2}{(1 - \omega^2/\omega_0^2)^2 + (\sqrt{2}\omega/\omega_0)^2}}$
Second order	$\dfrac{\hat{x}}{x^*} = \dfrac{1 + 2s/\omega_0 + 2s^2/\omega_0^2}{1 + 2s/\omega_0 + 2s^2/\omega_0^2 + s^3/\omega_0^3}$	$\left\|\dfrac{\hat{x}}{x^*}\right\| = \sqrt{\dfrac{(1 - 2\omega^2/\omega_0^2)^2 + (2\omega/\omega_0)^2}{(1 - 2\omega^2/\omega_0^2)^2 + (2\omega/\omega_0 - \omega^3/\omega_0^3)^2}}$

Listing 6.4 Program to calculate magnitudes of Kalman-filter transfer functions

```
IMPLICIT REAL*8 (A-H)
IMPLICIT REAL*8 (O-Z)
OPEN(1,STATUS='UNKNOWN',FILE='DATFIL')
W0=10.
DO 10 W=1.,100.
XMAG1=1./SQRT(1.+(W/W0)**2)
TOP1=1.+2.*(W/W0)**2
BOT1=(1.-(W*W/(W0*W0)))**2+2.*(W/W0)**2
XMAG2=SQRT(TOP1/(BOT1+.00001))
TOP2=(1.-2.*W*W/(W0*W0))**2+(2.*W/W0)**2
TEMP1=(1.-2.*W*W/(W0*W0))**2
TEMP2=(2.*W/W0-(W/W0)**3)**2
XMAG3=SQRT(TOP2/(TEMP1+TEMP2+.00001))
WRITE(9,*)W,XMAG1,XMAG2,XMAG3
WRITE(1,*)W,XMAG1,XMAG2,XMAG3
10 CONTINUE
CLOSE(1)
PAUSE
END
```

the filter natural frequency. If the filter natural frequency is high, the filter will have a wider bandwidth. This means that the filter will be more responsive, but it also means that the filter will pass more noise. Again, we have seen in preceding chapters that when we increase or add process noise the filter is able to better track the signal (i.e., less lag or higher bandwidth) but has more noise transmission (i.e., noisier estimates).

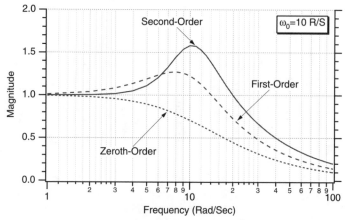

Fig. 6.25 Higher-order filters have less attenuation after filter natural frequency.

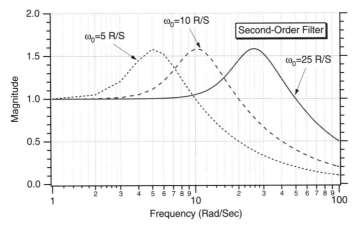

Fig. 6.26   Increasing filter natural frequency increases filter bandwidth.

Summary

In this chapter we have investigated different-order continuous polynomial Kalman filters. Although these filters are generally not implemented, they can be useful for helping us to understand the properties of the more popular discrete Kalman filters. Steady-state solutions for the gains and covariances of the various-order polynomial Kalman filters were obtained. From these solutions we also were able to derive transfer functions for the Kalman filter. We saw from the resultant transfer functions that the polynomial Kalman filter acted as a low-pass filter. The bandwidth of the various-order polynomial Kalman filters depended on the ratio of the spectral density of the process noise to the spectral density of the measurement noise. Increasing the spectral density ratio, by either increasing the process noise or decreasing the measurement noise, tended to increase the bandwidth of the Kalman filter. All results were derived analytically and confirmed by simulation.

References

[1]Kalman, R. E. and Bucy, R. S., "New Results in Linear Filtering and Prediction Theory," *Journal of Basic Engineering*, Vol. 83, No. 3, 1961, pp. 95–108.

[2]Gelb., A., *Applied Optimal Estimation*, Massachusetts Inst. of Technology Press, Cambridge, MA, 1974, pp. 119–127.

[3]Nesline, F. W. and Zarchan, P., "A New Look at Classical Versus Modern Homing Guidance," *Journal of Guidance and Control*, Vol. 4, No. 1, 1981, pp. 78–85.

[4]Karni, S., *Network Theory: Analysis and Synthesis*, Allyn and Bacon, London, 1966, pp. 344–349.

Chapter 7

Extended Kalman Filtering

Introduction

S O FAR we have seen how linear Kalman filters are designed and how they perform when the real world can be described by linear differential equations expressed in state-space form and when the measurements are linear functions of the states. In most realistic problems the real world may be described by nonlinear differential equations, and the measurements may not be a function of those states. In this chapter we will introduce extended Kalman filtering for a sample problem in which the measurements are a linear function of the states, but the model of the real world is nonlinear.

Theoretical Equations[1]

To apply extended Kalman-filtering techniques, it is first necessary to describe the real world by a set of nonlinear differential equations. These equations also can be expressed in nonlinear state-space form as a set of first-order nonlinear differential equations or

$$\dot{x} = f(x) + w$$

where x is a vector of the system states, $f(x)$ is a nonlinear function of those states, and w is a random zero-mean process. The process-noise matrix describing the random process w for the preceding model is given by

$$Q = E(ww^T)$$

Finally, the measurement equation, required for the application of extended Kalman filtering, is considered to be a nonlinear function of the states according to

$$z = h(x) + v$$

where v is a zero-mean random process described by the measurement noise matrix R, which is defined as

$$R = E(vv^T)$$

For systems in which the measurements are discrete, we can rewrite the nonlinear measurement equation as

$$z_k = h(x_k) + v_k$$

The discrete measurement noise matrix R_k consists of a matrix of variances representing each measurement noise source. Because the system and measurement equations are nonlinear, a first-order approximation is used in the continuous Riccati equations for the systems dynamics matrix F and the measurement matrix H. The matrices are related to the nonlinear system and measurement equations according to

$$F = \frac{\partial f(x)}{\partial x}\bigg|_{x=\hat{x}}$$

$$H = \frac{\partial h(x)}{\partial x}\bigg|_{x=\hat{x}}$$

The fundamental matrix, required for the discrete Riccati equations, can be approximated by the Taylor-series expansion for $\exp(FT_s)$ and is given by

$$\Phi_k = I + FT_s + \frac{F^2 T_s^2}{2!} + \frac{F^3 T_s^3}{3!} + \cdots$$

where T_s is the sampling time and I is the identity matrix. Often the series is approximated by only the first two terms or

$$\Phi_k \approx I + FT_s$$

In our applications of extended Kalman filtering, the fundamental matrix will only be used in the calculation of the Kalman gains. Because the fundamental matrix will not be used in the propagation of the states, we have already demonstrated in Chapter 5 that adding more terms to the Taylor-series expansion for the fundamental matrix will not generally improve the performance of the overall filter.

For linear systems, the systems dynamics matrix, measurement matrix, and fundamental matrix are all linear. However, these same matrices for the extended Kalman filter may be nonlinear because they depend on the system state estimates. The Riccati equations, required for the computation of the Kalman gains, are identical to those of the linear case and are repeated here for convenience:

$$M_k = \Phi_k P_{k-1} \Phi_k^T + Q_k$$
$$K_k = M_k H^T (H M_k H^T + R_k)^{-1}$$
$$P_k = (I - K_k H) M_k$$

where P_k is a covariance matrix representing errors in the state estimates after an update and M_k is the covariance matrix representing errors in the state estimates before an update. Because Φ_k and H are nonlinear functions of the state estimates, the Kalman gains cannot be computed off line as is possible with a linear Kalman filter. The discrete process-noise matrix Q_k can still be found from the continuous process-noise matrix according to

$$Q_k = \int_0^{T_s} \Phi(\tau) Q \Phi^T(\tau) \, d\tau$$

If the dynamical model of a linear Kalman filter is matched to the real world, the covariance matrix P_k can not only be used to calculate Kalman gains but can also provide predictions of the errors in the state estimates. The extended Kalman filter offers no such guarantees, and, in fact, the extended Kalman filter covariance matrix may indicate excellent performance when the filter is performing poorly or is even broken.

As was already mentioned, the preceding approximations for the fundamental and measurement matrices only have to be used in the computation of the Kalman gains. The actual extended Kalman-filtering equations do not have to use those approximations but instead can be written in terms of the nonlinear state and measurement equations. With an extended Kalman filter the new state estimate is the old state estimate projected forward to the new sampling or measurement time plus a gain times a residual or

$$\hat{x}_k = \bar{x}_k + K_k[z_k - h(\bar{x}_k)]$$

In the preceding equation the residual is the difference between the actual measurement and the nonlinear measurement equation. The old estimates that have to be propagated forward do not have to be done with the fundamental matrix but instead can be propagated directly by integrating the actual nonlinear differential equations forward at each sampling interval. For example, Euler integration can be applied to the nonlinear system differential equations yielding

$$\bar{x}_k = \hat{x}_{k-1} + \dot{\hat{x}}_{k-1} T_s$$

where the derivative is obtained from

$$\dot{\hat{x}}_{k-1} = f(\hat{x}_{k-1})$$

In the preceding equation the sampling time T_s is used as an integration interval. In problems where the sampling time is large, T_s would have to be replaced by a smaller integration interval, or possibly a more accurate method of integration has to be used. The best way to show how an extended Kalman filter is implemented and performs is by doing an example.

Drag Acting on Falling Object

Let us again reconsider the one-dimensional example of a high-velocity object falling on a tracking radar, as was shown in Fig. 4.26. Recall that the drag-free object was initially at 400,000 ft above the radar and had a velocity of 6000 ft/s toward the radar that was located on the surface of a flat Earth. In that example only gravity g (i.e., $g = 32.2\,\text{ft/s}^2$) acted on the object. Now we will add one degree of complexity to the problem by also considering the influence of drag. Drag will make the resultant equation describing the acceleration of the object nonlinear. However, we will assume that the amount of drag acting on the object is known. As was the case in Chapter 4, let us pretend that the radar measures the range from the radar to the object (i.e., altitude of the object) with a 1000-ft standard deviation measurement accuracy. The radar takes altitude measurements 10 times a second for 30 s. In this example the object will initially be at 200,000 ft above the radar, as is shown in Fig. 7.1. We would like to build a filter to estimate the altitude and velocity of the object.

If x is the distance from the radar to the object, then the acceleration acting on the object consists of gravity and drag[1,2] or

$$\ddot{x} = \text{Drag} - g = \frac{Q_p g}{\beta} - g$$

where β is the ballistic coefficient of the object and Q_p is the dynamic pressure. The ballistic coefficient, which is considered to be a constant in this problem, is a term describing the amount of drag acting on the object. Small values of β indicate high drag, and high values of β indicate low drag. Setting β equal to infinity eliminates the drag, and we end up with the problem of Chapter 4, where only gravity acts on the object. The dynamic pressure Q_p is given by

$$Q_p = 0.5 \rho \dot{x}^2$$

where ρ is the air density and $\dot{x}$ is the velocity of the object. For this sample problem we can assume that the air density is an exponential function of altitude or[2]

$$\rho = 0.0034 e^{-x/22000}$$

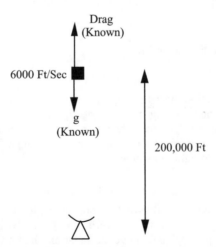

Fig. 7.1   Radar tracking falling object in presence of drag.

where x is altitude. Therefore, the equation expressing the acceleration acting on the object can now be expressed as the nonlinear second-order differential equation

$$\ddot{x} = \frac{Q_p g}{\beta} - g = \frac{0.5 g \rho \dot{x}^2}{\beta} - g = \frac{0.0034 g e^{-x/22000} \dot{x}^2}{2\beta} - g$$

A program was written to integrate numerically the preceding nonlinear differential equation using the second-order Runge–Kutta numerical integration technique described in Chapter 1. Listing 7.1 shows that the object is initially at 200,000-ft altitude traveling downward with an initial velocity of 6000 ft/s, as was shown in Fig. 7.1. The listing also shows that the ballistic coefficient β is made a parameter. The position X, velocity XD, and acceleration XDD of the object are printed out every 0.1 s for 30 s.

Cases were run with Listing 7.1 in which the ballistic coefficient BETA was varied from infinity to 500 lb/ft². We can see from Fig. 7.2 that there is not much difference in the final location of the object after 30 s for all ballistic coefficients considered. If there is no drag, the object descends from 200,000 to 5,500 ft in 30 s. If the ballistic coefficient is 500 lb/ft² (i.e., the most drag considered in this example), the object only descends to 25,400 ft in the same 30 s. In other words, increasing the drag only decreases the lowest altitude that can be reached in 30 sec.

As expected, Fig. 7.3 shows that objects with the most drag slow up the most. If there is no drag, we can see that the object actually speeds up from its initial value of 6000 ft/s to nearly 7000 ft/s as a result of gravity. With a ballistic coefficient of 2000 lb/ft² the object slows up to approximately 5000 ft/s. Reducing the ballistic coefficient to 1000 lb/ft² causes the object to slow up to 4100 ft/s. Finally, reducing the ballistic coefficient even further to 500 lb/ft² reduces the speed of the object from 6000 ft/s at the beginning of the flight to only 3300 ft/s at the end of the 30-s flight.

Figure 7.4 shows that, as expected, if there is no drag, the maximum acceleration experienced by the object is that of gravity (i.e., -32.2 ft/s²). However, the presence of drag can cause the object to go through decelerations in excess of 10 g (i.e., 322 ft/s²). We can see from Fig. 7.4 that the maximum acceleration increases with increasing drag (i.e., decreasing ballistic coefficient).

First Attempt at Extended Kalman Filter

We now have enough of a feel for the dynamics of the falling high-speed body to proceed with the design of an extended Kalman filter. Let us assume that the ballistic coefficient of the object we are tracking is known in advance. Under these circumstances it is only necessary to estimate the altitude and velocity of the object in order to track it. Thus, the proposed states for the filter design are given by

$$x = \begin{bmatrix} x \\ \dot{x} \end{bmatrix}$$

**Listing 7.1   Simulation of falling object under influence of
drag and gravity**

```
IMPLICIT REAL*8 (A-H)
IMPLICIT REAL*8 (O-Z)
G=32.2
X=200000.
XD=-6000.
BETA=500.
OPEN(1,STATUS='UNKNOWN',FILE='DATFIL')
TS=.1
TF=30.
T=0.
S=0.
H=.001
WHILE(T<=TF)
        XOLD=X
        XDOLD=XD
        XDD=.0034*G*XD*XD*EXP(-X/22000.)/(2.*BETA)-G
        X=X+H*XD
        XD=XD+H*XDD
        T=T+H
        XDD=.0034*G*XD*XD*EXP(-X/22000.)/(2.*BETA)-G
        X=.5*(XOLD+X+H*XD)
        D=.5*(XDOLD+XD+H*XDD)
        S=S+H
        IF(S>=(TS-.00001))THEN
                S=0.
                WRITE(9,*)T,X,XD,XDD
                WRITE(1,*)T,X,XD,XDD
        ENDIF
END  DO
PAUSE
CLOSE(1)
END
```

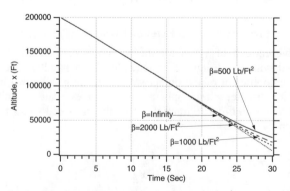

Fig. 7.2   Increasing drag decreases lowest altitude that can be reached in 30 s.

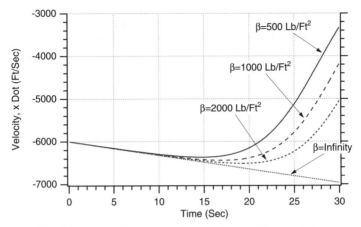

Fig. 7.3   Increasing drag reduces velocity at lower altitudes.

Because we have already shown that the acceleration acting on the object is a nonlinear function of the states according to

$$\ddot{x} = \frac{0.0034ge^{-x/22000}\dot{x}^2}{2\beta} - g$$

we can linearize the preceding equation by saying that

$$\begin{bmatrix} \Delta\dot{x} \\ \Delta\ddot{x} \end{bmatrix} = \begin{bmatrix} \dfrac{\partial\dot{x}}{\partial x} & \dfrac{\partial\dot{x}}{\partial\dot{x}} \\ \dfrac{\partial\ddot{x}}{\partial x} & \dfrac{\partial\ddot{x}}{\partial\dot{x}} \end{bmatrix} \begin{bmatrix} \Delta x \\ \Delta\dot{x} \end{bmatrix} + \begin{bmatrix} 0 \\ u_s \end{bmatrix}$$

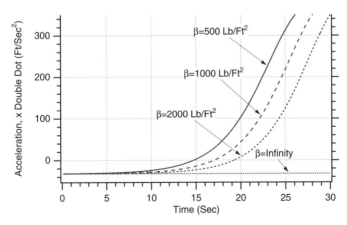

Fig. 7.4   Drag causes high decelerations.

In the preceding equation u_s is a white process noise that has been added to the acceleration equation for possible future protection. We can see from the preceding linearized equation that the systems dynamics matrix is the matrix of partial derivatives or

$$F = \begin{bmatrix} \dfrac{\partial \dot{x}}{\partial x} & \dfrac{\partial \dot{x}}{\partial \dot{x}} \\[2mm] \dfrac{\partial \ddot{x}}{\partial x} & \dfrac{\partial \ddot{x}}{\partial \dot{x}} \end{bmatrix}_{x=\hat{x}}$$

Here we can see that the partial derivatives are evaluated at the current state estimates. We can also see from the linearized state-space equation that the continuous process-noise matrix is given by

$$Q = E(ww^T)$$

Therefore, by inspection of the linearized state-space equation, we can say for this example that the continuous process-noise matrix is

$$Q(t) = \begin{bmatrix} 0 & 0 \\ 0 & \Phi_s \end{bmatrix}$$

where Φ_s is the spectral density of the white noise sources assumed to be on acceleration. To evaluate the systems dynamics matrix, we must take the necessary partial derivatives or

$$\frac{\partial \dot{x}}{\partial x} = 0$$

$$\frac{\partial \dot{x}}{\partial \dot{x}} = 1$$

$$\frac{\partial \ddot{x}}{\partial x} = -\frac{0.0034 e^{-x/22000} \dot{x}^2 g}{2\beta(22,000)} = \frac{-\rho g \dot{x}^2}{44,000\beta}$$

$$\frac{\partial \ddot{x}}{\partial \dot{x}} = -\frac{2*0.0034 e^{-x/22000} \dot{x} g}{2\beta} = \frac{\rho g \dot{x}}{\beta}$$

Therefore, the systems dynamics matrix turns out to be

$$F(t) = \begin{bmatrix} 0 & 1 \\[2mm] \dfrac{-\hat{\rho} g \hat{x}^2}{44,000\beta} & \dfrac{\hat{\rho} \hat{x} g}{\beta} \end{bmatrix}$$

where gravity g and the ballistic coefficient β are assumed to be known perfectly and the estimated air density is a function of the estimated altitude and is given by

$$\hat{\rho} = 0.0034 e^{-\hat{x}/22000}$$

If we assume that the systems dynamics matrix is approximately constant between sampling instants, the fundamental matrix can be found by the infinite Taylor-series approximation

$$\Phi(t) = I + Ft + \frac{F^2 t^2}{2!} + \frac{F^3 t^3}{3!} + \cdots$$

If we define

$$f_{21} = \frac{-\hat{\rho} g \hat{x}^2}{44{,}000 \beta}$$

$$f_{22} = \frac{\hat{\rho} \hat{x} g}{\beta}$$

then the systems dynamics matrix can be written as

$$F(t) = \begin{bmatrix} 0 & 1 \\ f_{21} & f_{22} \end{bmatrix}$$

If we only take the first two terms of the Taylor-series expansion for the fundamental matrix, we get

$$\Phi(t) = I + Ft = \begin{bmatrix} 1 & 0 \\ 0 & 1 \end{bmatrix} + \begin{bmatrix} 0 & 1 \\ f_{21} & f_{22} \end{bmatrix} t = \begin{bmatrix} 1 & t \\ f_{21} t & 1 + f_{22} t \end{bmatrix}$$

or more simply

$$\Phi(t) = \begin{bmatrix} 1 & t \\ f_{21} t & 1 + f_{22} t \end{bmatrix}$$

Therefore, the discrete fundamental matrix can be found by substituting T_s for t and is given by

$$\Phi_k = \begin{bmatrix} 1 & T_s \\ f_{21} T_s & 1 + f_{22} T_s \end{bmatrix}$$

In this example the altitude measurement is a linear function of the states and is given by

$$x_k^* = \begin{bmatrix} 1 & 0 \end{bmatrix} \begin{bmatrix} x_k \\ \dot{x}_k \end{bmatrix} + v_k$$

where v_k is the measurement noise. We can see from the preceding equation that the measurement matrix is given by

$$H = \begin{bmatrix} 1 & 0 \end{bmatrix}$$

The discrete measurement noise matrix is given by

$$R_k = E(v_k v_k^T)$$

Therefore, we can say that for this problem the discrete measurement noise matrix turns out to be a scalar and is simply

$$R_k = \sigma_v^2$$

where σ_v^2 is the variance of the measurement noise. Similarly, we have already shown that the continuous process-noise matrix is given by

$$Q(t) = \begin{bmatrix} 0 & 0 \\ 0 & \Phi_s \end{bmatrix}$$

where Φ_s is the spectral density of the white noise sources assumed to be on acceleration. The discrete process-noise matrix can be derived from the continuous process-noise matrix according to

$$Q_k = \int_0^{T_s} \Phi(\tau) Q \Phi^T(\tau) \, d\tau$$

Therefore, substitution of the appropriate matrices into the preceding equation yields

$$Q_k = \Phi_s \int_0^{T_s} \begin{bmatrix} 1 & \tau \\ f_{21}\tau & 1 + f_{22}\tau \end{bmatrix} \begin{bmatrix} 0 & 0 \\ 0 & 1 \end{bmatrix} \begin{bmatrix} 1 & f_{21}\tau \\ \tau & 1 + f_{22}\tau \end{bmatrix} d\tau$$

After multiplying out the matrices, we get

$$Q_k = \Phi_s \int_0^{T_s} \begin{bmatrix} \tau^2 & \tau + f_{22}\tau^2 \\ \tau + f_{22}\tau^2 & 1 + 2f_{22}\tau + f_{22}^2\tau^2 \end{bmatrix} d\tau$$

Finally, after integration we obtain the final expression for the discrete process-noise matrix as

$$
Q_k = \Phi_s \begin{bmatrix} \dfrac{T_s^3}{3} & \dfrac{T_s^2}{2} + f_{22}\dfrac{T_s^3}{3} \\[2ex] \dfrac{T_s^2}{2} + f_{22}\dfrac{T_s^3}{3} & T_s + f_{22}T_s^2 + f_{22}^2\dfrac{T_s^3}{3} \end{bmatrix}
$$

We now have defined all of the matrices required to solve the matrix Riccati equations. The next step is to write down the equations for the Kalman-filter section. First we must be able to propagate the states from the present sampling time to the next sampling time. This cannot be done in this example exactly with the fundamental matrix because the fundamental matrix is approximate (i.e., only two terms of a Taylor-series expansion were used), and the fundamental matrix was also based on a linearization of the problem. Therefore, we will use Euler numerical integration of the nonlinear differential equations until the next update, using the sampling time T_s as the integration interval. Therefore, given the nonlinear acceleration equation

$$
\ddot{\bar{x}}_{k-1} = \frac{0.0034 g e^{-\hat{x}_{k-1}/22000} \dot{\hat{x}}_{k-1}^2}{2\beta} - g
$$

the projected velocity and position are determined by a one-step Euler integration to the next sampling interval as

$$
\dot{\bar{x}}_k = \dot{\hat{x}}_{k-1} + T_s \ddot{\bar{x}}_{k-1}
$$
$$
\bar{x}_k = \hat{x}_{k-1} + T_s \dot{\bar{x}}_{k-1}
$$

Now the Kalman-filtering equations can be written as

$$
\hat{x}_k = \bar{x}_k + K_{1_k}(x_k^* - \bar{x}_k)
$$
$$
\dot{\hat{x}}_k = \dot{\bar{x}}_k + K_{2_k}(x_k^* - \bar{x}_k)
$$

where x_k^* is the noisy measurement of altitude. Again, note that the matrices required by the Riccati equations in order to obtain the Kalman gains were based on linearized equations, while the Kalman filter made use of the nonlinear equations.

The preceding equations for the Kalman filter, along with the Riccati equations, were programmed and are shown in Listing 7.2, along with a simulation of the real world. We can see that the process-noise matrix is set to zero in this example (i.e., $\Phi_s = 0$). We have initialized the states of the filter close to the true values. The initial filter estimate of altitude XH is in error by 100 ft, and the initial filter estimate of velocity XDH is in error by 10 ft/s. We can see from Listing 7.2 that the initial covariance matrix reflects those errors. We can also see that nominally there is 1000 ft of measurement noise (i.e., SIGN-OISE = 1000).

Listing 7.2   First attempt at extended Kalman filter for tracking a falling object under the influence of both drag and gravity

```
C THE FIRST THREE STATEMENTS INVOKE THE ABSOFT RANDOM
  NUMBER GENERATOR ON THE MACINTOSH
      GLOBAL DEFINE
              INCLUDE 'quickdraw.inc'
      END
      IMPLICIT REAL*8 (A-H)
      IMPLICIT REAL*8 (O-Z)
      REAL*8 PHI(2,2),P(2,2),M(2,2),PHIP(2,2),PHIPPHIT(2,2),GAIN(2,1)
      REAL*8 Q(2,2),HMAT(1,2),HM(1,2),MHT(2,1)
      REAL*8 PHIT(2,2)
      REAL*8 HMHT(1,1),HT(2,1),KH(2,2),IDN(2,2),IKH(2,2)
      INTEGER ORDER
      SIGNOISE=1000.
      X=200000.
      XD=-6000.
      BETA=500.
      XH=200025.
      XDH=-6150.
      OPEN(1,STATUS='UNKNOWN',FILE='DATFIL')
      OPEN(2,STATUS='UNKNOWN',FILE='COVFIL')
      ORDER=2
      TS=.1
      TF=30.
      PHIS=0.
      T=0.
      S=0.
      H=.001
      DO 1000 I=1,ORDER
      DO 1000 J=1,ORDER
              PHI(I,J)=0.
              P(I,J)=0.
              Q(I,J)=0.
              IDN(I,J)=0.
1000 CONTINUE
      IDN(1,1)=1.
      IDN(2,2)=1.
      P(1,1)=SIGNOISE*SIGNOISE
      P(2,2)=20000.
      DO 1100 I=1,ORDER
              HMAT(1,I)=0.
              HT(I,1)=0.
1100 CONTINUE
      HMAT(1,1)=1.
      HT(1,1)=1.
      WHILE(T<=TF)
              XOLD=X
              XDOLD=XD
              XDD=.0034*32.2*XD*XD*EXP(-X/22000.)/(2.*BETA)-32.2
              X=X+H*XD
```

(continued)

Listing 7.2   (*Continued*)

```
XD=XD+H*XDD
T=T+H
XDD=.0034*32.2*XD*XD*EXP(-X/22000.)/(2.*BETA)-32.2
X=.5*(XOLD+X+H*XD)
XD=.5*(XDOLD+XD+H*XDD)
S=S+H
IF(S>=(TS-.00001))THEN
S=0.
RHOH=.0034*EXP(-XH/22000.)
F21=-32.2*RHOH*XDH*XDH/(44000.*BETA)
F22=RHOH*32.2*XDH/BETA
PHI(1,1)=1.
PHI(1,2)=TS
PHI(2,1)=F21*TS
PHI(2,2)=1.+F22*TS
Q(1,1)=PHIS*TS*TS*TS/3.
Q(1,2)=PHIS*(TS*TS/2.+F22*TS*TS*TS/3.)
Q(2,1)=Q(1,2)
Q(2,2)=PHIS*(TS+F22*TS*TS+F22*F22*TS*TS*TS/3.)
CALL  MATTRN(PHI,ORDER,ORDER,PHIT)
CALL  MATMUL(PHI,ORDER,ORDER,P,ORDER,ORDER,
     PHIP)
CALL  MATMUL(PHIP,ORDER,ORDER,PHIT,ORDER,
     ORDER,PHIPPHIT)
CALL  MATADD(PHIPPHIT,ORDER,ORDER,Q,M)
CALL  MATMUL(HMAT,1,ORDER,M,ORDER,ORDER,HM)
CALL  MATMUL(HM,1,ORDER,HT,ORDER,1,HMHT)
HMHTR=HMHT(1,1)+SIGNOISE*SIGNOISE
HMHTRINV=1./HMHTR
CALL  MATMUL(M,ORDER,ORDER,HT,ORDER,1,MHT)
DO  150  I=1,ORDER
          GAIN(I,1)=MHT(I,1)*HMHTRINV
CONTINUE
CALL  MATMUL(GAIN,ORDER,1,HMAT,1,ORDER,KH)
CALL  MATSUB(IDN,ORDER,ORDER,KH,IKH)
CALL  MATMUL(IKH,ORDER,ORDER,M,ORDER,
     ORDER,P)
CALL  GAUSS(XNOISE,SIGNOISE)
XDDB=.0034*32.2*XDH*XDH*EXP(-XH/22000.)/
     (2.*BETA)-32.2
XDB=XDH+XDDB*TS
XB=XH+TS*XDB
RES=X+XNOISE-XB
XH=XB+GAIN(1,1)*RES
XDH=XDB+GAIN(2,1)*RES
ERRX=X-XH
SP11=SQRT(P(1,1))
ERRXD=XD-XDH
SP22=SQRT(P(2,2))
```

(*continued*)

Listing 7.2 *(Continued)*

```
                  WRITE(9,*)T,X,XH,XD,XDH,RES,GAIN(1,1),GAIN(2,1)
                  WRITE(1,*)T,X,XH,XD,XDH,RES,GAIN(1,1),GAIN(2,1)
                  WRITE(2,*)T,ERRX,SP11,-SP11,ERRXD,SP22,-SP22
        ENDIF
      END DO
      PAUSE
      CLOSE(1)
      END


C SUBROUTINE GAUSS IS SHOWN IN LISTING 1.8
C SUBROUTINE MATTRN IS SHOWN IN LISTING 1.3
C SUBROUTINE MATMUL IS SHOWN IN LISTING 1.4
C SUBROUTINE MATADD IS SHOWN IN LISTING 1.1
C SUBROUTINE MATSUB IS SHOWN IN LISTING 1.2
```

The nominal case of Listing 7.2 was run where the measurement noise standard deviation was 1000 ft and the process noise was set to zero. We can see from Fig. 7.5 that the error in the estimate of altitude appears to be within the covariance matrix predictions (i.e., square root of P_{11}) approximately 68% of the time. Figure 7.6 also shows that the error in the estimate of velocity also appears to be within the covariance matrix bounds (i.e., square root of P_{22}) the right percentage of time. Thus, at first glance, it appears that our extended Kalman filter appears to be working properly.

Another case was run with Listing 7.2 in which the measurement noise standard deviation SIGNOISE was reduced from 1000 to 25 ft. Although we can see from Fig. 7.7 that the error in the estimate of altitude has been significantly reduced (i.e., see Fig. 7.5), we can also see that the error in the estimate of altitude

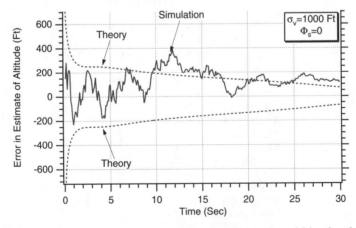

Fig. 7.5 Error in the estimate of position appears to be within the theoretical bounds.

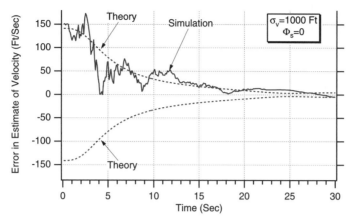

Fig. 7.6   Error in the estimate of velocity appears to be within the theoretical bounds.

is no longer within the theoretical bounds and is in fact diverging from the theoretical bounds. Similarly, Fig. 7.8 also shows that, although the error in the estimate of velocity has been reduced (i.e., see Fig. 7.6), we can no longer say that the error in the estimate of velocity is within the theoretical bounds. It, too, appears to be diverging for most of the flight. Clearly something is wrong with the design of the extended Kalman filter in this example.

Recall that the filter residual is the difference between the actual measurement and the nonlinear measurement equation. Figure 7.9 shows how the residual behaves as a function of time. Here we can see that the filter residual starts to drift from its theoretical value of zero (we will derive the fact that the residual should have zero mean in Chapter 14) after approximately 20 s. To prevent the residual from drifting, the filter should pay more attention to the measurements after 20 s. The only way for the filter to pay attention to the measurements is through the

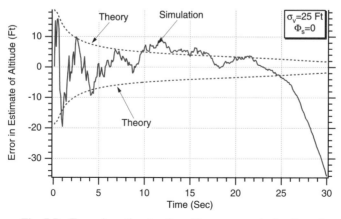

Fig. 7.7   Error in estimate of position appears to be diverging.

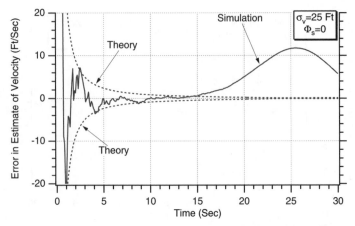

Figure 7.8   Error in estimate of velocity appears to be growing for most of the flight.

Kalman gains. A high set of gains will enable the filter to make use of the measurements, whereas a low set of gains will cause the filter to ignore the measurements. We can see from Fig. 7.10 that the first Kalman gain is very small and the second Kalman gain is approximately zero after 20 s. Thus, we can conclude that after a while the filter is ignoring the measurements, which causes the residual to drift.

We have seen in preceding chapters that the addition of process noise can often help account for the fact that our filter may not be matched to the real world. A case was run in which the spectral density of the process noise was increased from 0 to 100 when the measurement noise standard deviation was 25 ft. No attempt was made to optimize the amount of process noise required. We can see

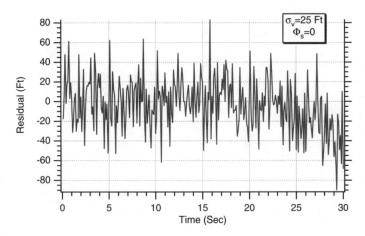

Fig. 7.9   Residual drifts from zero after 20 s.

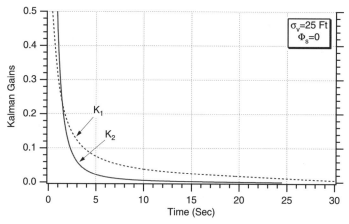

Fig. 7.10 Both Kalman gains approach zero.

from Fig. 7.11 that the addition of process noise has now made the residual zero mean. The reason for this is shown in Fig. 7.12, where the Kalman gains no longer approach zero. The larger Kalman gains (i.e., compared to the case in which there was no process noise) enable the filter to pay more attention to the measurements and thus prevent the residual from drifting off.

The elimination of the divergence problem can also be seen when we monitor the errors in the estimates of both states. We can see from Figs. 7.13 and 7.14 that the addition of process noise eliminated filter divergence. Comparing Fig. 7.7 with Fig. 7.13 shows that the error in the estimate of position has been reduced from 30 to approximately 10 ft. Comparing Fig. 7.8 with Fig. 7.14 shows that the error in the estimate of velocity remains at approximately 10 ft/s, except that now the error in the estimate of velocity is no longer behaving strangely. We also can

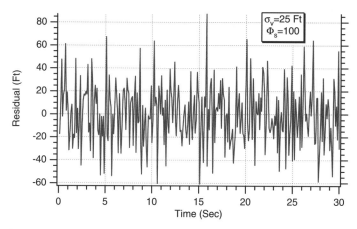

Fig. 7.11 Adding process noise prevents residual from drifting away from zero.

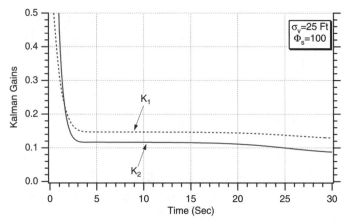

Fig. 7.12 Process noise prevents Kalman gains from going to zero.

see from Figs. 7.13 and 7.14 that the addition of process noise causes the errors in the estimates to no longer approach zero as more measurements are taken but simply to approach a steady-state value.

Second Attempt at Extended Kalman Filter

We saw in the preceding section that without process noise the extended Kalman filter's errors in the estimates diverged rather than got smaller as more measurements were taken. Adding process noise appeared to be the engineering fix for making the divergence problem go away. However, in the model of the real world for this problem there was no process noise. Therefore, the extended Kalman filter was apparently matched to the real world, and there should not have

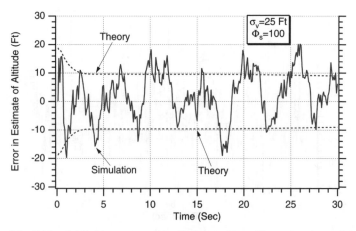

Fig. 7.13 Adding process noise eliminates filter divergence in position.

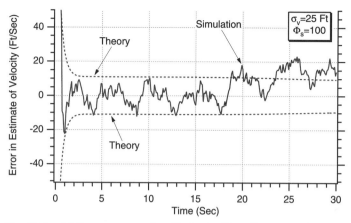

Fig. 7.14 Adding process noise eliminates filter divergence in velocity.

been any divergence. In this section we shall attempt to make the extended Kalman filter even better in an attempt to remove the divergence in the errors in the state estimates without resorting to the addition of process noise.

One possible cause of the divergence is that we only used two terms to compute the fundamental matrix. In reality, the fundamental matrix can be expressed by the infinite Taylor-series expansion

$$\Phi_k = I + FT_s + \frac{F^2 T_s^2}{2!} + \frac{F^3 T_s^3}{3!} + \cdots$$

For this example we have shown that the systems dynamics matrix is given by

$$F(t) = \begin{bmatrix} 0 & 1 \\ f_{21} & f_{22} \end{bmatrix}$$

where f_{21} and f_{22} can be written in terms of the state estimates as

$$f_{21} = \frac{-\hat{\rho} g \hat{x}^2}{44,000 \beta}$$

$$f_{22} = \frac{\hat{\rho} \hat{x} g}{\beta}$$

We previously showed that, assuming the systems dynamics matrix to be approximately constant between sampling times, the two-term Taylor-series approximation to the continuous fundamental matrix was

$$\Phi(t) = I + Ft = \begin{bmatrix} 1 & 0 \\ 0 & 1 \end{bmatrix} + \begin{bmatrix} 0 & 1 \\ f_{21} & f_{22} \end{bmatrix} t = \begin{bmatrix} 1 & t \\ f_{21}t & 1 + f_{22}t \end{bmatrix}$$

or more simply

$$\Phi(t) = \begin{bmatrix} 1 & t \\ f_{21}t & 1 + f_{22}t \end{bmatrix}$$

Therefore, the discrete fundamental matrix can be found by substituting T_s for t and is given by

$$\Phi_k = \begin{bmatrix} 1 & T_s \\ f_{21}T_s & 1 + f_{22}T_2 \end{bmatrix}$$

To get better approximations to the fundamental matrix, we first have to find F^2 and F^3 or

$$F^2 = \begin{bmatrix} 0 & 1 \\ f_{21} & f_{22} \end{bmatrix}\begin{bmatrix} 0 & 1 \\ f_{21} & f_{22} \end{bmatrix} = \begin{bmatrix} f_{21} & f_{22} \\ f_{22}f_{21} & f_{21} + f_{22}^2 \end{bmatrix}$$

$$F^3 = \begin{bmatrix} 0 & 1 \\ f_{21} & f_{22} \end{bmatrix}\begin{bmatrix} f_{21} & f_{22} \\ f_{22}f_{21} & f_{21} + f_{22}^2 \end{bmatrix} = \begin{bmatrix} f_{22}f_{21} & f_{21} + f_{22}^2 \\ f_{21}^2 + f_{22}^2 f_{21} & 2f_{22}f_{21} + f_{22}^3 \end{bmatrix}$$

Therefore, the three-term Taylor-series approximation to the fundamental matrix is given by

$$\Phi_{k_{3\,\mathrm{Term}}} = \begin{bmatrix} 1 & T_s \\ f_{21}T_s & 1 + f_{22}T_s \end{bmatrix} + \begin{bmatrix} f_{21} & f_{22} \\ f_{22}f_{21} & (f_{21} + f_{22}^2) \end{bmatrix}\frac{T_s^2}{2}$$

or more simply

$$\Phi_{k_{3\,\mathrm{Term}}} = \begin{bmatrix} 1 + f_{21}\dfrac{T_s^2}{2} & T_s + f_{22}\dfrac{T_s^2}{2} \\ f_{21}T_s + f_{22}f_{21}\dfrac{T_s^2}{2} & 1 + f_{22}T_s + (f_{21} + f_{22}^2)\dfrac{T_s^2}{2} \end{bmatrix}$$

The four-term approximation to the fundamental matrix can be found from

$$\Phi_{k_{4\,\mathrm{Term}}} = \begin{bmatrix} 1 + f_{21}\dfrac{T_s^2}{2} & T_s + f_{22}\dfrac{T_s^2}{2} \\ f_{21}T_s + f_{22}f_{21}\dfrac{T_s^2}{2} & 1 + f_{22}T_s + (f_{21} + f_{22}^2)\dfrac{T_s^2}{2} \end{bmatrix}$$
$$+ \begin{bmatrix} f_{22}f_{21} & f_{21} + f_{22}^2 \\ f_{21}^2 + f_{22}^2 f_{21} & 2f_{22}f_{21} + f_{22}^3 \end{bmatrix}\frac{T_s^3}{6}$$

Listing 7.3   Second attempt at extended Kalman filter in which more terms are added to fundamental matrix calculation

```
C THE FIRST THREE STATEMENTS INVOKE THE ABSOFT RANDOM
  NUMBER GENERATOR ON THE MACINTOSH
      GLOBAL DEFINE
              INCLUDE 'quickdraw.inc'
      END
      IMPLICIT REAL*8 (A-H)
      IMPLICIT REAL*8 (O-Z)
      REAL*8 PHI(2,2),P(2,2),M(2,2),PHIP(2,2),PHIPPHIT(2,2),GAIN(2,1)
      REAL*8 Q(2,2),HMAT(1,2),HM(1,2),MHT(2,1)
      REAL*8 PHIT(2,2)
      REAL*8 HMHT(1,1),HT(2,1),KH(2,2),IDN(2,2),IKH(2,2)
      INTEGER ORDER
      ITERM=4
      SIGNOISE=25.
      X=200000.
      XD=-6000.
      BETA=500.
      XH=200025.
      XDH=-6150.
      OPEN(1,STATUS='UNKNOWN',FILE='DATFIL')
      OPEN(2,STATUS='UNKNOWN',FILE='COVFIL')
      ORDER=2
      TS=.1
      TF=30.
      PHIS=0./TF
      T=0.
      S=0.
      H=.001
      DO 1000 I=1,ORDER
      DO 1000 J=1,ORDER
              PHI(I,J)=0.
              P(I,J)=0.
              Q(I,J)=0.
              IDN(I,J)=0.
1000  CONTINUE
      IDN(1,1)=1.
      IDN(2,2)=1.
      P(1,1)=SIGNOISE*SIGNOISE
      P(2,2)=20000.
      DO 1100 I=1,ORDER
              HMAT(1,I)=0.
              HT(I,1)=0.
1100  CONTINUE
      HMAT(1,1)=1.
      HT(1,1)=1.
      WHILE(T<=TF)
              XOLD=X
              XDOLD=XD
              XDD=.0034*32.2*XD*XD*EXP(-X/22000.)/(2.*BETA)-32.2
```

(continued)

Listing 7.3 *(Continued)*

```
X=X+H*XD
XD=XD+H*XDD
T=T+H
XDD=.0034*32.2*XD*XD*EXP(-X/22000.)/(2.*BETA)-32.2
X=.5*(XOLD+X+H*XD)
XD=.5*(XDOLD+XD+H*XDD)
S=S+H
IF(S>=(TS-.00001))THEN
S=0.
RHOH=.0034*EXP(-XH/22000.)
F21=-32.2*RHOH*XDH*XDH/(44000.*BETA)
F22=RHOH*32.2*XDH/BETA
RHOH=.0034*EXP(-XH/22000.)
F21=-32.2*RHOH*XDH*XDH/(44000.*BETA)
F22=RHOH*32.2*XDH/BETA
IF(ITERM.EQ.2)THEN
        PHI(1,1)=1.
        PHI(1,2)=TS
        PHI(2,1)=F21*TS
        PHI(2,2)=1.+F22*TS
ELSEIF(ITERM.EQ.3)THEN
        PHI(1,1)=1.+.5*TS*TS*F21
        PHI(1,2)=TS+.5*TS*TS*F22
        PHI(2,1)=F21*TS+.5*TS*TS*F21*F22
        PHI(2,2)=1.+F22*TS+.5*TS*TS*(F21+F22*F22)
ELSE
        PHI(1,1)=1.+.5*TS*TS*F21+TS*TS*TS*F22*F21/6.

        PHI(1,2)=TS+.5*TS*TS*F22+TS*TS*TS*
1           (F21+F22*F22)/6.
        PHI(2,1)=F21*TS+.5*TS*TS*F21*F22+TS*TS
1           *TS*(F21*F21+F22*F22*F21)/6.
        PHI(2,2)=1.+F22*TS+.5*TS*TS*(F21+F22
1           *F22)+TS*TS*TS*(2.*F21*F22
2           +F22**3)/6.
ENDIF
C Q Matrix Only Valid For Two Term Expansion For Fundamental Matrix
Q(1,1)=PHIS*TS*TS*TS/3.
Q(1,2)=PHIS*(TS*TS/2.+F22*TS*TS*TS/3.)
Q(2,1)=Q(1,2)
Q(2,2)=PHIS*(TS+F22*TS*TS+F22*F22*TS*TS*TS/3.)
CALL MATTRN(PHI,ORDER,ORDER,PHIT)
CALL MATMUL(PHI,ORDER,ORDER,P,ORDER,ORDER,
    PHIP)
CALL MATMUL(PHIP,ORDER,ORDER,PHIT,ORDER,
1       ORDER,PHIPPHIT)
CALL MATADD(PHIPPHIT,ORDER,ORDER,Q,M)
CALL MATMUL(HMAT,1,ORDER,M,ORDER,ORDER,HM)
CALL MATMUL(HM,1,ORDER,HT,ORDER,1,HMHT)
```

(continued)

Listing 7.3 (*Continued*)

```
                    HMHTR=HMHT(1,1)+SIGNOISE*SIGNOISE
                    HMHTRINV=1./HMHTR
                    CALL  MATMUL(M,ORDER,ORDER,HT,ORDER,1,MHT)
                    DO  150  I=1,ORDER
                       GAIN(I,1)=MHT(I,1)*HMHTRINV
150                 CONTINUE
                    CALL  MATMUL(GAIN,ORDER,1,HMAT,1,ORDER,KH)
                    CALL  MATSUB(IDN,ORDER,ORDER,KH,IKH)
                    CALL  MATMUL(IKH,ORDER,ORDER,M,ORDER,
                       ORDER,P)
                    CALL  GAUSS(XNOISE,SIGNOISE)
                    XDDB=.0034*32.2*XDH*XDH*EXP(-XH/22000.)/
1                      (2.*BETA)-32.2
                    XDB=XDH+XDDB*TS
                    XB=XH+TS*XDB
                    RES=X+XNOISE-XB
                    XH=XB+GAIN(1,1)*RES
                    XDH=XDB+GAIN(2,1)*RES
                    ERRX=X-XH
                    SP11=SQRT(P(1,1))
                    ERRXD=XD-XDH
                    SP22=SQRT(P(2,2))
                    WRITE(9,*)T,X,XH,XD,XDH
                    WRITE(1,*)T,X,XH,XD,XDH
                    WRITE(2,*)T,ERRX,SP11,-SP11,ERRXD,SP22,-SP22
          ENDIF
       END  DO
       PAUSE
       CLOSE(1)
       END

C SUBROUTINE GAUSS IS SHOWN IN LISTING 1.8
C SUBROUTINE MATTRN IS SHOWN IN LISTING 1.3
C SUBROUTINE MATMUL IS SHOWN IN LISTING 1.4
C SUBROUTINE MATADD IS SHOWN IN LISTING 1.1
C SUBROUTINE MATSUB IS SHOWN IN LISTING 1.2
```

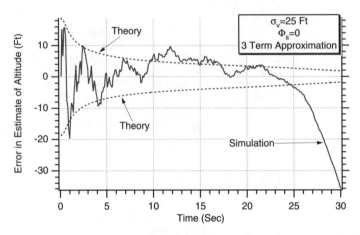

Fig. 7.15 Having three-term approximation for fundamental matrix does not remove filter divergence when there is no process noise.

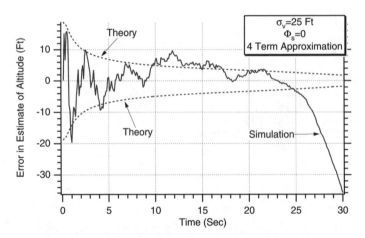

Fig. 7.16 Having four-term approximation for fundamental matrix does not remove filter divergence when there is no process noise.

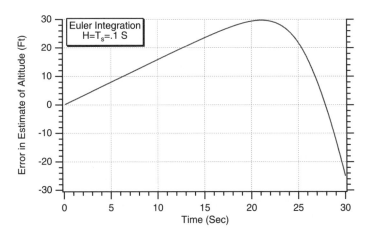

Fig. 7.17 Euler integration with an integration interval of 0.1 s is not adequate for
eliminating altitude errors.

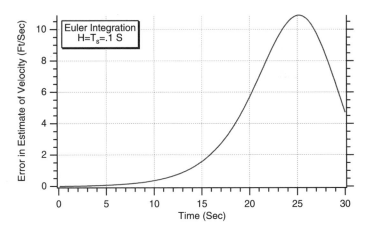

Fig. 7.18 Euler integration with an integration interval of .1 s is not adequate for
eliminating velocity errors.

Listing 7.4   Simulation to test state propagation

```
IMPLICIT REAL*8 (A-H)
IMPLICIT REAL*8 (O-Z)
X=200000.
XD=-6000.
BETA=500.
XH=X
XDH=XD
OPEN(1,STATUS='UNKNOWN',FILE='DATFIL')
TS=.1
TF=30.
T=0.
S=0.
H=.001
HP=.1
WHILE(T<=TF)
        XOLD=X
        XDOLD=XD
        XDD=.0034*32.2*XD*XD*EXP(-X/22000.)/(2.*BETA)-32.2
        X=X+H*XD
        XD=XD+H*XDD
        T=T+H
        XDD=.0034*32.2*XD*XD*EXP(-X/22000.)/(2.*BETA)-32.2
        X=.5*(XOLD+X+H*XD)
        XD=.5*(XDOLD+XD+H*XDD)
        S=S+H
        IF(S>=(TS-.00001))THEN
                S=0.
                CALL PROJECT(T,TS,XH,XDH,BETA,XB,XDB,XDDB,HP)
                XH=XB
                XDH=XDB
                ERRX=X-XH
                ERRXD=XD-XDH
                WRITE(9,*)T,ERRX,ERRXD
                WRITE(1,*)T,ERRX,ERRXD
        ENDIF
END DO
PAUSE
CLOSE(1)
END

SUBROUTINE PROJECT(TP,TS,XP,XDP,BETA,XH,XDH,XDDH,HP)
IMPLICIT REAL*8 (A-H)
IMPLICIT REAL*8 (O-Z)
T=0.
X=XP
XD=XDP
H=HP
WHILE(T<=(TS-.0001))
        XDD=.0034*32.2*XD*XD*EXP(-X/22000.)/(2.*BETA)-32.2
        XD=XD+H*XDD
```

(*continued*)

Listing 7.4 (*Continued*)

```
        X=X+H*XD
        T=T+H
END  DO
XH=X
XDH=XD
XDDH=XDD
RETURN
END
```

or more simply

$$\Phi_{k_{4\text{Term}}} = \begin{bmatrix} 1 + f_{21}\frac{T_s^2}{2} + f_{22}f_{21}\frac{T_s^3}{6} & T_s + f_{22}\frac{T_s^2}{2} + (f_{21} + f_{22}^2)\frac{T_s^3}{6} \\ f_{21}T_s + f_{22}f_{21}\frac{T_s^2}{2} + (f_{21}^2 + f_{22}^2 f_{21})\frac{T_s^3}{6} & 1 + f_{22}T_s + (f_{21} + f_{22}^2)\frac{T_s^2}{2} + (2f_{22}f_{21} + f_{22}^3)\frac{T_s^3}{6} \end{bmatrix}$$

Listing 7.2 was modified to account for the fact that we might want to have more terms in the Taylor-series expansion to approximate the fundamental matrix, and the resultant simulation appears in Listing 7.3. All changes from the original simulation are highlighted in bold. Because we are only going to run with zero

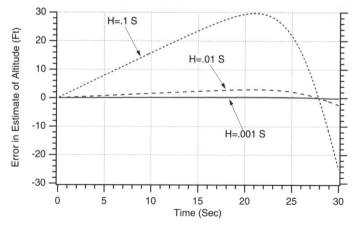

Fig. 7.19 Integration step size in propagation subroutine must be reduced to 0.001 s to keep errors in estimate of altitude near zero.

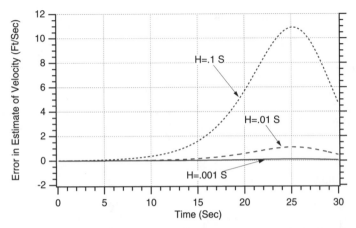

Fig. 7.20   Integration step size in propagation subroutine must be reduced to 0.001 s to keep errors in estimate of velocity near zero.

process noise, the discrete process matrix was not modified in this experiment. If ITERM = 2, we have a two-term approximation; if ITERM = 3, we have a three-term approximation; and, finally, if ITERM = 4, we have a four-term approximation to the fundamental matrix.

Cases were rerun with Listing 7.3 in which there was no process noise and 25 ft of measurement noise. We can see from Figs. 7.15 and 7.16 that having more terms in the approximation for the fundamental matrix has absolutely no influence on the filter divergence in the error in the estimate on altitude. In fact these curves are virtually identical to the results of the two-term approximation to the fundamental matrix (see Fig. 7.7). Therefore, we can conclude for this problem that the number of terms used in the series approximation for the fundamental matrix is not important and that something else must be causing the filter estimates to diverge.

Third Attempt at Extended Kalman Filter

Another possibility for the filter divergence is that perhaps our method of numerical integration in propagating the states forward from the nonlinear differential equation is not sufficiently accurate. The original program of Listing 7.2 was modified so that the initial state estimates were perfect and the Kalman gains were set to zero. Under these conditions the filter estimates should be perfect if a perfect method of integration was used because there are no errors. Therefore, the errors in the state estimates should be zero. However, we can see from Figs. 7.17 and 7.18 that the errors in the estimates of altitude and velocity are not zero under these circumstances. Therefore, the method of integration or state propagation needs improvement.

Another program was written in which the integration interval of the propagation method could be varied, and it is shown in Listing 7.4. We can see that Euler integration is used in subroutine PROJECT, which integrates the

Listing 7.5   Extended Kalman filter with accurate propagation subroutine

```
C THE FIRST THREE STATEMENTS INVOKE THE ABSOFT RANDOM
  NUMBER GENERATOR ON THE MACINTOSH
        GLOBAL DEFINE
               INCLUDE 'quickdraw.inc'
        END
        IMPLICIT REAL*8 (A-H)
        IMPLICIT REAL*8 (O-Z)
        REAL*8 PHI(2,2),P(2,2),M(2,2),PHIP(2,2),PHIPPHIT(2,2),GAIN(2,1)
        REAL*8 Q(2,2),HMAT(1,2),HM(1,2),MHT(2,1)
        REAL*8 PHIT(2,2)
        REAL*8 HMHT(1,1),HT(2,1),KH(2,2),IDN(2,2),IKH(2,2)
        INTEGER ORDER
        SIGNOISE=25.
        X=200000.
        XD=-6000.
        BETA=500.
        XH=200025.
        XDH=-6150.
        OPEN(1,STATUS='UNKNOWN',FILE='DATFIL')
        OPEN(2,STATUS='UNKNOWN',FILE='COVFIL')
        ORDER=2
        TS=.1
        TF=30.
        PHIS=0.
        T=0.
        S=0.
        H=.001

        DO 1000 I=1,ORDER
        DO 1000 J=1,ORDER
               PHI(I,J)=0.
               P(I,J)=0.
               Q(I,J)=0.
               IDN(I,J)=0.
1000 CONTINUE
        IDN(1,1)=1.
        IDN(2,2)=1.
        P(1,1)=SIGNOISE*SIGNOISE
        P(2,2)=20000.
        DO 1100 I=1,ORDER
               HMAT(1,I)=0.
               HT(I,1)=0.
1100 CONTINUE
        HMAT(1,1)=1.
        HT(1,1)=1.
        WHILE(T<=TF)
               XOLD=X
               XDOLD=XD
               XDD=.0034*32.2*XD*XD*EXP(-X/22000.)/(2.*BETA)-32.2
               X=X+H*XD
```

(continued)

Listing 7.5 (*Continued*)

```
        XD=XD+H*XDD
        T=T+H
        XDD=.0034*32.2*XD*XD*EXP(-X/22000.)/(2.*BETA)-32.2
        X=.5*(XOLD+X+H*XD)
        XD=.5*(XDOLD+XD+H*XDD)
        S=S+H
        IF(S>=(TS-.00001))THEN
                S=0.
                RHOH=.0034*EXP(-XH/22000.)
                F21=-32.2*RHOH*XDH*XDH/(44000.*BETA)
                F22=RHOH*32.2*XDH/BETA
                PHI(1,1)=1.
                PHI(1,2)=TS
                PHI(2,1)=F21*TS
                PHI(2,2)=1.+F22*TS
                Q(1,1)=PHIS*TS*TS*TS/3.
                Q(1,2)=PHIS*(TS*TS/2.+F22*TS*TS*TS/3.)
                Q(2,1)=Q(1,2)
                Q(2,2)=PHIS*(TS+F22*TS*TS+F22*F22*TS*TS*TS/3.)
                CALL MATTRN(PHI,ORDER,ORDER,PHIT)
                CALL MATMUL(PHI,ORDER,ORDER,P,ORDER,ORDER,
                  PHIP)
                CALL MATMUL(PHIP,ORDER,ORDER,PHIT,ORDER,
1                   ORDER,PHIPPHIT)
                CALL MATADD(PHIPPHIT,ORDER,ORDER,Q,M)
                CALL MATMUL(HMAT,1,ORDER,M,ORDER,ORDER,HM)
                CALL MATMUL(HM,1,ORDER,HT,ORDER,1,HMHT)
                HMHTR=HMHT(1,1)+SIGNOISE*SIGNOISE
                HMHTRINV=1./HMHTR
                CALL MATMUL(M,ORDER,ORDER,HT,ORDER,1,MHT)
                DO 150 I=1,ORDER
                        GAIN(I,1)=MHT(I,1)*HMHTRINV
150             CONTINUE
                CALL MATMUL(GAIN,ORDER,1,HMAT,1,ORDER,KH)
                CALL MATSUB(IDN,ORDER,ORDER,KH,IKH)
                CALL MATMUL(IKH,ORDER,ORDER,M,ORDER,
                  ORDER,P)
                CALL GAUSS(XNOISE,SIGNOISE)
                CALL PROJECT(T,TS,XH,XDH,BETA,XB,XDB,XDDB)
                RES=X+XNOISE-XB
                XH=XB+GAIN(1,1)*RES
                XDH=XDB+GAIN(2,1)*RES
                ERRX=X-XH
                SP11=SQRT(P(1,1))
                ERRXD=XD-XDH
                SP22=SQRT(P(2,2))
                WRITE(9,*)T,X,XH,XD,XDH
                WRITE(1,*)T,X,XH,XD,XDH
                WRITE(2,*)T,ERRX,SP11,-SP11,ERRXD,SP22,-SP22
        ENDIF
```

(*continued*)

Listing 7.5   (*Continued*)

```
END DO
PAUSE
CLOSE(1)
END

SUBROUTINE PROJECT(TP,TS,XP,XDP,BETA,XH,XDH,XDDH)
IMPLICIT REAL*8 (A-H)
IMPLICIT REAL*8 (O-Z)
T=0.
X=XP
XD=XDP
H=.001
WHILE(T<=(TS-.0001))
        XDD=.0034*32.2*XD*XD*EXP(-X/22000.)/(2.*BETA)-32.2
        XD=XD+H*XDD
        X=X+H*XD
        T=T+H
END DO
XH=X
XDH=XD
XDDH=XDD
RETURN
END
C SUBROUTINE GAUSS IS SHOWN IN LISTING 1.8
C SUBROUTINE MATTRN IS SHOWN IN LISTING 1.3
C SUBROUTINE MATMUL IS SHOWN IN LISTING 1.4
C SUBROUTINE MATADD IS SHOWN IN LISTING 1.1
C SUBROUTINE MATSUB IS SHOWN IN LISTING 1.2
```

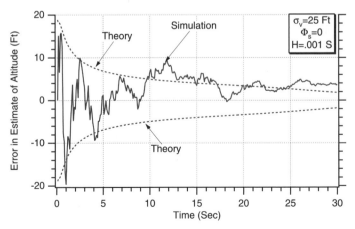

Fig. 7.21 Divergence has been eliminated in altitude estimate by use of more accurate state propagation methods.

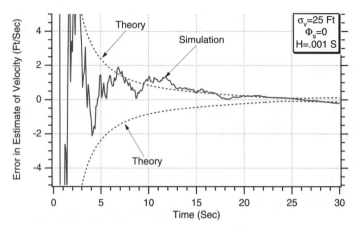

Fig. 7.22 Divergence has been eliminated in velocity estimate by use of more accurate state propagation methods.

nonlinear differential equations forward one sampling interval to produce the estimates. Second-order Runge–Kutta numerical integration with an integration step size of 0.001 s is used to propagate the actual states forward. The estimates are initially set equal to the actual states so that a perfect propagation subroutine would yield perfect state estimates. The parameter HP determines the integration interval used by the subroutine. When the integration interval HP is small enough the errors in the estimates of altitude ERRX and velocity ERRXD should go to zero.

Cases were run with Listing 7.4 in which the subroutine integration interval HP was made a parameter and varied from 0.1 to 0.001 s. We can see from Figs. 7.19 and 7.20 that a value of 0.001 s must be used for the subroutine integration interval to keep the errors in the estimates of altitude and velocity near zero.

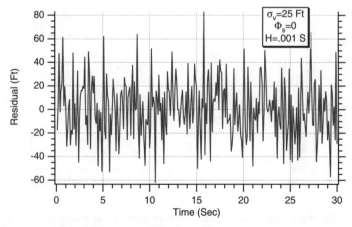

Fig. 7.23 More accurate state propagation ensures residual has zero mean even though Kalman gains approach zero.

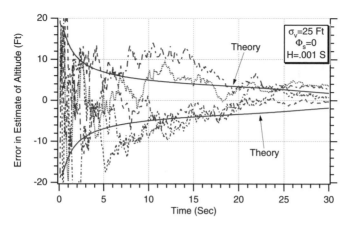

Fig. 7.24 Monte Carlo results are within theoretical bounds for error in estimate of position.

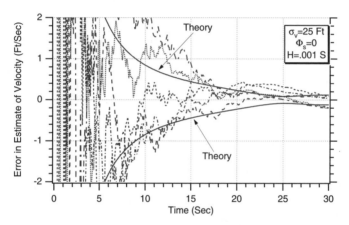

Fig. 7.25 Monte Carlo results are within theoretical bounds for error in estimate of velocity.

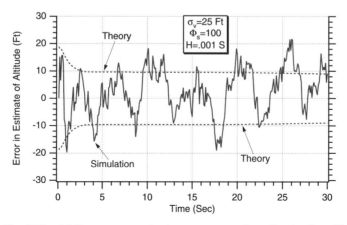

Fig. 7.26 Adding process noise increases error in estimate of position.

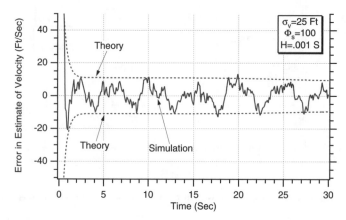

Fig. 7.27   Adding process noise increases error in estimate of velocity.

Therefore, to avoid additional errors in our extended Kalman filter, we will use a value of 0.001 s for the integration interval in the propagation subroutine.

The extended Kalman filter simulation of Listing 7.2 was modified to account for the more accurate method of integration required to propagate the states between measurements and appears in Listing 7.5. The changes to the original simulation are highlighted in bold.

The nominal case of Listing 7.5 was run in which there was 25 ft of measurement noise and no process noise. The integration interval was set to 0.001 s in the integration method for projecting the states (see subroutine PROJECT). We can see from Figs. 7.21–7.22 that the preceding divergence problem in the errors in the estimates of altitude and velocity has now been eliminated. This simple experiment shows that the accuracy of the propagation method for the filter states is extremely important when there is little or no process noise.

We can also see from Fig. 7.23 that the more accurate state propagation ensures that the residual has an average value of zero. This is now true even though the Kalman gains approach zero (see Fig. 7.10).

Although it was clear (at least to the authors) in Figs. 7.21 and 7.22 that the divergence problem disappeared when accurate state propagation was used, some might not be convinced from those single-run results. Listing 7.5 was modified so that it could be run in the Monte Carlo mode for this particular example. The results of Figs. 7.21 and 7.22 were repeated for five different runs, and the error in the estimate results appears in Figs. 7.24 and 7.25. From these results it is clearer that the simulated errors in the estimates of altitude and velocity are within the theoretical bounds approximately 68% of the time. In particular, note that the scale on Fig. 7.25 has been expanded to see how close the error in the estimate of velocity is to the theoretical values near the end of the measurements at 30 s.

Another case was run to see how the new filter with the more accurate state propagation methods performed in the presence of process noise. The process noise used was $\Phi_s = 100$, which is identical to what was used before to solve the divergence problem. We can see from Figs. 7.26 and 7.27 that the errors in the estimates of altitude and velocity are approximately what they were before (see Figs. 7.13 and 7.14) when process noise was used to solve the divergence problem. However, decreasing the integration step size from 0.1 to 0.001 s for the state propagation is considerably more costly in terms of computer throughput than simply using a step size of 0.1 s with process noise.

Summary

The extended Kalman-filtering equations were presented in this chapter, and a numerical example was used to illustrate the operation and characteristics of the filter. The extended Kalman filter was intentionally designed and tested in different ways to illustrate things that can go wrong. A divergence problem was presented, and it was shown that either more accurate integration in the state propagation or the addition of process noise were legitimate methods for improving filter performance. Also, adding more terms to the Taylor-series expansion for the fundamental matrix not only did not solve the divergence problem but, for this example, hardly had any influence on the results.

References

[1] Gelb., A., *Applied Optimal Estimation*, Massachusetts Inst. of Technology Press, Cambridge, MA, 1974, pp. 180–228.

[2] Zarchan, P., *Tactical and Strategic Missile Guidance*, 3rd ed., Progress in Astronautics and Aeronautics, AIAA, Reston, VA, 1998, pp. 373–387.

Chapter 8

Drag and Falling Object

Introduction

IN THE preceding chapter we introduced extended Kalman filtering by trying to estimate the altitude and velocity of a falling object under the influence of drag. It was assumed in Chapter 7 that the amount of drag acting on the object was known in advance via knowledge of the object's ballistic coefficient. In that example we derived a two-state extended Kalman filter in which it was assumed that there were noisy measurements of the object's altitude and that the altitude and velocity of the object had to be estimated. In this chapter we will add further complexity to the problem by assuming that the drag or ballistic coefficient is unknown. Therefore, in this problem we desire to build another extended Kalman filter that will estimate the object's altitude, velocity, and ballistic coefficient based on noisy measurements of the object's altitude.

Problem Setup

Let us again reconsider the one-dimensional example of an object falling on a tracking radar, as was shown in Fig. 7.1 but is now redrawn for convenience in Fig. 8.1. This time the problem is different because we are assuming that the drag is unknown, which means that we have to estimate the ballistic coefficient in order to get a good estimate of the position and velocity of the object. Recall that the object was initially at 200,000 ft above the radar and had a velocity of 6000 ft/s toward the radar, which is located on the surface of a flat Earth. As was the case in Chapters 4 and 7, let us pretend that the radar measures the range from the radar to the object (i.e., altitude of the object) with a 25-ft standard deviation measurement accuracy. The radar takes measurements 10 times a second for 30 s. We would like to build a filter to estimate the altitude, velocity, and ballistic coefficient of the object.

As was the case in Chapter 7, if x is the distance from the radar to the object, then the acceleration acting on the object consists of gravity and drag or

$$\ddot{x} = \text{Drag} - g = \frac{Q_p g}{\beta} - g$$

where β is the ballistic coefficient of the object. Recall that the ballistic coefficient, which is considered to be a constant in this problem, is a term

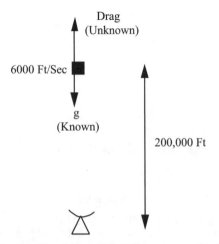

Fig. 8.1 Radar tracking falling object in presence of unknown drag.

describing the amount of drag acting on the object. In Chapter 7 we assumed that β was known in advance. The dynamic pressure Q_p in the preceding equation is given by

$$Q_p = 0.5\rho\dot{x}^2$$

where ρ is the air density. As was the case in Chapter 7, we can still assume that the air density is an exponential function of altitude or

$$\rho = 0.0034e^{-x/22000}$$

Therefore, the acceleration acting on the object can be expressed as the nonlinear second-order differential equation

$$\ddot{x} = \frac{Q_p g}{\beta} - g = \frac{0.5g\rho\dot{x}^2}{\beta} - g = \frac{0.0034ge^{-x/22000}\dot{x}^2}{2\beta} - g$$

Because the ballistic coefficient of the object we are tracking is unknown, we can make it a state. Under these circumstances it is necessary to estimate the altitude, velocity, and ballistic coefficient of the object in order to track it.[1,2] Thus, the states for the proposed filter are given by

$$x = \begin{bmatrix} x \\ \dot{x} \\ \beta \end{bmatrix}$$

Therefore, the acceleration acting on the object is a nonlinear function of the states according to

$$\ddot{x} = \frac{0.0034ge^{-x/22000}\dot{x}^2}{2\beta} - g$$

In addition, if we believe that the ballistic coefficient is constant, that means its derivative must be zero or

$$\dot{\beta} = 0$$

However, to make the resultant filter more robust we can add process noise to the derivative of the ballistic coefficient (i.e., the ballistic coefficient may not be a constant). Therefore, the equation for the derivative of the ballistic coefficient becomes

$$\dot{\beta} = u_s$$

where u_s is white process noise. We can linearize the preceding two differential equations by saying that

$$
\begin{bmatrix} \Delta\dot{x} \\ \Delta\ddot{x} \\ \Delta\dot{\beta} \end{bmatrix} =
\begin{bmatrix}
\dfrac{\partial\dot{x}}{\partial x} & \dfrac{\partial\dot{x}}{\partial \dot{x}} & \dfrac{\partial\dot{x}}{\partial \beta} \\[2mm]
\dfrac{\partial\ddot{x}}{\partial x} & \dfrac{\partial\ddot{x}}{\partial \dot{x}} & \dfrac{\partial\ddot{x}}{\partial \beta} \\[2mm]
\dfrac{\partial\dot{\beta}}{\partial x} & \dfrac{\partial\dot{\beta}}{\partial \dot{x}} & \dfrac{\partial\dot{\beta}}{\partial \beta}
\end{bmatrix}
\begin{bmatrix} \Delta x \\ \Delta\dot{x} \\ \Delta\beta \end{bmatrix} +
\begin{bmatrix} 0 \\ 0 \\ u_s \end{bmatrix}
$$

We can see from the preceding state-space differential equation that the systems dynamics matrix is the matrix of partial derivatives or

$$
F =
\begin{bmatrix}
\dfrac{\partial\dot{x}}{\partial x} & \dfrac{\partial\dot{x}}{\partial \dot{x}} & \dfrac{\partial\dot{x}}{\partial \beta} \\[2mm]
\dfrac{\partial\ddot{x}}{\partial x} & \dfrac{\partial\ddot{x}}{\partial \dot{x}} & \dfrac{\partial\ddot{x}}{\partial \beta} \\[2mm]
\dfrac{\partial\dot{\beta}}{\partial x} & \dfrac{\partial\dot{\beta}}{\partial \dot{x}} & \dfrac{\partial\dot{\beta}}{\partial \beta}
\end{bmatrix}_{x=\hat{x}}
$$

Here we can see that the partial derivatives are evaluated at the current state estimates. We can also see from the linearized state-space equation that the continuous process-noise matrix is given by

$$Q = E(ww^T)$$

Therefore, we can say that for this example the continuous process-noise matrix is

$$
Q(t) =
\begin{bmatrix}
0 & 0 & 0 \\
0 & 0 & 0 \\
0 & 0 & \Phi_s
\end{bmatrix}
$$

where Φ_s is the spectral density of the white noise source assumed to be on the derivative of the ballistic coefficient. To evaluate the systems dynamics matrix, we must take the necessary partial derivatives of the nonlinear state-space equation. The first row of partial derivatives yields

$$\frac{\partial \dot{x}}{\partial x} = 0$$

$$\frac{\partial \dot{x}}{\partial \dot{x}} = 1$$

$$\frac{\partial \dot{x}}{\partial \beta} = 0$$

Whereas the second row of partial derivatives yields

$$\frac{\partial \ddot{x}}{\partial x} = \frac{-0.0034e^{-x/22000}\dot{x}^2 g}{2\beta(22,000)} = \frac{-\rho g \dot{x}^2}{44,000\beta}$$

$$\frac{\partial \ddot{x}}{\partial \dot{x}} = \frac{2*0.0034e^{-x/22000}\dot{x}g}{2\beta} = \frac{\rho g \dot{x}}{\beta}$$

$$\frac{\partial \ddot{x}}{\partial \beta} = \frac{-0.0034e^{-x/22000}\dot{x}^2 g}{2\beta^2} = \frac{-\rho g \dot{x}^2}{2\beta^2}$$

and, finally, the last row of partial derivatives yields

$$\frac{\partial \dot{\beta}}{\partial x} = 0$$

$$\frac{\partial \dot{\beta}}{\partial \dot{x}} = 0$$

$$\frac{\partial \dot{\beta}}{\partial \beta} = 0$$

Therefore, the systems dynamics matrix turns out to be

$$F = \begin{bmatrix} 0 & 1 & 0 \\ \dfrac{-\hat{\rho} g \hat{\dot{x}}^2}{44,000\hat{\beta}} & \dfrac{\hat{\rho} g \hat{\dot{x}}}{\hat{\beta}} & \dfrac{-\hat{\rho} g \hat{\dot{x}}^2}{2\hat{\beta}^2} \\ 0 & 0 & 0 \end{bmatrix}$$

where gravity g is assumed to be known perfectly, but this time the ballistic coefficient β must be estimated. The estimated air density in the preceding expression depends on the estimated altitude and is given by

$$\hat{\rho} = 0.0034e^{-\hat{x}/22000}$$

The fundamental matrix can be found by a Taylor-series approximation, yielding

$$\Phi(t) = I + Ft + \frac{F^2 t^2}{2!} + \frac{F^3 t^3}{3!} + \cdots$$

If we define

$$f_{21} = \frac{-\hat{\rho}g\hat{x}^2}{44{,}000\beta}$$

$$f_{22} = \frac{\hat{\rho}\hat{x}g}{\beta}$$

$$f_{23} = \frac{-\hat{\rho}g\hat{x}^2}{2\hat{\beta}^2}$$

then the systems dynamics matrix can be written as

$$F(t) = \begin{bmatrix} 0 & 1 & 0 \\ f_{21} & f_{22} & f_{23} \\ 0 & 0 & 0 \end{bmatrix}$$

Because the fundamental matrix will only be used in the Riccati equations but will not be used in state propagation, we will only take the first two terms of the Taylor-series expansion for the fundamental matrix and obtain

$$\Phi(t) \approx I + Ft = \begin{bmatrix} 1 & 0 & 0 \\ 0 & 1 & 0 \\ 0 & 0 & 1 \end{bmatrix} + \begin{bmatrix} 0 & 1 & 0 \\ f_{21} & f_{22} & f_{23} \\ 0 & 0 & 0 \end{bmatrix} t$$

$$= \begin{bmatrix} 1 & t & 0 \\ f_{21}t & 1 + f_{22}t & f_{23}t \\ 0 & 0 & 1 \end{bmatrix}$$

or, more simply,

$$\Phi(t) \approx \begin{bmatrix} 1 & t & 0 \\ f_{21}t & 1 + f_{22}t & f_{23}t \\ 0 & 0 & 1 \end{bmatrix}$$

Therefore, the discrete fundamental matrix can be found by substituting T_s for t and is given by

$$\Phi_k \approx \begin{bmatrix} 1 & T_s & 0 \\ f_{21}T_s & 1 + f_{22}T_s & f_{23}T_s \\ 0 & 0 & 1 \end{bmatrix}$$

In this example the altitude measurement is a linear function of the states and is given by

$$x^* = \begin{bmatrix} 1 & 0 & 0 \end{bmatrix} \begin{bmatrix} x \\ \dot{x} \\ \beta \end{bmatrix} + v$$

where v is the measurement noise. From the preceding equation the measurement noise matrix can be written by inspection as

$$H = \begin{bmatrix} 1 & 0 & 0 \end{bmatrix}$$

Because the discrete measurement noise matrix is by definition

$$R_k = E(v_k v_k^T)$$

we can say that, for this problem, the discrete measurement noise matrix turns out to be a scalar and is given by

$$R_k = \sigma_v^2$$

where σ_v^2 is the variance of the altitude noise. Recall that the continuous process-noise matrix is

$$Q(t) = \begin{bmatrix} 0 & 0 & 0 \\ 0 & 0 & 0 \\ 0 & 0 & \Phi_s \end{bmatrix}$$

Therefore, the discrete process-noise matrix can be derived from the continuous process-noise matrix according to

$$Q_k = \int_0^{T_s} \Phi(\tau) Q \Phi^T(\tau) \, d\tau$$

Substitution of the appropriate matrices into the preceding expression yields

$$Q_k = \Phi_s \int_0^{T_s} \begin{bmatrix} 1 & \tau & 0 \\ f_{21}\tau & 1+f_{22}\tau & f_{23}\tau \\ 0 & 0 & 1 \end{bmatrix} \begin{bmatrix} 0 & 0 & 0 \\ 0 & 0 & 0 \\ 0 & 0 & 1 \end{bmatrix} \begin{bmatrix} 1 & f_{21}\tau & 0 \\ \tau & 1+f_{22}\tau & 0 \\ 0 & f_{23}\tau & 1 \end{bmatrix} d\tau$$

After some algebra we obtain

$$Q_k = \Phi_s \int_0^{T_s} \begin{bmatrix} 0 & 0 & 0 \\ 0 & f_{23}^2\tau^2 & f_{23}\tau \\ 0 & f_{23}\tau & 1 \end{bmatrix} d\tau$$

Finally, after integration we get the final expression for the discrete process-noise matrix:

$$Q_k = \Phi_s \begin{bmatrix} 0 & 0 & 0 \\ 0 & f_{23}^2 \dfrac{T_s^3}{3} & f_{23} \dfrac{T_s^2}{2} \\ 0 & f_{23} \dfrac{T_s^2}{2} & T_s \end{bmatrix}$$

We now have defined all of the matrices required to solve the Riccati equations. The next step is to write down the equations for the Kalman-filter section. First we must be able to propagate the states from the present sampling time to the next sampling time. We learned in Chapter 7 that the best way to do this is to integrate, using Euler integration with a small step size, the nonlinear differential equations until the next update. The results of the integration will yield the projected position and velocity. In this example the nonlinear differential equation to be integrated is

$$\ddot{\bar{x}}_{k-1} = \frac{0.0034 g e^{-\hat{x}_{k-1}/22000} \hat{\dot{x}}_{k-1}^2}{2\hat{\beta}_{k-1}} - g$$

Therefore, the Kalman-filtering equations can be written as

$$\hat{x}_k = \bar{x}_k + K_{1_k}(x_k^* - \bar{x}_k)$$
$$\hat{\dot{x}}_k = \bar{\dot{x}}_k + K_{2_k}(x_k^* - \bar{x}_k)$$
$$\hat{\beta}_k = \hat{\beta}_{k-1} + K_{3_k}(x_k^* - \bar{x}_k)$$

where x_k^* is the noisy measurement of altitude and the barred quantities are to be obtained by using Euler numerical integration with a small integration step size.

The preceding equations for the three-state extended Kalman filter and Riccati equations were programmed and are shown in Listing 8.1, along with a simulation of the real world. We can see that the process-noise matrix is set to zero in this listing (i.e., PHIS = 0) and that the third diagonal element of the initial covariance matrix has been set to 300^2 to reflect the uncertainty in our initial guess at the estimated ballistic coefficient. We have initialized the states of the filter close to the true values. The position states are in error by 25 ft, and the velocity states are in error by 150 ft/s. The initial covariance matrix reflects those errors. As was the case near the end of Chapter 7, we now have the subroutine PROJECT to numerically integrate the nonlinear acceleration equation over the sampling interval. We can see that this subroutine uses Euler integration with an integration step size of 0.001 s (i.e., HP = 0.001).

A nominal case was run with Listing 8.1 in which the process noise was zero and the single-run simulation results for the errors in the estimates of altitude, velocity, and ballistic coefficient appear in Figs. 8.2–8.4. We can see from Fig. 8.2 that the error in the estimate of altitude is within the theoretical bounds.

Listing 8.1   Simulation of extended Kalman filter to estimate ballistic coefficient

```
C THE FIRST THREE STATEMENTS INVOKE THE ABSOFT RANDOM
  NUMBER GENERATOR ON THE MACINTOSH
      GLOBAL  DEFINE
              INCLUDE 'quickdraw.inc'
      END
      IMPLICIT REAL*8 (A-H)
      IMPLICIT REAL*8 (O-Z)
      REAL*8 PHI(3,3),P(3,3),M(3,3),PHIP(3,3),PHIPPHIT(3,3),GAIN(3,1)
      REAL*8 Q(3,3),HMAT(1,3),HM(1,3),MHT(3,1)
      REAL*8 PHIT(3,3)
      REAL*8 HMHT(1,1),HT(3,1),KH(3,3),IDN(3,3),IKH(3,3)
      INTEGER ORDER
      ITERM=1
      SIGNOISE=25.
      X=200000.
      XD=-6000.
      BETA=500.
      XH=200025.
      XDH=-6150.
      BETAH=800.
      OPEN(2,STATUS='UNKNOWN',FILE='COVFIL')
      OPEN(1,STATUS='UNKNOWN',FILE='DATFIL')
      ORDER=3
      TS=.1
      TF=30.
      PHIS=0.
      T=0.
      S=0.
      H=.001
      HP=.001
      DO 1000 I=1,ORDER
      DO 1000 J=1,ORDER
              PHI(I,J)=0.
              P(I,J)=0.
              Q(I,J)=0.
              IDN(I,J)=0.
1000 CONTINUE
      IDN(1,1)=1.
      IDN(2,2)=1.
      IDN(3,3)=1.
      P(1,1)=SIGNOISE*SIGNOISE
      P(2,2)=20000.
      P(3,3)=300.**2
      DO 1100 I=1,ORDER
              HMAT(1,I)=0.
              HT(I,1)=0.
1100 CONTINUE
      HMAT(1,1)=1.
      HT(1,1)=1.
```

(continued)

Listing 8.1 (*Continued*)

```
WHILE(T<=TF)
XOLD=X
XDOLD=XD
XDD=.0034*32.2*XD*XD*EXP(-X/22000.)/(2.*BETA)-32.2
X=X+H*XD
XD=XD+H*XDD
T=T+H
XDD=.0034*32.2*XD*XD*EXP(-X/22000.)/(2.*BETA)-32.2
X=.5*(XOLD+X+H*XD)
XD=.5*(XDOLD+XD+H*XDD)
S=S+H
IF(S>=(TS-.00001))THEN
        S=0.
        RHOH=.0034*EXP(-XH/22000.)
        F21=-32.2*RHOH*XDH/(44000.*BETAH)
        F22=RHOH*32.2*XDH/BETAH
        F23=-RHOH*32.2*XDH*XDH/(2.*BETAH*BETAH)
        IF(ITERM.EQ.1)THEN
                PHI(1,1)=1.
                PHI(1,2)=TS
                PHI(2,1)=F21*TS
                PHI(2,2)=1.+F22*TS
                PHI(2,3)=F23*TS
                PHI(3,3)=1.
        ELSE
                PHI(1,1)=1.+.5*TS*TS*F21
                PHI(1,2)=TS+.5*TS*TS*F22
                PHI(1,3)=.5*TS*TS*F23
                PHI(2,1)=F21*TS+.5*TS*TS*F22*F21
                PHI(2,2)=1.+F22*TS+.5*TS*TS*(F21+F22*F22)
                PHI(2,3)=F23*TS+.5*TS*TS*F22*F23
                PHI(3,3)=1.
        ENDIF
C THIS PROCESS NOISEMATRIX ONLY VALID FOR TWO TERM
APPROXIMATION TO FUNDAMNETAL MATRIX
                Q(2,2)=F23*F23*PHIS*TS*TS*TS/3.
                Q(2,3)=F23*PHIS*TS*TS/2.
                Q(3,2)=Q(2,3)
                Q(3,3)=PHIS*TS
                CALL  MATTRN(PHI,ORDER,ORDER,PHIT)
                CALL  MATMUL(PHI,ORDER,ORDER,P,ORDER,ORDER,
                PHIP)
                CALL  MATMUL(PHIP,ORDER,ORDER,PHIT,ORDER,ORDER,
      1          PHIPPHIT)
                CALL  MATADD(PHIPPHIT,ORDER,ORDER,Q,M)
                CALL  MATMUL(HMAT,1,ORDER,M,ORDER,ORDER,HM)
                CALL  MATMUL(HM,1,ORDER,HT,ORDER,1,HMHT)
                HMHTR=HMHT(1,1)+SIGNOISE*SIGNOISE
                HMHTRINV=1./HMHTR
```

(*continued*)

Listing 8.1 *(Continued)*

```
                CALL  MATMUL(M,ORDER,ORDER,HT,ORDER,1,MHT)
                DO  150  I=1,ORDER
                    GAIN(I,1)=MHT(I,1)*HMHTRINV
150             CONTINUE
                CALL  MATMUL(GAIN,ORDER,1,HMAT,1,ORDER,KH)
                CALL  MATSUB(IDN,ORDER,ORDER,KH,IKH)
                CALL  MATMUL(IKH,ORDER,ORDER,M,ORDER,ORDER,P)
                CALL  GAUSS(XNOISE,SIGNOISE)
                CALL  PROJECT(T,TS,XH,XDH,BETAH,XB,XDB,XDDB,HP)
                RES=X+XNOISE-XB
                XH=XB+GAIN(1,1)*RES
                XDH=XDB+GAIN(2,1)*RES
                BETAH=BETAH+GAIN(3,1)*RES

                ERRX=X-XH
                SP11=SQRT(P(1,1))
                ERRXD=XD-XDH
                SP22=SQRT(P(2,2))
                ERRBETA=BETA-BETAH
                SP33=SQRT(P(3,3))
                WRITE(9,*)T,X,XH,XD,XDH,BETA,BETAH
                WRITE(1,*)T,X,XH,XD,XDH,BETA,BETAH
                WRITE(2,*)T,ERRX,SP11,-SP11,ERRXD,SP22,-SP22,
1                   ERRBETA,SP33,-SP33
            ENDIF
        END  DO
        PAUSE
        CLOSE(1)
        END

        SUBROUTINE  PROJECT(TP,TS,XP,XDP,BETAP,XH,XDH,XDDH,HP)
        IMPLICIT  REAL*8 (A-H)
        IMPLICIT  REAL*8 (O-Z)
        T=0.
        X=XP
        XD=XDP
        BETA=BETAP
        H=HP
        WHILE(T<=(TS-.0001))
            XDD=.0034*32.2*XD*XD*EXP(-X/22000.)/(2.*BETA)-32.2
            XD=XD+H*XDD
            X=X+H*XD
            T=T+H
        END  DO
        XH=X
        XDH=XD
        XDDH=XDD
        RETURN
        END
```

(continued)

Listing 8.1 *(Continued)*

```
C SUBROUTINE GAUSS IS SHOWN IN LISTING 1.8
C SUBROUTINE MATTRN IS SHOWN IN LISTING 1.3
C SUBROUTINE MATMUL IS SHOWN IN LISTING 1.4
C SUBROUTINE MATADD IS SHOWN IN LISTING 1.1
C SUBROUTINE MATSUB IS SHOWN IN LISTING 1.2
```

However, Figs. 8.3 and 8.4 indicate that there appear to be hangoff errors in the error in the estimates of velocity and ballistic coefficient. This is not a filter divergence problem because the error is not growing with time. However, we can see that there is a difference between what the covariance matrix is indicating and what is actually being achieved. Theory says that if there is no process noise, the errors in the estimates should continually decrease as more measurements are taken. We can see from Figs. 8.3 and 8.4 that there appears to be a lower limit on how small we can make the errors in the estimates. In addition, Fig. 8.4 indicates that it takes approximately 10 s before the errors in the estimates of ballistic coefficient start to decrease. The reason for this is that at the high altitudes (i.e., 200 kft) there is very little drag making the ballistic coefficient unobservable. Only as the object descends in altitude (i.e., time is increasing) and there is more drag are we able to get more accurate estimates of the ballistic coefficient.

Figure 8.5 expands the scales of Fig. 8.4 to get a more detailed understanding of how the errors in the estimate of ballistic coefficient reduce as time increases (i.e., altitude decreases). We can see from Fig. 8.5 that, although the errors reduce, there is also definitely hangoff error in this state (i.e., error does not go to zero). We can see that there is a hangoff error of approximately $3 \, lb/ft^2$ in the ballistic coefficient estimate.

An alternative hypothesis is that perhaps the reason for the hangoff error in the velocity and ballistic coefficient estimates was that there were an insufficient

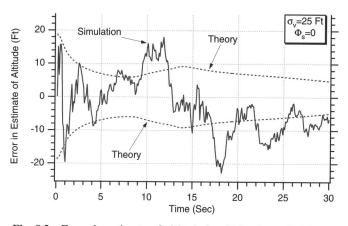

Fig. 8.2   Error in estimate of altitude is within theoretical bounds.

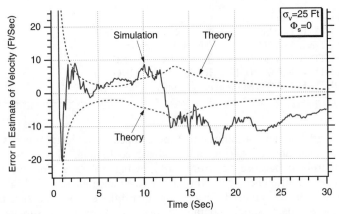

Fig. 8.3   Error in estimate of velocity appears to have hangoff error.

number of terms in the Taylor-series expansion used in obtaining the fundamental matrix. Recall that the systems dynamics matrix was given by

$$F(t) = \begin{bmatrix} 0 & 1 & 0 \\ f_{21} & f_{22} & f_{23} \\ 0 & 0 & 0 \end{bmatrix}$$

where the terms f_{21}, f_{22}, and f_{23} have already been defined. If we now use the first three terms of a Taylor series involving the systems dynamics matrix to define the fundamental matrix, we get

$$\Phi(t) = I + Ft + \frac{F^2 t^2}{2!}$$

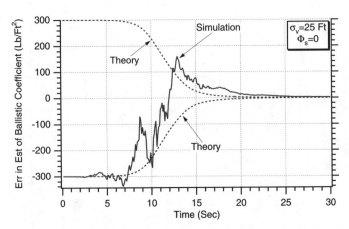

Fig. 8.4   It takes more than 10 s to reduce errors in estimating ballistic coefficient.

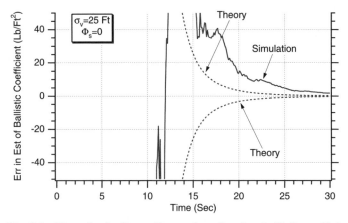

Fig. 8.5 There is also hangoff error in estimating ballistic coefficient.

If we define the two-term fundamental matrix as Φ_2, then the three-term fundamental matrix can be written as

$$\Phi_3 = \Phi_2 + \frac{F^2 t^2}{2}$$

or

$$\Phi_3 = \begin{bmatrix} 1 & t & 0 \\ f_{21}t & 1+f_{22}t & f_{23}t \\ 0 & 0 & 1 \end{bmatrix} + \frac{F^2 t^2}{2}$$

We can find the square of the systems dynamics matrix from multiplication as

$$F^2 = \begin{bmatrix} 0 & 1 & 0 \\ f_{21} & f_{22} & f_{23} \\ 0 & 0 & 0 \end{bmatrix}\begin{bmatrix} 0 & 1 & 0 \\ f_{21} & f_{22} & f_{23} \\ 0 & 0 & 0 \end{bmatrix} = \begin{bmatrix} f_{21} & f_{22} & f_{23} \\ f_{22}f_{21} & f_{21}+f_{22}^2 & f_{22}f_{23} \\ 0 & 0 & 0 \end{bmatrix}$$

Substitution of the preceding expression into the definition of the three-term Taylor-series expansion yields

$$\Phi_3 = \begin{bmatrix} 1 & t & 0 \\ f_{21}t & 1+f_{22}t & f_{23}t \\ 0 & 0 & 1 \end{bmatrix} + \begin{bmatrix} f_{21}\dfrac{t^2}{2} & f_{22}\dfrac{t^2}{2} & f_{23}\dfrac{t^2}{2} \\ f_{22}f_{21}\dfrac{t^2}{2} & (f_{21}+f_{22}^2)\dfrac{t^2}{2} & f_{22}f_{23}\dfrac{t^2}{2} \\ 0 & 0 & 0 \end{bmatrix}$$

After simplification we get

$$
\Phi_3 =
\begin{bmatrix}
1 + f_{21}\dfrac{t^2}{2} & t + f_{22}\dfrac{t^2}{2} & f_{23}\dfrac{t^2}{2} \\[2ex]
f_{21}t + f_{22}f_{21}\dfrac{t^2}{2} & 1 + f_{22}t + (f_{21} + f_{22}^2)\dfrac{t^2}{2} & f_{23}t + f_{22}f_{23}\dfrac{t^2}{2} \\[2ex]
0 & 0 & 1
\end{bmatrix}
$$

To find the new discrete fundamental matrix, we simply substitute the sampling time for time or

$$
\Phi_{3_k} =
\begin{bmatrix}
1 + f_{21}\dfrac{T_s^2}{2} & T_s + f_{22}\dfrac{T_s^2}{2} & f_{23}\dfrac{T_s^2}{2} \\[2ex]
f_{21}T_s + f_{22}f_{21}\dfrac{T_s^2}{2} & 1 + f_{22}T_s + (f_{21} + f_{22}^2)\dfrac{T_s^2}{2} & f_{23}T_s + f_{22}f_{23}\dfrac{T_s^2}{2} \\[2ex]
0 & 0 & 1
\end{bmatrix}
$$

In theory we should also modify the discrete process-noise matrix because the fundamental matrix has changed. However, because we will run with zero process noise, this will not be necessary. A case was run with the new fundamental matrix by setting ITERM = 2 in Listing 8.1. Figure 8.6 displays the resultant error in the estimate of velocity as a function of time. We can see that adding an extra term to the approximation for the fundamental matrix had no influence on removing the hangoff error.

It was then hypothesized that perhaps the integration interval used to project the states forward was not small enough. The integration step size HP was reduced in subroutine PROJECT from 0.001 to 0.0001 s in Listing 8.1. We see from Fig. 8.7 that reducing the integration interval for the method of state propagation also has no effect in reducing the hangoff error.

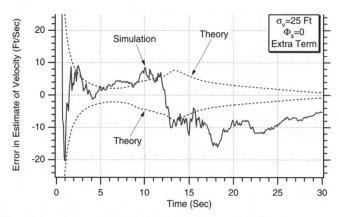

Fig. 8.6 Adding extra term to fundamental matrix does not remove hangoff error.

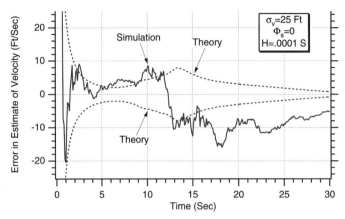

Fig. 8.7   Making integration interval 10 times smaller in PROJECT subroutine does not remove hangoff error in velocity estimate.

As the object descends in altitude, there is more drag, and the object becomes more observable from a filtering point of view. However, because there is no process noise the filter gains will go to zero. This means that the filter will stop paying attention to the measurements (i.e., when the ballistic coefficient is most observable) and hangoff error will result, as can be seen in Fig. 8.5. Finally, process noise was added to the extended Kalman filter. Figures 8.8–8.10 show how the errors in the estimates change with the addition of process noise. For example, we can see from Fig. 8.8 that using a value of 3000 (i.e., $\Phi_s = 3000$) increases slightly the error in the estimate of altitude. Figures 8.9 and 8.10 show how the addition of process noise eliminates the hangoff error in the velocity and

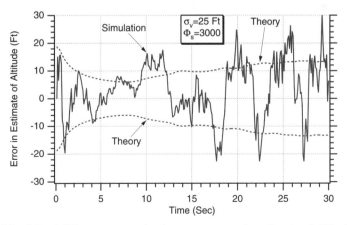

Fig. 8.8   Adding process noise increases errors in estimate of altitude.

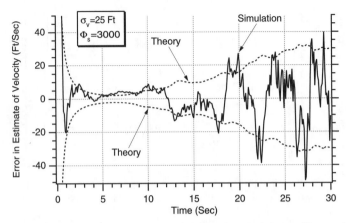

Fig. 8.9   Adding process noise removes hangoff error at expense of increasing errors in estimate of velocity.

ballistic coefficient estimate. By comparing Figs. 8.7 with 8.9, we see that the price we paid for adding the process noise was slightly higher errors in the estimates of velocity. By comparing Figs. 8.4 and 8.10, we can see that adding process noise removed the hangoff error in the ballistic coefficient at the expense of increased errors in that estimate. Adding process noise is the normal engineering fix to situations in which the errors in the estimates do not quite match theory. We are simply accounting for our imperfect knowledge of the real world. In general, adding process noise will make a filter more robust by giving it a wider bandwidth and is thus the prudent thing to do.

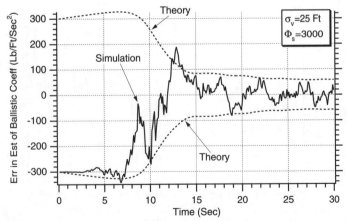

Fig. 8.10   Adding process noise removes hangoff error at expense of increasing errors in estimate of ballistic coefficient.

Changing Filter States

The choice of states for this problem is not unique. Unlike a linear Kalman filter, where the choice of states should make no difference provided that the new states are linear functions of the old states, the choice of states in an extended Kalman filter can make a major difference. Recall that the nonlinear differential equation describing the acceleration on the object is given by

$$\ddot{x} = \frac{0.0034ge^{-x/22000}\dot{x}^2}{2\beta} - g = \left(\frac{1}{\beta}\right)\frac{0.0034ge^{-x/22000}\dot{x}^2}{2} - g$$

Because the preceding equation is inversely proportional to the ballistic coefficient, it seems that a reasonable alternate set of states would be

$$x = \begin{bmatrix} x \\ \dot{x} \\ 1/\beta \end{bmatrix}$$

because the original differential equation looks less nonlinear. In addition, if we believe that the ballistic coefficient is constant that means its derivative must be zero. Therefore, the derivative of the inverse ballistic coefficient must also be zero or

$$(\dot{1/\beta}) = 0$$

We can linearize the preceding differential equations by saying that

$$\begin{bmatrix} \Delta\dot{x} \\ \Delta\ddot{x} \\ \Delta(\dot{1/\beta}) \end{bmatrix} = \begin{bmatrix} \dfrac{\partial\dot{x}}{\partial x} & \dfrac{\partial\dot{x}}{\partial\dot{x}} & \dfrac{\partial\dot{x}}{\partial(1/\beta)} \\ \dfrac{\partial\ddot{x}}{\partial x} & \dfrac{\partial\ddot{x}}{\partial\dot{x}} & \dfrac{\partial\ddot{x}}{\partial(1/\beta)} \\ \dfrac{\partial(\dot{1/\beta})}{\partial x} & \dfrac{\partial(\dot{1/\beta})}{\partial\dot{x}} & \dfrac{\partial(\dot{1/\beta})}{\partial(1/\beta)} \end{bmatrix} \begin{bmatrix} \Delta x \\ \Delta\dot{x} \\ \Delta(1/\beta) \end{bmatrix} + \begin{bmatrix} 0 \\ 0 \\ u_s \end{bmatrix}$$

where u_s is white process noise that has been added to the derivative of the inverse ballistic coefficient rate equation for possible future protection (i.e., the ballistic coefficient may not be a constant). We can see from the preceding state-space differential equation that the systems dynamics matrix is the matrix of partial derivatives or

$$F = \begin{bmatrix} \dfrac{\partial\dot{x}}{\partial x} & \dfrac{\partial\dot{x}}{\partial\dot{x}} & \dfrac{\partial\dot{x}}{\partial(1/\beta)} \\ \dfrac{\partial\ddot{x}}{\partial x} & \dfrac{\partial\ddot{x}}{\partial\dot{x}} & \dfrac{\partial\ddot{x}}{\partial(1/\beta)} \\ \dfrac{\partial(\dot{1/\beta})}{\partial x} & \dfrac{\partial(\dot{1/\beta})}{\partial\dot{x}} & \dfrac{\partial(\dot{1/\beta})}{\partial(1/\beta)} \end{bmatrix}_{x=\hat{x}}$$

Note that in the preceding systems dynamics matrix the partial derivatives are evaluated at the current state estimates. To evaluate the systems dynamics matrix, we must take the necessary partial derivatives. The first row of partial derivatives yields

$$\frac{\partial \dot{x}}{\partial x} = 0$$

$$\frac{\partial \dot{x}}{\partial \dot{x}} = 1$$

$$\frac{\partial \dot{x}}{\partial (1/\beta)} = 0$$

while the second row of partial derivatives becomes

$$\frac{\partial \ddot{x}}{\partial x} = \frac{-0.0034e^{-x/22000}\dot{x}^2 g}{2\beta(22,000)} = \frac{-\rho g \dot{x}^2}{44,000\beta} = \left(\frac{1}{\beta}\right)\left(\frac{-\rho g \dot{x}^2}{44,000}\right)$$

$$\frac{\partial \ddot{x}}{\partial \dot{x}} = \frac{2*0.0034e^{-x/22000}\dot{x}g}{2\beta} = \frac{\rho g \dot{x}}{\beta} = \left(\frac{1}{\beta}\right)\rho g \dot{x}$$

$$\frac{\partial \ddot{x}}{\partial (1/\beta)} = \frac{0.0034e^{-x/22000}\dot{x}^2 g}{2} = \frac{\rho g \dot{x}^2}{2}$$

Finally, the last row of partial derivatives yields

$$\frac{\partial (1/\dot{\beta})}{\partial x} = 0$$

$$\frac{\partial (1/\dot{\beta})}{\partial \dot{x}} = 0$$

$$\frac{\partial (1/\dot{\beta})}{\partial (1/\beta)} = 0$$

Therefore, the systems dynamics matrix turns out to be

$$F(t) = \begin{bmatrix} 0 & 1 & 0 \\ f_{21} & f_{22} & f_{23} \\ 0 & 0 & 0 \end{bmatrix}$$

where

$$f_{21} = \frac{-\hat{\rho}g\hat{x}^2}{44,000}\widehat{(1/\beta)}$$

$$f_{22} = \hat{\rho}\hat{\dot{x}}g\widehat{(1/\beta)}$$

$$f_{23} = \frac{-\hat{\rho}g\hat{x}^2}{2}$$

As was the case before, the two-term approximation to the fundamental is given by

$$\Phi_k = \begin{bmatrix} 1 & T_s & 0 \\ f_{21}T_s & 1+f_{22}T_s & f_{23}T_s \\ 0 & 0 & 1 \end{bmatrix}$$

The simulation of the extended Kalman filter, for which $1/\beta$ is a state, appears in Listing 8.2. We can see that the inputs are identical to the other extended Kalman filter, which appears in Listing 8.1. The simulation is set up so that nominally there is no process noise (i.e., PHIS = 0).

The nominal case of Listing 8.2 was run when there was no process noise. We can see from Fig. 8.11 that the error in the estimate of altitude is approximately the same as it was for the other extended Kalman filter (see Fig. 8.2). However, we also can see from Fig. 8.12 that there is no longer any hangoff error in the error in the estimate of velocity when there is no process noise.

Figure 8.13 seems also to be saying that there is also no hangoff error in the error of the estimate for the inverse ballistic coefficient. We also can see that the inverse ballistic coefficient is not observable for about the same period of time as the other extended Kalman filter (see Fig. 8.4). However, Fig. 8.14, which simply expands the scales of the preceding plot for more clarity, also says that there is no hangoff error.

Because most people do not have a feeling for the inverse ballistic coefficient, the actual estimate of the ballistic coefficient (i.e., reciprocal of inverse ballistic coefficient) appears in Fig. 8.15 along with the actual ballistic coefficient. We can see that after 10 s we get extremely accurate estimates.

Thus, we can see that the hangoff error of the original extended Kalman filter was eliminated by changing the filter states. The choice of $1/\beta$ as a state made the state-space equations appear less strongly nonlinear, thus eliminating the need for process noise to correct matters. In principle, using smaller amounts of process noise (or none at all) will reduce the errors in the estimates as long as the filter model is perfectly matched to the real world.

Why Process Noise Is Required

It would appear from the work presented in this chapter that it is desirable to set the process noise to zero because it reduces the effect of the measurement noise. We have seen that if the errors in the estimate (i.e., truth minus estimate)

Listing 8.2   Simulation of extended Kalman filter with inverse ballistic coefficient as state

```
C THE FIRST THREE STATEMENTS INVOKE THE ABSOFT RANDOM
  NUMBER GENERATOR ON THE MACINTOSH
     GLOBAL DEFINE
               INCLUDE 'quickdraw.inc'
     END
     IMPLICIT REAL*8 (A-H)
     IMPLICIT REAL*8 (O-Z)
     REAL*8 PHI(3,3),P(3,3),M(3,3),PHIP(3,3),PHIPPHIT(3,3),GAIN(3,1)
     REAL*8 Q(3,3),HMAT(1,3),HM(1,3),MHT(3,1)
     REAL*8 PHIT(3,3)
     REAL*8 HMHT(1,1),HT(3,1),KH(3,3),IDN(3,3),IKH(3,3)
     INTEGER ORDER
     ITERM=1
     G=32.2
     SIGNOISE=25.
     X=200000.
     XD=-6000.
     BETA=500.
     XH=200025.
     XDH=-6150.
     BETAH=800.
     BETAINV=1./BETA
     BETAINVH=1./BETAH
     OPEN(2,STATUS='UNKNOWN',FILE='COVFIL')
     OPEN(1,STATUS='UNKNOWN',FILE='DATFIL')
     ORDER=3
     TS=.1
     TF=30.
     PHIS=0.
     T=0.
     S=0.
     H=.001
     HP=.001
     DO 1000 I=1,ORDER
     DO 1000 J=1,ORDER
               PHI(I,J)=0.
               P(I,J)=0.
               Q(I,J)=0.
               IDN(I,J)=0.
1000 CONTINUE
     IDN(1,1)=1.
     IDN(2,2)=1.
     IDN(3,3)=1.
     P(1,1)=SIGNOISE*SIGNOISE
     P(2,2)=20000.
     P(3,3)=(BETAINV-BETAINVH)**2
     DO 1100 I=1,ORDER
               HMAT(1,I)=0.
               HT(I,1)=0.
```

(continued)

Listing 8.2 *(Continued)*

```
1100 CONTINUE
     HMAT(1,1)=1.
     HT(1,1)=1.
     WHILE(T<=TF)
          XOLD=X
          XDOLD=XD
          XDD=.0034*32.2*XD*XD*EXP(-X/22000.)/(2.*BETA)-32.2
          X=X+H*XD
          XD=XD+H*XDD
          T=T+H
          XDD=.0034*32.2*XD*XD*EXP(-X/22000.)/(2.*BETA)-32.2
          X=.5*(XOLD+X+H*XD)
          XD=.5*(XDOLD+XD+H*XDD)
          S=S+H
          IF(S>=(TS-.00001))THEN
               S=0.
               RHOH=.0034*EXP(-XH/22000.)
               F21=-G*RHOH*XDH*XDH*BETAINVH/44000.
               F22=RHOH*G*XDH*BETAINVH
               F23=.5*RHOH*XDH*XDH*G
               PHI(1,1)=1.
               PHI(1,2)=TS
               PHI(2,1)=F21*TS
               PHI(2,2)=1.+F22*TS
               PHI(2,3)=F23*TS
               PHI(3,3)=1.
               Q(2,2)=F23*F23*PHIS*TS*TS*TS/3.
               Q(2,3)=F23*PHIS*TS*TS/2.
               Q(3,2)=Q(2,3)
               Q(3,3)=PHIS*TS
               CALL MATTRN(PHI,ORDER,ORDER,PHIT)
               CALL MATMUL(PHI,ORDER,ORDER,P,ORDER,ORDER,
                 PHIP)
               CALL MATMUL(PHIP,ORDER,ORDER,PHIT,ORDER,
                 ORDER,PHIPPHIT)
               CALL MATADD(PHIPPHIT,ORDER,ORDER,Q,M)
               CALL MATMUL(HMAT,1,ORDER,M,ORDER,ORDER,HM)
               CALL MATMUL(HM,1,ORDER,HT,ORDER,1,HMHT)
               HMHTR=HMHT(1,1)+SIGNOISE*SIGNOISE
               HMHTRINV=1./HMHTR
               CALL MATMUL(M,ORDER,ORDER,HT,ORDER,1,MHT)
               DO 150 I=1,ORDER
                    GAIN(I,1)=MHT(I,1)*HMHTRINV
150            CONTINUE
               CALL MATMUL(GAIN,ORDER,1,HMAT,1,ORDER,KH)
               CALL MATSUB(IDN,ORDER,ORDER,KH,IKH)
               CALL MATMUL(IKH,ORDER,ORDER,M,ORDER,
                 ORDER,P)
               CALL GAUSS(XNOISE,SIGNOISE)
               BETAH=1./BETAINVH
```

(continued)

Listing 8.2 *(Continued)*

```
                  CALL  PROJECT(T,TS,XH,XDH,BETAH,XB,XDB,
                     XDDB,HP)
                  RES=X+XNOISE-XB
                  XH=XB+GAIN(1,1)*RES
                  XDH=XDB+GAIN(2,1)*RES
                  BETAINVH=BETAINVH+GAIN(3,1)*RES

                  ERRX=X-XH
                  SP11=SQRT(P(1,1))
                  ERRXD=XD-XDH
                  SP22=SQRT(P(2,2))
                  ERRBETAINV=1./BETA-BETAINVH
                  SP33=SQRT(P(3,3))
                  BETAH=1./BETAINVH
                  WRITE(9,*)T,X,XH,XD,XDH,BETA,BETAH
                  WRITE(1,*)T,X,XH,XD,XDH,BETA,BETAH
                  WRITE(2,*)T,ERRX,SP11,-SP11,ERRXD,SP22,-SP22,
1                    ERRBETAINV,SP33,-SP33
          ENDIF
      END DO
      PAUSE
      CLOSE(1)
      END

      SUBROUTINE  PROJECT(TP,TS,XP,XDP,BETAP,XH,XDH,XDDH,HP)
      IMPLICIT  REAL*8  (A-H)
      IMPLICIT  REAL*8  (O-Z)
      T=0.
      X=XP
      XD=XDP
      BETA=BETAP
      H=HP
      WHILE(T<=(TS-.0001))
          XDD=.0034*32.2*XD*XD*EXP(-X/22000.)/(2.*BETA)-32.2
          XD=XD+H*XDD
          X=X+H*XD
          T=T+H
      END DO
      XH=X
      XDH=XD
      XDDH=XDD
      RETURN
      END

C SUBROUTINE GAUSS IS SHOWN IN LISTING 1.8
C SUBROUTINE MATTRN IS SHOWN IN LISTING 1.3
C SUBROUTINE MATMUL IS SHOWN IN LISTING 1.4
C SUBROUTINE MATADD IS SHOWN IN LISTING 1.1
C SUBROUTINE MATSUB IS SHOWN IN LISTING 1.2
```

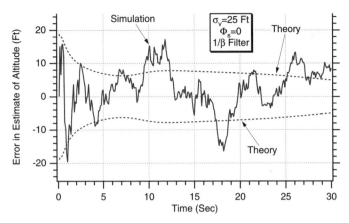

Fig. 8.11 **Error in the estimate of altitude is approximately the same for both extended Kalman filters.**

have no hangoff error, then having the process noise set to zero enables the error in the estimate to approach zero as more measurements are taken. Although increasing the process noise often removes hangoff errors, we have also seen that the errors in the estimates increase with increasing process noise because of the higher Kalman gains, which admit more measurement noise. The real purpose of the process noise is to account for the fact that the filter's knowledge of the real world is often in error.

Consider the case in which we are still tracking the same object, but after 25 s of tracking the object breaks in half. This causes the ballistic coefficient to reduce by a factor of two (i.e., from 500 to 250 lb/ft^2). By using zero process noise, the extended Kalman-filter model was assuming that the actual ballistic coefficient

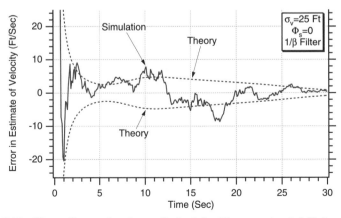

Fig. 8.12 **Hangoff error has been eliminated with new extended Kalman filter.**

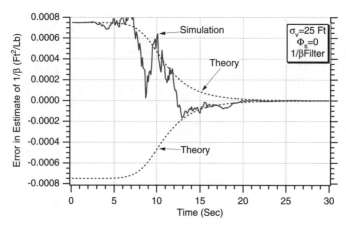

Fig. 8.13   It still takes approximately 10 s to reduce error in estimate of inverse ballistic coefficient.

was a constant. Figure 8.16 shows that before the object breaks in half (i.e., first 25 s of simulation) the extended Kalman filter's estimate of the ballistic coefficient was excellent. However, after the actual ballistic coefficient changes by a factor of two (i.e., object breaks in half), the filter's estimate of the ballistic coefficient considerably lags the actual ballistic coefficient. This is because setting the process noise to zero eventually causes the filter to stop paying attention to new measurements (i.e., filter bandwidth is reduced or the filter gain is too small for incorporating more measurements). In other words, a zero process noise extended Kalman filter eventually goes to sleep!

Although the amount of process noise to use is often determined experimentally, a good starting point is that the amount of process noise that a Kalman filter

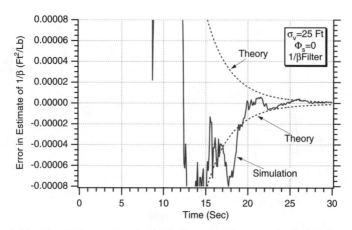

Fig. 8.14   Hangoff error has been eliminated with new extended Kalman filter.

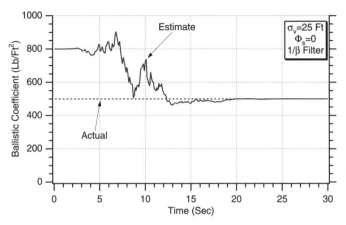

Fig. 8.15   It takes 10 s to get accurate estimates of ballistic coefficient.

requires should reflect our estimate of our lack of knowledge of the real world. One method for determining process noise is to square the uncertainty in our expected initial error in the estimate and divide by the amount of filtering time.[2] For example, because we initially thought that the ballistic coefficient was $800 \, \text{lb/ft}^2$ rather than $500 \, \text{lb/ft}^2$ and the amount of filtering time was 40 s (i.e., see Fig. 8.16) we could say that a good value of the process noise for the inverse ballistic coefficient filter should be

$$\Phi_s = \frac{(1/500 - 1/800)^2}{40}$$

Figure 8.17 shows that when the process noise is set to the value of the preceding equation the track quality of the extended Kalman filter is still excellent

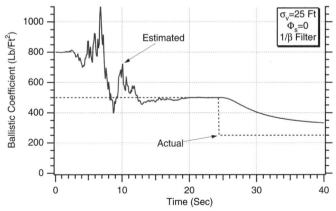

Fig. 8.16   Filter cannot track changes in ballistic coefficient if there is no process noise.

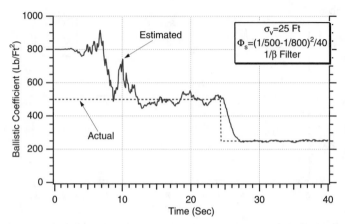

Fig. 8.17 Adding process noise enables filter to track changes in the ballistic coefficient.

for the first 25 s (i.e., where the ballistic coefficient is constant) but somewhat noisier than the case in which there was no process noise. More importantly, when the ballistic coefficient changes, the extended Kalman filter is now able to keep track of the motion.

We have now seen that it is probably prudent to include process noise in an extended Kalman filter trying to estimate the ballistic coefficient or its inverse. However, there is also a price that must be paid when process noise is included. We have just observed from Fig. 8.17 that the estimates will be noisier than they were without the inclusion of process noise. Figures 8.18–8.20 show how the errors in the estimates change when process noise is included. Figure 8.18

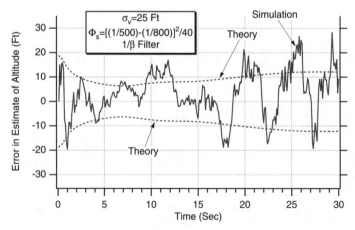

Fig. 8.18 Nominal errors in estimate of altitude deteriorate slightly when process noise is added.

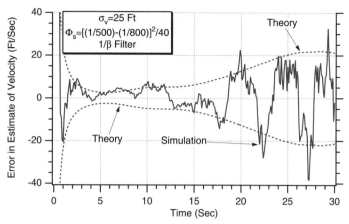

Fig. 8.19 **Nominal errors in estimate of velocity deteriorate slightly when process noise is added.**

presents the single-run simulation results and theoretical errors in the estimate of altitude for the inverse ballistic coefficient extended Kalman filter. By comparing Fig. 8.18 with Fig. 8.11, we can see that after 30 s the filter with process-noise yields errors in the estimate of altitude that are approximately twice as large as when the process noise is zero (i.e. 10 vs 5 ft). By comparing Fig. 8.19 with Fig. 8.12, we can see that after 30 s, the filter with process noise yields errors in the estimate of velocity that are approximately an order of magnitude larger than when the process noise is zero (i.e., 20 vs. 2 ft/s). We can see from Fig. 8.20 that when process noise is included the error in the estimate of the inverse ballistic coefficient no longer goes to zero as it did in Fig. 8.13, but it now approaches a

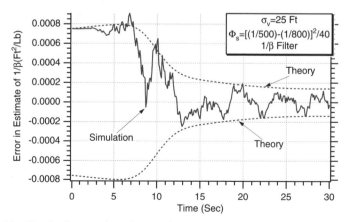

Fig. 8.20 **Nominal errors in estimate of ballistic coefficient deteriorate slightly when process noise is added.**

steady-state value. This is not a hangoff error because the error in the estimate of the inverse ballistic coefficient appears to have zero mean.

Linear Polynomial Kalman Filter

We have seen so far in this chapter that an extended Kalman filter could be used to track a rapidly decelerating falling object by estimating its ballistic coefficient. It is reasonable to ask how a linear polynomial Kalman filter would perform for the same task. Because the object is decelerating, at least a second-order or three-state linear polynomial Kalman filter will be required to track the object without building up excessive truncation error. For a linear polynomial Kalman-filter design we will not make use of any a priori information concerning the object's dynamics. We will simply assume that because the object's acceleration is not constant we can say that the derivative of acceleration or jerk is equal to white noise or

$$\dddot{x} = u_s$$

If we assume the states are altitude, velocity, and acceleration, the state-space equation, upon which the linear Kalman filter will be designed, can be written as

$$\begin{bmatrix} \dot{x} \\ \ddot{x} \\ \dddot{x} \end{bmatrix} = \begin{bmatrix} 0 & 1 & 0 \\ 0 & 0 & 1 \\ 0 & 0 & 0 \end{bmatrix} \begin{bmatrix} x \\ \dot{x} \\ \ddot{x} \end{bmatrix} + \begin{bmatrix} 0 \\ 0 \\ u_s \end{bmatrix}$$

where u_s is white noise with spectral density Φ_s. The continuous process-noise matrix for this example can be obtained from the preceding equation as

$$Q = E\left[\begin{bmatrix} 0 & 0 & u_s \end{bmatrix}\begin{bmatrix} 0 \\ 0 \\ u_s \end{bmatrix}\right] = \begin{bmatrix} 0 & 0 & 0 \\ 0 & 0 & 0 \\ 0 & 0 & \Phi_s \end{bmatrix}$$

whereas the systems dynamics matrix can be written by inspection as

$$F = \begin{bmatrix} 0 & 1 & 0 \\ 0 & 0 & 1 \\ 0 & 0 & 0 \end{bmatrix}$$

In this example, we can assume that the measurement x^* is linearly related to the states according to

$$x^* = \begin{bmatrix} 1 & 0 & 0 \end{bmatrix}\begin{bmatrix} x \\ \dot{x} \\ \ddot{x} \end{bmatrix} + v$$

where v is white measurement noise. From the preceding equation we can see that the measurement matrix is given by

$$H = \begin{bmatrix} 1 & 0 & 0 \end{bmatrix}$$

For the second-order or three-state linear polynomial Kalman filter, we summarized in Chapter 4 (see Table 4.1) the discrete form of the fundamental and measurement noise matrices for the three-state linear polynomial Kalman filter, which are given by

$$\Phi_k = \begin{bmatrix} 1 & T_s & 0.5T_s^2 \\ 0 & 1 & T_s \\ 0 & 0 & 1 \end{bmatrix}$$

$$R_k = \sigma_n^2$$

where T_s is the sampling time and σ_n^2 is the variance of the measurement noise v. The discrete process-noise matrix for the second-order or three-state linear polynomial Kalman filter was also shown in Table 4.2 of Chapter 4 to be

$$Q_k = \Phi_s \begin{bmatrix} \dfrac{T_s^5}{20} & \dfrac{T_s^4}{8} & \dfrac{T_s^3}{6} \\ \dfrac{T_s^4}{8} & \dfrac{T_s^3}{3} & \dfrac{T_s^2}{2} \\ \dfrac{T_s^3}{6} & \dfrac{T_s^2}{2} & T_s \end{bmatrix}$$

We now have enough information to build the linear polynomial Kalman filter to estimate the altitude, velocity, and acceleration of the falling object based on noisy measurements of the object's altitude. However, if we would also like to estimate the object's ballistic coefficient, which is not required for tracking the object, then a little more work must be done.

At the beginning of this chapter, we showed that the acceleration of the object could be expressed in terms of the drag and gravity acting on the object as

$$\ddot{x} = \frac{g\rho \dot{x}^2}{2\beta} - g$$

We can invert the preceding expression to solve for the ballistic coefficient to obtain

$$\beta = \frac{g\rho \dot{x}^2}{2(\ddot{x} + g)}$$

where the air density is a function of altitude according to

$$\rho = 0.0034 e^{-x/22000}$$

Therefore, we can use the altitude, velocity, and acceleration estimates from the three-state linear polynomial Kalman filter to estimate the ballistic coefficient according to

$$\hat{\beta} = \frac{g\hat{\rho}\hat{\dot{x}}^2}{2(\hat{\ddot{x}} + g)}$$

where the estimated air density is a function of the estimated altitude from

$$\hat{\rho} = 0.0034 e^{-\hat{x}/22000}$$

Listing 8.3 is a simulation of the falling object and the three-state linear polynomial Kalman filter for estimating the object's altitude, velocity, acceleration, and ballistic coefficient. We can see from Listing 8.3 that the initial conditions for the state estimates and covariance matrix are the same as they were in Listing 8.2. This filter has process noise, whose spectral density is given by

$$\Phi_s = \frac{322^2}{30}$$

to account for the fact that the object's deceleration might be as large as 10 g (i.e., 322 ft/s²) over a 30-s flight. The third diagonal element of the initial covariance matrix also has an initial value to reflect the uncertainty in the object's deceleration (i.e., $P_{33} = 322*322$). We can see from Listing 8.3 that the object's acceleration state of the filter has initially been set to zero, which means we are not assuming any a priori information regarding the object's initial acceleration.

The nominal case of Listing 8.3 was run, and the error in the estimate results are displayed in Figs. 8.21–8.23. First, we can see from Figs. 8.21–8.23 that because the single-run errors in the estimates are within the theoretical bounds the linear second-order polynomial Kalman filter appears to be working correctly. We can also see from Fig. 8.21 that the error in the estimate of altitude is approximately 15 ft, which is slightly larger than the results of the extended Kalman filter with the inverse ballistic coefficient (see Fig. 8.18, where the error in the estimate of altitude is approximately 10 ft when process noise is present). Using smaller values for the process noise in the linear filter probably would have yielded nearly identical results to the extended filter. We can see from Fig. 8.22 that the error in the estimate of velocity is approximately 30 ft/s, which is slightly larger than the results of the extended Kalman filter with the inverse ballistic coefficient (see Fig. 8.20, where the error in the estimate of velocity is approximately 20 ft/s when process noise is present). Again, using smaller values for the process noise in the linear filter probably would have yielded nearly identical results to the extended filter. We can see from Fig. 8.23 that the

Listing 8.3 Using three-state linear polynomial Kalman filter to track falling object under the influence of drag

```
C THE FIRST THREE STATEMENTS INVOKE THE ABSOFT RANDOM
  NUMBER GENERATOR ON THE MACINTOSH
    GLOBAL DEFINE
            INCLUDE 'quickdraw.inc'
    END
    IMPLICIT REAL*8 (A-H)
    IMPLICIT REAL*8 (O-Z)
    REAL*8 PHI(3,3),P(3,3),M(3,3),PHIP(3,3),PHIPPHIT(3,3),GAIN(3,1)
    REAL*8 Q(3,3),HMAT(1,3),HM(1,3),MHT(3,1)
    REAL*8 PHIT(3,3)
    REAL*8 HMHT(1,1),HT(3,1),KH(3,3),IDN(3,3),IKH(3,3)
    INTEGER ORDER
    ITERM=1
    G=32.2
    SIGNOISE=25.
    X=200000.
    XD=-6000.
    BETA=500.
    XH=200025.
    XDH=-6150.
    XDDH=0.
    XNT=322.
    OPEN(2,STATUS='UNKNOWN',FILE='COVFIL')
    OPEN(1,STATUS='UNKNOWN',FILE='DATFIL')
    ORDER=3
    TS=.1
    TF=30.
    PHIS=XNT*XNT/TF
    T=0.
    S=0.
    H=.001
    HP=.001
    TS2=TS*TS
    TS3=TS2*TS
    TS4=TS3*TS
    TS5=TS4*TS
    DO 1000 I=1,ORDER
    DO 1000 J=1,ORDER
            PHI(I,J)=0.
            P(I,J)=0.
            Q(I,J)=0.
            IDN(I,J)=0.
1000 CONTINUE
    PHI(1,1)=1.
    PHI(1,2)=TS
    PHI(1,3)=.5*TS*TS
    PHI(2,2)=1.
    PHI(2,3)=TS
```

(continued)

Listing 8.3 (*Continued*)

```
        PHI(3,3)=1.
        CALL  MATTRN(PHI,ORDER,ORDER,PHIT)
        Q(1,1)=TS5*PHIS/20.
        Q(1,2)=TS4*PHIS/8.
        Q(1,3)=PHIS*TS3/6.
        Q(2,1)=Q(1,2)
        Q(2,2)=PHIS*TS3/3.
        Q(2,3)=.5*TS2*PHIS
        Q(3,1)=Q(1,3)
        Q(3,2)=Q(2,3)
        Q(3,3)=PHIS*TS
        IDN(1,1)=1.
        IDN(2,2)=1.
        IDN(3,3)=1.
        P(1,1)=SIGNOISE*SIGNOISE
        P(2,2)=20000.
        P(3,3)=XNT*XNT
        DO 1100 I=1,ORDER
                HMAT(1,I)=0.
                HT(I,1)=0.
1100 CONTINUE
        HMAT(1,1)=1.
        HT(1,1)=1.
        WHILE(T<=TF)
                XOLD=X
                XDOLD=XD
                XDD=.0034*32.2*XD*XD*EXP(-X/22000.)/(2.*BETA)-32.2
                X=X+H*XD
                XD=XD+H*XDD
                T=T+H
                XDD=.0034*32.2*XD*XD*EXP(-X/22000.)/(2.*BETA)-32.2
                X=.5*(XOLD+X+H*XD)
                XD=.5*(XDOLD+XD+H*XDD)
                S=S+H
                IF(S>=(TS-.00001))THEN
                        S=0.
                        CALL  MATMUL(PHI,ORDER,ORDER,P,ORDER,ORDER,
                        PHIP)
                        CALL  MATMUL(PHIP,ORDER,ORDER,PHIT,ORDER,
1                       ORDER,PHIPPHIT)
                        CALL  MATADD(PHIPPHIT,ORDER,ORDER,Q,M)
                        CALL  MATMUL(HMAT,1,ORDER,M,ORDER,ORDER,HM)
                        CALL  MATMUL(HM,1,ORDER,HT,ORDER,1,HMHT)
                        HMHTR=HMHT(1,1)+SIGNOISE*SIGNOISE
                        HMHTRINV=1./HMHTR
                        CALL  MATMUL(M,ORDER,ORDER,HT,ORDER,1,MHT)
                        DO 150 I=1,ORDER
                                GAIN(I,1)=MHT(I,1)*HMHTRINV
150                     CONTINUE
                        CALL  MATMUL(GAIN,ORDER,1,HMAT,1,ORDER,KH)
                                                          (continued)
```

Listing 8.3 (*Continued*)

```
              CALL  MATSUB(IDN,ORDER,ORDER,KH,IKH)
              CALL  MATMUL(IKH,ORDER,ORDER,M,ORDER,
                ORDER,P)
              CALL  GAUSS(XNOISE,SIGNOISE)
              RES=X+XNOISE-XH-TS*XDH-.5*TS*TS*XDDH
              XH=XH+TS*XDH+.5*TS*TS*XDDH+GAIN(1,1)*RES
              XDH=XDH+TS*XDDH+GAIN(2,1)*RES
              XDDH=XDDH+GAIN(3,1)*RES
              RHOH=.0034*EXP(-XH/22000.)
              BETAH=16.1*RHOH*XDH*XDH/(XDDH+32.2)
              ERRX=X-XH
              SP11=SQRT(P(1,1))
              ERRXD=XD-XDH
              SP22=SQRT(P(2,2))
              ERRXDD=XDD-XDDH
              SP33=SQRT(P(3,3))
              WRITE(9,*)T,X,XH,XD,XDH,XDD,XDDH,BETA,BETAH
              WRITE(1,*)T,X,XH,XD,XDH,XDD,XDDH,BETA,BETAH
              WRITE(2,*)T,ERRX,SP11,-SP11,ERRXD,SP22,-SP22,
     1          ERRXDD,SP33,-SP33
         ENDIF
      END DO
      PAUSE
      CLOSE(1)
      END


C SUBROUTINE GAUSS IS SHOWN IN LISTING 1.8
C SUBROUTINE MATTRN IS SHOWN IN LISTING 1.3
C SUBROUTINE MATMUL IS SHOWN IN LISTING 1.4
C SUBROUTINE MATADD IS SHOWN IN LISTING 1.1
C SUBROUTINE MATSUB IS SHOWN IN LISTING 1.2
```

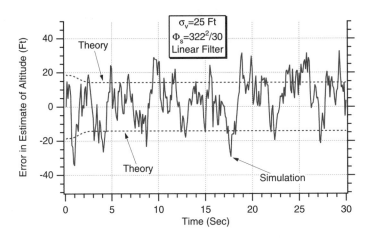

Fig. 8.21 Altitude estimates are slightly worse with linear filter.

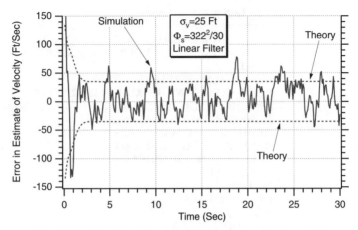

Fig. 8.22 Velocity estimates slightly worse with linear filter.

error in the estimate of acceleration is approximately $50 \, \text{ft/s}^2$ in the steady state. We cannot compare these results to the extended filter results because the extended Kalman filter that estimated the inverse ballistic coefficient did not estimate the object's acceleration directly.

To clarify the filtering results even further, Fig. 8.24 compares the acceleration estimate of the three-state linear polynomial Kalman filter to the actual acceleration of the object. We can see that the linear filter is able to track the highly changing acceleration characteristics of the object. There does not appear to be any significant lag between the actual acceleration and the estimated acceleration. In addition, we can see from Fig. 8.24 that the estimates are not too noisy, indicating that just about the right amount of process noise was used.

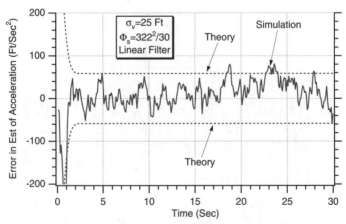

Fig. 8.23 Second-order linear polynomial Kalman filter is able to track object's acceleration.

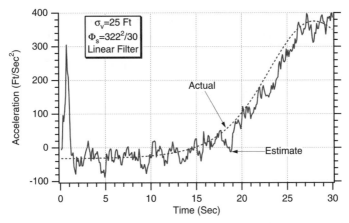

Fig. 8.24   Second-order linear polynomial Kalman filter is able to track object's acceleration without too much lag or noise propagation.

From the same computer run where the acceleration estimates of the second-order linear polynomial Kalman filter were fairly decent, an attempt was also made to estimate the object's ballistic coefficient using the formulas already derived

$$\hat{\beta} = \frac{g\hat{\rho}\hat{\dot{x}}^2}{2(\hat{\ddot{x}} + g)}$$

$$\hat{\rho} = 0.0034e^{-\hat{x}/22000}$$

We can see from Fig. 8.25 that using the linear filter to estimate ballistic coefficient is not a good idea because the estimate is far too noisy for most of the

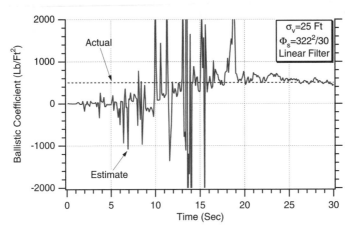

Fig. 8.25   It takes approximately 20 s for the second-order linear polynomial Kalman filter to accurately estimate the ballistic coefficient.

flight. Only after about 20 s are we able to estimate the ballistic coefficient. The accuracy to which the ballistic coefficient can be estimated with the linear polynomial Kalman filter is far worse than the accuracy results of both extended Kalman filters. Of course, an estimate of the ballistic coefficient is not required for the second-order linear polynomial Kalman filter to estimate the altitude, velocity, and acceleration of the falling object, and so this type of filter may be perfectly suitable for this application.

To understand why the ballistic coefficient estimate of the linear filter was so terrible, another experiment was conducted. Because the ballistic coefficient is inversely proportional to the acceleration, it was hypothesized that the division by the noisy acceleration estimate effectively multiplied the noise. To check the hypothesis, Listing 8.3 was modified so that the ballistic coefficient estimate made use of perfect acceleration information, rather than estimated acceleration from the filter, or

$$\hat{\beta} = \frac{g\hat{\rho}\hat{\dot{x}}^2}{2(\ddot{x}+g)}$$

We can now see from Fig. 8.26 that the linear filter's ballistic coefficient estimate is now near perfect, indicating that there was too much noise on the acceleration estimate to accurately provide an estimate of the ballistic coefficient for most of the flight.

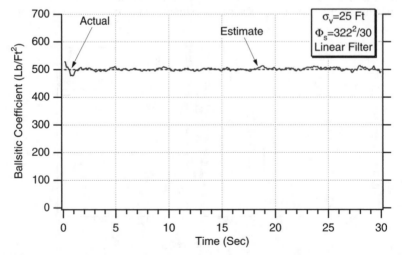

Fig. 8.26 If acceleration known exactly, ballistic coefficient estimate is near perfect.

Summary

In this chapter we have worked on several extended Kalman filters for another version of the falling-object problem. We have again seen the utility of adding process noise to the extended Kalman filter to eliminate hangoff errors in the estimates and to make the filter more responsive to unanticipated changes in the object's dynamics. We have also seen that changing the states of an extended Kalman filter can sometimes reduce estimation errors because certain states make the problem look less nonlinear. Finally, we also have seen that for this application a second-order linear polynomial Kalman filter could be made to work if the only goal was to track the object and estimate altitude, velocity, and acceleration. On the other hand, we also saw that the linear filter had certain deficiencies if we attempted to also estimate the object's ballistic coefficient.

References

[1]Gelb., A., *Applied Optimal Estimation*, Massachusetts Inst. of Technology Press, Cambridge, MA, 1974, pp. 194–198.

[2]Zarchan, P., *Tactical and Strategic Missile Guidance*, 3rd ed. Progress in Astronautics and Aeronautics, AIAA, Reston, VA, 1998, pp. 373–387.

Chapter 9

Cannon-Launched Projectile Tracking Problem

Introduction

IN ALL of the applications so far, there was only one sensor measurement from which we built our Kalman filters. In this chapter we will attempt to track a cannon-launched projectile with a sensor that measures both the range and angle to the projectile. We will first see how the extra sensor measurement is incorporated into an extended Kalman filter. In this example the real world is linear in the Cartesian coordinate system, but the measurements are nonlinear. The extended Kalman filter will initially be designed in the Cartesian coordinate system to establish a benchmark for tracking performance. We will then see if performance can be improved by redesigning the filter in the polar coordinate system, where the measurements are linear but the model of the real world is nonlinear. Finally, we will also see if linear coupled and decoupled polynomial Kalman filters also can be made to work for this problem.

Problem Statement

Consider another simple example in which a radar tracks a cannon-launched projectile traveling in a two-dimensional, drag-free environment, as shown in Fig. 9.1. After the projectile is launched at an initial velocity, only gravity acts on the projectile. The tracking radar, which is located at coordinates x_R, y_R of Fig. 9.1, measures the range r and angle θ from the radar to the projectile.

In this example the projectile is launched at an initial velocity of 3000 ft/s at a 45-deg angle with respect to the horizontal axis in a zero-drag environment (i.e., atmosphere is neglected). The radar is located on the ground and is 100,000 ft downrange from where the projectile is initially launched. The radar is considered to have an angular measurement accuracy of 0.01 rad and a range measurement accuracy of 100 ft.

Before we can proceed with the development of a filter for this application, it is important to first get a numerical and analytical feel for all aspects of the problem. Because only gravity acts on the projectile, we can say that in the downrange or x direction there is no acceleration, whereas in the altitude or y direction there is the acceleration of gravity g or[1]

$$\ddot{x}_T = 0$$

$$\ddot{y}_T = -g$$

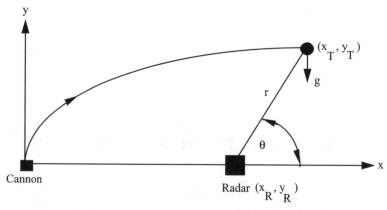

Fig. 9.1 Radar tracking cannon-launched projectile.

where the acceleration of gravity has a value of 32.2 ft/s^2 in the English system of units. Using trigonometry, the angle and range from the radar to the projectile can be obtained by inspection of Fig. 9.1 to be

$$\theta = \tan^{-1}\left(\frac{y_T - y_R}{x_T - x_R}\right)$$

$$r = \sqrt{(x_T - x_R)^2 + (y_T - y_R)^2}$$

Before we build an extended Kalman filter to track the projectile and estimate its position and velocity at all times based on noisy angle and range measurements, let us see what can be done without any filtering at all. By inspection of Fig. 9.1, we can see that we can express the location of the cannon-launched projectile (i.e., x_T, y_T) in terms of the radar range and angle as

$$x_T = r\cos\theta + x_R$$

$$y_T = r\sin\theta + y_R$$

If the radar coordinates are known precisely and r^* and θ^* are the noisy radar range and angle measurements, then the location of the projectile at any time can be calculated directly or estimated without any filtering at all as

$$\hat{x}_T = r^*\cos\theta^* + x_R$$

$$\hat{y}_T = r^*\sin\theta^* + y_R$$

The estimates of the velocity components of the projectile can be obtained from the noisy position estimates by using the definition of a derivative from calculus. In other words, we simply subtract the old position estimate from the new position estimate and divide by the sampling time or the time between measurements to get the estimated projectile velocity components or

$$\hat{\dot{x}}_{T_k} = \frac{\hat{x}_{T_k} - \hat{x}_{t_{k-1}}}{T_s}$$

$$\hat{\dot{y}}_{T_k} = \frac{\hat{y}_{T_k} - \hat{y}_{t_{k-1}}}{T_s}$$

Figure 9.2 displays the true projectile trajectory and the estimated projectile trajectory based on the raw radar measurements. We can see that the estimated trajectory appears to be quite close to the actual trajectory, so one might wonder why filtering techniques are required in this example. On the other hand, Figs. 9.3 and 9.4 show that when the raw measurements are used to estimate the projectile velocity the results are very poor. For example, the actual downrange velocity is constant and is approximately 2000 ft/s. However, we can see from Fig. 9.3 that the estimated downrange velocity varies from 500 to 5000 ft/s. Similarly, Fig. 9.4 shows that the actual altitude velocity varies from 2000 ft/s at cannon ball launch to −2000 ft/s near impact. However, the figure also shows that the estimated altitude velocity is often in error by several thousand feet per second. Clearly, filtering is required if we desire a better way of estimating the projectile's velocity.

It might appear from Fig. 9.2 that the position estimates of the projectile were satisfactory because they appeared to be so close to the actual trajectory. We have seen in the earlier filtering examples of this text that we often compared the error in the estimates to the theoretical predictions of the covariance matrix. Although in this example there is no covariance matrix, we can look at the errors in the estimates of downrange and altitude. For this problem the initial downrange and altitude estimates of the projectile were in error by 1000 ft. We can see from Figs. 9.5 and 9.6 that the errors in the estimates of the projectile's downrange and altitude are never substantially below 1000 ft. In other words, we are not reducing

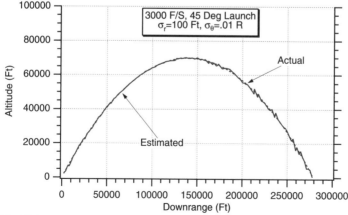

Fig. 9.2   Using raw measurements to estimate trajectory appears to be satisfactory.

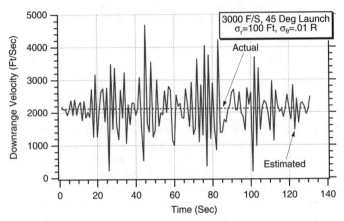

Fig. 9.3   Using raw measurements yields terrible downrange velocity estimates.

the initial errors. This is to be expected because we are not really doing any filtering but are simply using the raw measurements to obtain the location of the projectile.

Extended Cartesian Kalman Filter

If we desire to keep the model describing the real world linear in the cannon-launched projectile tracking problem, we can choose as states projectile location and velocity in the downrange or x direction and projectile location and velocity in the altitude or y direction. Thus, the proposed states are given by

$$x = \begin{bmatrix} x_T \\ \dot{x}_T \\ y_T \\ \dot{y}_T \end{bmatrix}$$

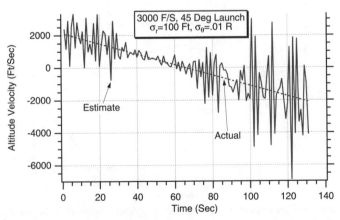

Fig. 9.4   Using raw measurements also yields very poor altitude velocity estimates.

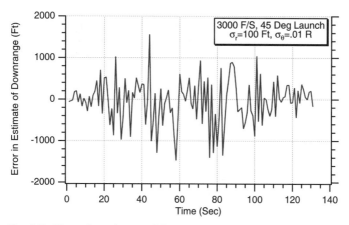

Fig. 9.5   Error in estimate of downrange is often greater than 1000 ft.

Therefore, when the preceding Cartesian states are chosen the state-space differential equation describing projectile motion becomes

$$
\begin{bmatrix} \dot{x}_T \\ \ddot{x}_T \\ \dot{y}_T \\ \ddot{y}_T \end{bmatrix} = \begin{bmatrix} 0 & 1 & 0 & 0 \\ 0 & 0 & 0 & 0 \\ 0 & 0 & 1 & 0 \\ 0 & 0 & 0 & 0 \end{bmatrix} \begin{bmatrix} x_T \\ \dot{x}_T \\ y_T \\ \dot{y}_T \end{bmatrix} + \begin{bmatrix} 0 \\ 0 \\ 0 \\ -g \end{bmatrix} + \begin{bmatrix} 0 \\ u_s \\ 0 \\ u_s \end{bmatrix}
$$

Notice that in the preceding equation gravity g is not a state that has to be estimated but is assumed to be known in advance. We have also added process noise u_s to the acceleration portion of the equations as protection for effects that may not be considered by the Kalman filter. However, in our initial experiments

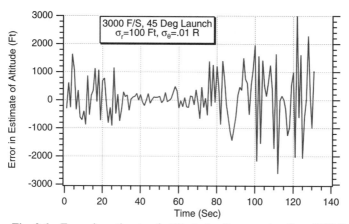

Fig. 9.6   Error in estimate of altitude is often greater than 1000 ft.

we will assume the process noise to be zero because our state-space equations upon which the Kalman filter is based will be a perfect match to the real world.

From the preceding state-space equation we can see that systems dynamics matrix is given by

$$F = \begin{bmatrix} 0 & 1 & 0 & 0 \\ 0 & 0 & 0 & 0 \\ 0 & 0 & 0 & 1 \\ 0 & 0 & 0 & 0 \end{bmatrix}$$

Because the fundamental matrix for a time-invariant system is given by

$$\Phi(t) = \mathcal{L}^{-1}[(sI - F)^{-1}]$$

we could find the fundamental matrix from the preceding expression. However, the required inversion of a 4×4 matrix might prove very tedious. A faster way to compute the fundamental matrix in this example is to simply use the Taylor-series approximation, which is given by

$$\Phi(t) = I + Ft + \frac{F^2 t^2}{2!} + \frac{F^3 t^3}{3!} + \cdots$$

Because

$$F^2 = \begin{bmatrix} 0 & 1 & 0 & 0 \\ 0 & 0 & 0 & 0 \\ 0 & 0 & 0 & 1 \\ 0 & 0 & 0 & 0 \end{bmatrix} \begin{bmatrix} 0 & 1 & 0 & 0 \\ 0 & 0 & 0 & 0 \\ 0 & 0 & 0 & 1 \\ 0 & 0 & 0 & 0 \end{bmatrix} = \begin{bmatrix} 0 & 0 & 0 & 0 \\ 0 & 0 & 0 & 0 \\ 0 & 0 & 0 & 0 \\ 0 & 0 & 0 & 0 \end{bmatrix}$$

all of the higher-order terms of the Taylor-series expansion must be zero, and the fundamental matrix becomes

$$\Phi(t) = I + Ft = \begin{bmatrix} 1 & 0 & 0 & 0 \\ 0 & 1 & 0 & 0 \\ 0 & 0 & 1 & 0 \\ 0 & 0 & 0 & 1 \end{bmatrix} + \begin{bmatrix} 0 & 1 & 0 & 0 \\ 0 & 0 & 0 & 0 \\ 0 & 0 & 0 & 1 \\ 0 & 0 & 0 & 0 \end{bmatrix} t$$

or, more simply,

$$\Phi(t) = \begin{bmatrix} 0 & t & 0 & 0 \\ 0 & 1 & 0 & 0 \\ 0 & 0 & 1 & t \\ 0 & 0 & 0 & 1 \end{bmatrix}$$

Therefore, the discrete fundamental matrix can be found by substituting the sampling time T_s for time t and is given by

$$\Phi_k = \begin{bmatrix} 0 & T_s & 0 & 0 \\ 0 & 1 & 0 & 0 \\ 0 & 0 & 1 & T_s \\ 0 & 0 & 0 & 1 \end{bmatrix}$$

Because our states have been chosen to be Cartesian, the radar measurements r and θ will automatically be nonlinear functions of those states. Therefore, we must write the linearized measurement equation as

$$\begin{bmatrix} \Delta\theta^* \\ \Delta r^* \end{bmatrix} = \begin{bmatrix} \dfrac{\partial\theta}{\partial x_T} & \dfrac{\partial\theta}{\partial \dot{x}_T} & \dfrac{\partial\theta}{\partial y_T} & \dfrac{\partial\theta}{\partial \dot{y}_T} \\ \dfrac{\partial r}{\partial x_T} & \dfrac{\partial r}{\partial \dot{x}_T} & \dfrac{\partial r}{\partial y_T} & \dfrac{\partial r}{\partial \dot{y}_T} \end{bmatrix} \begin{bmatrix} \Delta x_T \\ \Delta \dot{x}_T \\ \Delta y_T \\ \Delta \dot{y}_T \end{bmatrix} + \begin{bmatrix} v_\theta \\ v_r \end{bmatrix}$$

where v_θ and v_r represent the measurement noise on angle and range, respectively. Because the angle from the radar to the projectile is given by

$$\theta = \tan^{-1}\left(\frac{y_T - y_R}{x_T - x_R}\right)$$

the four partial derivatives of the angle with respect to each of the states can be computed as

$$\frac{\partial\theta}{\partial x_T} = \frac{1}{1 + [(y_T - y_R)^2/(x_T - x_R)^2]} \frac{(x_T - x_R)*0 - (y_T - y_R)*1}{(x_T - x_R)^2} = \frac{-(y_T - y_R)}{r^2}$$

$$\frac{\partial\theta}{\partial \dot{x}_T} = 0$$

$$\frac{\partial\theta}{\partial y_T} = \frac{1}{1 + [(y_T - y_R)^2/(x_T - x_R)^2]} \frac{(x_T - x_R)*1 - (y_T - y_R)*0}{(x_T - x_R)^2} = \frac{(x_T - x_R)}{r^2}$$

$$\frac{\partial\theta}{\partial \dot{y}_T} = 0$$

Similarly, because the range from the radar to the projectile is given by

$$r = \sqrt{(x_T - x_R)^2 + (y_T - y_R)^2}$$

the four partial derivatives of the range with respect to each of the states can be computed as

$$\frac{\partial r}{\partial x_T} = \frac{1}{2}[(x_T - x_R)^2 + (y_T - y_R)^2]^{-1/2}2(x_T - x_R) = \frac{(x_T - x_R)}{r}$$

$$\frac{\partial r}{\partial \dot{x}_T} = 0$$

$$\frac{\partial r}{\partial y_T} = \frac{1}{2}[(x_T - x_R)^2 + (y_T - y_R)^2]^{-1/2}2(y_T - y_R) = \frac{(y_T - y_R)}{r}$$

$$\frac{\partial r}{\partial \dot{y}_T} = 0$$

Because the linearized measurement matrix is given by

$$H = \begin{bmatrix} \dfrac{\partial \theta}{\partial x_T} & \dfrac{\partial \theta}{\partial \dot{x}_T} & \dfrac{\partial \theta}{\partial y_T} & \dfrac{\partial \theta}{\partial \dot{y}_T} \\ \dfrac{\partial r}{\partial x_T} & \dfrac{\partial r}{\partial \dot{x}_T} & \dfrac{\partial r}{\partial y_T} & \dfrac{\partial r}{\partial \dot{y}_T} \end{bmatrix}$$

we can use substitution of the already derived partial derivatives to yield the linearized measurement matrix for this example to be

$$H = \begin{bmatrix} \dfrac{-(y_T - y_R)}{r^2} & 0 & \dfrac{x_T - x_R}{r^2} & 0 \\ \dfrac{x_T - x_R}{r} & 0 & \dfrac{y_T - y_R}{r} & 0 \end{bmatrix}$$

For this problem it is assumed that we know where the radar is so that x_R and y_R are known and do not have to be estimated. The states required for the discrete linearized measurement matrix will be based on the projected state estimate or

$$H_k = \begin{bmatrix} \dfrac{-(\bar{y}_{T_k} - y_R)}{\bar{r}_k^2} & 0 & \dfrac{\bar{x}_{T_k} - x_R}{\bar{r}^2} & 0 \\ \dfrac{\bar{x}_{T_k} - x_R}{\bar{r}_k} & 0 & \dfrac{\bar{y}_{T_k} - y_R}{\bar{r}_k} & 0 \end{bmatrix}$$

Because the discrete measurement noise matrix is given by

$$R_k = E(v_k v_k^T)$$

we can say that for this problem the discrete measurement noise matrix is

$$R_k = \begin{bmatrix} \sigma_\theta^2 & 0 \\ 0 & \sigma_r^2 \end{bmatrix}$$

where σ_θ^2 and σ_r^2 are the variances of the angle noise and range noise measurements, respectively. Similarly because the continuous process-noise matrix is given by

$$Q = E(ww^T)$$

we can easily show from the linear state-space equation for this example that the continuous process-noise matrix is

$$Q(t) = \begin{bmatrix} 0 & 0 & 0 & 0 \\ 0 & \Phi_s & 0 & 0 \\ 0 & 0 & 0 & 0 \\ 0 & 0 & 0 & \Phi_s \end{bmatrix}$$

where Φ_s is the spectral density of the white noise sources assumed to be on the downrange and altitude accelerations acting on the projectile. The discrete process-noise matrix can be derived from the continuous process-noise matrix according to

$$Q_k = \int_0^{T_s} \Phi(\tau)Q\Phi^T(\tau)\, dt$$

Therefore, substitution of the appropriate matrices into the preceding expression yields

$$Q_k = \int_0^{T_s} \begin{bmatrix} 1 & \tau & 0 & 0 \\ 0 & 1 & 0 & 0 \\ 0 & 0 & 1 & \tau \\ 0 & 0 & 0 & 1 \end{bmatrix} \begin{bmatrix} 0 & 0 & 0 & 0 \\ 0 & \Phi_s & 0 & 0 \\ 0 & 0 & 0 & 0 \\ 0 & 0 & 0 & \Phi_s \end{bmatrix} \begin{bmatrix} 1 & 0 & 0 & 0 \\ \tau & 1 & 0 & 0 \\ 0 & 0 & 1 & 0 \\ 0 & 0 & \tau & 1 \end{bmatrix} d\tau$$

After some algebra we get

$$Q_k = \int_0^{T_s} \begin{bmatrix} \tau^2\Phi_s & \tau\Phi_s & 0 & 0 \\ \tau\Phi_s & \Phi_s & 0 & 0 \\ 0 & 0 & \tau^2\Phi_s & \tau\Phi_s \\ 0 & 0 & \tau\Phi_s & \Phi_s \end{bmatrix} d\tau$$

Finally, after integration we obtain the final expression for the discrete process-noise matrix to be

$$
Q_k = \begin{bmatrix}
\dfrac{T_s^3 \Phi_s}{3} & \dfrac{T_s^2 \Phi_s}{2} & 0 & 0 \\[2ex]
\dfrac{T_s^2 \Phi_s}{2} & T_s \Phi_s & 0 & 0 \\[2ex]
0 & 0 & \dfrac{T_s^3 \Phi_s}{2} & \dfrac{T_s^2 \Phi_s}{2} \\[2ex]
0 & 0 & \dfrac{T_s^2 \Phi_s}{2} & T_s \Phi_s
\end{bmatrix}
$$

We now have defined all of the matrices required to solve the Riccati equations. The next step is to write down the equations for the Kalman-filter section. First, we must be able to propagate the states from the present sampling time to the next sampling time. This can be done in this example precisely with the fundamental matrix because the fundamental matrix is exact. In other nonlinear examples, where the fundamental matrix was approximate, we have already shown that the numerical integration of the nonlinear differential equations until the next update (i.e., using Euler integration) was required to eliminate potential filtering problems (i.e., hangoff error or even filter divergence).

Recall that for the linear filtering problem the real world was represented by the state-space equation

$$
\dot{x} = Fx + Gu + w
$$

where G is a matrix multiplying a known disturbance or control vector u that does not have to be estimated. We can show that the discrete linear Kalman-filtering equation is given by

$$
\hat{x}_k = \Phi_k \hat{x}_{k-1} + G_k u_{k-1} + K_k(z_k - H\Phi_k \hat{x}_{k-1} - HG_k u_{k-1})
$$

where G_k is obtained from

$$
G_k = \int_0^{T_s} \Phi(\tau) G \, d\tau
$$

If we forget about the gain times the residual portion of the filtering equation, we can see that the projected state is simply

$$
\bar{x}_k = \Phi_k \hat{x}_{k-1} + G_k u_{k-1}
$$

For this problem

$$
G = Gu = \begin{bmatrix} 0 \\ 0 \\ 0 \\ -g \end{bmatrix}
$$

Therefore, the G_k becomes

$$G_k = \int_0^{T_s} \begin{bmatrix} 1 & \tau & 0 & 0 \\ 0 & 1 & 0 & 0 \\ 0 & 0 & 1 & \tau \\ 0 & 0 & 0 & 1 \end{bmatrix} \begin{bmatrix} 0 \\ 0 \\ 0 \\ -g \end{bmatrix} d\tau = \begin{bmatrix} 0 \\ 0 \\ -\dfrac{gT_s^2}{2} \\ -gT_s \end{bmatrix}$$

and our projected state is determined from

$$\bar{x}_k = \begin{bmatrix} 1 & T_s & 0 & 0 \\ 0 & 1 & 0 & 0 \\ 0 & 0 & 1 & T_s \\ 0 & 0 & 0 & 1 \end{bmatrix} \hat{x}_{k-1} + \begin{bmatrix} 0 \\ 0 \\ -\dfrac{gT_s^2}{2} \\ -gT_s \end{bmatrix}$$

When we convert the preceding matrix equation for the projected states to four scalar equations, we obtain

$$\bar{x}_{T_k} = \hat{x}_{T_{k-1}} + T_s \hat{\dot{x}}_{T_{k-1}}$$

$$\overline{\dot{x}}_{T_k} = \hat{\dot{x}}_{T_{k-1}}$$

$$\bar{y}_{T_k} = \hat{y}_{T_{k-1}} + T_s \hat{\dot{y}}_{T_{k-1}} - 0.5gT_s^2$$

$$\overline{\dot{y}}_{T_k} = \hat{\dot{y}}_{T_{k-1}} - gT_s$$

The next portion of the Kalman filter uses gains times residuals. Because the measurements are nonlinear, the residuals are simply the measurements minus the projected values of the measurements (i.e., we do not want to use the linearized measurement matrix). Therefore, the projected value of the angle and range from the radar to the projectile must be based upon the projected state estimates or

$$\bar{\theta}_k = \tan^{-1}\left(\frac{\bar{y}_{T_{k-1}} - y_R}{\bar{x}_{T_{k-1}} - x_R}\right)$$

$$\bar{r}_k = \sqrt{(\bar{x}_{T_{k-1}} - x_R)^2 + (\bar{y}_{T_{k-1}} - y_R)^2}$$

Now the extended Kalman-filtering equations can be written simply as

$$\hat{x}_{T_k} = \bar{x}_{T_k} + K_{11_k}(\theta_k^* - \bar{\theta}_k) + K_{12_k}(r_k^* - \bar{r}_k)$$

$$\hat{\dot{x}}_{T_k} = \bar{\dot{x}}_{T_k} + K_{21_k}(\theta_k^* - \bar{\theta}_k) + K_{22_k}(r_k^* - \bar{r}_k)$$

$$\hat{y}_{T_k} = \bar{y}_{T_k} + K_{31_k}(\theta_k^* - \bar{\theta}_k) + K_{32_k}(r_k^* - \bar{r}_k)$$

$$\hat{\dot{y}}_{T_k} = \bar{\dot{y}}_{T_k} + K_{41_k}(\theta_k^* - \bar{\theta}_k) + K_{42_k}(r_k^* - \bar{r}_k)$$

where θ_k^* and r_k^* are the noisy measurements of radar angle and range. Again, notice that we are using the actual nonlinear measurement equations in the extended Kalman filter.

The preceding equations for the Kalman filter and Riccati equations were programmed and are shown in Listing 9.1, along with a simulation of the real world. We can see that the process-noise matrix is set to zero in this example (i.e., $\Phi_s = 0$). In addition, we have initialized the states of the filter close to the true values. The filter's position states are in error by 1000 ft, and the velocity states are in error by 100 ft/s. The initial covariance matrix reflects those errors. Also, because we have two independent measurements, the Riccati equation requires the inverse of a 2×2 matrix. This inverse is done exactly using the matrix inverse formula for a 2×2 matrix from Chapter 1.

The extended Kalman filter of Listing 9.1 was run for the nominal conditions described. Figure 9.7 compares the error in the estimate of the projectile's

Listing 9.1   Cartesian extended Kalman-filter simulation for projectile tracking problem

```
C THE FIRST THREE STATEMENTS INVOKE THE ABSOFT RANDOM
NUMBER GENERATOR ON THE MACINTOSH
        GLOBAL DEFINE
                INCLUDE 'quickdraw.inc'
        END
        IMPLICIT REAL*8(A-H,O-Z)
        REAL*8 P(4,4),Q(4,4),M(4,4),PHI(4,4),HMAT(2,4),HT(4,2),PHIT(4,4)
        REAL*8 RMAT(2,2),IDN(4,4),PHIP(4,4),PHIPPHIT(4,4),HM(2,4)
        REAL*8 HMHT(2,2),HMHTR(2,2),HMHTRINV(2,2),MHT(4,2),K(4,2)
        REAL*8 KH(4,4),IKH(4,4)
        INTEGER ORDER
        TS=1.
        ORDER=4
        PHIS=0.
        SIGTH=.01
        SIGR=100.
        VT=3000.
        GAMDEG=45.
        G=32.2
        XT=0.
        YT=0.
```

(continued)

Listing 9.1   (*Continued*)

```
        XTD=VT*COS(GAMDEG/57.3)
        YTD=VT*SIN(GAMDEG/57.3)
        XR=100000.
        YR=0.
        OPEN(1,STATUS='UNKNOWN',FILE='DATFIL')
        OPEN(2,STATUS='UNKNOWN',FILE='COVFIL')
        T=0.
        S=0.
        H=.001
        DO  14  I=1,ORDER
        DO  14  J=1,ORDER
        PHI(I,J)=0.
        P(I,J)=0.
        Q(I,J)=0.
        IDN(I,J)=0.
14      CONTINUE
        TS2=TS*TS
        TS3=TS2*TS
        Q(1,1)=PHIS*TS3/3.
        Q(1,2)=PHIS*TS2/2.
        Q(2,1)=Q(1,2)
        Q(2,2)=PHIS*TS
        Q(3,3)=Q(1,1)
        Q(3,4)=Q(1,2)
        Q(4,3)=Q(3,4)
        Q(4,4)=Q(2,2)
        PHI(1,1)=1.
        PHI(1,2)=TS
        PHI(2,2)=1.
        PHI(3,3)=1.
        PHI(3,4)=TS
        PHI(4,4)=1.
        CALL  MATTRN(PHI,ORDER,ORDER,PHIT)
        RMAT(1,1)=SIGTH**2
        RMAT(1,2)=0.
        RMAT(2,1)=0.
        RMAT(2,2)=SIGR**2
        IDN(1,1)=1.
        IDN(2,2)=1.
        IDN(3,3)=1.
        IDN(4,4)=1.
        P(1,1)=1000.**2
        P(2,2)=100.**2
        P(3,3)=1000.**2
        P(4,4)=100.**2
        XTH=XT+1000.
        XTDH=XTD-100.
        YTH=YT-1000.
        YTDH=YTD+100.
```

(*continued*)

Listing 9.1   (*Continued*)

```
WHILE(YT>=0.)
        XTOLD=XT
        XTDOLD=XTD
        YTOLD=YT
        YTDOLD=YTD
        XTDD=0.
        YTDD=-G
        XT=XT+H*XTD
        XTD=XTD+H*XTDD
        YT=YT+H*YTD
        YTD=YTD+H*YTDD
        T=T+H
        XTDD=0.
        YTDD=-G
        XT=.5*(XTOLD+XT+H*XTD)
        XTD=.5*(XTDOLD+XTD+H*XTDD)
        YT=.5*(YTOLD+YT+H*YTD)
        YTD=.5*(YTDOLD+YTD+H*YTDD)
        S=S+H
        IF(S>=(TS-.00001))THEN
                S=0.
                XTB=XTH+TS*XTDH
                XTDB=XTDH
                YTB=YTH+TS*YTDH-.5*G*TS*TS
                YTDB=YTDH-G*TS
                RTB=SQRT((XTB-XR)**2+(YTB-YR)**2)
                HMAT(1,1)=-(YTB-YR)/RTB**2
                HMAT(1,2)=0.
                HMAT(1,3)=(XTB-XR)/RTB**2
                HMAT(1,4)=0.
                HMAT(2,1)=(XTB-XR)/RTB
                HMAT(2,2)=0.
                HMAT(2,3)=(YTB-YR)/RTB
                HMAT(2,4)=0.
                CALL  MATTRN(HMAT,2,ORDER,HT)
                CALL  MATMUL(PHI,ORDER,ORDER,P,ORDER,
                    ORDER,PHIP)
                CALL  MATMUL(PHIP,ORDER,ORDER,PHIT,ORDER,
                    ORDER,PHIPPHIT)
                CALL  MATADD(PHIPPHIT,ORDER,ORDER,Q,M)
                CALL  MATMUL(HMAT,2,ORDER,M,ORDER,
                    ORDER,HM)
                CALL  MATMUL(HM,2,ORDER,HT,ORDER,2,HMHT)
                CALL  MATADD(HMHT,ORDER,ORDER,RMAT,
                    HMHTR)
                DET=HMHTR(1,1)*HMHTR(2,2)-HMHTR(1,2)*
                    HMHTR(2,1)
                HMHTRINV(1,1)=HMHTR(2,2)/DET
                HMHTRINV(1,2)=-HMHTR(1,2)/DET
```

1

(*continued*)

Listing 9.1   (*Continued*)

```
HMHTRINV(2,1)=-HMHTR(2,1)/DET
HMHTRINV(2,2)=HMHTR(1,1)/DET
CALL  MATMUL(M,ORDER,ORDER,HT,ORDER,2,MHT)
CALL  MATMUL(MHT,ORDER,2,HMHTRINV,2,2,K)
CALL  MATMUL(K,ORDER,2,HMAT,2,ORDER,KH)
CALL  MATSUB(IDN,ORDER,ORDER,KH,IKH)
CALL  MATMUL(IKH,ORDER,ORDER,M,ORDER,
     ORDER,P)
CALL  GAUSS(THETNOISE,SIGTH)
CALL  GAUSS(RTNOISE,SIGR)
THET=ATAN2((YT-YR),(XT-XR))
RT=SQRT((XT-XR)**2+(YT-YR)**2)
THETMEAS=THET+THETNOISE
RTMEAS=RT+RTNOISE
THETB=ATAN2((YTB-YR),(XTB-XR))
RTB=SQRT((XTB-XR)**2+(YTB-YR)**2)
RES1=THETMEAS-THETB
RES2=RTMEAS-RTB
XTH=XTB+K(1,1)*RES1+K(1,2)*RES2
XTDH=XTDB+K(2,1)*RES1+K(2,2)*RES2
YTH=YTB+K(3,1)*RES1+K(3,2)*RES2
YTDH=YTDB+K(4,1)*RES1+K(4,2)*RES2
ERRX=XT-XTH
SP11=SQRT(P(1,1))
ERRXD=XTD-XTDH
SP22=SQRT(P(2,2))
ERRY=YT-YTH
SP33=SQRT(P(3,3))
ERRYD=YTD-YTDH
SP44=SQRT(P(4,4))
WRITE(9,*)T,XT,YT,THET*57.3,RT
WRITE(1,*)T,XT,XTH,XTD,XTDH,YT,YTH,YTD,YTDH
WRITE(2,*)T,ERRX,SP11,-SP11,ERRXD,SP22,-SP22,
     ERRY,SP33,-SP33,ERRYD,SP44,-SP44
1          ENDIF
      END DO
      PAUSE
      CLOSE(1)
      END

C SUBROUTINE GAUSS IS SHOWN IN LISTING 1.8
C SUBROUTINE MATTRN IS SHOWN IN LISTING 1.3
C SUBROUTINE MATMUL IS SHOWN IN LISTING 1.4
C SUBROUTINE MATADD IS SHOWN IN LISTING 1.1
C SUBROUTINE MATSUB IS SHOWN IN LISTING 1.2
```

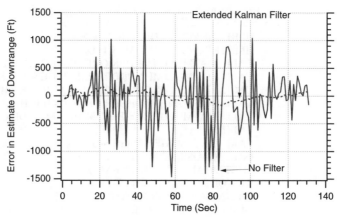

Fig. 9.7 Filtering dramatically reduces position error over using raw measurements.

downrange position for a single-run in which the extended Kalman filter is used in the preceding example in which there was no filtering at all (i.e., projectile's position reconstructed directly from range and angle measurements). The solid curve represents the case in which no filtering is used, whereas the dashed curve represents the error in the estimate when the extended Kalman filter is used. We can see that filtering dramatically reduces the error in the estimate of the projectile's position from more than 1000 ft to less than 100 ft.

To get a better intuitive feel for how the extended Kalman filter is performing, it is best to compare errors in the estimate of the projectile's position and velocity to the theoretical predictions of the covariance matrix. Figures 9.8 and 9.9 show how the single simulation run errors in the estimates of downrange position (i.e.,

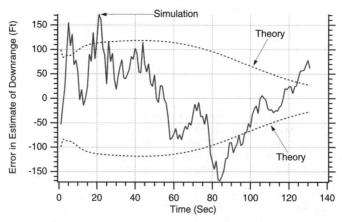

Fig. 9.8 Extended Cartesian Kalman filter's projectile's downrange estimates appear to agree with theory.

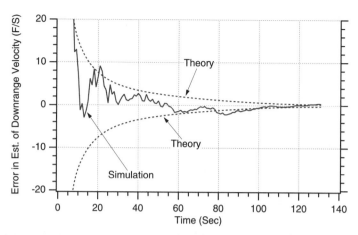

Fig. 9.9   Extended Cartesian Kalman filter's projectile's downrange velocity estimates appear to agree with theory.

x_T) and velocity (i.e., $\dot{x}_T$) compare with the theoretical predictions of the Riccati equation covariance matrix (i.e., square root of $P11$ and $P22$, respectively). We can see that the single flight simulation results lie within the theoretical bounds approximately 68% of the time, giving us a good feeling that the extended Cartesian Kalman filter is performing properly. Because there is no process noise in this example, the errors in the estimates continue to get smaller as more measurements are taken.

Similarly, Figs. 9.10 and 9.11 show how the single-run simulation errors in the estimates of the projectile's altitude (i.e., y_T) and altitude velocity (i.e., $\dot{y}_T$) compare with the theoretical predictions of the covariance matrix (i.e., square root of $P33$ and $P44$). Again, we can see that the single-run simulation results lie

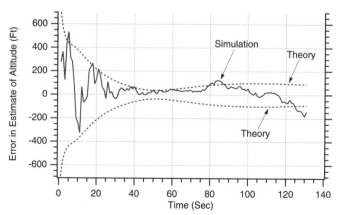

Fig. 9.10   Extended Cartesian Kalman filter's projectile's altitude estimates appear to agree with theory.

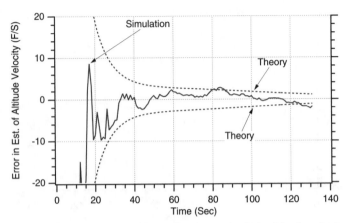

Fig. 9.11   Extended Cartesian Kalman filter's projectile's altitude velocity estimates appear to agree with theory.

within the theoretical bounds 68% of the time, giving us a good indication that the filter is performing properly.

In summary, we can say that for this example both the projectile's downrange and altitude position estimates appeared to be good to within 100 ft. The projectile's downrange and altitude velocity estimates appear to be good to within a few feet per second of the actual velocity. The results were for the case in which there was zero process noise. Adding process noise for practical reasons (i.e., prevent filter divergence when projectile does something that is not anticipated) will make the errors in the estimates larger. Recall that the process-noise matrix for the Cartesian extended Kalman filter of this example is given by

$$
Q_k = \begin{bmatrix}
\dfrac{T_s^3 \Phi_s}{3} & \dfrac{T_s^2 \Phi_s}{2} & 0 & 0 \\[2mm]
\dfrac{T_s^2 \Phi_s}{2} & T_s \Phi_s & 0 & 0 \\[2mm]
0 & 0 & \dfrac{T_s^3 \Phi_s}{3} & \dfrac{T_s^2 \Phi_s}{2} \\[2mm]
0 & 0 & \dfrac{T_s^2 \Phi_s}{2} & T_s \Phi_s
\end{bmatrix}
$$

Therefore, we can see that the amount of process noise is determined by Φ_s. Several runs were made in which the amount of process noise was varied, and the square root of the second diagonal element of the covariance matrix was saved. This covariance matrix element represents the theoretical error in the estimate of the projectile's downrange velocity. We can see from Fig. 9.12 that when there is no process noise (i.e., $\Phi_s = 0$) the projectile's velocity errors get smaller as time goes on (i.e., more measurements are taken). However, with process noise the

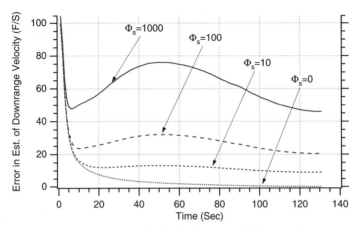

Fig. 9.12   Errors in estimate of velocity increase with increasing process noise.

error in the estimate of the projectile's downrange velocity approaches a steady state. In other words, errors in the estimate of the projectile's velocity do not decrease as more measurements are taken. We can see that the steady-state value of the error in the estimate of the projectile's velocity increases with increasing values of the process noise Φ_s.

Polar Coordinate System

In the preceding section we derived an extended Kalman filter, whose states were in a Cartesian coordinate system and whose measurements were in a polar coordinate system. With this methodology the model of the real world could easily be represented by linear differential equations, while the measurements were nonlinear functions of the states. In this section we will derive an extended Kalman filter based on a different set of states. These states will be for a polar coordinate system, which in turn will lead to measurements that are linear functions of the states. However, the resulting polar differential equations representing the real world will now be more complex and nonlinear. Eventually we want to see if an extended polar Kalman filter, based on this alternative representation, will perform better than the extended Cartesian Kalman filter. To accomplish this task, we must first derive and verify the polar equations for the cannon-launched projectile tracking problem.

Recall from Fig. 9.1 that the angle and range from the radar to the cannon ball is given by

$$\theta = \tan^{-1}\left(\frac{y_T - y_R}{x_T - x_R}\right)$$

$$r = \sqrt{(x_T - x_R)^2 + (y_T - y_R)^2}$$

Taking the first derivative of the angle yields

$$\dot\theta = \frac{1}{1 + (y_T - y_R)^2/(x_T - x_R)^2} \frac{(x_T - x_R)\dot y_T - (y_T - y_R)\dot x_T}{(x_T - x_R)^2}$$

After some simplification we can substitute the formula for range into the preceding expression to obtain

$$\dot\theta = \frac{(x_T - x_R)\dot y_T - (y_T - y_R)\dot x_T}{r^2}$$

Using calculus, we can also take the first derivative of range

$$\dot r = \frac{1}{2}[(x_T - x_R)^2 + (y_7 - y_R)^2]^{-1/2}[2(x_T - x_R)\dot x_T + 2(y_T - y_R)\dot y_T]$$

which simplifies to

$$\dot r = \frac{(x_T - x_R)\dot x_T + (y_T - y_R)\dot y_T}{r}$$

We can also find the angular acceleration by taking the derivative of the angular velocity or

$$\ddot\theta = \frac{r^2[\dot x_T \dot y_T + (x_T - x_R)\ddot y_t - \dot x_T \dot y_T - (y_T - y_R)\ddot x_T] - [(x_T - x_R)\dot y_T - (y_T - y_R)\dot y_T]2r\dot r}{r^4}$$

After some simplification we obtain

$$\ddot\theta = \frac{(x_T - x_R)\ddot y_T - (y_T - y_R)\ddot x_T - 2r\dot r\dot\theta}{r^2}$$

Recognizing that

$$\ddot x_T = 0$$

$$\ddot y_T = -g$$

$$\cos\theta = \frac{x_T - x_R}{r}$$

$$\sin\theta = \frac{y_T - y_R}{r}$$

the formula for angular acceleration simplifies even further to

$$\ddot\theta = \frac{-g\cos\theta - 2\dot r\dot\theta}{r}$$

Finally, the second derivative of range yields

$$\ddot{r} = \frac{r[\dot{x}_T\dot{x}_T + (x_T - x_R)\ddot{x}_T + \dot{y}_T\dot{y}_T + (y_T - y_R)\ddot{y}_T] - [(x_T - x_R)\dot{x}_T + (y_T - y_R)\dot{y}_T]\dot{r}}{r^2}$$

After a great deal of algebraic manipulation and simplification, we obtain

$$\ddot{r} = \frac{r^2\dot{\theta}^2 - gr\sin\theta}{r}$$

In summary, we are saying that the nonlinear polar differential equations representing the cannon-launched projectile are given by

$$\ddot{\theta} = \frac{-g\cos\theta - 2\dot{r}\dot{\theta}}{r}$$

$$\ddot{r} = \frac{r^2\dot{\theta}^2 - gr\sin\theta}{r}$$

The preceding two nonlinear differential equations are completely equivalent to the linear Cartesian differential equations

$$\ddot{x}_T = 0$$

$$\ddot{y}_T = -g$$

if they both have the same initial conditions. For example, if the initial conditions in the Cartesian equations are denoted

$$x_T(0), \quad y_T(0), \quad \dot{x}_T(0), \quad \dot{y}_T(0)$$

then the initial conditions on the polar equations must be

$$\theta(0) = \tan^{-1}\left[\frac{y_T(0) - y_R}{x_T(0) - x_R}\right]$$

$$r(0) = \sqrt{[x_T(0) - x_R]^2 + [y_T(0) - y_R]^2}$$

$$\dot{\theta}(0) = \frac{[x_T(0) - x_R]\dot{y}_T(0) - [y_T(0) - y_R]\dot{x}_T(0)}{r(0)^2}$$

$$\dot{r}(0) = \frac{[x_T(0) - x_R]\dot{x}_T(0) + [y_T(0) - y_R]\dot{y}_T(0)}{r(0)}$$

When we have integrated the polar equations, we must also then find a way to convert the resultant answers back to Cartesian coordinates in order to perform a direct comparison with the Cartesian differential equations. The conversion can be easily accomplished by recognizing that

$$\hat{x}_T = r\cos\theta + x_R$$

$$\hat{y}_T = r\sin\theta + y_R$$

where the caret denotes the estimates based on the conversion. Taking the derivative of the preceding two equations yields the estimated velocity components, which are also based on the conversion or

$$\hat{\dot{x}}_T = \dot{r}\cos\theta - r\dot{\theta}\sin\theta$$

$$\hat{\dot{y}}_T = \dot{r}\sin\theta + r\dot{\theta}\cos\theta$$

The simulation that compares the flight of the cannon-launched projectile in both the polar and Cartesian coordinate systems appears in Listing 9.2. We can see that the simulation consists of both Cartesian and polar differential equations for the projectile. Both sets of differential equations have the same initial conditions and integration step size. The location and velocity of the projectile, based on the Cartesian and polar differential equations, are printed out every second.

The nominal case of Listing 9.2 was run, and we can see from Fig. 9.13 that the projectile locations for both the Cartesian and polar formulation of the

Listing 9.2   Simulation that compares polar and Cartesian differential equations of cannon-launched projectile

```
IMPLICIT REAL*8(A-H,O-Z)
VT=3000.
GAMDEG=45.
G=32.2
TS=1.
XT=0.
YT=0.
XTD=VT*COS(GAMDEG/57.3)
YTD=VT*SIN(GAMDEG/57.3)
XR=100000.
YR=0.
OPEN(1,STATUS='UNKNOWN',FILE='DATFIL')
T=0.
S=0.
H=.001
THET=ATAN2((YT-YR),(XT-XR))
RT=SQRT((XT-XR)**2+(YT-YR)**2)
THETD=((XT-XR)*YTD-(YT-YR)*XTD)/RT**2
RTD=((XT-XR)*XTD+(YT-YR)*YTD)/RT
WHILE(YT>=0.)
        XTOLD=XT
        XTDOLD=XTD
        YTOLD=YT
        YTDOLD=YTD
        THETOLD=THET
```

(continued)

Listing 9.2   (*Continued*)

```
THETDOLD=THETD
RTOLD=RT
RTDOLD=RTD
XTDD=0.
YTDD=-G
THETDD=(-G*COS(THET)-2.*RTD*THETD)/RT
RTDD=((RT*THETD)**2-G*RT*SIN(THET))/RT
XT=XT+H*XTD
XTD=XTD+H*XTDD
YT=YT+H*YTD
YTD=YTD+H*YTDD
THET=THET+H*THETD
THETD=THETD+H*THETDD
RT=RT+H*RTD
RTD=RTD+H*RTDD
T=T+H
XTDD=0.
YTDD=-G
THETDD=(-G*COS(THET)-2.*RTD*THETD)/RT
RTDD=((RT*THETD)**2-G*RT*SIN(THET))/RT
XT=.5*(XTOLD+XT+H*XTD)
XTD=.5*(XTDOLD+XTD+H*XTDD)
YT=.5*(YTOLD+YT+H*YTD)
YTD=.5*(YTDOLD+YTD+H*YTDD)
THET=.5*(THETOLD+THET+H*THETD)
THETD=.5*(THETDOLD+THETD+H*THETDD)
RT=.5*(RTOLD+RT+H*RTD)
RTD=.5*(RTDOLD+RTD+H*RTDD)
S=S+H
IF(S>=(TS-.00001))THEN
        S=0.
        XTH=RT*COS(THET)+XR
        YTH=RT*SIN(THET)+YR
        XTDH=RTD*COS(THET)-RT*SIN(THET)*THETD
        YTDH=RTD*SIN(THET)+RT*COS(THET)*THETD
        WRITE(9,*)T,XT,XTH,YT,YTH,XTD,XTDH,YTD,YTDH
        WRITE(1,*)T,XT,XTH,YT,YTH,XTD,XTDH,YTD,YTDH
    ENDIF
END DO
PAUSE
CLOSE(1)
END
```

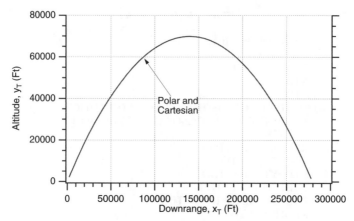

Fig. 9.13 Polar and Cartesian differential equations are identical in position.

differential equations are identical. In addition, Fig. 9.14 shows that the projectile component velocities are also identical in both coordinate systems, indicating the polar nonlinear differential equations have been derived correctly and are completely equivalent to the Cartesian linear differential equations.

Extended Polar Kalman Filter

We have already derived and validated the nonlinear polar differential equations for the cannon-launched projectile, which are given by

$$\ddot{\theta} = \frac{-g \cos \theta - 2\dot{r}\dot{\theta}}{r}$$

$$\ddot{r} = \frac{r^2 \dot{\theta}^2 - gr \sin \theta}{r}$$

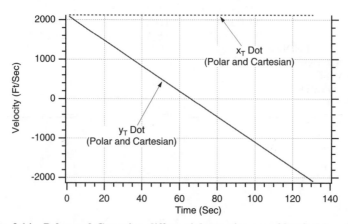

Fig. 9.14 Polar and Cartesian differential equations are identical in velocity.

In addition, we have already seen that, for this example, the extended Cartesian Kalman filter performed well without a process-noise matrix because the filter model of the real world was perfect. Let us assume, for purposes of comparison, that we also can neglect the process noise in polar coordinates. Under these circumstances we can express the nonlinear polar differential equations describing projectile motion in state-space form as

$$
\begin{bmatrix} \Delta\dot{\theta} \\ \Delta\ddot{\theta} \\ \Delta\dot{r} \\ \Delta\ddot{r} \end{bmatrix} =
\begin{bmatrix}
\dfrac{\partial\dot{\theta}}{\partial\theta} & \dfrac{\partial\dot{\theta}}{\partial\dot{\theta}} & \dfrac{\partial\dot{\theta}}{\partial r} & \dfrac{\partial\dot{\theta}}{\partial\dot{r}} \\
\dfrac{\partial\ddot{\theta}}{\partial\theta} & \dfrac{\partial\ddot{\theta}}{\partial\dot{\theta}} & \dfrac{\partial\ddot{\theta}}{\partial r} & \dfrac{\partial\ddot{\theta}}{\partial\dot{r}} \\
\dfrac{\partial\dot{r}}{\partial\theta} & \dfrac{\partial\dot{r}}{\partial\dot{\theta}} & \dfrac{\partial\dot{r}}{\partial r} & \dfrac{\partial\dot{r}}{\partial\dot{r}} \\
\dfrac{\partial\ddot{r}}{\partial\theta} & \dfrac{\partial\ddot{r}}{\partial\dot{\theta}} & \dfrac{\partial\ddot{r}}{\partial r} & \dfrac{\partial\ddot{r}}{\partial\dot{r}}
\end{bmatrix}
\begin{bmatrix} \Delta\theta \\ \Delta\dot{\theta} \\ \Delta r \\ \Delta\dot{r} \end{bmatrix} =
\begin{bmatrix}
0 & 1 & 0 & 0 \\
\dfrac{\partial\ddot{\theta}}{\partial\theta} & \dfrac{\partial\ddot{\theta}}{\partial\dot{\theta}} & \dfrac{\partial\ddot{\theta}}{\partial r} & \dfrac{\partial\ddot{\theta}}{\partial\dot{r}} \\
0 & 0 & 0 & 1 \\
\dfrac{\partial\ddot{r}}{\partial\theta} & \dfrac{\partial\ddot{r}}{\partial\dot{\theta}} & \dfrac{\partial\ddot{r}}{\partial r} & \dfrac{\partial\ddot{r}}{\partial\dot{r}}
\end{bmatrix}
\begin{bmatrix} \Delta\theta \\ \Delta\dot{\theta} \\ \Delta r \\ \Delta\dot{r} \end{bmatrix}
$$

In this case the system dynamics matrix, which is the preceding matrix of partial derivatives, can be written by inspection as

$$
F = \begin{bmatrix}
0 & 1 & 0 & 0 \\
\dfrac{g\sin\theta}{r} & \dfrac{-2\dot{r}}{r} & \dfrac{g\cos\theta + 2\dot{\theta}\dot{r}}{r^2} & \dfrac{-2\dot{\theta}}{r} \\
0 & 0 & 0 & 1 \\
-g\cos\theta & 2r\dot{\theta} & \dot{\theta}^2 & 0
\end{bmatrix}
$$

Therefore, if we use a two-term Taylor-series approximation for the fundamental matrix, we get

$$
\Phi(t) = I + Ft = \begin{bmatrix}
1 & 0 & 0 & 0 \\
0 & 1 & 0 & 0 \\
0 & 0 & 1 & 0 \\
0 & 0 & 0 & 1
\end{bmatrix} +
\begin{bmatrix}
0 & t & 0 & 0 \\
\dfrac{g\sin\theta}{r}t & \dfrac{-2\dot{r}}{r}t & \dfrac{g\cos\theta + 2\dot{\theta}\dot{r}}{r^2}t & \dfrac{-2\dot{\theta}}{r}t \\
0 & 0 & 0 & t \\
-g\cos\theta\, t & 2r\dot{\theta}t & \dot{\theta}^2 t & 0
\end{bmatrix}
$$

After some simplification we finally obtain

$$
\Phi(t) = \begin{bmatrix}
1 & t & 0 & 0 \\
\dfrac{g\sin\theta}{r}t & 1 - \dfrac{2\dot{r}}{r}t & \dfrac{g\cos\theta + 2\dot{\theta}\dot{r}}{r^2}t & \dfrac{-2\dot{\theta}}{r}t \\
0 & 0 & 1 & t \\
-g\cos\theta\, t & 2r\dot{\theta}t & \dot{\theta}^2 t & 1
\end{bmatrix}
$$

The discrete fundamental matrix can be found by substituting T_s for t in the preceding expression, thus obtaining

$$\Phi_k = \begin{bmatrix} 1 & T_s & 0 & 0 \\ \dfrac{g\sin\theta}{r}T_s & 1 - \dfrac{2\dot{r}}{r}T_s & \dfrac{g\cos\theta + 2\dot{\theta}\dot{r}}{r^2}T_s & \dfrac{-2\dot{\theta}}{r}T_s \\ 0 & 0 & 1 & T_s \\ -g\cos\theta T_s & 2r\dot{\theta}T_s & \dot{\theta}^2 T_s & 1 \end{bmatrix}$$

Because the states are polar, the measurement equation is linear and can be expressed as

$$\begin{bmatrix} \theta^* \\ r^* \end{bmatrix} = \begin{bmatrix} 1 & 0 & 0 & 0 \\ 0 & 0 & 1 & 0 \end{bmatrix} \begin{bmatrix} \theta \\ \dot{\theta} \\ r \\ \dot{r} \end{bmatrix} + \begin{bmatrix} v_\theta \\ v_r \end{bmatrix}$$

where the measurement matrix is

$$H = \begin{bmatrix} 1 & 0 & 0 & 0 \\ 0 & 0 & 1 & 0 \end{bmatrix}$$

Because the discrete measurement noise matrix is given by

$$R_k = E(v_k v_k^T)$$

we can say that for this problem the discrete measurement noise matrix turns out to be

$$R_k = \begin{bmatrix} \sigma_\theta^2 & 0 \\ 0 & \sigma_r^2 \end{bmatrix}$$

where σ_θ^2 and σ_r^2 are the variances of the angle noise and range noise measurements, respectively.

Because we are assuming zero process noise, we now have defined all of the matrices required to solve the Riccati equations. However, if we want to compare the resultant polar covariance matrices obtained from the Riccati equations with the preceding Cartesian covariance matrices, then we must have a way of going between both systems. Recall that we have shown in the preceding section that the four equations relating the polar description of the projectile to the Cartesian description of the projectile are given by

$$x_T = r \cos \theta + x_R$$

$$y_T = r \sin \theta + y_R$$

$$\dot{x}_T = \dot{r} \cos \theta - r\dot{\theta} \sin \theta$$

$$\dot{y}_T = \dot{r} \sin \theta + r\dot{\theta} \cos \theta$$

We can use the chain rule from calculus to get the total differentials

$$\Delta x_T = \frac{\partial x_T}{\partial \theta} \Delta\theta + \frac{\partial x_T}{\partial \dot{\theta}} \Delta\dot{\theta} + \frac{\partial x_T}{\partial r} \Delta r + \frac{\partial x_T}{\partial \dot{r}} \Delta\dot{r}$$

$$\Delta \dot{x}_T = \frac{\partial \dot{x}_T}{\partial \theta} \Delta\theta + \frac{\partial \dot{x}_T}{\partial \dot{\theta}} \Delta\dot{\theta} + \frac{\partial \dot{x}_T}{\partial r} \Delta r + \frac{\partial \dot{x}_T}{\partial \dot{r}} \Delta\dot{r}$$

$$\Delta y_T = \frac{\partial y_T}{\partial \theta} \Delta\theta + \frac{\partial y_T}{\partial \dot{\theta}} \Delta\dot{\theta} + \frac{\partial y_T}{\partial r} \Delta r + \frac{\partial y_T}{\partial \dot{r}} \Delta\dot{r}$$

$$\Delta \dot{y}_T = \frac{\partial \dot{y}_T}{\partial \theta} \Delta\theta + \frac{\partial \dot{y}_T}{\partial \dot{\theta}} \Delta\dot{\theta} + \frac{\partial \dot{y}_T}{\partial r} \Delta r + \frac{\partial \dot{y}_T}{\partial \dot{r}} \Delta\dot{r}$$

Evaluating each of the partial derivatives yields the following expressions for each of the total differentials:

$$\Delta x_T = -r \sin \theta \Delta\theta + \cos \theta \Delta r$$

$$\Delta \dot{x}_T = (-\dot{r} \sin \theta - r\dot{\theta} \cos \theta)\Delta\theta - r \sin \theta \Delta\dot{\theta} - \dot{\theta} \sin \theta \Delta r + \cos \theta \Delta\dot{r}$$

$$\Delta y_T = r \cos \theta \Delta\theta + \sin \theta \Delta r$$

$$\Delta \dot{y}_T = (\dot{r} \cos \theta - r\dot{\theta} \sin \theta)\Delta\theta + r \cos \theta \Delta\dot{\theta} + \dot{\theta} \cos \theta \Delta r + \sin \theta \Delta\dot{r}$$

Placing the preceding set of equations into state-space form yields a more compact expression for the total differential

$$\begin{bmatrix} \Delta x_T \\ \Delta \dot{x}_T \\ \Delta y_T \\ \Delta \dot{y}_T \end{bmatrix} = \begin{bmatrix} -r \sin \theta & 0 & \cos \theta & 0 \\ -\dot{r} \sin \theta - r\dot{\theta} \cos \theta & -r \sin \theta & -\dot{\theta} \sin \theta & \cos \theta \\ r \cos \theta & 0 & \sin \theta & 0 \\ \dot{r} \cos \theta & r \cos \theta & \dot{\theta} \cos \theta & \sin \theta \end{bmatrix} \begin{bmatrix} \Delta\theta \\ \Delta\dot{\theta} \\ \Delta r \\ \Delta\dot{r} \end{bmatrix}$$

If we define the transformation matrix A relating the polar total differentials to the Cartesian total differentials, we get

$$
A = \begin{bmatrix}
-r\sin\theta & 0 & \cos\theta & 0 \\
-\dot{r}\sin\theta - r\dot{\theta}\cos\theta & -r\sin\theta & -\dot{\theta}\sin\theta & \cos\theta \\
r\cos\theta & 0 & \sin\theta & 0 \\
\dot{r}\cos\theta & r\cos\theta & \dot{\theta}\cos\theta & \sin\theta
\end{bmatrix}
$$

It is easy to show that the Cartesian covariance matrix P_{CART} is related to the polar covariance matrix P_{POL} according to

$$
P_{CART} = A P_{POL} A^T
$$

The next step is to write down the equations for the Kalman-filter section. First, we must be able to propagate the states from the present sampling time to the next sampling time. This could be done exactly with the fundamental matrix when we were working in the Cartesian system because the fundamental matrix was exact. In the polar system the fundamental matrix is approximate, and so it is best to numerically integrate the nonlinear differential equations for a sampling interval to get the projected states. Now the Kalman-filtering equations can be written as

$$
\hat{\theta}_k = \bar{\theta}_k + K_{11_k}(\theta_k^* - \bar{\theta}_k) + K_{12_k}(r_k^* - \bar{r}_k)
$$

$$
\hat{\dot{\theta}}_k = \bar{\dot{\theta}}_k + K_{21_k}(\theta_k^* - \bar{\theta}_k) + K_{22_k}(r_k^* - \bar{r}_k)
$$

$$
\hat{r}_k = \bar{r}_k + K_{31_k}(\theta_k^* - \bar{\theta}_k) + K_{32_k}(r_k^* - \bar{r}_k)
$$

$$
\hat{\dot{r}}_k = \bar{\dot{r}}_k + K_{41_k}(\theta_k^* - \bar{\theta}_k) + K_{42_k}(r_k^* - \bar{r}_k)
$$

where θ_k^* and r_k^* are the noisy measurements of radar angle and range and the bar indicates the projected states that are obtained directly by numerical integration.

The preceding equations for the polar Kalman filter and Riccati equations were programmed and are shown in Listing 9.3, along with a simulation of the real world. There is no process-noise matrix in this formulation. We have initialized the states of the polar filter to have the same initialization of the Cartesian filter. The initial covariance matrix reflects the initial state estimate errors. The subroutine PROJECT handles the numerical integration of the nonlinear polar differential equations to get the projected value of the states T_s seconds in the future.

The nominal case of Listing 9.3 was run, and the polar and Cartesian errors in the estimates were computed. By comparing Fig. 9.15 with Fig. 9.8, we can see that the errors in the estimates of downrange are comparable for both filters. Initially, the error in the estimate of x is 100 ft, and after 130 s the error reduces to approximately 25 ft. Figure 9.15 also shows that the single-run simulation error in the estimate of x appears to be correct when compared to the theoretical

Listing 9.3   Polar extended Kalman filter simulation for projectile tracking problem

```
C THE FIRST THREE STATEMENTS INVOKE THE ABSOFT RANDOM
  NUMBER GENERATOR ON THE MACINTOSH
        GLOBAL DEFINE
                INCLUDE 'quickdraw.inc'
        END
        IMPLICIT REAL*8 (A-H)
        IMPLICIT REAL*8 (O-Z)
        REAL*8 P(4,4),M(4,4),GAIN(4,2),PHI(4,4),PHIT(4,4),PHIP(4,4)
        REAL*8 Q(4,4),HMAT(2,4),HM(2,4),MHT(4,2),PHIPPHIT(4,4),AT(4,4)
        REAL*8 RMAT(2,2),HMHTR(2,2),HMHTRINV(2,2),A(4,4),PNEW(4,4)
        REAL*8 HMHT(2,2),HT(4,2),KH(4,4),IDN(4,4),IKH(4,4),FTS(4,4)
        REAL*8 AP(4,4)
        INTEGER ORDER
106     CONTINUE
        OPEN(1,STATUS='UNKNOWN',FILE='DATFIL')
        OPEN(2,STATUS='UNKNOWN',FILE='COVFIL')
        OPEN(3,STATUS='UNKNOWN',FILE='COVFIL2')
        SIGTH=.01
        SIGR=100.
        TS=1.
        G=32.2
        VT=3000.
        GAMDEG=45.
        XT=0.
        YT=0.
        XTD=VT*COS(GAMDEG/57.3)
        YTD=VT*SIN(GAMDEG/57.3)
        XR=100000.
        YR=0.
        XTH=XT+1000.
        YTH=YT-1000.
        XTDH=XTD-100.
        YTDH=YTD+100.
        TH=ATAN2((YT-YR),(XT-XR)+.001)
        R=SQRT((XT-XR)**2+(YT-YR)**2)
        THD=((XT-XR)*YTD-(YT-YR)*XTD)/R**2
        RD=((XT-XR)*XTD+(YT-YR)*YTD)/R
        THH=ATAN2((YTH-YR),(XTH-XR)+.001)
        RH=SQRT((XTH-XR)**2+(YTH-YR)**2)
        THDH=((XTH-XR)*YTDH-(YTH-YR)*XTDH)/RH**2
        RDH=((XTH-XR)*XTDH+(YTH-YR)*YTDH)/RH
        ORDER=4
        TF=100.
        T=0.
        S=0.
        H=.001
        HP=.001
        DO 1000 I=1,ORDER
        DO 1000 J=1,ORDER
```

(*continued*)

Listing 9.3 *(Continued)*

```
                P(I,J)=0.
                Q(I,J)=0.
                IDN(I,J)=0.
                PHI(I,J)=0.
                FTS(I,J)=0.
                A(I,J)=0.
1000    CONTINUE
        IDN(1,1)=1.
        IDN(2,2)=1.
        IDN(3,3)=1.
        IDN(4,4)=1.
        P(1,1)=(TH-THH)**2
        P(2,2)=(THD-THDH)**2
        P(3,3)=(R-RH)**2
        P(4,4)=(RD-RDH)**2
        RMAT(1,1)=SIGTH**2
        RMAT(1,2)=0.
        RMAT(2,1)=0.
        RMAT(2,2)=SIGR**2
        HMAT(1,1)=1.
        HMAT(1,2)=0.
        HMAT(1,3)=0.
        HMAT(1,4)=0.
        HMAT(2,1)=0.
        HMAT(2,2)=0.
        HMAT(2,3)=1.
        HMAT(2,4)=0.
        WHILE(YT>=0.)
                THOLD=TH
                THDOLD=THD
                ROLD=R
                RDOLD=RD
                THDD=(-G*R*COS(TH)-2.*THD*R*RD)/R**2
                RDD=(R*R*THD*THD-G*R*SIN(TH))/R
                TH=TH+H*THD
                THD=THD+H*THDD
                R=R+H*RD
                RD=RD+H*RDD
                T=T+H
                THDD=(-G*R*COS(TH)-2.*THD*R*RD)/R**2
                RDD=(R*R*THD*THD-G*R*SIN(TH))/R
                TH=.5*(THOLD+TH+H*THD)
                THD=.5*(THDOLD+THD+H*THDD)
                R=.5*(ROLD+R+H*RD)
                RD=.5*(RDOLD+RD+H*RDD)
                S=S+H
                IF(S>=(TS-.00001))THEN
                        S=0.
                        FTS(1,2)=1.*TS
                        FTS(2,1)=G*SIN(THH)*TS/RH
```

(continued)

Listing 9.3   (*Continued*)

```
FTS(2,2)=-2.*RDH*TS/RH
FTS(2,3)=(G*COS(THH)+2.*THDH*RDH)*TS/RH**2
FTS(2,4)=-2.*THDH*TS/RH
FTS(3,4)=1.*TS
FTS(4,1)=-G*COS(THH)*TS
FTS(4,2)=2.*RH*THDH*TS
FTS(4,3)=(THDH**2)*TS
CALL  MATADD(FTS,ORDER,ORDER,IDN,PHI)
CALL  MATTRN(HMAT,2,ORDER,HT)
CALL  MATTRN(PHI,ORDER,ORDER,PHIT)
CALL  MATMUL(PHI,ORDER,ORDER,P,ORDER,
  ORDER,PHIP)
CALL  MATMUL(PHIP,ORDER,ORDER,PHIT,ORDER,
  ORDER, PHIPPHIT)
CALL  MATADD(PHIPPHIT,ORDER,ORDER,Q,M)
CALL  MATMUL(HMAT,2,ORDER,M,ORDER,
  ORDER,HM)
CALL  MATMUL(HM,2,ORDER,HT,ORDER,2,HMHT)
CALL  MATADD(HMHT,2,2,RMAT,HMHTR)
DET=HMHTR(1,1)*HMHTR(2,2)-HMHTR(1,2)*
  HMHTR(2,1)
HMHTRINV(1,1)=HMHTR(2,2)/DET
HMHTRINV(1,2)=-HMHTR(1,2)/DET
HMHTRINV(2,1)=-HMHTR(2,1)/DET
HMHTRINV(2,2)=HMHTR(1,1)/DET
CALL  MATMUL(M,ORDER,ORDER,HT,ORDER,
  2,MHT)
CALL  MATMUL(MHT,ORDER,2,HMHTRINV,2,
  2,GAIN)
CALL  MATMUL(GAIN,ORDER,2,HMAT,2,ORDER,KH)
CALL  MATSUB(IDN,ORDER,ORDER,KH,IKH)
CALL  MATMUL(IKH,ORDER,ORDER,M,ORDER,
  ORDER,P)
CALL  GAUSS(THNOISE,SIGTH)
CALL  GAUSS(RNOISE,SIGR)
CALL  PROJECT(T,TS,THH,THDH,RH,RDH,THB,
  THDB,RB,RDB,HP)
RES1=TH+THNOISE-THB
RES2=R+RNOISE-RB
THH=THB+GAIN(1,1)*RES1+GAIN(1,2)*RES2
THDH=THDB+GAIN(2,1)*RES1+GAIN(2,2)*RES2
RH=RB+GAIN(3,1)*RES1+GAIN(3,2)*RES2
RDH=RDB+GAIN(4,1)*RES1+GAIN(4,2)*RES2
ERRTH=TH-THH
SP11=SQRT(P(1,1))
ERRTHD=THD-THDH
SP22=SQRT(P(2,2))
ERRR=R-RH
```

(*continued*)

Listing 9.3 (*Continued*)

```
SP33=SQRT(P(3,3))
ERRRD=RD-RDH
SP44=SQRT(P(4,4))
XT=R*COS(TH)+XR
YT=R*SIN(TH)+YR
XTD=RD*COS(TH)-R*THD*SIN(TH)
YTD=RD*SIN(TH)+R*THD*COS(TH)
XTH=RH*COS(THH)+XR
YTH=RH*SIN(THH)+YR
XTDH=RDH*COS(THH)-RH*THDH*SIN(THH)
YTDH=RDH*SIN(THH)+RH*THDH*COS(THH)
A(1,1)=-RH*SIN(THH)
A(1,3)=COS(THH)
A(2,1)=-RDH*SIN(THH)-RH*THDH*COS(THH)
A(2,2)=-RH*SIN(THH)
A(2,3)=-THDH*SIN(THH)
A(2,4)=COS(THH)
A(3,1)=RH*COS(THH)
A(3,3)=SIN(THH)
A(4,1)=RDH*COS(THH)-RH*SIN(THH)*THDH
A(4,2)=RH*COS(THH)
A(4,3)=THDH*COS(THH)
A(4,4)=SIN(THH)
CALL  MATTRN(A,ORDER,ORDER,AT)
CALL  MATMUL(A,ORDER,ORDER,P,ORDER,
   ORDER,AP)
CALL  MATMUL(AP,ORDER,ORDER,AT,ORDER,
   ORDER,PNEW)
ERRXT=XT-XTH
SP11P=SQRT(PNEW(1,1))
ERRXTD=XTD-XTDH
SP22P=SQRT(PNEW(2,2))
ERRYT=YT-YTH
SP33P=SQRT(PNEW(3,3))
ERRYTD=YTD-YTDH
SP44P=SQRT(PNEW(4,4))
WRITE(9,*)T,R,RH,RD,RDH,TH,THH,THD,THDH
WRITE(1,*)T,R,RH,RD,RDH,TH,THH,THD,THDH
WRITE(2,*)T,ERRTH,SP11,-SP11,ERRTHD,SP22,-SP22,
1      ERRR,SP33,-SP33,ERRRD,SP44,-SP44
WRITE(3,*)T,ERRXT,SP11P,-SP11P,ERRXTD,SP22P,
1      -SP22P,ERRYT,SP33P,-SP33P,ERRYTD,
2      SP44P,-SP44P
            ENDIF
      END DO
      PAUSE
      CLOSE(1)
      CLOSE(2)
      CLOSE(3)
      END
```

(*continued*)

Listing 9.3   (*Continued*)

```
SUBROUTINE  PROJECT(TP,TS,THP,THDP,RP,RDP,THH,THDH,RH,
   RDH,HP)
IMPLICIT  REAL*8  (A-H)
IMPLICIT  REAL*8  (O-Z)
T=0.
G=32.2
TH=THP
THD=THDP
R=RP
RD=RDP
H=HP
WHILE(T<=(TS-.0001))
        THDD=(-G*R*COS(TH)-2.*THD*R*RD)/R**2
        RDD=(R*R*THD*THD-G*R*SIN(TH))/R
        THD=THD+H*THDD
        TH=TH+H*THD
        RD=RD+H*RDD
        R=R+H*RD
        T=T+H
END  DO
RH=R
RDH=RD
THH=TH
THDH=THD
RETURN
END
```

```
C SUBROUTINE  GAUSS  IS  SHOWN  IN  LISTING  1.8
C SUBROUTINE  MATTRN  IS  SHOWN  IN  LISTING  1.3
C SUBROUTINE  MATMUL  IS  SHOWN  IN  LISTING  1.4
C SUBROUTINE  MATADD  IS  SHOWN  IN  LISTING  1.1
C SUBROUTINE  MATSUB  IS  SHOWN  IN  LISTING  1.2
C SUBROUTINE  MATSCA  IS  SHOWN  IN  LISTING  1.5
```

predictions from the transformed covariance matrix in the sense that it appears to be within the theoretical error bounds approximately 68% of the time.

Figure 9.16 displays the error in the estimate of downrange velocity. By comparing Fig. 9.16 with Fig. 9.9, we can see that the errors in the estimate of downrange velocity are comparable for both the polar and Cartesian extended Kalman filters. Initially, the error in the estimate of velocity is approximately 20 ft/s at 10 s and diminishes to approximately 1 ft/s after 130 s. Figure 9.16 also shows that the single flight error in the estimate of $\dot{x}$ appears to be correct when compared to the theoretical predictions from the transformed covariance matrix.

Figure 9.17 displays the error in the estimate of altitude. By comparing Fig. 9.17 with Fig. 9.10, we can see that the errors in the estimate of altitude are comparable for both the polar and Cartesian extended Kalman filters. Initially, the

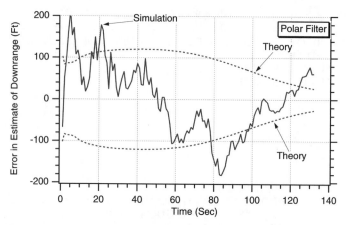

Fig. 9.15   Polar and Cartesian extended Kalman filters yield similar results for downrange estimates.

error in the estimate of altitude is approximately 600 ft and diminishes to approximately 100 ft after 130 s. Figure 9.17 also shows that the single-run simulation error in the estimate of y also appears to be correct when compared to the theoretical predictions from the transformed covariance matrix.

Figure 9.18 displays the error in the estimate of velocity in the altitude direction. By comparing Fig. 9.18 with Fig. 9.11, we can see that the errors in the estimate of the altitude velocity component are comparable for both the polar and Cartesian extended Kalman filters. Initially, the error in the estimate of altitude velocity is approximately 20 ft/s at 20 s and diminishes to approximately 2 ft/s after 130 s. Figure 9.18 also shows that the single-run simulation error in the estimate of $\dot{y}$ appears to be correct when compared to the theoretical predictions from the transformed covariance matrix.

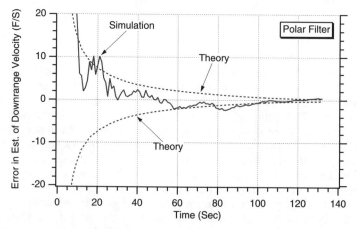

Fig. 9.16   Polar and Cartesian extended Kalman filters yield similar results for downrange velocity estimates.

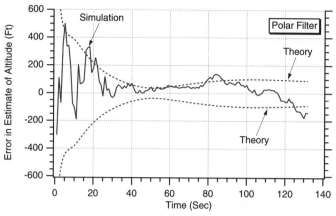

Fig. 9.17   Polar and Cartesian extended Kalman filters yield similar results for altitude estimates.

Recall that the polar filter states are angle, angle rate, distance, and distance rate or

$$\begin{bmatrix} \theta \\ \dot{\theta} \\ r \\ \dot{r} \end{bmatrix}$$

The covariance matrix resulting from the polar extended-Kalman-filter Riccati equations represents errors in the estimates of those states. Figures 9.19–9.22 display the single simulation run and theoretical (i.e., square root of appropriate diagonal elements of covariance matrix) error in the estimate of angle, angle rate,

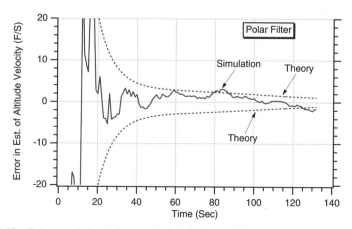

Fig. 9.18   Polar and Cartesian extended Kalman filters yield similar results for altitude velocity estimates.

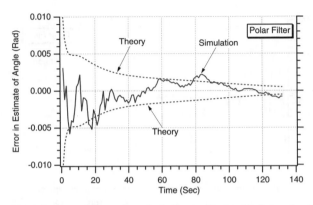

Fig. 9.19   Error in the estimate of angle indicates that extended polar Kalman filter appears to be working properly.

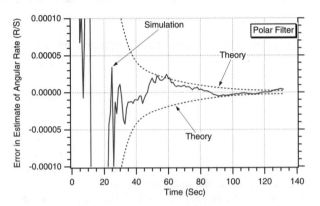

Fig. 9.20   Error in the estimate of angle rate indicates that extended polar Kalman filter appears to be working properly.

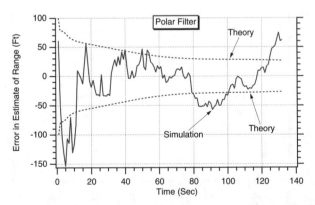

Fig. 9.21   Error in the estimate of range indicates that extended polar Kalman filter appears to be working properly.

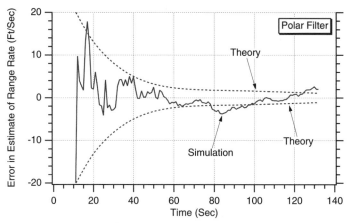

Fig. 9.22   Error in the estimate of range rate indicates that extended polar Kalman filter appears to be working properly.

range, and range rate, respectively. We can see from each of the plots that the single-run simulation error in the estimate is within the theoretical predictions of the covariance matrix approximately 68% of the time, indicating that the extended polar Kalman filter appears to be working properly.

In summary, we can say that there does not appear to be any advantage in using a polar extended Kalman filter for this example. The performance results of the filter are comparable to those of the Cartesian extended Kalman filter, although the computational burden is significantly greater. The increase in computational burden for the polar filter is primarily caused by the states having to be propagated forward at each sampling instant by numerical integration. With the Cartesian extended Kalman filter the fundamental matrix was exact, and numerical integration was not required for state propagation (i.e., multiplying the states by the fundamental matrix propagates the states).

Using Linear Decoupled Polynomial Kalman Filters

The cannon-launched projectile problem was ideal for an extended Kalman filter because either the measurements were nonlinear (if viewed in a Cartesian frame) or the system equations were nonlinear (if viewed in a polar frame). The problem also could have been manipulated to make it appropriate to apply two linear decoupled two-state polynomial Kalman filters—one in the downrange direction and the other in the altitude direction. This can be accomplished by pretending that the filter measures downrange and altitude (i.e., x_T^* and y_T^*), rather than angle and range (i.e., θ^* and r^*).

To find the equivalent noise on downrange and altitude, we have to go through the following procedure. We have already shown that downrange and altitude can be expressed in terms of range and angle according to

$$x_T = r \cos \theta + x_R$$
$$y_T = r \sin \theta + y_R$$

Using calculus, we can find the total differential of the preceding two equations as

$$\Delta x_T = \frac{\partial x_T}{\partial r}\Delta r + \frac{\partial x_T}{\partial \theta}\Delta\theta = \cos\theta\Delta r - r\sin\theta\Delta\theta$$

$$\Delta y_T = \frac{\partial y_T}{\partial r}\Delta r + \frac{\partial y_T}{\partial \theta}\Delta\theta = \sin\theta\Delta r + r\cos\theta\Delta\theta$$

We can square each of the preceding equations to obtain

$$\Delta x_T^2 = \cos^2\theta\Delta r^2 - 2r\sin\theta\cos\theta\Delta r\Delta\theta + r^2\sin^2\theta\Delta\theta^2$$
$$\Delta y_T^2 = \sin^2\theta\Delta r^2 + 2r\sin\theta\cos\theta\Delta r\Delta\theta + r^2\cos^2\theta\Delta\theta^2$$

To find the variance of the pseudonoise on downrange and altitude, we can take expectations of both sides of the equations, assuming that the actual range and angle noise are not correlated [i.e., $E(\Delta r\Delta\theta) = 0$). Therefore, we can say that

$$E(\Delta x_T^2) = \cos^2\theta E(\Delta r^2) + r^2\sin^2\theta E(\Delta\theta^2)$$

$$E(\Delta y_T^2) = \sin^2\theta E(\Delta r^2) + r^2\cos^2\theta E(\Delta\theta^2)$$

Because

$$\sigma_{x_T}^2 = E(\Delta x_T^2)$$

$$\sigma_{y_T}^2 = E(\Delta y_T^2)$$

$$\sigma_r^2 = E(\Delta r^2)$$

$$\sigma_\theta^2 = E(\Delta\theta^2)$$

we can also say that

$$\sigma_{x_T}^2 = \cos^2\theta\sigma_r^2 + r^2\sin^2\theta\sigma_\theta^2$$
$$\sigma_{y_T}^2 = \sin^2\theta\sigma_r^2 + r^2\cos^2\theta\sigma_\theta^2$$

In other words, we are pretending that we are measuring downrange and altitude with measurement accuracies described by the two preceding equations. We now have enough information for the design of the filters in each channel. In downrange there is no acceleration acting on the projectile. The model of the real world in this channel can be represented in state-space form as

$$\begin{bmatrix} \dot{x}_T \\ \ddot{x}_T \end{bmatrix} = \begin{bmatrix} 0 & 1 \\ 0 & 0 \end{bmatrix}\begin{bmatrix} x_T \\ \dot{x}_T \end{bmatrix} + \begin{bmatrix} 0 \\ u_s \end{bmatrix}$$

where u_s is white noise, which has been added to acceleration for possible protection as a result of uncertainties in modeling. Therefore, the continuous process-noise matrix is given by

$$Q = \begin{bmatrix} 0 & 0 \\ 0 & \Phi_s \end{bmatrix}$$

where Φ_s is the spectral density of the white process noise. The systems dynamics matrix is obtained from the state-space equation by inspection and is given by

$$F = \begin{bmatrix} 0 & 1 \\ 0 & 0 \end{bmatrix}$$

We have already shown in Chapter 4 that the fundamental matrix can be obtained exactly from this systems dynamics matrix and is given by

$$\Phi(t) = \begin{bmatrix} 0 & t \\ 0 & 1 \end{bmatrix}$$

which means that the discrete fundamental matrix is

$$\Phi_k = \begin{bmatrix} 0 & T_s \\ 0 & 1 \end{bmatrix}$$

The discrete process-noise matrix is related to the continuous process-noise matrix according to

$$Q_k \int_0^{T_s} \Phi(\tau) Q \Phi^T(\tau) \, dt$$

We have already derived the discrete process-noise matrix for this type of example in Chapter 4, and the results are given by

$$Q_k = \Phi_s \begin{bmatrix} \dfrac{T_s^3}{3} & \dfrac{T_s^2}{2} \\ \dfrac{T_s^2}{2} & T_s \end{bmatrix}$$

The measurement equation is linear because we have assumed a pseudomeasurement and it is given by

$$x_T^* = \begin{bmatrix} 1 & 0 \end{bmatrix} \begin{bmatrix} x_T \\ \dot{x}_T \end{bmatrix} + v_x$$

Therefore, the measurement matrix is given by

$$H = \begin{bmatrix} 1 & 0 \end{bmatrix}$$

In this case the measurement noise matrix is simply a scalar and is given by

$$R_k = \sigma_{x_T}^2$$

We now have enough information to solve the Riccati equations for the Kalman gains in the downrange channel. Because the fundamental matrix is exact in this example, we can also use the preceding matrices in the linear Kalman-filtering equation

$$\hat{x}_k = \Phi_k \hat{x}_{k-1} + K_k(z_k - H\Phi_k \hat{x}_{k-1})$$

to obtain

$$\begin{bmatrix} x_{T_k} \\ \hat{\dot{x}}_{T_k} \end{bmatrix} = \begin{bmatrix} 1 & T_s \\ 0 & 1 \end{bmatrix} \begin{bmatrix} x_{T_{k-1}} \\ \hat{\dot{x}}_{T_{k-1}} \end{bmatrix} + \begin{bmatrix} K_{1_k} \\ K_{2_k} \end{bmatrix} \left(x^*_{T_k} - \begin{bmatrix} 1 & 0 \end{bmatrix} \begin{bmatrix} 1 & T_s \\ 0 & 1 \end{bmatrix} \begin{bmatrix} x_{T_{k-1}} \\ \hat{\dot{x}}_{T_{k-1}} \end{bmatrix} \right)$$

Multiplying out the terms of the preceding matrix equation yields the two scalar equations, which represent the linear polynomial Kalman filter in the downrange channel or

$$\hat{x}_{T_k} = \hat{x}_{t_{k-1}} + T_s\hat{\dot{x}}_{T_{k-1}} + K_{1_k}(x^*_{T_k} - \hat{x}_{T_{k-1}} - T_s\hat{\dot{x}}_{T_{k-1}})$$

$$\hat{\dot{x}}_{T_k} = \hat{\dot{x}}_{T_{k-1}} + K_{2_k}(x^*_{T_k} - \hat{x}_{T_{k-1}} - T_s\hat{\dot{x}}_{T_{k-1}})$$

In the altitude channel there is gravity, and so the equations will be slightly different. The state-space equation describing the real world is now given by

$$\begin{bmatrix} \dot{y}_T \\ \ddot{y}_T \end{bmatrix} = \begin{bmatrix} 0 & 1 \\ 0 & 0 \end{bmatrix} \begin{bmatrix} y_T \\ \dot{y}_T \end{bmatrix} + \begin{bmatrix} 0 \\ -g \end{bmatrix} + \begin{bmatrix} 0 \\ u_s \end{bmatrix}$$

where g is gravity and is assumed to be known in advance. Therefore, the continuous known disturbance or control vector in this equation is given by

$$G = \begin{bmatrix} 0 \\ -g \end{bmatrix}$$

Because the systems dynamics matrix in the altitude channel is identical to the one in the downrange channel, the fundamental matrix is also the same. Therefore, the discrete control vector can be found from the continuous one from the relationship

$$G_k = \int_0^{T_s} \Phi G \, d\tau = \int_0^{T_s} \begin{bmatrix} 0 & \tau \\ 0 & 1 \end{bmatrix} \begin{bmatrix} 0 \\ -g \end{bmatrix} d\tau = \begin{bmatrix} -0.5 T_s^2 g \\ -T_s g \end{bmatrix}$$

The measurement equation is given by

$$y_T^* = \begin{bmatrix} 1 & 0 \end{bmatrix} \begin{bmatrix} y_T \\ \dot{y}_T \end{bmatrix} + v_y$$

so that the measurement matrix in the altitude channel is the same as it was in the downrange channel. The measurement noise matrix is also simply a scalar in the altitude channel and is given by

$$R_k = \sigma_{y_T}^2$$

We now have enough information to solve the Riccati equations for the Kalman gains in the altitude channel. Because the fundamental matrix is exact in this example, we can also use the preceding matrices in the Kalman-filtering equation

$$\hat{x}_k = \Phi_k \hat{x}_{k-1} + G_k u_{k-1} + K_k(z_k - H\Phi_k \hat{x}_{k-1} - HG_k u_{k-1})$$

to obtain

$$\begin{bmatrix} y_{Y_k} \\ \hat{\dot{y}}_{T_k} \end{bmatrix} = \begin{bmatrix} 1 & T_s \\ 0 & 1 \end{bmatrix} \begin{bmatrix} y_{T_{k-1}} \\ \hat{\dot{y}}_{T_{k-1}} \end{bmatrix} - \begin{bmatrix} 0.5gT_s^2 \\ gT_s \end{bmatrix}$$
$$+ \begin{bmatrix} K_{1_k} \\ K_{2_k} \end{bmatrix} \left[y_{T_k}^* - \begin{bmatrix} 1 & 0 \end{bmatrix} \begin{bmatrix} 1 & T_s \\ 0 & 1 \end{bmatrix} \begin{bmatrix} y_{T_{k-1}} \\ \hat{\dot{y}}_{T_{k-1}} \end{bmatrix} + \begin{bmatrix} 1 & 0 \end{bmatrix} \begin{bmatrix} 0.5gT_s^2 \\ gT_s \end{bmatrix} \right]$$

Multiplying out the terms of the preceding matrix equation yields the two scalar equations, which represent the linear polynomial Kalman filter in the downrange channel or

$$\hat{y}_{T_k} = \hat{y}_{T_{k-1}} + T_s \hat{\dot{y}}_{T_{k-1}} - 0.5gT_s^2 + K_{1_k}(x_{T_k}^* - \hat{x}_{T_{k-1}} - T_s \hat{\dot{x}}_{T_{k-1}} + 0.5gT_s^2)$$

$$\hat{\dot{y}}_{T_k} = \hat{\dot{y}}_{T_{k-1}} - gT_s + K_{2_k}(x_{T_k}^* - \hat{x}_{T_{k-1}} - T_s \hat{\dot{x}}_{T_{k-1}} + 0.5gT_s^2)$$

The preceding equations for the two linear decoupled polynomial Kalman filters and associated Riccati equations were programmed and are shown in Listing 9.4, along with a simulation of the real world. Again, we can see that the process-noise matrix is set to zero in this example (i.e., $\Phi_s = 0$). As before, we have initialized the states of the filter close to the true values. The position states are in error by 1000 ft, and the velocity states are in error by 100 ft/s, while the

Listing 9.4 Two decoupled polynomial linear Kalman filters for tracking projectile

```
C THE FIRST THREE STATEMENTS INVOKE THE ABSOFT RANDOM
  NUMBER GENERATOR ON THE MACINTOSH
      GLOBAL DEFINE
              INCLUDE 'quickdraw.inc'
      END
      IMPLICIT REAL*8(A-H,O-Z)
      REAL*8  P(2,2),Q(2,2),M(2,2),PHI(2,2),HMAT(1,2),HT(2,1),PHIT(2,2)
      REAL*8  RMAT(1,1),IDN(2,2),PHIP(2,2),PHIPPHIT(2,2),HM(1,2)
      REAL*8  HMHT(1,1),HMHTR(1,1),HMHTRINV(1,1),MHT(2,1),K(2,1)
      REAL*8  KH(2,2),IKH(2,2)
      REAL*8  RMATY(1,1),PY(2,2),PHIPY(2,2),PHIPPHITY(2,2),MY(2,2)
      REAL*8  HMY(1,2),HMHTY(1,1),HMHTRY(1,1),HMHTRINVY(1,1)
      REAL*8  MHTY(2,1),KY(2,1),KHY(2,2),IKHY(2,2)
      INTEGER ORDER
      TS=1.
      ORDER=2
      PHIS=0.
      SIGTH=.01
      SIGR=100.
      VT=3000.
      GAMDEG=45.
      G=32.2
      XT=0.
      YT=0.
      XTD=VT*COS(GAMDEG/57.3)
      YTD=VT*SIN(GAMDEG/57.3)
      XR=100000.
      YR=0.
      OPEN(1,STATUS='UNKNOWN',FILE='DATFIL')
      OPEN(2,STATUS='UNKNOWN',FILE='COVFIL')
      T=0.
      S=0.
      H=.001
      DO 14 I=1,ORDER
      DO 14 J=1,ORDER
      PHI(I,J)=0.
      P(I,J)=0.
      Q(I,J)=0.
      IDN(I,J)=0.
14    CONTINUE
      PHI(1,1)=1.
      PHI(1,2)=TS
      PHI(2,2)=1.
      HMAT(1,1)=1.
      HMAT(1,2)=0.
      CALL MATTRN(PHI,ORDER,ORDER,PHIT)
      CALL MATTRN(HMAT,1,ORDER,HT)
      IDN(1,1)=1.
      IDN(2,2)=1.
      Q(1,1)=PHIS*TS*TS*TS/3.
```

(continued)

Listing 9.4   (*Continued*)

```
Q(1,2)=PHIS*TS*TS/2.
Q(2,1)=Q(1,2)
Q(2,2)=PHIS*TS
P(1,1)=1000.**2
P(2,2)=100.**2
PY(1,1)=1000.**2
PY(2,2)=100.**2
XTH=XT+1000.
XTDH=XTD-100.
YTH=YT-1000.
YTDH=YTD-100.
WHILE(YT>=0.)
        XTOLD=XT
        XTDOLD=XTD
        YTOLD=YT
        YTDOLD=YTD
        XTDD=0.
        YTDD=-G
        XT=XT+H*XTD
        XTD=XTD+H*XTDD
        YT=YT+H*YTD
        YTD=YTD+H*YTDD
        T=T+H
        XTDD=0.
        YTDD=-G
        XT=.5*(XTOLD+XT+H*XTD)
        XTD=.5*(XTDOLD+XTD+H*XTDD)
        YT=.5*(YTOLD+YT+H*YTD)
        YTD=.5*(YTDOLD+YTD+H*YTDD)
        S=S+H
        IF(S>=(TS-.00001))THEN
                S=0.
                THETH=ATAN2((YTH-YR),(XTH-XR))
                RTH=SQRT((XTH-XR)**2+(YTH-YR)**2)
                RMAT(1,1)=(COS(THETH)*SIGR)**2+(RTH*SIN(THETH)
1                   *SIGTH)**2
                CALL MATMUL(PHI,ORDER,ORDER,P,ORDER,
                    ORDER,PHIP)
                CALL MATMUL(PHIP,ORDER,ORDER,PHIT,ORDER,
1                   ORDER,PHIPPHIT)
                CALL MATADD(PHIPPHIT,ORDER,ORDER,Q,M)
                CALL MATMUL(HMAT,1,ORDER,M,ORDER,
                    ORDER,HM)
                CALL MATMUL(HM,1,ORDER,HT,ORDER,1,HMHT)
                CALL MATADD(HMHT,1,1,RMAT,HMHTR)
                HMHTRINV(1,1)=1./HMHTR(1,1)
                CALL MATMUL(M,ORDER,ORDER,HT,ORDER,1,MHT)
                CALL MATMUL(MHT,ORDER,1,HMHTRINV,1,1,K)
                CALL MATMUL(K,ORDER,1,HMAT,1,ORDER,KH)
```

(*continued*)

Listing 9.4   (*Continued*)

```
            CALL  MATSUB(IDN,ORDER,ORDER,KH,IKH)
            CALL  MATMUL(IKH,ORDER,ORDER,M,ORDER,
            ORDER,P)
            CALL  GAUSS(THETNOISE,SIGTH)
            CALL  GAUSS(RTNOISE,SIGR)
            THET=ATAN2((YT-YR),(XT-XR))
            RT=SQRT((XT-XR)**2+(YT-YR)**2)
            THETMEAS=THET+THETNOISE
            RTMEAS=RT+RTNOISE
            XTMEAS=RTMEAS*COS(THETMEAS)+XR
            RES1=XTMEAS-XTH-TS*XTDH
            XTH=XTH+TS*XTDH+K(1,1)*RES1
            XTDH=XTDH+K(2,1)*RES1
            RMATY(1,1)=(SIN(THETH)*SIGR)**2+
1              (RTH*COS(THETH)*SIGTH)**2
            CALL  MATMUL(PHI,ORDER,ORDER,PY,ORDER,
1              ORDER,PHIPY)
            CALL  MATMUL(PHIPY,ORDER,ORDER,PHIT,ORDER,
1              ORDER,PHIPPHITY)
            CALL  MATADD(PHIPPHITY,ORDER,ORDER,Q,MY)
            CALL  MATMUL(HMAT,1,ORDER,MY,ORDER,ORDER,
            HMY)
            CALL  MATMUL(HMY,1,ORDER,HT,ORDER,1,HMHTY)
            CALL  MATADD(HMHTY,1,1,RMATY,HMHTRY)
            HMHTRINVY(1,1)=1./HMHTRY(1,1)
            CALL  MATMUL(MY,ORDER,ORDER,HT,ORDER,1,
            MHTY)
            CALL  MATMUL(MHTY,ORDER,1,HMHTRINVY,1,
            1,KY)
            CALL  MATMUL(KY,ORDER,1,HMAT,1,ORDER,KHY)
            CALL  MATSUB(IDN,ORDER,ORDER,KHY,IKHY)
            CALL  MATMUL(IKHY,ORDER,ORDER,MY,ORDER,
            ORDER,PY)
            YTMEAS=RTMEAS*SIN(THETMEAS)+YR
            RES2=YTMEAS-YTH-TS*YTDH+.5*TS*TS*G
            YTH=YTH+TS*YTDH-.5*TS*TS*G+KY(1,1)*RES2
            YTDH=YTDH-TS*G+KY(2,1)*RES2
            ERRX=XT-XTH
            SP11=SQRT(P(1,1))
            ERRXD=XTD-XTDH
            SP22=SQRT(P(2,2))
            ERRY=YT-YTH
            SP11Y=SQRT(PY(1,1))
            ERRYD=YTD-YTDH
            SP22Y=SQRT(PY(2,2))
            WRITE(9,*)T,XT,XTH,XTD,XTDH,YT,YTH,YTD,YTDH
            WRITE(1,*)T,XT,XTH,XTD,XTDH,YT,YTH,YTD,YTDH
            WRITE(2,*)T,ERRX,SP11,-SP11,ERRXD,SP22,-SP22,
1              ERRY,SP11Y,-SP11Y,ERRYD,SP22Y,-SP22Y
      ENDIF
```

(*continued*)

Listing 9.4   (*Continued*)

```
END DO
PAUSE
CLOSE(1)
CLOSE(2)
END

C SUBROUTINE GAUSS IS SHOWN IN LISTING 1.8
C SUBROUTINE MATTRN IS SHOWN IN LISTING 1.3
C SUBROUTINE MATMUL IS SHOWN IN LISTING 1.4
C SUBROUTINE MATADD IS SHOWN IN LISTING 1.1
C SUBROUTINE MATSUB IS SHOWN IN LISTING 1.2
```

initial covariance matrix reflects those errors. The simulation running time of Listing 9.4 is much faster than that of the extended Cartesian Kalman filter of Listing 9.1 because there are effectively fewer equations with two two-state filters than there are for those with one four-state filter.

The nominal case was run, and the resultant errors in the estimates for position and velocity in the downrange and altitude direction appear in Figs. 9.23–9.26. We can see from the four figures that the linear Kalman filters appear to be working correctly because the single-run simulation errors in the estimates appear to lie between the theoretical predictions of the associated covariance matrices approximately 68% of the time. However, if we compare these results to the errors in the estimates of the Cartesian extended Kalman filter shown in Figs. 9.8–9.12, we can see that these new results appear to be slightly worse than before. In other words, the errors in the estimates of the two two-state linear

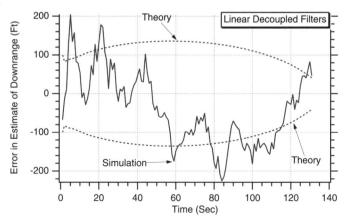

Fig. 9.23   Linear decoupled Kalman filter downrange error in the estimate of position is larger than that of extended Kalman filter.

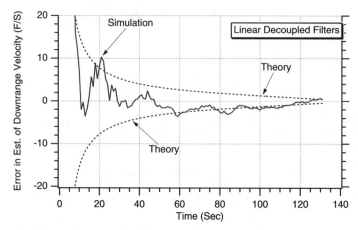

Fig. 9.24   Linear decoupled Kalman filter downrange error in the estimate of velocity is larger than that of extended Kalman filter.

decoupled polynomial Kalman filters appear to be larger than those of the four-state Cartesian extended Kalman filter.

Using Linear Coupled Polynomial Kalman Filters

In the preceding section we pretended that the noise in the altitude and downrange channels was not correlated, which allowed us to decouple the linear Kalman filters. Let us now remove that restriction in order to see what happens.

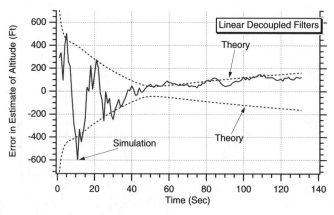

Fig. 9.25   Linear decoupled Kalman filter altitude error in the estimate of position is larger than that of extended Kalman filter.

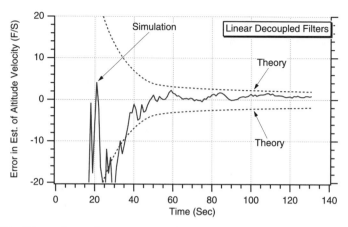

Fig. 9.26   Linear decoupled Kalman filter altitude error in the estimate of velocity is larger than that of extended Kalman filter.

Recall that we have already shown that downrange and altitude can be expressed in terms of range and angle according to

$$x_T = r \cos \theta + x_R$$
$$y_T = r \sin \theta + y_R$$

Using calculus, we can find the total differential of the preceding two equations as

$$\Delta x_T = \frac{\partial x_T}{\partial r} \Delta r + \frac{\partial x_T}{\partial \theta} \Delta \theta = \cos \theta \Delta r - r \sin \theta \Delta \theta$$

$$\Delta y_T = \frac{\partial y_T}{\partial r} \Delta r + \frac{\partial y_T}{\partial \theta} \Delta \theta = \sin \theta \Delta r + r \cos \theta \Delta \theta$$

Before, we squared each of the preceding equations to obtain

$$\Delta x_T^2 = \cos^2 \theta \Delta r^2 - 2r \sin \theta \cos \theta \Delta r \Delta \theta + r^2 \sin^2 \theta \Delta \theta^2$$
$$\Delta y_T^2 = \sin^2 \theta \Delta r^2 + 2r \sin \theta \cos \theta \Delta r \Delta \theta + r^2 \cos^2 \theta \Delta \theta^2$$

but now we also can find

$$\Delta x_T \Delta y_T = \sin \theta \cos \theta \Delta r^2 + r \sin^2 \theta \Delta r \Delta \theta + r \cos^2 \theta \Delta r \Delta \theta - r^2 \sin \theta \cos \theta \Delta \theta^2$$

To find the variance of the pseudonoise, we can take expectations of both sides of the equations, assuming that the actual range and angle noise are not correlated [i.e., $E(\Delta r \Delta \theta) = 0$]. Therefore, we can say that

$$E(\Delta x_T^2) = \cos^2 \theta E(\Delta r^2) + r^2 \sin^2 \theta E(\Delta \theta^2)$$

$$E(\Delta y_T^2) = \sin^2 \theta E(\Delta r^2) + r^2 \cos^2 \theta E(\Delta \theta^2)$$

$$E(\Delta x_T \Delta y_T) = \sin \theta \cos \theta E(\Delta r^2) - r^2 \sin \theta \cos \theta E(\Delta \theta^2)$$

Because

$$\sigma_{x_T}^2 = E(\Delta x_T^2)$$

$$\sigma_{y_T}^2 = E(\Delta y_T^2)$$

$$\sigma_{x_T y_T}^2 = E(\Delta x_T \Delta y_T)$$

$$\sigma_r^2 = E(r^2)$$

$$\sigma_\theta^2 = E(\Delta \theta^2)$$

we can say that

$$\sigma_{x_T}^2 = \cos^2 \theta \theta \sigma_r^2 + r^2 \sin^2 \theta \sigma_\theta^2$$

$$\sigma_{y_T}^2 = \sin^2 \theta \theta \sigma_r^2 + r^2 \cos^2 \theta \sigma_\theta^2$$

$$\sigma_{x_T y_T}^2 = \sin \theta \cos \theta \sigma_r^2 - r^2 \sin \theta \cos \theta \sigma_\theta^2$$

Therefore, unlike the preceding section, where we had a downrange and altitude scalar noise matrix, we now have a fully coupled 2×2 noise matrix given by

$$\boldsymbol{R}_k = \begin{bmatrix} \cos^2 \theta \sigma_r^2 + r^2 \sin^2 \theta \sigma_\theta^2 & \sin \theta \cos \theta \sigma_r^2 - r^2 \sin \theta \cos \theta \sigma_\theta^2 \\ \sin \theta \cos \theta \sigma_r^2 - r^2 \sin \theta \cos \theta \sigma_\theta^2 & \sin^2 \theta \sigma_r^2 + r^2 \cos^2 \theta \sigma_\theta^2 \end{bmatrix}$$

Because we are no longer decoupling the channels, the states upon which the new filter will be based are given by

$$\boldsymbol{x} = \begin{bmatrix} x_T \\ \dot{x}_T \\ y_T \\ \dot{y}_T \end{bmatrix}$$

and the pseudomeasurements are now linearly related to the states according to

$$
\begin{bmatrix} x_T^* \\ y_T^* \end{bmatrix} = \begin{bmatrix} 1 & 0 & 0 & 0 \\ 0 & 0 & 1 & 0 \end{bmatrix} \begin{bmatrix} x_T \\ \dot{x}_T \\ y_T \\ \dot{y}_T \end{bmatrix} + \begin{bmatrix} v_x \\ v_y \end{bmatrix}
$$

From the preceding equation we can see that the measurement matrix is given by

$$
H = \begin{bmatrix} 1 & 0 & 0 & 0 \\ 0 & 0 & 1 & 0 \end{bmatrix}
$$

Therefore, when the preceding Cartesian states are chosen, we have already shown that the state-space differential equation describing projectile motion is also linear and given by

$$
\begin{bmatrix} \dot{x}_T \\ \ddot{x}_T \\ \dot{y}_T \\ \ddot{y}_T \end{bmatrix} = \begin{bmatrix} 0 & 1 & 0 & 0 \\ 0 & 0 & 0 & 0 \\ 0 & 0 & 0 & 1 \\ 0 & 0 & 0 & 0 \end{bmatrix} \begin{bmatrix} x_T \\ \dot{x}_T \\ y_T \\ \dot{y}_T \end{bmatrix} + \begin{bmatrix} 0 \\ 0 \\ 0 \\ -g \end{bmatrix} + \begin{bmatrix} 0 \\ u_s \\ 0 \\ u_s \end{bmatrix}
$$

Again, notice that in the preceding equation gravity g is not a state that has to be estimated but is assumed to be known in advance. We also have added white process noise u_s to the acceleration portion of the equations as protection for effects that may not considered by the Kalman filter. The preceding equation is equivalent to the model used in deriving the extended Cartesian Kalman filter. The only difference in this formulation is that the actual measurements have been simplified to pseudomeasurements for purposes of making the filter linear.

From the preceding state-space equation we can see that systems dynamics matrix is given by

$$
F = \begin{bmatrix} 0 & 1 & 0 & 0 \\ 0 & 0 & 0 & 0 \\ 0 & 0 & 0 & 1 \\ 0 & 0 & 0 & 0 \end{bmatrix}
$$

and we have already shown in this chapter that the fundamental matrix for this system of equations is

$$
\Phi(t) = \begin{bmatrix} 0 & t & 0 & 0 \\ 0 & 1 & 0 & 0 \\ 0 & 0 & 1 & t \\ 0 & 0 & 0 & 1 \end{bmatrix}
$$

Therefore, the discrete fundamental matrix can be found by substituting T_s for t and is given by

$$\Phi_k = \begin{bmatrix} 0 & T_s & 0 & 0 \\ 0 & 1 & 0 & 0 \\ 0 & 0 & 1 & T_s \\ 0 & 0 & 0 & 1 \end{bmatrix}$$

Because the continuous process-noise matrix is given by

$$Q = E(ww^T)$$

we can say that, for this example, the continuous process-noise matrix is

$$Q(t) = \begin{bmatrix} 0 & 0 & 0 & 0 \\ 0 & \Phi_s & 0 & 0 \\ 0 & 0 & 0 & 0 \\ 0 & 0 & 0 & \Phi_s \end{bmatrix}$$

where Φ_s is the spectral density of the white noise sources assumed to be on the downrange and altitude accelerations acting on the projectile. The discrete process-noise matrix can be derived from the continuous process-noise matrix according to

$$Q_k = \int_0^{T_s} \Phi(\tau) Q \Phi^T(\tau) \, dt$$

and we have already shown that the discrete process-noise matrix turns out to be

$$Q_k = \begin{bmatrix} \dfrac{T_s^3 \Phi_s}{3} & \dfrac{T_s^2 \Phi_s}{2} & 0 & 0 \\[2ex] \dfrac{T_s^2 \Phi_s}{2} & T_s \Phi_s & 0 & 0 \\[2ex] 0 & 0 & \dfrac{T_s^3 \Phi_s}{3} & \dfrac{T_s^2 \Phi_s}{2} \\[2ex] 0 & 0 & \dfrac{T_s^2 \Phi_s}{2} & T_s \Phi_s \end{bmatrix}$$

From the state-space equation we can see that the continuous control vector in this equation is given by

$$G = \begin{bmatrix} 0 \\ 0 \\ 0 \\ -g \end{bmatrix}$$

Therefore, the discrete control vector can be found from the continuous one from the relationship

$$G_k = \int_0^{T_s} \Phi G \, d\tau = \int_0^{T_s} \begin{bmatrix} 0 & \tau & 0 & 0 \\ 0 & 1 & 0 & 0 \\ 0 & 0 & 1 & \tau \\ 0 & 0 & 0 & 1 \end{bmatrix} \begin{bmatrix} 0 \\ 0 \\ 0 \\ -g \end{bmatrix} d\tau = \begin{bmatrix} 0 \\ 0 \\ -0.5gT_s^2 \\ -gT_s \end{bmatrix}$$

We now have defined all of the matrices required to solve the Riccati equations. Because the fundamental matrix is exact in this example, we also can use the preceding matrices in the Kalman-filtering equation

$$\hat{x}_k = \Phi_k \hat{x}_{k-1} + G_k u_{k-1} + K_k(z_k - H\Phi_k \hat{x}_{k-1} - HG_k u_{k-1})$$

to obtain

$$\begin{bmatrix} \hat{x}_{T_k} \\ \hat{\dot{x}}_{T_k} \\ \hat{y}_{T_k} \\ \hat{\dot{y}}_{T_k} \end{bmatrix} = \begin{bmatrix} 0 & T_s & 0 & 0 \\ 0 & 1 & 0 & 0 \\ 0 & 0 & 1 & T_s \\ 0 & 0 & 0 & 1 \end{bmatrix} \begin{bmatrix} \hat{x}_{T_{k-1}} \\ \hat{\dot{x}}_{T_{k-1}} \\ \hat{y}_{T_{k-1}} \\ \hat{\dot{y}}_{T_{k-1}} \end{bmatrix} + \begin{bmatrix} 0 \\ 0 \\ -0.5gT_s^2 \\ -gT_s \end{bmatrix}$$

$$+ \begin{bmatrix} K_{11_k} & K_{12_k} \\ K_{21_k} & K_{22_k} \\ K_{31_k} & K_{32_k} \\ K_{41_k} & K_{42_k} \end{bmatrix} \left(\begin{bmatrix} 1 & 0 & 0 & 0 \\ 0 & 0 & 1 & 0 \end{bmatrix} \begin{bmatrix} 0 & T_s & 0 & 0 \\ 0 & 1 & 0 & 0 \\ 0 & 0 & 1 & T_s \\ 0 & 0 & 0 & 1 \end{bmatrix} \begin{bmatrix} \hat{x}_{T-1} \\ \hat{\dot{x}}_{T_{k-1}} \\ \hat{y}_{T_{k-1}} \\ \hat{\dot{y}}_{T_{k-1}} \end{bmatrix} \right.$$

$$\left. - \begin{bmatrix} 1 & 0 & 0 & 0 \\ 0 & 0 & 1 & 0 \end{bmatrix} \begin{bmatrix} 0 \\ 0 \\ -0.5gT_s^2 \\ -gT_s \end{bmatrix} \right) \right]$$

Multiplying out the terms yields the following scalar equations:

$$\text{Res}_1 = x_{T_k}^* - \hat{x}_{T_{k-1}} - T_s \hat{\dot{x}}_{T_{k-1}}$$

$$\text{Res}_2 = y_{T_k}^* - \hat{y}_{T_{k-1}} - T_s \hat{\dot{y}}_{T_{k-1}} + 0.5gT_s^2$$

$$\hat{x}_{T_k} = \hat{x}_{T_{k-1}} + T_s \hat{\dot{x}}_{T_{k-1}} + K_{11_k} \text{Res}_1 + K_{12_k} \text{Res}_2$$

$$\hat{\dot{x}}_{T_k} = \hat{\dot{x}}_{T_{k-1}} + K_{21_k} \text{Res}_1 + K_{22_k} \text{Res}_2$$

$$\hat{y}_{T_k} = \hat{y}_{T_{k-1}} + T_s \hat{\dot{y}}_{T_{k-1}} - 0.5gT_s^2 + K_{31_k} \text{Res}_1 + K_{32_k} \text{Res}_2$$

$$\hat{\dot{y}}_{T_k} = \hat{\dot{y}}_{T_{k-1}} - gT_s + K_{41_k} \text{Res}_1 + K_{42_k} \text{Res}_2$$

The preceding equations for the four-state linear polynomial Kalman filters and associated Riccati equations were programmed and are shown in Listing 9.5, along with a simulation of the real world. Again, we can see that the process-noise matrix is set to zero in this example (i.e., $\Phi_s = 0$). As before, we have initialized the states of the filter close to the true values. The position states are in

Listing 9.5 Linear coupled four-state polynomial Kalman filter

```
C THE FIRST THREE STATEMENTS INVOKE THE ABSOFT RANDOM
  NUMBER GENERATOR ON THE MACINTOSH
        GLOBAL DEFINE
                INCLUDE 'quickdraw.inc'
        END
        IMPLICIT REAL*8(A-H,O-Z)
        REAL*8 P(4,4),Q(4,4),M(4,4),PHI(4,4),HMAT(2,4),HT(2,4),PHIT(4,4)
        REAL*8 RMAT(2,2),IDN(4,4),PHIP(4,4),PHIPPHIT(4,4),HM(2,4)
        REAL*8 HMHT(2,2),HMHTR(2,2),HMHTRINV(2,2),MHT(4,2),K(4,2)
        REAL*8 KH(4,4),IKH(4,4)
        INTEGER ORDER
        TS=1.
        ORDER=4
        PHIS=0.
        SIGTH=.01
        SIGR=100.
        VT=3000.
        GAMDEG=45.
        G=32.2
        XT=0.
        YT=0.
        XTD=VT*COS(GAMDEG/57.3)
        YTD=VT*SIN(GAMDEG/57.3)
        XR=100000.
        YR=0.
        OPEN(1,STATUS='UNKNOWN',FILE='DATFIL')
        OPEN(2,STATUS='UNKNOWN',FILE='COVFIL')
        T=0.
        S=0.
        H=.001
        DO 14 I=1,ORDER
        DO 14 J=1,ORDER
        PHI(I,J)=0.
        P(I,J)=0.
        Q(I,J)=0.
        IDN(I,J)=0.
14      CONTINUE
        PHI(1,1)=1.
        PHI(1,2)=TS
        PHI(2,2)=1.
        PHI(3,3)=1.
```

(continued)

Listing 9.5   *(Continued)*

```
PHI(3,4)=TS
PHI(4,4)=1.
HMAT(1,1)=1.
HMAT(1,2)=0.
HMAT(1,3)=0.
HMAT(1,4)=0.
HMAT(2,1)=0.
HMAT(2,2)=0.
HMAT(2,3)=1.
HMAT(2,4)=0.
CALL  MATTRN(PHI,ORDER,ORDER,PHIT)
CALL  MATTRN(HMAT,2,ORDER,HT)
IDN(1,1)=1.
IDN(2,2)=1.
IDN(3,3)=1.
IDN(4,4)=1.
Q(1,1)=PHIS*TS*TS*TS/3.
Q(1,2)=PHIS*TS*TS/2.
Q(2,1)=Q(1,2)
Q(2,2)=PHIS*TS
Q(3,3)=PHIS*TS*TS*TS/3.
Q(3,4)=PHIS*TS*TS/2.
Q(4,3)=Q(3,4)
Q(4,4)=PHIS*TS
P(1,1)=1000.**2
P(2,2)=100.**2
P(3,3)=1000.**2
P(4,4)=100.**2
XTH=XT+1000.
XTDH=XTD-100.
YTH=YT-1000.
YTDH=YTD+100.
WHILE(YT>=0.)
        XTOLD=XT
        XTDOLD=XTD
        YTOLD=YT
        YTDOLD=YTD
        XTDD=0.
        YTDD=-G
        XT=XT+H*XTD
        XTD=XTD+H*XTDD
        YT=YT+H*YTD
        YTD=YTD+H*YTDD
        T=T+H
        XTDD=0.
        YTDD=-G
        XT=.5*(XTOLD+XT+H*XTD)
        XTD=.5*(XTDOLD+XTD+H*XTDD)
        YT=.5*(YTOLD+YT+H*YTD)
```

(continued)

Listing 9.5 (*Continued*)

```
YTD=.5*(YTDOLD+YTD+H*YTDD)
S=S+H
IF(S>=(TS-.00001))THEN
    S=0.
    THETH=ATAN2((YTH-YR),(XTH-XR))
    RTH=SQRT((XTH-XR)**2+(YTH-YR)**2)
    RMAT(1,1)=(COS(THETH)*SIGR)**2+(RTH*SIN
        (THETH)*SIGTH)**2
    RMAT(2,2)=(SIN(THETH)*SIGR)**2+(RTH*COS
        (THETH)*SIGTH)**2
    RMAT(1,2)=SIN(THETH)*COS(THETH)*(SIGR**2-
        (RTH*SIGTH)**2)
    RMAT(2,1)=RMAT(1,2)
    CALL  MATMUL(PHI,ORDER,ORDER,P,ORDER,
        ORDER,PHIP)
    CALL  MATMUL(PHIP,ORDER,ORDER,PHIT,ORDER,
        ORDER,PHIPPHIT)
    CALL  MATADD(PHIPPHIT,ORDER,ORDER,Q,M)
    CALL  MATMUL(HMAT,2,ORDER,M,ORDER,
        ORDER,HM)
    CALL  MATMUL(HM,2,ORDER,HT,ORDER,2,HMHT)
    CALL  MATADD(HMHT,2,2,RMAT,HMHTR)
    DET=HMHTR(1,1)*HMHTR(2,2)-
        HMHTR(1,2)*HMHTR(2,1)
    HMHTRINV(1,1)=HMHTR(2,2)/DET
    HMHTRINV(1,2)=-HMHTR(1,2)/DET
    HMHTRINV(2,1)=-HMHTR(2,1)/DET
    HMHTRINV(2,2)=HMHTR(1,1)/DET
    CALL  MATMUL(M,ORDER,ORDER,HT,ORDER,
        2,MHT)
    CALL  MATMUL(MHT,ORDER,2,HMHTRINV,2,2,K)
    CALL  MATMUL(K,ORDER,2,HMAT,2,ORDER,KH)
    CALL  MATSUB(IDN,ORDER,ORDER,KH,IKH)
    CALL  MATMUL(IKH,ORDER,ORDER,M,ORDER,
        ORDER,P)
    CALL  GAUSS(THETNOISE,SIGTH)
    CALL  GAUSS(RTNOISE,SIGR)
    THET=ATAN2((YT-YR),(XT-XR))
    RT=SQRT((XT-XR)**2+(YT-YR)**2)
    THETMEAS=THET+THETNOISE
    RTMEAS=RT+RTNOISE
    XTMEAS=RTMEAS*COS(THETMEAS)+XR
    YTMEAS=RTMEAS*SIN(THETMEAS)+YR
    RES1=XTMEAS-XTH-TS*XTDH
    RES2=YTMEAS-YTH-TS*YTDH+.5*TS*TS*G
    XTH=XTH+TS*XTDH+K(1,1)*RES1+K(1,2)*RES2
    XTDH=XTDH+K(2,1)*RES1+K(2,2)*RES2
    YTH=YTH+TS*YTDH-.5*TS*TS*G+K(3,1)*RES1+
        K(3,2)*RES2
    YTDH=YTDH-TS*G+K(4,1)*RES1+K(4,2)*RES2
```

(continued)

Listing 9.5   *(Continued)*

```
                ERRX=XT-XTH
                SP11=SQRT(P(1,1))
                ERRXD=XTD-XTDH
                SP22=SQRT(P(2,2))
                ERRY=YT-YTH
                SP33=SQRT(P(3,3))
                ERRYD=YTD-YTDH
                SP44=SQRT(P(4,4))
                WRITE(9,*)T,XT,XTH,XTD,XTDH,YT,YTH,YTD,YTDH
                WRITE(1,*)T,XT,XTH,XTD,XTDH,YT,YTH,YTD,YTDH
                WRITE(2,*)T,ERRX,SP11,-SP11,ERRXD,SP22,-SP22,
1               ERRY,SP33,-SP33,ERRYD,SP44,-SP44
        ENDIF
      END DO
      PAUSE
      CLOSE(1)
      CLOSE(2)
      END

C SUBROUTINE GAUSS IS SHOWN IN LISTING 1.8
C SUBROUTINE MATTRN IS SHOWN IN LISTING 1.3
C SUBROUTINE MATMUL IS SHOWN IN LISTING 1.4
C SUBROUTINE MATADD IS SHOWN IN LISTING 1.1
C SUBROUTINE MATSUB IS SHOWN IN LISTING 1.2
```

error by 1000 ft, and the velocity states are in error by 100 ft/s, while the initial covariance matrix reflects those errors.

The nominal case of Listing 9.5 was run in which there was no process noise. Figures 9.27–9.30 display the errors in the estimates of the downrange and altitude states. By comparing these figures with those of the Cartesian extended Kalman filter (see Figs. 9.8–9.11), we can see that the results are now virtually identical. In other words, in the cannon-launched projectile example there is no degradation in performance by using a linear coupled filter with pseudomeasurements as opposed to the extended Cartesian Kalman filter with the actual measurements. However, there was some degradation in performance when we used decoupled linear filters, as was done in the preceding section.

Robustness Comparison of Extended and Linear Coupled Kalman Filters

So far we have seen that both the extended and linear coupled Kalman filters appear to have approximately the same performance. However, these results were based on work in which the initialization errors in the filter states were relatively modest. Let us first see how the extended Kalman filter reacts to larger initialization errors.

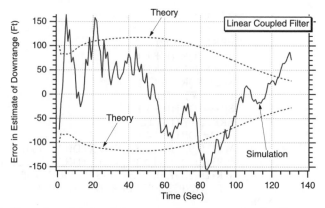

Fig. 9.27 Error in estimate of downrange is the same for the linear coupled and extended Kalman filters.

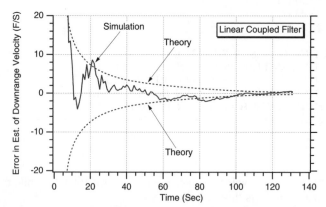

Fig. 9.28 Error in estimate of downrange velocity is the same for the linear coupled and extended Kalman filters.

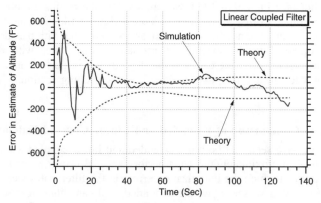

Fig. 9.29 Error in estimate of altitude is the same for the linear coupled and extended Kalman filters.

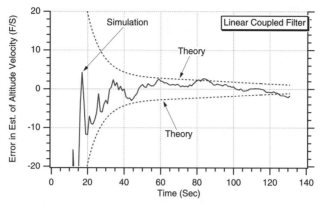

Fig. 9.30   Error in estimate of altitude velocity is the same for the linear coupled and extended Kalman filters.

In all of the experiments of this chapter, the initial state estimates of the filter were the true states plus an error of 1000 ft on position and 100 ft/s on velocity, as indicated here:

$$\begin{bmatrix} \hat{x}_T(0) \\ \hat{\dot{x}}_T(0) \\ \hat{y}_T(0) \\ \hat{\dot{y}}_T(0) \end{bmatrix} = \begin{bmatrix} x_T(0) \\ \dot{x}_T(0) \\ y_T(0) \\ \dot{y}_T(0) \end{bmatrix} + \begin{bmatrix} 1000 \\ -100 \\ -1000 \\ 100 \end{bmatrix}$$

The initial covariance matrix of both the linear and extended Kalman filters reflected the initialization errors by having each diagonal element be the square of the appropriate initialization error or

$$P_0 = \begin{bmatrix} 1000^2 & 0 & 0 & 0 \\ 0 & 100^2 & 0 & 0 \\ 0 & 0 & 1000^2 & 0 \\ 0 & 0 & 0 & 100^2 \end{bmatrix}$$

To see how sensitive the extended Kalman filter is to initialization errors, let us first double those errors so that the initial estimates of the extended Kalman filter and its covariance matrix are now given by

$$\begin{bmatrix} \hat{x}_T(0) \\ \hat{\dot{x}}_T(0) \\ \hat{y}_T(0) \\ \hat{\dot{y}}_T(0) \end{bmatrix} = \begin{bmatrix} x_T(0) \\ \dot{x}_T(0) \\ y_T(0) \\ \dot{y}_T(0) \end{bmatrix} + \begin{bmatrix} 2000 \\ -200 \\ -2000 \\ 200 \end{bmatrix}$$

$$P_0 = \begin{bmatrix} 2000^2 & 0 & 0 & 0 \\ 0 & 200^2 & 0 & 0 \\ 0 & 0 & 2000^2 & 0 \\ 0 & 0 & 0 & 200^2 \end{bmatrix}$$

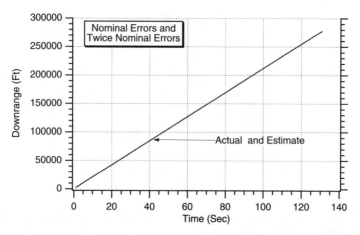

Fig. 9.31 Extended Kalman filter appears to yield good estimates even when initialization errors are twice as large as nominal.

Listing 9.1, which represents the extended Cartesian Kalman filter, was rerun with the preceding initialization errors. We can see from Fig. 9.31 that the estimate of the downrange location of the projectile appears to be independent of initialization error because the estimates do not change when the initialization errors are doubled. In addition, the estimates of the projectile location appear to be very close to the actual projectile location.

Next, the initialization errors were made five times larger than the nominal values so that the initial state estimates and covariance matrix are now given by

$$
\begin{bmatrix} \hat{x}_T(0) \\ \hat{\dot{x}}_T(0) \\ \hat{y}_T(0) \\ \hat{\dot{y}}_T(0) \end{bmatrix} = \begin{bmatrix} x_T(0) \\ \dot{x}_T(0) \\ y_T(0) \\ \dot{y}_T(0) \end{bmatrix} + \begin{bmatrix} 5000 \\ -500 \\ -5000 \\ 500 \end{bmatrix}
$$

$$
P_0 = \begin{bmatrix} 5000^2 & 0 & 0 & 0 \\ 0 & 500^2 & 0 & 0 \\ 0 & 0 & 5000^2 & 0 \\ 0 & 0 & 0 & 500^2 \end{bmatrix}
$$

Listing 9.1 was rerun with the new initialization errors. We can see from Fig. 9.32 that with the larger initialization errors the extended Kalman filter's estimate of projectile downrange degrades severely. In addition, we can see that for the first 10 s there is a filter transient caused by the initialization errors. However, after the transient settles out the filter estimate of the projectile downrange position severely lags the actual downrange location.

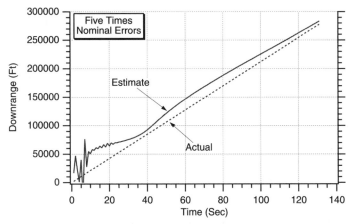

Fig. 9.32   Estimates from extended Kalman filter degrade severely when initialization errors are five time larger than nominal.

Next, the extended Kalman filter's initialization errors were made 10 times larger than the nominal values. The new initial state estimates and initial covariance matrix are now given by

$$
\begin{bmatrix} \hat{x}_T(0) \\ \hat{\dot{x}}_T(0) \\ \hat{y}_T(0) \\ \hat{\dot{y}}_T(0) \end{bmatrix} = \begin{bmatrix} x_T(0) \\ \dot{x}_T(0) \\ y_T(0) \\ \dot{y}_T(0) \end{bmatrix} + \begin{bmatrix} 10000 \\ -1000 \\ -10000 \\ 1000 \end{bmatrix}
$$

$$
P_0 = \begin{bmatrix} 10{,}000^2 & 0 & 0 & 0 \\ 0 & 1{,}000^2 & 0 & 0 \\ 0 & 0 & 10{,}000^2 & 0 \\ 0 & 0 & 0 & 1{,}000^2 \end{bmatrix}
$$

Listing 9.1 was rerun with the preceding initialization conditions. We can see from Fig. 9.33 that when the initialization errors of the filter estimate and the initial covariance matrix are 10 times the nominal value the estimate of projectile, downrange is worthless. We can see that the filter estimate of the downrange location of the projectile is monotonically decreasing, whereas the actual location is monotonically increasing.

All of the initialization experiments conducted so far in this section had zero process noise (i.e., $\Phi_s = 0$). We know from Chapter 6 that the addition of process noise will widen the bandwidth of the Kalman filter, and hopefully this also will improve the extended Kalman filter's transient response. The case in which the initialization errors for the extended Kalman were 10 times larger than nominal and the filter's estimates were worthless was reexamined when process-noise was present. Figure 9.34 shows that the addition of process noise (i.e., $\Phi_s = 1$) now

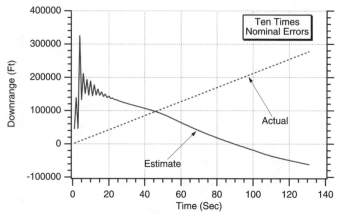

Fig. 9.33   Estimates from extended Kalman filter are worthless when initialization errors are 10 times larger than nominal.

enables the extended Kalman filter to estimate the downrange location of the projectile. Figures 9.34 and 9.35 show that increasing the process noise enables the filter to eliminate the estimation transient error associated with the large initialization errors more rapidly. This is not surprising because we already know from Chapter 6 that increasing the process noise will increase the bandwidth of the Kalman filter and thus improve its transient response.

For more precise work we must also see how the actual errors in the estimates compare to the covariance matrix predictions for the errors in the estimates. Figure 9.36 examines the error in the estimate of the projectile's downrange location. We can see that the extended Kalman filter with process noise (i.e., $\Phi_s = 1000$) now appears to be working correctly because, after an initial transient

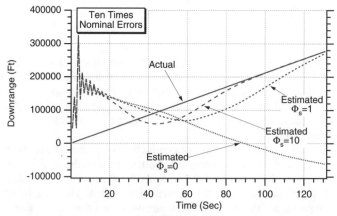

Fig. 9.34   Addition of process noise enables extended Kalman filter to better estimate downrange in presence of large initialization errors.

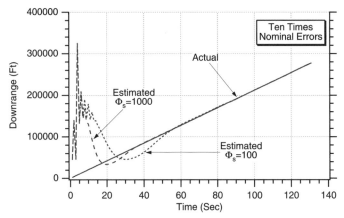

Fig. 9.35   Making process noise larger further improves extended Kalman filter's ability to estimate downrange in presence of large initialization errors.

period, the error in the estimate of projectile downrange position agrees with the results of the covariance matrix. In other words, the simulated error in the estimate appears to lie within the theoretical bounds (i.e., square root of P_{11}) approximately 68% of the time.

We now ask ourselves how the linear coupled polynomial Kalman filter would have performed under similar circumstances. Listing 9.5, which we have already shown has the linear coupled polynomial Kalman filter, was rerun for the case in which the initialization errors were 10 times their nominal value. The linear coupled polynomial Kalman filter was run *without* process-noise. We can see from Fig. 9.37 that under these adverse conditions the linear filter does not appear to have a problem in estimating the downrange location of the projectile. Again,

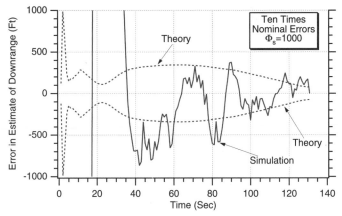

Fig. 9.36   After an initial transient period simulation results agree with covariance matrix predictions.

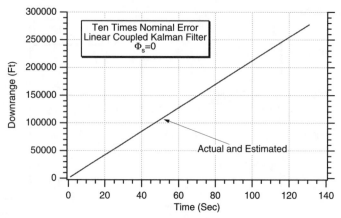

Fig. 9.37   Linear coupled polynomial Kalman filter does not appear to be sensitive to large initialization errors.

we can examine filter performance more critically by examining the error in the estimate from simulation and comparing it to the covariance matrix prediction. We can see from Fig. 9.38 that the single-run simulation results of the error in the estimate of the projectile downrange location appear to lie within the theoretical bounds approximately 68% of the time. Therefore, the linear coupled polynomial Kalman filter *without* process-noise appears to be working in the presence of large initialization errors. If we compare these results to the case in which the initialization errors were their nominal values (see Fig. 9.8), we can see that the performance of the linear filter does not appear to depend on the size of the initialization error.

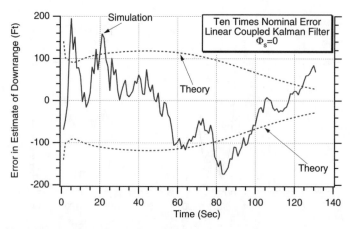

Fig. 9.38   Linear coupled polynomial Kalman filter appears to be working correctly in presence of large initialization errors.

If we compare the preceding results to those of the extended Kalman filter (see Fig. 9.36), we can see that the performance of the linear Kalman filter is far superior to that of the extended Kalman filter. When the initialization errors were 10 times the nominal value, the extended filter yielded estimates for projectile downrange location with errors on the order of 300 ft, while the linear filter errors were on the order of 100 ft.

To further demonstrate the robustness of the linear coupled polynomial Kalman filter to large initialization errors, another experiment was conducted. In this case the initial state estimates were set to zero or

$$\begin{bmatrix} \hat{x}_T(0) \\ \hat{\dot{x}}_T(0) \\ \hat{y}_T(0) \\ \hat{\dot{y}}_T(0) \end{bmatrix} = \begin{bmatrix} 0 \\ 0 \\ 0 \\ 0 \end{bmatrix}$$

The diagonal elements of the initial covariance matrix were then set to infinity to account for the large uncertainties in the initial errors in the estimates or

$$\boldsymbol{P}_0 = \begin{bmatrix} \infty & 0 & 0 & 0 \\ 0 & \infty & 0 & 0 \\ 0 & 0 & \infty & 0 \\ 0 & 0 & 0 & \infty \end{bmatrix}$$

Figure 9.39 shows sample results for the linear coupled polynomial Kalman filter. We can see from Fig. 9.39 that the errors in the estimate of downrange are well behaved, even when the filter is extremely poorly initialized. Recall that when the diagonal elements of the initial covariance matrix are infinite and the process-noise matrix is zero we have a recursive least-squares filter. By comparing Fig 9.39 with Fig. 9.8, we can see that the filter performance is virtually identical, indicating once again that initialization is not an important issue for a linear Kalman filter. However, good initialization is critical for an extended Kalman filter.

Therefore, if we have a problem in which both a linear and extended Kalman filter work and yield comparable performance, it is safer to use the linear Kalman filter because it is more robust when large initialization errors are present.

Summary

In this chapter we have shown how an extended Kalman filter could be built in either the Cartesian or polar coordinate systems. For the cannon-launched projectile problem, in which the filter was in the Cartesian system, the state-space equation was linear, but the measurement equation was nonlinear. If the filter were in the polar system, the state-space equation was nonlinear, but the measurement equation was linear. We saw that for the example considered the performance of both extended Kalman filters was approximately the same. In other words, if computational requirements were not an issue the preferred

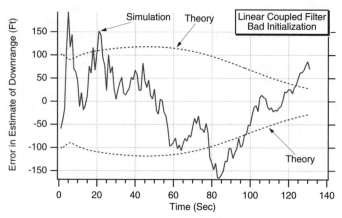

Fig. 9.39 Linear coupled Kalman filter performance appears to be independent of initialization errors.

coordinate system for the extended Kalman filter was a matter of personal preference and not related to performance issues. The polar filter was more computationally expensive than the Cartesian filter because numerical integration was required for state propagation. We also saw that a linear coupled Kalman filter that had near identical performance to that of the extended Kalman filters could also be built. The advantage of the linear filter was that it was more robust, as illustrated by an experiment in which the filter was improperly initialized. The linear coupled filter worked fine under such adverse conditions, whereas the extended Kalman filter failed. Process noise could be used to fix the extended Kalman filter, but the price paid was much larger estimation errors when compared to the linear Kalman filter. We also saw that if computation was an issue the linear filter also could be decoupled. The computer throughput requirements were better in the resultant filter pair because a pair of two-state linear filters require significantly less computation than one linear four-state filter. However, the price for reduced computation was a slight degradation in filter performance as measured by the estimation errors.

Reference

[1]Richards, J. A., Sears, F. W., Wehr, M. R., and Zemansky, M. W., *Modern University Physics, Part 1: Mechanics and Thermodynamics*, Addison Wesley Longman, Reading, MA, 1960, pp. 54–74.

Chapter 10

Tracking a Sine Wave

Introduction

IN THIS chapter we will attempt to apply extended Kalman filtering to a problem we briefly investigated with a linear Kalman filter. We will revisit the problem of tracking a sinusoidal signal measurement corrupted by noise. We have already shown that if the frequency of the sinusoid was known in advance the signal could be tracked quite well with a linear Kalman filter. However, if the frequency of the sinusoidal signal is unknown either our model of the real world or our measurement model becomes nonlinear, and we must resort to an extended Kalman filter. Several possible extended Kalman filters for this application, each of which has different states, will be explored in this chapter.

Extended Kalman Filter

In Chapter 5 we attempted to estimate the states (i.e., derivatives) of a sinusoidal signal based on noisy measurement of the sinusoidal signal. We showed that a linear polynomial Kalman filter was adequate for estimation purposes, but a much better linear Kalman filter, which made use of the fact that we knew that the true signal was sinusoidal, could be constructed. However, it was also demonstrated that if our a priori information was in error (i.e., knowledge of the frequency of the sinusoid is in error) the performance of the better linear Kalman filter deteriorated to the point where the estimates were no better, and possibly worse, than that of the linear polynomial Kalman filter (i.e., linear Kalman filter required no a priori information at all). In this section we will attempt to build an extended Kalman filter that operates on the sinusoidal measurement but also takes into account that the frequency of the sinusoidal signal is unknown and must also be estimated.[1,2]

In some of the one-dimensional work that follows, we will make use of the concept of negative and positive frequencies. Some of these frequency concepts may be difficult to understand from a physical point of view. These concepts are much easier to visualize in two dimensions. For example, in the two-dimensional x-y plane, circular motion can be described by the equations

$$x = A \sin \omega t$$
$$y = A \cos \omega t$$

In this case a positive frequency means that circular motion will be clockwise, and a negative frequency will describe circular motion in the counterclockwise direction.

In this chapter we consider the one-dimensional sinusoidal signal given by

$$x = A \sin \omega t$$

We can define a new variable ϕ to be

$$\phi = \omega t$$

If the frequency of the sinusoid is constant, we can take the derivative of the preceding equation to obtain

$$\dot{\phi} = \omega$$

Similarly, by assuming that the frequency and amplitude of the sinusoid are constant we are also saying that

$$\dot{\omega} = 0$$
$$\dot{A} = 0$$

We can express the preceding set of scalar equations that model our version of the real world in state-space form as

$$\begin{bmatrix} \dot{\phi} \\ \dot{\omega} \\ \dot{A} \end{bmatrix} = \begin{bmatrix} 0 & 1 & 0 \\ 0 & 0 & 0 \\ 0 & 0 & 0 \end{bmatrix} \begin{bmatrix} \phi \\ \omega \\ A \end{bmatrix} + \begin{bmatrix} 0 \\ u_{s1} \\ u_{s2} \end{bmatrix}$$

where u_{s1} and u_{s2} are white process noise sources that have been added to the derivatives of frequency and amplitude. These white noise sources may be required later to get the filter to work if we encounter problems. The process noise has been added to the derivatives of the states that are the least certain because we really do not know if the frequency and amplitude of the sinusoid will be constant. The preceding state-space equation is linear for this particular formulation. From the preceding state-space equation the continuous process-noise matrix can be written by inspection as

$$Q = \begin{bmatrix} 0 & 0 & 0 \\ 0 & \Phi_{s1} & 0 \\ 0 & 0 & \Phi_{s2} \end{bmatrix}$$

where Φ_{s1} and Φ_{s2} are the spectral densities of the white noise sources u_{s1} and u_{s2}. The systems dynamics matrix also can be written by inspection of the state-space equation as

$$F = \begin{bmatrix} 0 & 1 & 0 \\ 0 & 0 & 0 \\ 0 & 0 & 0 \end{bmatrix}$$

Because F^2 is zero as a result of

$$F^2 = \begin{bmatrix} 0 & 1 & 0 \\ 0 & 0 & 0 \\ 0 & 0 & 0 \end{bmatrix} \begin{bmatrix} 0 & 1 & 0 \\ 0 & 0 & 0 \\ 0 & 0 & 0 \end{bmatrix} = \begin{bmatrix} 0 & 0 & 0 \\ 0 & 0 & 0 \\ 0 & 0 & 0 \end{bmatrix}$$

the fundamental matrix can be expressed exactly as the two-term Taylor-series expansion

$$\Phi = I + Ft = \begin{bmatrix} 1 & 0 & 0 \\ 0 & 1 & 0 \\ 0 & 0 & 1 \end{bmatrix} + \begin{bmatrix} 0 & 1 & 0 \\ 0 & 0 & 0 \\ 0 & 0 & 0 \end{bmatrix} t = \begin{bmatrix} 1 & t & 0 \\ 0 & 1 & 0 \\ 0 & 0 & 1 \end{bmatrix}$$

Therefore, the exact discrete fundamental matrix can be obtained by replacing t with T_s, yielding

$$\Phi_k = \begin{bmatrix} 1 & T_s & 0 \\ 0 & 1 & 0 \\ 0 & 0 & 1 \end{bmatrix}$$

Although the state-space equation is linear, the measurement equation is nonlinear for this particular formulation of the problem. In fact, information concerning the sinusoidal nature of the signal is buried in the measurement equation. We have constructed the problem so that the linearized measurement equation (i.e., we are actually measuring x, which is not a state, thus making the measurement nonlinear) is given by

$$\Delta x^* = \begin{bmatrix} \dfrac{\partial x}{\partial \phi} & \dfrac{\partial x}{\partial \omega} & \dfrac{\partial x}{\partial A} \end{bmatrix} \begin{bmatrix} \Delta \phi \\ \Delta \omega \\ \Delta A \end{bmatrix} + v$$

where v is white measurement noise. To calculate the partial derivatives of the preceding equation, we first recall that

$$x = A \sin \omega t = A \sin \phi$$

Therefore, the partial derivatives in the measurement matrix are easily evaluated as

$$\frac{\partial x}{\partial \phi} = A \cos \phi$$

$$\frac{\partial x}{\partial \omega} = 0$$

$$\frac{\partial x}{\partial A} = \sin \phi$$

making the linerized measurement matrix (i.e., matrix of partial differentials)

$$\boldsymbol{H} = [A \cos \phi \quad 0 \quad \sin \phi]$$

The measurement matrix is evaluated at the projected estimates of A and ϕ. In this formulation the measurement noise is a scalar. Therefore, the discrete measurement noise matrix will also be a scalar given by

$$\boldsymbol{R}_k = \sigma_x^2$$

where σ_x^2 is the variance of the measurement noise.

Recall that the discrete process-noise matrix can be found from the continuous process-noise matrix according to

$$\boldsymbol{Q}_k = \int_0^{T_2} \Phi(\tau)\boldsymbol{Q}\Phi^T(\tau)\,\mathrm{d}t$$

Substitution of the appropriate matrices into the preceding expression yields

$$\boldsymbol{Q}_k = \int_0^{T_s} \begin{bmatrix} 1 & \tau & 0 \\ 0 & 1 & 0 \\ 0 & 0 & 1 \end{bmatrix} \begin{bmatrix} 0 & 0 & 0 \\ 0 & \Phi_{s1} & 0 \\ 0 & 0 & \Phi_{s2} \end{bmatrix} \begin{bmatrix} 1 & 0 & 0 \\ \tau & 1 & 0 \\ 0 & 0 & 1 \end{bmatrix} \mathrm{d}\tau$$

After multiplying out the three matrices we obtain

$$\boldsymbol{Q}_k = \int_0^{T_s} \begin{bmatrix} \tau^2 \Phi_{s1} & \tau \Phi_{s1} & 0 \\ \tau \Phi_{s1} & \Phi_{s1} & 0 \\ 0 & 0 & \Phi_{s2} \end{bmatrix} \mathrm{d}\tau$$

Finally, integrating the preceding expression shows the discrete process-noise matrix to be

$$Q_k = \begin{bmatrix} \dfrac{\Phi_{s1}T_s^3}{3} & \dfrac{\Phi_{s1}T_s^2}{2} & 0 \\[2mm] \dfrac{\Phi_{s1}T_s^2}{2} & \Phi_{s1}T_s & 0 \\[2mm] 0 & 0 & \Phi_{s2}T_s \end{bmatrix}$$

We now have enough information to solve the matrix Riccati equations for the Kalman gains.

Because the fundamental matrix is exact in this application, we can also use it to exactly propagate the state estimates in the extended Kalman filter over the sampling interval. In other words, by using the fundamental matrix we can propagate the states from time $k - 1$ to time k or in matrix form

$$\begin{bmatrix} \bar{\phi}_k \\ \bar{\omega}_k \\ \bar{A}_k \end{bmatrix} = \begin{bmatrix} 1 & T_s & 0 \\ 0 & 1 & 0 \\ 0 & 0 & 1 \end{bmatrix} \begin{bmatrix} \hat{\phi}_{k-1} \\ \hat{\omega}_{k-1} \\ \hat{A}_{k-1} \end{bmatrix}$$

We can multiply out the preceding matrix equation to get the three scalar equations

$$\bar{\phi}_k = \hat{\phi}_{k-1} + \hat{\omega}_{k-1}T_s$$
$$\bar{\omega}_k = \hat{\omega}_{k-1}$$
$$\bar{A}_k = \hat{A}_{k-1}$$

The linearized measurement matrix is used in the Riccati equations, but we do not have to use that matrix in the computation of the residual for the actual extended Kalman-filtering equations. We can do better by using the actual nonlinear equation for the residual or

$$\text{Res}_k = x_k^* - \bar{A}_k \sin \bar{\phi}_k$$

where the residual has been computed using the projected values of the states (i.e., barred quantities in the preceding equation). Now the extended Kalman-filtering equations can be written as

$$\hat{\phi}_k = \bar{\phi}_k + K_{1_k}\text{Res}_k$$
$$\hat{\omega}_k = \bar{\omega}_k + K_{2_k}\text{Res}_k$$
$$\hat{A}_k = \bar{A}_k + K_{3_k}\text{Res}_k$$

Listing 10.1 presents the extended Kalman filter for estimating the states of a noisy sinusoidal signal whose frequency is unknown. We can see that the

Listing 10.1   Extended Kalman filter for sinusoidal signal with unknown frequency

```
C THE FIRST THREE STATEMENTS INVOKE THE ABSOFT RANDOM
  NUMBER GENERATOR ON THE MACINTOSH
        GLOBAL DEFINE
                INCLUDE 'quickdraw.inc'
        END
        IMPLICIT REAL*8(A-H,O-Z)
        REAL*8 P(3,3),Q(3,3),M(3,3),PHI(3,3),HMAT(1,3),HT(3,1),PHIT(3,3)
        REAL*8 RMAT(1,1),IDN(3,3),PHIP(3,3),PHIPPHIT(3,3),HM(1,3)
        REAL*8 HMHT(1,1),HMHTR(1,1),HMHTRINV(1,1),MHT(3,1),K(3,1)
        REAL*8 KH(3,3),IKH(3,3)
        INTEGER ORDER
        A=1.
        W=1.
        TS=.1
        ORDER=3
        PHIS1=0.
        PHIS2=0.
        SIGX=1.
        OPEN(1,STATUS='UNKNOWN',FILE='DATFIL')
        OPEN(2,STATUS='UNKNOWN',FILE='COVFIL')
        T=0.
        S=0.
        H=.001
        DO 14 I=1,ORDER
        DO 14 J=1,ORDER
        PHI(I,J)=0.
        P(I,J)=0.
        Q(I,J)=0.
        IDN(I,J)=0.
14      CONTINUE
        RMAT(1,1)=SIGX**2
        IDN(1,1)=1.
        IDN(2,2)=1.
        IDN(3,3)=1.
        PHIH=0.
        WH=2.
        AH=3.
        P(1,1)=0.
        P(2,2)=(W-WH)**2
        P(3,3)=(A-AH)**2
        XT=0.
        XTD=A*W
        WHILE(T<=20.)
                XTOLD=XT
                XTDOLD=XTD
                XTDD=-W*W*XT
                XT=XT+H*XTD
                XTD=XTD+H*XTDD
                T=T+H
                XTDD=-W*W*XT
```

(continued)

Listing 10.1 (*Continued*)

```
XT=.5*(XTOLD+XT+H*XTD)
XTD=.5*(XTDOLD+XTD+H*XTDD)
S=S+H
IF(S>=(TS-.00001))THEN
        S=0.
        PHI(1,1)=1.
        PHI(1,2)=TS
        PHI(2,2)=1.
        PHI(3,3)=1.
        Q(1,1)=TS*TS*TS*PHIS1/3.
        Q(1,2)=.5*TS*TS*PHIS1
        Q(2,1)=Q(1,2)
        Q(2,2)=PHIS1*TS
        Q(3,3)=PHIS2*TS
        PHIB=PHIH+WH*TS
        HMAT(1,1)=AH*COS(PHIB)
        HMAT(1,2)=0.
        HMAT(1,3)=SIN(PHIB)
        CALL MATTRN(PHI,ORDER,ORDER,PHIT)
        CALL MATTRN(HMAT,1,ORDER,HT)
        CALL MATMUL(PHI,ORDER,ORDER,P,ORDER,
            ORDER,PHIP)
        CALL MATMUL(PHIP,ORDER,ORDER,PHIT,ORDER,
            ORDER,PHIPPHIT)
        CALL MATADD(PHIPPHIT,ORDER,ORDER,Q,M)
        CALL MATMUL(HMAT,1,ORDER,M,ORDER,
            ORDER,HM)
        CALL MATMUL(HM,1,ORDER,HT,ORDER,1,HMHT)
        CALL MATADD(HMHT,ORDER,ORDER,RMAT,
            HMHTR)HMHTRINV(1,1)=1./HMHTR(1,1)
        CALL MATMUL(M,ORDER,ORDER,HT,ORDER,1,
            MHT)
        CALL MATMUL(MHT,ORDER,1,HMHTRINV,1,1,K)
        CALL MATMUL(K,ORDER,1,HMAT,1,ORDER,KH)
        CALL MATSUB(IDN,ORDER,ORDER,KH,IKH)
        CALL MATMUL(IKH,ORDER,ORDER,M,ORDER,
            ORDER,P)
        CALL GAUSS(XTNOISE,SIGX)
        XTMEAS=XT+XTNOISE
        RES=XTMEAS-AH*SIN(PHIB)
        PHIH=PHIB+K(1,1)*RES
        WH=WH+K(2,1)*RES
        AH=AH+K(3,1)*RES
        ERRPHI=W*T-PHIH
        SP11=SQRT(P(1,1))
        ERRW=W-WH
        SP22=SQRT(P(2,2))
        ERRA=A-AH
        SP33=SQRT(P(3,3))
```

(*continued*)

Listing 10.1   *(Continued)*

```
                        XTH=AH*SIN(PHIH)
                        XTDH=AH*WH*COS(PHIH)
                        WRITE(9,*)T,XT,XTH,XTD,XTDH,W,WH,A,AH,W*T,
                          PHIH
                        WRITE(1,*)T,XT,XTH,XTD,XTDH,W,WH,A,AH,W*T,
                          PHIH
                        WRITE(2,*)T,ERRPHI,SP11,-SP11,ERRW,SP22,-SP22,
    1                     ERRA,SP33,-SP33
               ENDIF
          END DO
          PAUSE
          CLOSE(1)
          CLOSE(2)
          END


C SUBROUTINE GAUSS IS SHOWN IN LISTING 1.8
C SUBROUTINE MATTRN IS SHOWN IN LISTING 1.3
C SUBROUTINE MATMUL IS SHOWN IN LISTING 1.4
C SUBROUTINE MATADD IS SHOWN IN LISTING 1.1
C SUBROUTINE MATSUB IS SHOWN IN LISTING 1.2
```

simulation is initially set up to run without process noise (i.e., PHIS1 = PHIS2 = 0). The actual sinusoidal signal has unity amplitude, and the standard deviation of the measurement noise is also unity. The frequency of the actual sinusoidal signal is unity. The filter's initial state estimates of ϕ is set to zero, whereas the estimate of A has been set to three (rather than unity). The initial estimate of the frequency is set to two (rather than unity). Values are used for the initial covariance matrix to reflect the uncertainties in our initial state estimates. Because the fundamental matrix is exact in this filter formulation, a special subroutine is *not* required in this simulation to integrate the state equations over the sampling interval to obtain the state projections.

The nominal case of Listing 10.1 (i.e., in which there is no process noise) was run. We can see from Figs. 10.1 and 10.2 that for this run the extended Kalman filter is able to estimate the frequency and amplitude of the sinusoidal signal quite well when the actual frequency of the sinusoid is 1 rad/s and the filter's initial frequency estimate is 2 rad/s. After approximately 5 s the frequency and amplitude estimates are near perfect.

To test the filter's robustness, another case was run in which the actual frequency of the sinusoid is negative, but the filter is initialized with a positive frequency. We can see from Fig. 10.3 that now the filter is unable to estimate the negative frequency. In fact the final frequency estimate of the extended Kalman filter is approximately zero. Figure 10.4 indicates that under these circumstances the filter is also unable to estimate the signal amplitude. The figure indicates that the extended Kalman filter estimates the amplitude to be three, rather than the true value of unity.

Even though at this point we know the filter is not satisfactory in meeting our original goals, we would like to examine it further to see if we can learn

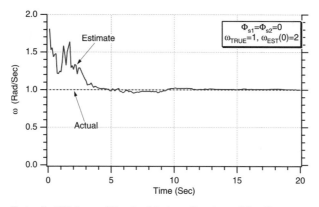

Fig. 10.1   Extended Kalman filter is able to estimate positive frequency when initial frequency estimate is also positive.

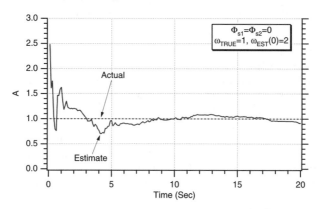

Fig. 10.2   Extended Kalman filter is able to estimate amplitude when actual frequency is positive and initial frequency estimate is also positive.

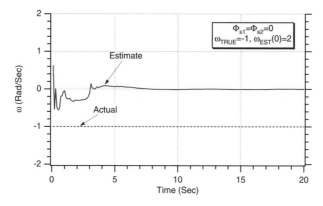

Fig. 10.3   Extended Kalman filter is unable to estimate negative frequency when initial frequency estimate is positive.

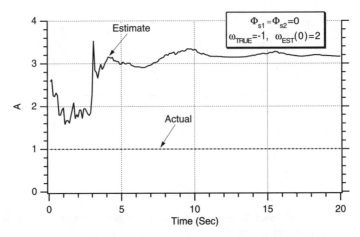

Fig. 10.4 Extended Kalman filter is unable to estimate amplitude when actual frequency is negative and initial frequency estimate is positive.

something that will be of future use. To further test the filter's properties, another case was run in which the actual frequency of the sinusoid is negative, but this time the filter is initialized with a negative frequency. We can see from Fig. 10.5 that the filter is now able to estimate the negative frequency fairly rapidly. Figure 10.6 also indicates that under these circumstances the filter is also able to estimate the signal amplitude accurately.

As a final test of the filter's properties, another case was run in which the actual frequency of the sinusoid is positive while the filter is initialized with a negative frequency. We can see from Fig. 10.7 that after approximately 10 s the filter is now able to estimate the positive frequency. However, Fig. 10.8 indicates that

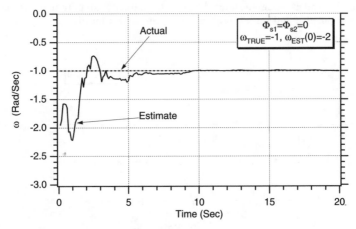

Fig. 10.5 Extended Kalman filter is now able to estimate negative frequency when initial frequency estimate is also negative.

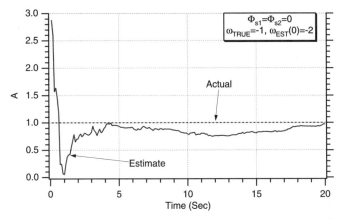

Fig. 10.6 Extended Kalman filter is now able to estimate amplitude when actual frequency is negative and initial frequency estimate is also negative.

under these circumstances the filter is unable to estimate the signal amplitude. The actual amplitude is 1, and the estimated amplitude is approximately 0.6.

From the preceding set of experiments, it appears that the extended Kalman filter only works if the initial estimate of the signal frequency is of the same sign as the actual frequency. We have already seen that when the signs of the actual frequency and initial frequency estimate are mismatched either the signal frequency or amplitude or both could not be estimated.

Sometimes adding process noise is the engineering fix for getting a Kalman filter to perform properly and robustly. Figures 10.9 and 10.10 reexamine the case where the actual frequency of the sinusoid is positive but the initial frequency estimate of the filter is negative. If we now add process noise (i.e.,

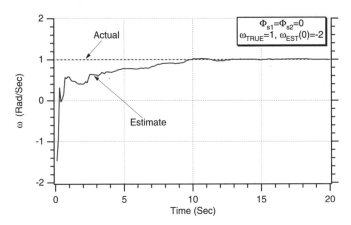

Fig. 10.7 Extended Kalman filter is able to estimate positive frequency when initial frequency estimate is negative.

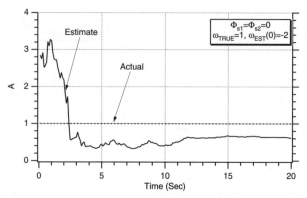

Fig. 10.8 Extended Kalman filter is not able to estimate amplitude when actual frequency is positive and initial frequency estimate is negative.

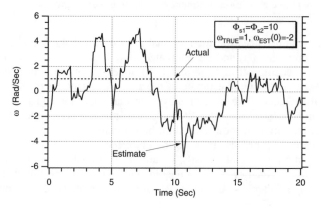

Fig. 10.9 Addition of process noise is not the engineering fix to enable filter to estimate frequency.

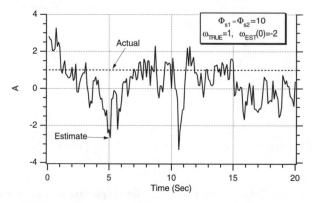

Fig. 10.10 Addition of process noise is not the engineering fix to enable filter to estimate amplitude.

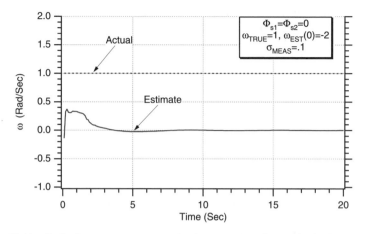

Fig. 10.11 Reducing measurement noise by an order of magnitude does not enable Kalman filter to estimate positive frequency when initial frequency estimate is negative.

$\Phi_{s1} = 10$, $\Phi_{s2} = 10$), the extended Kalman filter's frequency and amplitude estimates are considerably worse than they were when there was no process noise at all (see Figs. 10.7–10.8). Therefore, it appears that the extended Kalman filter is not able to estimate frequency and amplitude under all circumstances, even in the presence of process noise.

It also was hypothesized that perhaps there was too much measurement noise for the extended Kalman filter to work properly. Figure 10.11 indicates that even if we reduce the measurement noise by an order of magnitude the extended Kalman filter is still unable to estimate the positive frequency of the sinusoid when the initial filter frequency estimate is negative. In addition, from Fig. 10.12

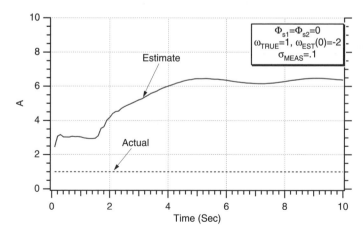

Fig. 10.12 Reducing measurement noise by an order of magnitude does not enable Kalman filter to estimate amplitude when actual frequency is positive and initial frequency estimate is negative.

we can see that the filter is also unable to estimate the amplitude under the same conditions. In other words, the filter's inability to estimate frequency and amplitude properly under a variety of initial conditions is not caused by the fact that the measurement noise is too large.

The conclusion reached is that the extended Kalman filter we have formulated in this section does not appear to be working satisfactorily if the filter is not properly initialized. Is it possible that, for this problem, the frequency of the sinusoid is unobservable?

Two-State Extended Kalman Filter with a Priori Information

So far we have shown that the extended Kalman filter of the preceding section was unable to estimate the frequency and amplitude of the sinusoidal signal unless the filter was properly initialized. With perfect hindsight this now seems reasonable because if we have a signal given by

$$x = A \sin \omega t$$

it is possible to determine if a negative value of x is a result of either a negative frequency ω or a negative amplitude A. Therefore, to gain a deeper understanding of the problem it is hypothesized that if the amplitude of the sinusoid is known in advance we should be able to estimate the frequency based on measurements of x. For the academic purpose of testing our hypothesis, we will derive another extended Kalman filter, which assumes that the amplitude of the sinusoidal signal is known precisely. Under these circumstances the model of the real world can be simplified from that of the preceding section to

$$\begin{bmatrix} \dot{\phi} \\ \dot{\omega} \end{bmatrix} = \begin{bmatrix} 0 & 1 \\ 0 & 0 \end{bmatrix} \begin{bmatrix} \phi \\ \omega \end{bmatrix} + \begin{bmatrix} 0 \\ u_s \end{bmatrix}$$

where u_s is white process noise that has been added to the derivative of frequency to account for the fact that we may not be modeling the real world perfectly. Therefore, the continuous process-noise matrix can be written by inspection of the preceding state-space equation as

$$Q = \begin{bmatrix} 0 & 0 \\ 0 & \Phi_s \end{bmatrix}$$

where Φ_s is the spectral density of the white process noise. The systems dynamics matrix also can be obtained by inspection of the state-space equation representing the real world as

$$F = \begin{bmatrix} 0 & 1 \\ 0 & 0 \end{bmatrix}$$

Because F^2 is zero, the fundamental matrix can be represented exactly by the two-term Taylor-series expansion

$$\Phi = I + Ft = \begin{bmatrix} 1 & 0 \\ 0 & 1 \end{bmatrix} + \begin{bmatrix} 0 & 1 \\ 0 & 0 \end{bmatrix} t = \begin{bmatrix} 1 & t \\ 0 & 1 \end{bmatrix}$$

which means that the discrete fundamental matrix is given by

$$\Phi_k = \begin{bmatrix} 1 & T_s \\ 0 & 1 \end{bmatrix}$$

Although the state-space equation representing our model of the real world is linear, the measurement equation (i.e., we are actually measuring x, which is not a state) is nonlinear. Therefore, the linearized measurement equation is given by

$$\Delta x^* = \begin{bmatrix} \dfrac{\partial x}{\partial \phi} & \dfrac{\partial x}{\partial \omega} \end{bmatrix} \begin{bmatrix} \Delta\phi \\ \Delta\omega \end{bmatrix} + v$$

where v is white measurement noise. We have already shown in the preceding section that

$$x = A \sin \omega t = A \sin \phi$$

the partial derivatives of the measurement matrix can be easily evaluated as

$$\frac{\partial x}{\partial \phi} = A \cos \phi$$

$$\frac{\partial x}{\partial \omega} = 0$$

making the linearized measurement matrix

$$H = [A \cos \phi \quad 0]$$

The measurement matrix will be evaluated at the projected estimate of ϕ. In this example, because the measurement noise is a scalar, the discrete measurement noise matrix will also be a scalar given by

$$R_k = \sigma_x^2$$

where σ_x^2 is the variance of the measurement noise. Recall that the discrete process-noise matrix can be found from the continuous process-noise matrix according to

$$Q_k = \int_0^{T_2} \Phi(\tau) Q \Phi^T(\tau) \, dt$$

Substitution of the appropriate matrices into the preceding expression yields

$$
Q_k = \int_0^{T_s} \begin{bmatrix} 1 & \tau \\ 0 & 1 \end{bmatrix} \begin{bmatrix} 0 & 0 \\ 0 & \Phi_s \end{bmatrix} \begin{bmatrix} 1 & 0 \\ \tau & 1 \end{bmatrix} d\tau
$$

After multiplying out the three matrices, we get

$$
Q_k = \int_0^{T_s} \begin{bmatrix} \tau^2 \Phi_s & \tau \Phi_s \\ \tau \Phi_s & \Phi_s \end{bmatrix} d\tau
$$

Finally, after integration we obtain for the discrete process-noise matrix

$$
Q_k = \begin{bmatrix} \dfrac{\Phi_s T_s^3}{3} & \dfrac{\Phi_s T_s^2}{2} \\ \dfrac{\Phi_s T_s^2}{2} & \Phi_s T_s \end{bmatrix}
$$

We now have enough information to solve the matrix Riccati equations for the Kalman gains.

Because the fundamental matrix is also exact in this application, we can also use it to propagate the state estimates in the extended Kalman filter over the sampling interval. Using the fundamental matrix, we can propagate the states from time $k - 1$ to time k or in matrix form

$$
\begin{bmatrix} \bar{\phi}_k \\ \bar{\omega}_k \end{bmatrix} = \begin{bmatrix} 1 & T_s \\ 0 & 1 \end{bmatrix} \begin{bmatrix} \hat{\phi}_{k-1} \\ \hat{\omega}_{k-1} \end{bmatrix}
$$

We can multiply out the preceding matrix equation to get the two scalar equations

$$
\bar{\phi}_k = \hat{\phi}_{k-1} + \hat{\omega}_{k-1} T_s
$$
$$
\bar{\omega}_k = \hat{\omega}_{k-1}
$$

Again, as was the case in the preceding section, because the measurement matrix is linearized we do not have to use it in the calculation of the residual for the actual extended Kalman-filtering equations. Instead, we can do better by using the actual nonlinear equation for the residual or

$$
\text{Res}_k = x_k^* - A \sin \bar{\phi}_k
$$

In the preceding equation the amplitude A is not estimated but is assumed to be known a prior. The extended Kalman-filtering equations can now be written as

$$
\hat{\phi}_k = \bar{\phi}_k + K_{1_k} \text{Res}_k
$$
$$
\hat{\omega}_k = \bar{\omega}_k + K_{2_k} \text{Res}_k
$$

where the barred quantities represent projected states that have been already defined.

Listing 10.2 presents the two-state extended Kalman filter for estimating the states of a noisy sinusoidal signal whose frequency is unknown but whose amplitude is known. As was the case with Listing 10.1, we can see that the simulation is initially set up to run without process noise (i.e., PHIS = 0). The actual sinusoidal signal still has unity amplitude, and the standard deviation of the measurement noise is also unity. We can see from Listing 10.2 that the frequency of the actual sinusoidal signal is also unity. The filter's initial state estimate of ϕ is set to zero because time is initially zero (i.e., $\phi = \omega t$), while the initial estimate of the frequency is set to two (rather than unity). Values are used for the initial covariance matrix to reflect the uncertainties in our initial state estimates. Again, because the fundamental matrix is exact, a special subroutine is not required in this simulation to integrate the state equations over the sampling interval in order to obtain the state projections.

Listing 10.2   Two-state extended Kalman filter with a priori information for estimating frequency of sinusoidal signal

```
C THE FIRST THREE STATEMENTS INVOKE THE ABSOFT RANDOM
  NUMBER GENERATOR ON THE MACINTOSH
          GLOBAL DEFINE
                  INCLUDE 'quickdraw.inc'
          END
          IMPLICIT REAL*8(A-H,O-Z)
          REAL*8 P(2,2),Q(2,2),M(2,2),PHI(2,2),HMAT(1,2),HT(2,1),PHIT(2,2)
          REAL*8 RMAT(1,1),IDN(2,2),PHIP(2,2),PHIPPHIT(2,2),HM(1,2)
          REAL*8 HMHT(1,1),HMHTR(1,1),HMHTRINV(1,1),MHT(2,1),K(2,1)
          REAL*8 KH(2,2),IKH(2,2)
          INTEGER ORDER
          TS=.1
          A=1.
          W=1.
          PHIS=0.
          SIGX=1.
          ORDER=2
          OPEN(1,STATUS='UNKNOWN',FILE='DATFIL')
          OPEN(2,STATUS='UNKNOWN',FILE='COVFIL')
          T=0.
          S=0.
          H=.001
          DO 14 I=1,ORDER
          DO 14 J=1,ORDER
          PHI(I,J)=0.
          P(I,J)=0.
          Q(I,J)=0.
          IDN(I,J)=0.
14        CONTINUE
          RMAT(1,1)=SIGX**2
```

(continued)

Listing 10.2 (*Continued*)

```
IDN(1,1)=1.
IDN(2,2)=1.
PHIH=0.
WH=2.
P(1,1)=0.**2
P(2,2)=(W-WH)**2
XT=0.
XTD=A*W
WHILE(T<=20.)
        XTOLD=XT
        XTDOLD=XTD
        XTDD=-W*W*XT
        XT=XT+H*XTD
        XTD=XTD+H*XTDD
        T=T+H
        XTDD=-W*W*XT
        XT=.5*(XTOLD+XT+H*XTD)
        XTD=.5*(XTDOLD+XTD+H*XTDD)
        S=S+H
        IF(S>=(TS-.00001))THEN
                S=0.
                PHI(1,1)=1.
                PHI(1,2)=TS
                PHI(2,2)=1.
                Q(1,1)=TS*TS*TS*PHIS/3.
                Q(1,2)=.5*TS*TS*PHIS
                Q(2,1)=Q(1,2)
                Q(2,2)=PHIS*TS
                PHIB=PHIH+WH*TS
                HMAT(1,1)=COS(PHIB)
                HMAT(1,2)=0.
                CALL MATTRN(PHI,ORDER,ORDER,PHIT)
                CALL MATTRN(HMAT,1,ORDER,HT)
                CALL MATMUL(PHI,ORDER,ORDER,P,ORDER,
                    ORDER,PHIP)
                CALL MATMUL(PHIP,ORDER,ORDER,PHIT,ORDER,
                    ORDER,PHIPPHIT)
                CALL MATADD(PHIPPHIT,ORDER,ORDER,Q,M)
                CALL MATMUL(HMAT,1,ORDER,M,ORDER,ORDER,
                    HM)
                CALL MATMUL(HM,1,ORDER,HT,ORDER,1,HMHT)
                CALL MATADD(HMHT,ORDER,ORDER,RMAT,
                    HMHTR)
                HMHTRINV(1,1)=1./HMHTR(1,1)
                CALL MATMUL(M,ORDER,ORDER,HT,ORDER,1,
                    MHT)
                CALL MATMUL(MHT,ORDER,1,HMHTRINV,1,1,K)
                CALL MATMUL(K,ORDER,1,HMAT,1,ORDER,KH)
                CALL MATSUB(IDN,ORDER,ORDER,KH,IKH)
                CALL MATMUL(IKH,ORDER,ORDER,M,ORDER,
```
                                                        (*continued*)

1

Listing 10.2 *(Continued)*

```
                        ORDER,P)
                        CALL  GAUSS(XTNOISE,SIGX)
                        XTMEAS=XT+XTNOISE
                        RES=XTMEAS-A*SIN(PHIB)
                        PHIH=PHIB+K(1,1)*RES
                        WH=WH+K(2,1)*RES
                        PHIREAL=W*T
                        ERRPHI=PHIREAL-PHIH
                        SP11=SQRT(P(1,1))
                        ERRW=W-WH
                        SP22=SQRT(P(2,2))
                        XTH=A*SIN(PHIH)
                        XTDH=A*WH*COS(PHIH)
                        WRITE(9,*)T,XT,XTH,XTD,XTDH,W,WH,PHI,PHIH
                        WRITE(1,*)T,XT,XTH,XTD,XTDH,W,WH,PHI,PHIH
                        WRITE(2,*)T,ERRPHI,SP11,-SP11,ERRW,SP22,-SP22
            ENDIF
      END  DO
      PAUSE
      CLOSE(1)
      CLOSE(2)
      END


C  SUBROUTINE GAUSS IS SHOWN IN LISTING 1.8
C  SUBROUTINE MATTRN IS SHOWN IN LISTING 1.3
C  SUBROUTINE MATMUL IS SHOWN IN LISTING 1.4
C  SUBROUTINE MATADD IS SHOWN IN LISTING 1.1
C  SUBROUTINE MATSUB IS SHOWN IN LISTING 1.2
```

The nominal case of Listing 10.2 was run, and we can see from Fig. 10.13 that the new two-state extended Kalman filter appears to be working because it is able to estimate the positive frequency when the filter frequency estimate is initialized positive. However, under these idealized initialization conditions the preceding extended Kalman filter also had similar performance (see Fig 10.1).

Another case was run in which the actual frequency was negative and the initial frequency estimate was positive. We can see from Fig. 10.14 that now the new two-state extended Kalman filter is able to estimate the negative frequency quite well under these circumstances, whereas the preceding three-state extended Kalman filter was unable to estimate the negative frequency under similar circumstances (see Fig. 10.3). Thus, it appears that when the amplitude of the sinusoid is known in advance it is possible to estimate the frequency of the sinusoid.

Figures 10.15 and 10.16 complete further experiments in which we are attempting to estimate the frequency for various filter frequency initialization conditions. As expected, Fig. 10.15 shows that a negative frequency can be easily estimated when the initial filter frequency estimate is negative. This is not

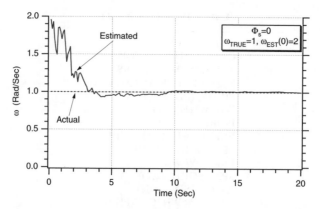

Fig. 10.13 Two-state extended Kalman filter estimates positive frequency when initial frequency estimate is also positive.

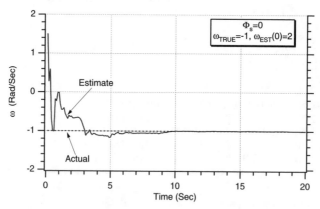

Fig. 10.14 New two-state extended Kalman filter estimates negative frequency when initialized positive.

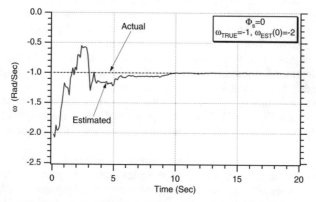

Fig. 10.15 New two-state extended Kalman filter estimates negative frequency when initialized negative.

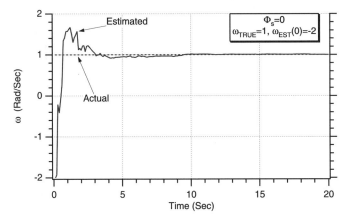

Fig. 10.16 **New two-state extended Kalman filter estimates positive frequency when initialized negative.**

surprising because the preceding three-state extended Kalman filter also could have accomplished this task. However, Fig. 10.16 shows that a positive frequency also can be estimated when the initial filter frequency estimate is negative. The preceding three-state extended Kalman filter would have failed at this attempt.

To see if the new two-state extended Kalman filter was really working all of the time, even more runs were made. Listing 10.2 was modified slightly so that it could be run in the Monte Carlo mode. Code was changed so that 10 runs could be made, one after another, each with a different noise stream. The first case that was examined in detail was the one in which the actual frequency of the sinusoid was negative and the filter's initial estimate of the frequency was positive. The single-run simulation results of Fig. 10.14 indicated that the filter was working. However, we can see from Fig. 10.17 that one case out of 10 failed to estimate the

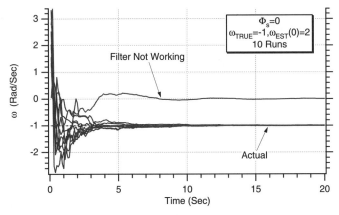

Fig. 10.17 **New two-state extended Kalman filter does not estimate negative frequency all of the time when filter is initialized positive.**

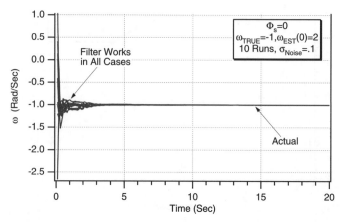

Fig. 10.18 **New two-state extended Kalman filter works all of the time when the measurement noise is reduced by an order of magnitude.**

negative frequency. In that particular case the filter thought the frequency was zero.

Because the filter should work all of the time, it was hypothesized that perhaps there was too much measurement noise. The standard deviation of the measurement noise was unity, and the amplitude of the sinusoidal signal was also unity. Figure 10.18 shows that when the measurement noise is reduced by an order of magnitude to 0.1 the filter is now successful in estimating the negative frequency in all 10 cases.

In the preceding figure the frequency estimates looked excellent. However, the real performance of the two-state extended Kalman filter is determined by the error in the estimate of frequency. Figure 10.19 displays the error in the estimate

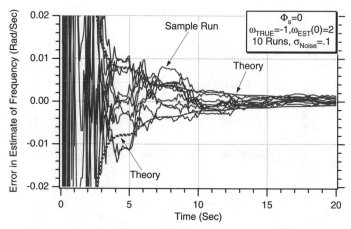

Fig. 10.19 **Error in the estimate results indicate that the new two-state extended Kalman filter is able to estimate frequency.**

of frequency for the case in which the actual frequency is negative, but the filter's initial estimate of frequency is positive when the measurement noise is 0.1. The theoretical error bounds for the error in the estimate of frequency, obtained by taking the square root of the second diagonal element of the covariance matrix, are also displayed in Fig. 10.19. We can see that for all 10 runs the simulated error in the estimate of frequency lies within the theoretical bounds most of the time, indicating that the new two-state extended Kalman filter is working properly.

The two-state extended Kalman filter derived in this section was an academic exercise in explaining the failure of the three-state extended Kalman filter of the preceding section. We still want to find out if an extended Kalman filter can be designed to estimate the signal states and frequency under poor initialization conditions.

Alternate Extended Kalman Filter for Sinusoidal Signal

In this section we will try again to build an extended Kalman filter that makes use of the fact that the signal is sinusoidal but also takes into account that both the frequency and amplitude of the sinusoid are also unknown.

Recall that the actual signal is given by

$$x = A \sin \omega t$$

As was just mentioned, in this model we will assume that both the amplitude A and the frequency ω of the sinusoid are unknown. If we take the derivative of the preceding equation, we also get a sinusoid or

$$\dot{x} = A\omega \cos \omega t$$

Taking the derivative again yields

$$\ddot{x} = -A\omega^2 \sin \omega t$$

Thus, we can see that the preceding second-order differential equation can be expressed in terms of the first equation or

$$\ddot{x} = -\omega^2 x$$

In other words, integrating the preceding equation twice, with the appropriate initial conditions, will yield the original sinusoidal signal. The preceding equation is especially useful because the sinusoidal term and signal amplitude have been eliminated and the second derivative of the signal or third state has been expressed in terms of the signal or first state. This is precisely what we desire for expressing the problem in state-space notation.

If we assume that the sinusoidal frequency is an unknown constant, its derivative must be zero. To account for the fact that the sinusoidal frequency may not be a constant, it is prudent to add process noise to the derivative of frequency. Therefore, the two differential equations representing the real world in this example are

$$\ddot{x} = -\omega^2 x$$
$$\dot{\omega} = u_s$$

where u_s is white process noise with spectral density Φ_s. The preceding differential equations can be expressed as three first-order differential equations (i.e., one of the equations is redundant) in state-space form as

$$
\begin{bmatrix} \dot{x} \\ \ddot{x} \\ \dot{\omega} \end{bmatrix} =
\begin{bmatrix} 0 & 1 & 0 \\ -\omega^2 & 0 & 0 \\ 0 & 0 & 0 \end{bmatrix}
\begin{bmatrix} x \\ \dot{x} \\ \omega \end{bmatrix} +
\begin{bmatrix} 0 \\ 0 \\ u_s \end{bmatrix}
$$

The preceding equation is nonlinear because a state also shows up in the 3×3 matrix multiplying the state vector. From the preceding equation we can see that the systems dynamics matrix turns out to be

$$
F = \frac{\partial f(x)}{\partial x} =
\begin{bmatrix}
\dfrac{\partial \dot{x}}{\partial x} & \dfrac{\partial \dot{x}}{\partial \dot{x}} & \dfrac{\partial \dot{x}}{\partial \omega} \\
\dfrac{\partial \ddot{x}}{\partial x} & \dfrac{\partial \ddot{x}}{\partial \dot{x}} & \dfrac{\partial \ddot{x}}{\partial \omega} \\
\dfrac{\partial \dot{\omega}}{\partial x} & \dfrac{\partial \dot{\omega}}{\partial \dot{x}} & \dfrac{\partial \dot{\omega}}{\partial \omega}
\end{bmatrix}
$$

where the partial derivatives are evaluated at the current estimates. Taking the partial derivatives in this example can be done by inspection, and the resultant systems dynamics matrix is given by

$$
F =
\begin{bmatrix} 0 & 1 & 0 \\ -\hat{\omega}^2 & 0 & -2\hat{\omega}\hat{x} \\ 0 & 0 & 0 \end{bmatrix}
$$

where the terms in the systems dynamics matrix have been evaluated at the current state estimates. In this example the exact fundamental matrix will be difficult, if not impossible, to find. If we assume that the elements of the systems dynamics matrix are approximately constant between sampling instants, then we use a two-term Taylor-series approximation for the fundamental matrix, yielding

$$
\Phi(t) \approx I + Ft =
\begin{bmatrix} 1 & t & 0 \\ -\hat{\omega}^2 t & 1 & -2\hat{\omega}\hat{x}t \\ 0 & 0 & 1 \end{bmatrix}
$$

Therefore, the discrete fundamental matrix can be obtained by substituting T_s for t or

$$\Phi_k \approx \begin{bmatrix} 1 & T_s & 0 \\ -\hat{\omega}_{k-1}^2 T_s & 1 & -2\hat{\omega}_{k-1}\hat{x}_{k-1}T_s \\ 0 & 0 & 1 \end{bmatrix}$$

The continuous process-noise matrix can by found from the original state-space equation to be

$$Q = E(ww^T) = E\begin{bmatrix} \begin{bmatrix} 0 \\ 0 \\ u_s \end{bmatrix} \begin{bmatrix} 0 & 0 & u_s \end{bmatrix} \end{bmatrix} = \begin{bmatrix} 0 & 0 & 0 \\ 0 & 0 & 0 \\ 0 & 0 & \Phi_s \end{bmatrix}$$

Recall that the discrete process-noise matrix can be found from the continuous process-noise matrix according to

$$Q_k = \int_0^{T_s} \Phi(\tau)Q\Phi^T(\tau)\,dt$$

By substitution of the appropriate matrices into the preceding equation, we obtain

$$Q_k = \int_0^{T_s} \begin{bmatrix} 1 & \tau & 0 \\ -\hat{\omega}^2\tau & 1 & -2\hat{\omega}\hat{x}\tau \\ 0 & 0 & 1 \end{bmatrix} \begin{bmatrix} 0 & 0 & 0 \\ 0 & 0 & 0 \\ 0 & 0 & \Phi_s \end{bmatrix} \begin{bmatrix} 1 & -\hat{\omega}^2\tau & 0 \\ \tau & 1 & 0 \\ 0 & -2\hat{\omega}\hat{x}\tau & 1 \end{bmatrix} dt$$

Multiplying out the three matrices yields

$$Q_k = \int_0^{T_s} \begin{bmatrix} 0 & 0 & 0 \\ 0 & 4\hat{\omega}^2\hat{x}^2\tau^2\Phi_s & -2\hat{\omega}\hat{x}\tau\Phi_s \\ 0 & -2\hat{\omega}\hat{x}\tau\Phi_s & \Phi_s \end{bmatrix} dt$$

If we again assume that the states are approximately constant between the sampling intervals, then the preceding integral can easily be evaluated as

$$Q_k = \begin{bmatrix} 0 & 0 & 0 \\ 0 & 1.333\hat{\omega}^2\hat{x}^2 T_s^3\Phi_s & -\hat{\omega}\hat{x}T_s^2\Phi_s \\ 0 & -\hat{\omega}\hat{x}T_s^2\Phi_s & T_s\Phi_s \end{bmatrix}$$

In this problem we are assuming that the measurement is of the first state plus noise or

$$x_k^* = x_k + v_k$$

Therefore, the measurement is linearly related to the states according to

$$x_k^* = [1 \quad 0 \quad 0] \begin{bmatrix} x \\ \dot{x} \\ \omega \end{bmatrix} + v_k$$

The measurement matrix can be obtained from the preceding equation by inspection as

$$H = [1 \quad 0 \quad 0]$$

In this example the discrete measurement noise matrix is a scalar and is given by

$$R_k = E(v_k v_k^T) = \sigma_x^2$$

We now have enough information to solve the matrix Riccati equations for the Kalman gains.

For this example the projected states in the actual extended Kalman-filtering equations do not have to use the approximation for the fundamental matrix. Instead the state projections, indicated by an overbar, can be obtained by numerically integrating the nonlinear differential equations over the sampling interval. Therefore, the extended Kalman-filtering equations can then be written as

$$\hat{x}_k = \bar{x}_k + K_{1_k}(x_k^* - \bar{x}_k)$$
$$\hat{\dot{x}}_k = \bar{\dot{x}}_k + K_{2_k}(x_k^* - \bar{x}_k)$$
$$\hat{\omega}_k = \hat{\omega}_{k-1} + K_{3_k}(x_k^* - \bar{x}_k)$$

Listing 10.3 presents the alternate three-state extended Kalman filter for estimating the states of a noisy sinusoidal signal whose frequency is unknown. We can see once again that the simulation is initially set up to run without process noise (i.e., PHIS = 0). The actual sinusoidal signal has unity amplitude, and the standard deviation of the measurement noise is also unity. We can see from Listing 10.3 that the frequency of the actual sinusoidal signal is also unity. The filter's initial state estimates of x and $\dot{x}$ are set to zero, while the initial estimate of the frequency is set to two (rather than unity). In other words, the second and third states are mismatched from the real world. Values are used for the initial covariance matrix to reflect the uncertainties in our initial state estimates. As was the case in Chapter 9, subroutine PROJECT is used to integrate the nonlinear equations using Euler integration over the sampling interval to obtain the state projections for x and $\dot{x}$.

A case was run using the nominal values of Listing 10.3 (i.e., no process noise), and we can see from Figs. 10.20–10.22 that the actual quantities and their estimates are quite close. An additional comparison of the results of our alternate three-state extended Kalman filter with the results of the first- and second-order polynomial Kalman filters of Chapter 5 indicates that we are estimating the states more accurately than those imperfect linear filters. However, our state estimates

Listing 10.3 **Three-state alternate extended Kalman filter for sinusoidal signal with unknown frequency**

```
C THE FIRST THREE STATEMENTS INVOKE THE ABSOFT RANDOM
  NUMBER GENERATOR ON THE MACINTOSH
        GLOBAL DEFINE
              INCLUDE 'quickdraw.inc'
        END
        IMPLICIT REAL*8(A-H,O-Z)
        REAL*8  P(3,3),Q(3,3),M(3,3),PHI(3,3),HMAT(1,3),HT(3,1),PHIT(3,3)
        REAL*8  RMAT(1,1),IDN(3,3),PHIP(3,3),PHIPPHIT(3,3),HM(1,3)
        REAL*8  HMHT(1,1),HMHTR(1,1),HMHTRINV(1,1),MHT(3,1),K(3,1),
     F(3,3)
        REAL*8  KH(3,3),IKH(3,3)
        INTEGER ORDER
        HP=.001
        W=1.
        A=1.
        TS=.1
        ORDER=3
        PHIS=0.
        SIGX=1.
        OPEN(1,STATUS='UNKNOWN',FILE='DATFIL')
        OPEN(2,STATUS='UNKNOWN',FILE='COVFIL')
        T=0.
        S=0.
        H=.001
        DO 14 I=1,ORDER
        DO 14 J=1,ORDER
        F(I,J)=0.
        PHI(I,J)=0.
        P(I,J)=0.
        Q(I,J)=0.
        IDN(I,J)=0.
14      CONTINUE
        RMAT(1,1)=SIGX**2
        IDN(1,1)=1.
        IDN(2,2)=1.
        IDN(3,3)=1.
        P(1,1)=SIGX**2
        P(2,2)=2.**2
        P(3,3)=2.**2
        XTH=0.
        XTDH=0.
        WH=2.
        XT=0.
        XTD=A*W
        WHILE(T<=20.)
              XTOLD=XT
              XTDOLD=XTD
```

(*continued*)

Listing 10.3 (*Continued*)

```
XTDD=-W*W*XT
XT=XT+H*XTD
XTD=XTD+H*XTDD
T=T+H
XTDD=-W*W*XT
XT=.5*(XTOLD+XT+H*XTD)
XTD=.5*(XTDOLD+XTD+H*XTDD)
S=S+H
IF(S>=(TS-.00001))THEN
        S=0.
        F(1,2)=1.
        F(2,1)=-WH**2
        F(2,3)=-2.*WH*XTH
        PHI(1,1)=1.
        PHI(1,2)=TS
        PHI(2,1)=-WH*WH*TS
        PHI(2,2)=1.
        PHI(2,3)=-2.*WH*XTH*TS
        PHI(3,3)=1.
        Q(2,2)=4.*WH*WH*XTH*XTH*TS*TS*TS*PHIS/3.
        Q(2,3)=-2.*WH*XTH*TS*TS*PHIS/2.
        Q(3,2)=Q(2,3)
        Q(3,3)=PHIS*TS
        HMAT(1,1)=1.
        HMAT(1,2)=0.
        HMAT(1,3)=0.
        CALL MATTRN(PHI,ORDER,ORDER,PHIT)
        CALL MATTRN(HMAT,1,ORDER,HT)
        CALL MATMUL(PHI,ORDER,ORDER,P,ORDER,
          ORDER,PHIP)
        CALL MATMUL(PHIP,ORDER,ORDER,PHIT,ORDER,
          ORDER,PHIPPHIT)
        CALL MATADD(PHIPPHIT,ORDER,ORDER,Q,M)
        CALL MATMUL(HMAT,1,ORDER,M,ORDER,ORDER,
          HM)
        CALL MATMUL(HM,1,ORDER,HT,ORDER,1,HMHT)
        CALL MATADD(HMHT,ORDER,ORDER,RMAT,
          HMHTR)
        HMHTRINV(1,1)=1./HMHTR(1,1)
        CALL MATMUL(M,ORDER,ORDER,HT,ORDER,1,
          MHT)
        CALL MATMUL(MHT,ORDER,1,HMHTRINV,1,1,K)
        CALL MATMUL(K,ORDER,1,HMAT,1,ORDER,KH)
        CALL MATSUB(IDN,ORDER,ORDER,KH,IKH)
        CALL MATMUL(IKH,ORDER,ORDER,M,ORDER,
          ORDER,P)
        CALL GAUSS(XTNOISE,SIGX)
        XTMEAS=XT+XTNOISE
        CALL PROJECT(T,TS,XTH,XTDH,XTB,XTDB,HP,WH)
        RES=XTMEAS-XTB
```

1

(*continued*)

Listing 10.3   *(Continued)*

```
                    XTH=XTB+K(1,1)*RES
                    XTDH=XTDB+K(2,1)*RES
                    WH=WH+K(3,1)*RES
                    ERRX=XT-XTH
                    SP11=SQRT(P(1,1))
                    ERRXD=XTD-XTDH
                    SP22=SQRT(P(2,2))
                    ERRW=W-WH
                    SP33=SQRT(P(3,3))
                    WRITE(9,*)T,XT,XTH,XTD,XTDH,W,WH
                    WRITE(1,*)T,XT,XTH,XTD,XTDH,W,WH
                    WRITE(2,*)T,ERRX,SP11,-SP11,ERRXD,SP22,-SP22,
1                      ERRW,SP33,-SP33
             ENDIF
        END DO
        PAUSE
        CLOSE(1)
        CLOSE(2)
        END

        SUBROUTINE PROJECT(TP,TS,XTP,XTDP,XTH,XTDH,HP,W)
        IMPLICIT REAL*8 (A-H)
        IMPLICIT REAL*8 (O-Z)
        T=0.
        XT=XTP
        XTD=XTDP
        H=HP
        WHILE(T<=(TS-.0001))
                XTDD=-W*W*XT
                XTD=XTD+H*XTDD
                XT=XT+H*XTD
                T=T+H
        END DO
        XTH=XT
        XTDH=XTD
        RETURN
        END

C SUBROUTINE GAUSS IS SHOWN IN LISTING 1.8
C SUBROUTINE MATTRN IS SHOWN IN LISTING 1.3
C SUBROUTINE MATMUL IS SHOWN IN LISTING 1.4
C SUBROUTINE MATADD IS SHOWN IN LISTING 1.1
C SUBROUTINE MATSUB IS SHOWN IN LISTING 1.2
```

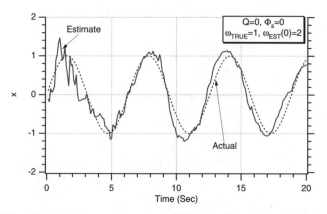

Fig. 10.20 Alternate three-state extended Kalman filter estimates first state well.

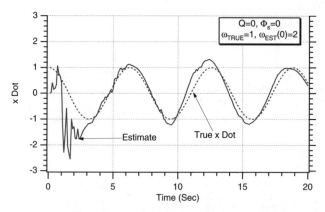

Fig. 10.21 Alternate three-state extended Kalman filter estimates second state well, even though filter is not initialized correctly.

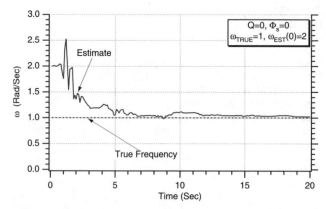

Fig. 10.22 Alternate three-state extended Kalman filter appears able to estimate the frequency of the sinusoid even though filter is not initialized correctly.

are not quite as good as those obtained from the linear sinusoidal Kalman filter of Chapter 5 (i.e., the one in which the fundamental matrix is exact) in which the frequency is known a priori. However, we are estimating the states far more accurately than the linear sinusoidal Kalman filter of Chapter 5, when the frequency is in error. From Fig. 10.22 we can see that it takes approximately 5 s to obtain an accurate estimate of the frequency for the parameters considered.

The fact that the new alternate three-state extended Kalman filter appears to be working under the benign initialization condition means that the filter is now ready for further testing. We will try to make the filter fail, as was done in the preceding two sections of this chapter. We will now consider the more difficult condition where the actual frequency is negative but the initial filter frequency estimate is positive. At first glance Fig. 10.23 appears to be telling us that the new filter also has failed in estimating the frequency. However, a closer examination of Fig. 10.23 indicates that the magnitude of the frequency has been estimated correctly, but the sign has not been estimated correctly.

To see if it is of any practical utility to only be able to estimate the magnitude of the frequency, more information is displayed in Figs. 10.24 and 10.25. Figure 10.24 compares the actual and estimated signal x, whereas Fig. 10.25 compares the actual and estimated derivative of the signal or $\dot{x}$. We can see that in both figures the estimates and actual quantities are very close even though the sign of the estimated frequency magnitude is not correct, indicating that the filter is working properly.

Another experiment was conducted in which the actual frequency of the signal was minus one, but the initial estimate of the frequency was minus two. Under these conditions we do not expect any problems, and Fig. 10.26 confirms that the alternate three-state extended Kalman filter estimates the frequency quite well.

Another experiment was conducted in which the actual frequency of the signal was plus one, but the initial estimate of the frequency was minus two. Under these conditions we expect problems, and Fig. 10.27 indicates that the alternate three-

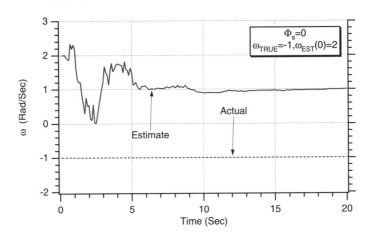

Fig. 10.23 Alternate three-state extended Kalman filter appears able to estimate the magnitude of the frequency but not its sign when the filter is not initialized correctly.

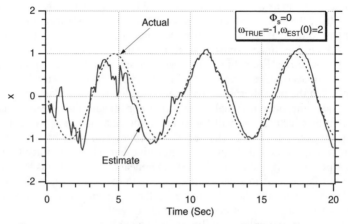

Fig. 10.24 Alternate three-state extended Kalman filter appears able to estimate the signal when the filter is not initialized correctly.

state extended Kalman filter estimates the magnitude of the frequency quite well but estimates the wrong sign.

To again see if it is of any practical utility to be able to estimate the magnitude of the frequency, more information is displayed in Figs. 10.28 and 10.29. Figure 10.28 compares the actual and estimated signal x, whereas Fig. 10.29 compares the actual and estimated derivative of the signal or $\dot{x}$. We can see that in both figures the estimates and actual quantities are very close, indicating that the filter is working properly even though we are not able to estimate the sign of the frequency.

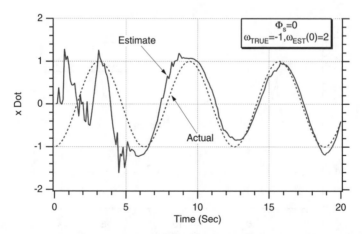

Fig. 10.25 Alternate three-state extended Kalman filter appears able to estimate the derivative of the signal when filter is not initialized correctly.

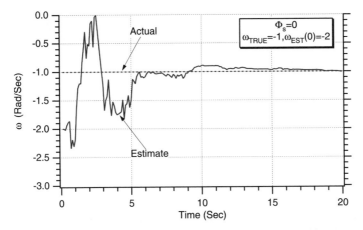

Fig. 10.26 **Alternate three-state extended Kalman filter appears able to estimate the frequency correctly when frequency and initial estimate are both negative.**

Figures 10.30–10.32 display the theoretical errors in the state estimates (i.e., obtained by taking the square root of the appropriate diagonal element of the covariance matrix) and the single-run simulation errors for each of the three states. The single-run errors in the estimates appear to be within the theoretical bounds approximately 68% of the time, indicating that the extended Kalman filter seems to be behaving properly. Because this example has zero process noise, the errors in the estimates diminish as more measurements are taken.

To verify that the filter was actually working all of the time under a variety of initialization conditions, Listing 10.3 was slightly modified to give it Monte Carlo

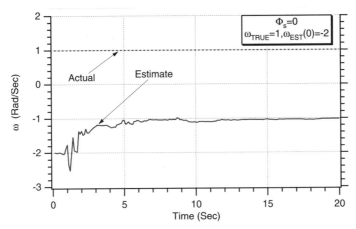

Fig. 10.27 **Alternate three-state extended Kalman filter appears able to estimate the magnitude of the frequency but not its sign when the filter is not initialized correctly.**

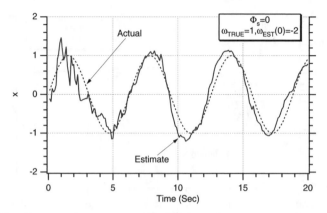

Fig. 10.28   Alternate three-state extended Kalman filter appears able to estimate the signal when the filter is not initialized correctly.

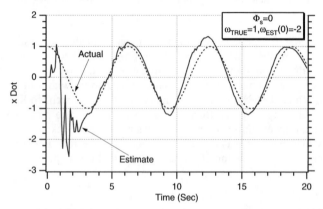

Fig. 10.29   Alternate three-state extended Kalman filter appears able to estimate the derivative of the signal when the filter is not initialized correctly.

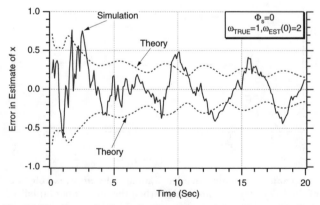

Fig. 10.30   Error in the estimate of first state agrees with theory.

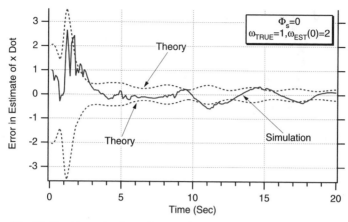

Fig. 10.31   Error in the estimate of second state agrees with theory.

capabilities. Ten-run Monte Carlo sets were run with the alternate three-state extended Kalman filter for the four different frequency and estimated frequency initialization conditions. The standard deviation of the measurement noise for these experiments was unity. The results, which are presented in Figs. 10.33–10.36, show that the correct magnitude of the frequency is always estimated accurately, regardless of initialization. We have already seen that when the correct frequency magnitude is estimated the other states also will be accurate. Thus, we can conclude that the alternate three-state extended Kalman filter is effective in estimating the states of a sinusoid.

We have seen that the alternate three-state extended Kalman filter is only able to estimate both the magnitude and sign of the frequency exactly when the initial

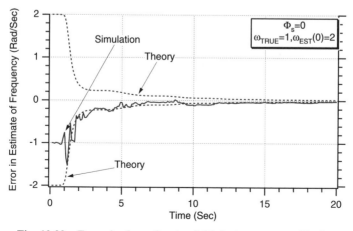

Fig. 10.32   Error in the estimate of third state agrees with theory.

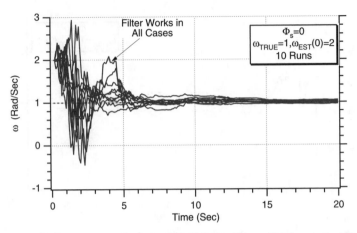

Fig. 10.33   Alternate three-state extended Kalman filter estimates correct frequency when initial frequency estimate is of correct sign.

frequency estimate of the filter is of the same sign as the actual frequency. When this condition is not met, the filter is only able to estimate the correct magnitude of the frequency but not the sign. It is hypothesized that the reason for the filter's lack of ability in always distinguishing between positive and negative frequencies is caused by the formulation of the state-space equations where an ω^2 term appears. Apparently, we cannot distinguish between positive and negative frequencies with only the ω^2 term (i.e., there is no ω term). However, if we are only interested in obtaining estimates of x and $\dot{x}$, knowing the magnitude of the frequency is sufficient (i.e., the sign of the frequency is not important).

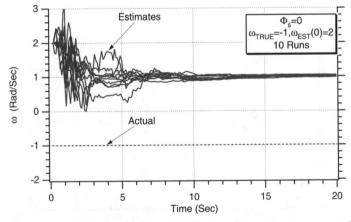

Fig. 10.34   Alternate three-state extended Kalman filter estimates correct frequency magnitude when initial frequency estimate is of the wrong sign.

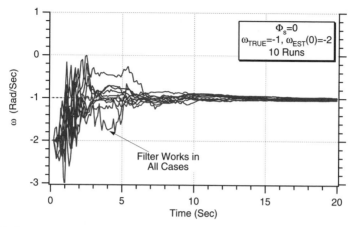

Fig. 10.35 Alternate three-state extended Kalman filter estimates correct frequency when initial frequency estimate is of correct sign.

Another Extended Kalman Filter for Sinusoidal Model

We have just derived an extended Kalman filter in the preceding section in which the sinusoidal frequency ω was a state. We demonstrated that in order to obtain accurate estimates of the states of a sinusoidal signal it was not important to estimate the sign of the sinusoidal frequency but only its magnitude. With that information we will see if we can do even better in the estimation process by reforming the problem to match our new experience base. Let us define a new state z, which is simply the square of the frequency or

$$z = \omega^2$$

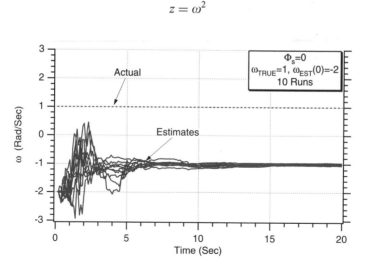

Fig. 10.36 Alternate three-state extended Kalman filter estimates correct frequency magnitude when initial frequency estimate is of the wrong sign.

Under these new circumstances the differential equations representing the real world in this new model are now given by

$$\ddot{x} = -zx$$

$$\dot{z} = u_s$$

where u_s is white process noise with spectral density Φ_s. The preceding differential equations can be expressed as three first-order differential equations (i.e., one of the equations is redundant) in state-space form as

$$\begin{bmatrix} \dot{x} \\ \ddot{x} \\ \dot{z} \end{bmatrix} = \begin{bmatrix} 0 & 1 & 0 \\ -z & 0 & 0 \\ 0 & 0 & 0 \end{bmatrix} \begin{bmatrix} x \\ \dot{x} \\ z \end{bmatrix} + \begin{bmatrix} 0 \\ 0 \\ u_s \end{bmatrix}$$

The preceding equation is also nonlinear because a state also shows up in the 3×3 matrix multiplying the state vector. From the preceding equation we know that the systems dynamics matrix is given by

$$F = \frac{\partial f(x)}{\partial x} = \begin{bmatrix} \dfrac{\partial \dot{x}}{\partial x} & \dfrac{\partial \dot{x}}{\partial \dot{x}} & \dfrac{\partial \dot{x}}{\partial z} \\ \dfrac{\partial \ddot{x}}{\partial x} & \dfrac{\partial \ddot{x}}{\partial \dot{x}} & \dfrac{\partial \ddot{x}}{\partial z} \\ \dfrac{\partial \dot{z}}{\partial x} & \dfrac{\partial \dot{z}}{\partial \dot{x}} & \dfrac{\partial \dot{z}}{\partial z} \end{bmatrix}$$

where the partial derivatives are evaluated at the current estimates. Taking the partial derivatives in this example can be done by inspection, and the resultant systems dynamics matrix is given by

$$F = \begin{bmatrix} 0 & 1 & 0 \\ -\hat{z} & 0 & -\hat{x} \\ 0 & 0 & 0 \end{bmatrix}$$

where the terms in the systems dynamics matrix have been evaluated at the current state estimates. As was the case in the last section, the exact fundamental matrix will be difficult, if not impossible, to find. If we assume that the elements of the systems dynamics matrix are approximately constant between sampling instants, then we use a two-term Taylor-series approximation for the fundamental matrix, yielding

$$\Phi(t) \approx I + Ft = \begin{bmatrix} 1 & t & 0 \\ -\hat{z}t & 1 & -\hat{x}t \\ 0 & 0 & 1 \end{bmatrix}$$

Therefore, the discrete fundamental matrix can be obtained by substituting T_s for t or

$$\Phi_k \approx \begin{bmatrix} 1 & T_s & 0 \\ -\hat{z}_{k-1}T_s & 1 & -\hat{x}_{k-1}T_s \\ 0 & 0 & 1 \end{bmatrix}$$

As was the case in the preceding section, the continuous process-noise matrix can by found from the original state-space equation to be

$$Q = E(ww^T) = E\left[\begin{bmatrix} 0 \\ 0 \\ u_s \end{bmatrix} \begin{bmatrix} 0 & 0 & u_s \end{bmatrix}\right] = \begin{bmatrix} 0 & 0 & 0 \\ 0 & 0 & 0 \\ 0 & 0 & \Phi_s \end{bmatrix}$$

Recall that the discrete process-noise matrix can be found from the continuous process-noise matrix according to

$$Q_k = \int_0^{T_s} \Phi(\tau)Q\Phi^T(\tau)\,d\tau$$

By substitution of the appropriate matrices into the preceding equation, we obtain

$$Q_k = \int_0^{T_s} \begin{bmatrix} 1 & \tau & 0 \\ -\hat{z}\tau & 1 & -\hat{x}\tau \\ 0 & 0 & 1 \end{bmatrix} \begin{bmatrix} 0 & 0 & 0 \\ 0 & 0 & 0 \\ 0 & 0 & \Phi_s \end{bmatrix} \begin{bmatrix} 1 & -\hat{z}\tau & 0 \\ \tau & 1 & 0 \\ 0 & -\hat{x}\tau & 1 \end{bmatrix} d\tau$$

Multiplying out the three matrices yields

$$Q_k = \int_0^{T_s} \begin{bmatrix} 0 & 0 & 0 \\ 0 & \hat{x}^2\tau^2\Phi_s & -\hat{x}\tau\Phi_s \\ 0 & -\hat{x}\tau\Phi_s & \Phi_s \end{bmatrix} d\tau$$

If we again assume that the states are approximately constant between the sampling intervals, then the preceding integral can easily be evaluated as

$$Q_k = \begin{bmatrix} 0 & 0 & 0 \\ 0 & 0.333\hat{x}^2 T_s^3\Phi_s & -0.5\hat{x}T_s^2\Phi_s \\ 0 & -0.5\hat{x}T_s^2\Phi_s & T_s\Phi_s \end{bmatrix}$$

In this problem we are also assuming that the measurement is of the first state plus noise or

$$x_k^* = x_k + v_k$$

Therefore, the measurement is linearly related to the states according to

$$x_k^* = \begin{bmatrix} 1 & 0 & 0 \end{bmatrix} \begin{bmatrix} x \\ \dot{x} \\ z \end{bmatrix} + v_k$$

The measurement matrix can be obtained from the preceding equation by inspection as

$$H = \begin{bmatrix} 1 & 0 & 0 \end{bmatrix}$$

In this example the discrete measurement noise matrix is a scalar and is given by

$$R_k = E(v_k v_k^T) = \sigma_x^2$$

We now have enough information to solve the matrix Riccati equations for the Kalman gains.

As was the case in the preceding section, the projected states in the actual extended Kalman-filtering equations do not have to use the approximation for the fundamental matrix. Instead, the state projections, indicated by an overbar, can be obtained by numerically integrating the nonlinear differential equations over the sampling interval. Therefore, the extended Kalman-filtering equations can then be written as

$$\hat{x}_k = \bar{x}_k + K_{1_k}(x_k^* - \bar{x}_k)$$
$$\hat{\dot{x}}_k = \bar{\dot{x}}_k + K_{2_k}(x_k^* - \bar{x}_k)$$
$$\hat{z}_k = \hat{z}_{k-1} + K_{3_k}(x_k^* - \bar{x}_k)$$

Listing 10.4 presents another three-state extended Kalman filter for estimating the states of a noisy sinusoidal signal whose frequency is unknown. As was the case in the last section, the simulation is initially set up to run without process noise (i.e., PHIS = 0). The actual sinusoidal signal has unity amplitude, and the standard deviation of the measurement noise is also unity. We can see from Listing 10.4 that the frequency of the actual sinusoidal signal is unity. The filter's initial state estimates of x and $\dot{x}$ are set to zero. Because the initial estimate of the frequency in the previous section was two (rather than unity), the initial value of the estimate of z is 4 (i.e., $z = \omega^2 = 2^2 = 4$). In other words, as was the case in the preceding section, the second and third states are mismatched from the real world. Values are used for the initial covariance matrix to reflect the uncertainties in our initial state estimates. Subroutine PROJECT is still used to integrate the nonlinear equations, using Euler integration over the sampling interval to obtain the state projections for x and $\dot{x}$.

A case was run using the nominal values of Listing 10.4 (i.e., no process noise), and we can see from a comparison that the actual quantities and their estimates in Figs. 10.37–10.39 are quite close. If we compare the estimates of x and $\dot{x}$ (i.e., Figs. 10.37 and 10.38) with the preceding results from our alternate

Listing 10.4   Another three-state extended Kalman filter for sinusoidal signal with unknown frequency

```
C THE FIRST THREE STATEMENTS INVOKE THE ABSOFT RANDOM
  NUMBER GENERATOR ON THE MACINTOSH
        GLOBAL DEFINE
               INCLUDE 'quickdraw.inc'
        END
        IMPLICIT REAL*8(A-H,O-Z)
        REAL*8 P(3,3),Q(3,3),M(3,3),PHI(3,3),HMAT(1,3),HT(3,1),PHIT(3,3)
        REAL*8 RMAT(1,1),IDN(3,3),PHIP(3,3),PHIPPHIT(3,3),HM(1,3)
        REAL*8 HMHT(1,1),HMHTR(1,1),HMHTRINV(1,1),MHT(3,1),K(3,1),
       F(3,3)
        REAL*8 KH(3,3),IKH(3,3)
        INTEGER ORDER
        HP=.001
        W=1.
        WH=2.
        A=1.
        TS=.1
        ORDER=3
        PHIS=0.
        SIGX=1.
        OPEN(1,STATUS='UNKNOWN',FILE='DATFIL')
        OPEN(2,STATUS='UNKNOWN',FILE='COVFIL')
        T=0.
        S=0.
        H=.001
        DO 14 I=1,ORDER
        DO 14 J=1,ORDER
        F(I,J)=0.
        PHI(I,J)=0.
        P(I,J)=0.
        Q(I,J)=0.
        IDN(I,J)=0.
14      CONTINUE
        RMAT(1,1)=SIGX**2
        IDN(1,1)=1.
        IDN(2,2)=1.
        IDN(3,3)=1.
        P(1,1)=SIGX**2
        P(2,2)=2.**2
        P(3,3)=4.**2
        XTH=0.
        XTDH=0.
        ZH=WH**2
        XT=0.
        XTD=A*W
        WHILE(T<=20.)
               XTOLD=XT
               XTDOLD=XTD
```

(continued)

Listing 10.4 (*Continued*)

```
XTDD=-W*W*XT
XT=XT+H*XTD
XTD=XTD+H*XTDD
T=T+H
XTDD=-W*W*XT
XT=.5*(XTOLD+XT+H*XTD)
XTD=.5*(XTDOLD+XTD+H*XTDD)
S=S+H
IF(S>=(TS-.00001))THEN
        S=0.
        F(1,2)=1.
        F(2,1)=-WH**2
        F(2,3)=-2.*WH*XTH
        PHI(1,1)=1.
        PHI(1,2)=TS
        PHI(2,1)=-ZH*TS
        PHI(2,2)=1.
        PHI(2,3)=-XTH*TS
        PHI(3,3)=1.
        Q(2,2)=XTH*XTH*TS*TS*TS*PHIS/3.
        Q(2,3)=-XTH*TS*TS*PHIS/2.
        Q(3,2)=Q(2,3)
        Q(3,3)=PHIS*TS
        HMAT(1,1)=1.
        HMAT(1,2)=0.
        HMAT(1,3)=0.
        CALL  MATTRN(PHI,ORDER,ORDER,PHIT)
        CALL  MATTRN(HMAT,1,ORDER,HT)
        CALL  MATMUL(PHI,ORDER,ORDER,P,ORDER,
          ORDER,PHIP)
        CALL  MATMUL(PHIP,ORDER,ORDER,PHIT,ORDER,
          ORDER,PHIPPHIT)
        CALL  MATADD(PHIPPHIT,ORDER,ORDER,Q,M)
        CALL  MATMUL(HMAT,1,ORDER,M,ORDER,ORDER,
          HM)
        CALL  MATMUL(HM,1,ORDER,HT,ORDER,1,HMHT)
        CALL  MATADD(HMHT,ORDER,ORDER,RMAT,
          HMHTR)HMHTRINV(1,1)=1./HMHTR(1,1)
        CALL  MATMUL(M,ORDER,ORDER,HT,ORDER,1,
          MHT)
        CALL  MATMUL(MHT,ORDER,1,HMHTRINV,1,1,K)
        CALL  MATMUL(K,ORDER,1,HMAT,1,ORDER,KH)
        CALL  MATSUB(IDN,ORDER,ORDER,KH,IKH)
        CALL  MATMUL(IKH,ORDER,ORDER,M,ORDER,
          ORDER,P)
        CALL  GAUSS(XTNOISE,SIGX)
        XTMEAS=XT+XTNOISE
        CALL  PROJECT(T,TS,XTH,XTDH,XTB,XTDB,HP,ZH)
        RES=XTMEAS-XTB
```

(*continued*)

Listing 10.4   *(Continued)*

```
                    XTH=XTB+K(1,1)*RES
                    XTDH=XTDB+K(2,1)*RES
                    ZH=ZH+K(3,1)*RES
                    ERRX=XT-XTH
                    SP11=SQRT(P(1,1))
                    ERRXD=XTD-XTDH
                    SP22=SQRT(P(2,2))
                    Z=W**2
                    ERRZ=Z-ZH
                    SP33=SQRT(P(3,3))
                    WH=SQRT(ZH)

                    WRITE(9,*)T,XT,XTH,XTD,XTDH ,Z,ZH
                    WRITE(1,*)T,XT,XTH,XTD,XTDH ,Z,ZH
                    WRITE(2,*)T,ERRX,SP11,-SP11,ERRXD,SP22,-SP22,
1                        ERRZ,SP33,-SP33
                ENDIF
         END DO
         PAUSE
         CLOSE(1)
         CLOSE(2)
         END

         SUBROUTINE PROJECT(TP,TS,XTP,XTDP,XTH,XTDH,HP,Z)
         IMPLICIT REAL*8 (A-H)
         IMPLICIT REAL*8 (O-Z)
         T=0.
         XT=XTP
         XTD=XTDP
         H=HP
         WHILE(T<=(TS-.0001))
                XTDD=-Z*XT
                XTD=XTD+H*XTDD
                XT=XT+H*XTD
                T=T+H
         END DO
         XTH=XT
         XTDH=XTD
         RETURN
         END

C SUBROUTINE GAUSS IS SHOWN IN LISTING 1.8
C SUBROUTINE MATTRN IS SHOWN IN LISTING 1.3
C SUBROUTINE MATMUL IS SHOWN IN LISTING 1.4
C SUBROUTINE MATADD IS SHOWN IN LISTING 1.1
C SUBROUTINE MATSUB IS SHOWN IN LISTING 1.2
```

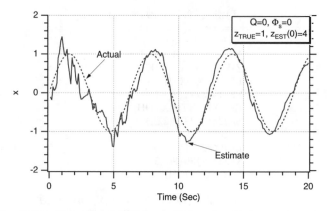

Fig. 10.37   New extended Kalman filter estimates first state quite well.

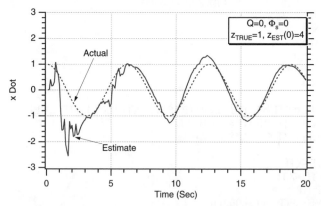

Fig. 10.38   New extended Kalman filter estimates second state well, even though that state is not correctly initialized.

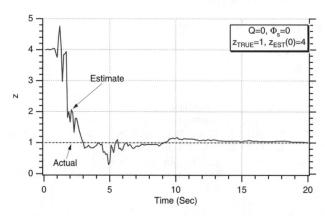

Fig. 10.39   After 5 s the new extended Kalman filter is able to estimate the square of frequency.

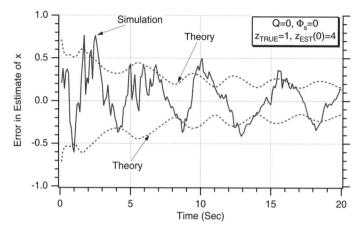

Fig. 10.40   Error in the estimate of first state agrees with theory.

extended Kalman filter (ie., Figs. 10.20 and 10.21), we can see that the new results are virtually identical to the preceding results. Our new filter does not appear to be any better or worse than the preceding filter. From Fig. 10.39 we can see that it takes approximately 5 s to obtain an accurate estimate of z (i.e., the square of the frequency) for the parameters considered.

Figures 10.40–10.42 display the theoretical errors in the state estimates (i.e., obtained by taking the square root of the appropriate diagonal element of the covariance matrix) and the single-run errors for each of the three states obtained from the simulation. The single-run errors in the estimates appear to be within the theoretical bounds approximately 68% of the time, indicating that the extended Kalman filter seems to be behaving properly. Because this example has zero process noise, the errors in the estimates diminish as more measurements are taken. The results of Figs. 10.40 and 10.41 are virtually identical to the results of

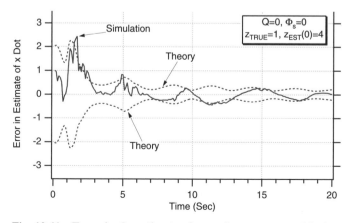

Fig. 10.41   Error in the estimate of second state agrees with theory.

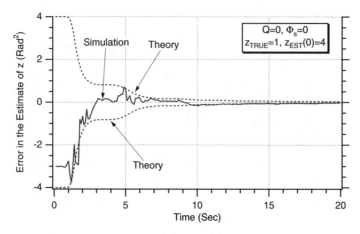

Fig. 10.42   Error in the estimate of third state agrees with theory.

Figs. 10.30 and 10.31. Thus, it appears that there is virtually no difference between the extended Kalman filter of this section and the one of the preceding section.

To verify that the filter was actually working all of the time, Listing 10.4 was slightly modified to give it Monte Carlo capabilities. A ten-run Monte Carlo set was run with the new three-state extended Kalman filter. Figure 10.43 shows that we are able to accurately estimate z in each of the 10 runs. Thus, we can conclude that the new three-state extended Kalman filter is also effective in estimating the states of a sinusoid.

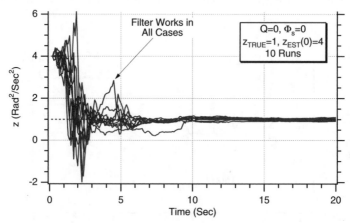

Fig. 10.43   New three-state extended Kalman filter is also effective in estimating the square of the frequency of a sinusoid.

Summary

We have developed several extended Kalman filters that attempt to estimate the states of a sinusoidal signal way based upon noisy measurements of the signal. We have demonstrated in this chapter that simply choosing states and building an extended Kalman filter does not guarantee that it will actually work as expected if programmed correctly. Sometimes, some states that are being estimated are not observable. Numerous experiments were conducted, and various extended Kalman filters were designed, each of which had different states, to highlight some of the issues. An extended Kalman filter was finally built that could estimate all of the states under a variety of initialization conditions, but it could not always distinguish between positive and negative frequencies (i.e., always a correct estimate of frequency magnitude but not always a correct estimate of the frequency sign). Monte Carlo experiments confirmed that this particular extended Kalman filter appeared to be very robust. An alternate form of the extended Kalman filter, in which the square of the frequency was being estimated, was shown to be equally as robust.

References

[1] Gelb., A., *Applied Optimal Estimation*, Massachusetts Inst. of Technology Press, Cambridge, MA, 1974, pp. 180–228.

[2] Zarchan, P., *Tactical and Strategic Missile Guidance* 3rd ed., Progress in Astronautics and Aeronautics, AIAA, Reston, VA, 1998, pp. 373–387.

Chapter 11

Satellite Navigation

Introduction

IN THIS chapter we will show how extended Kalman filtering can be used to locate a receiver based on range measurements from a number of satellites to the receiver. Although many simplifications are made, some aspects of problems of this type are similar to those encountered by a receiver processing measurements from the global positioning system (GPS).[1] We will first investigate the case in which filtering is not used at all, but we solve simply the noisy algebraic equations for the receiver location. Next, we will attempt to improve the accuracy in locating the receiver by filtering the range measurements. We will then show that even more accuracy can be achieved by using extended Kalman filtering. Once the filter is built, we will see what happens to accuracy and to the filter when the receiver is moving. All of the experiments will be conducted in two dimensions with either one or two satellites.

Problem with Perfect Range Measurements

Consider two satellites traveling in orbit in a two-dimensional world. If we are considering a flat Earth, then the satellites will appear as if they are each traveling at constant altitude. Let us assume that at any time the two satellite locations $[(x_{R1}, y_{R1})$ and $(x_{R2}, y_{R2})]$ are known perfectly. In addition, let us first consider the case where each satellite is able to take perfect range measurements r_1 and r_2 to a receiver at location (x, y), as shown in Fig. 11.1. It seems reasonable to believe that two range measurements are sufficient for determining the location of the receiver.

To calculate the location of the receiver numerically, let us first recall that the range from each of the satellites to the receiver can be written by inspection of Fig. 11.1 as

$$r_1 = \sqrt{(x_{R1} - x)^2 + (y_{R1} - y)^2}$$

$$r_2 = \sqrt{(x_{R2} - x)^2 + (y_{R2} - y)^2}$$

The preceding two equations have two unknowns (i.e., x and y are unknowns); because we are assuming that we know the location of both satellites perfectly (i.e., $x_{R1}, y_{R1}, x_{R2},$ and y_{R2} are known), we have enough information to solve for

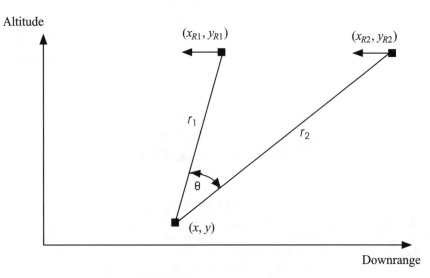

Fig. 11.1   Two satellites making range measurements to a receiver.

the location of the receiver. Squaring and multiplying out each of the terms of the two preceding equations yields

$$r_1^2 = x_{R1}^2 - 2x_{R1}x + x^2 + y_{R1}^2 - 2y_{R1}y + y^2$$

$$r_2^2 = x_{R2}^2 - 2x_{R2}x + x^2 + y_{R2}^2 - 2y_{R2}y + y^2$$

By subtracting the second equation from the first equation and combining like terms, we get

$$r_1^2 - r_2^2 = 2x(x_{R2} - x_{R1}) + 2y(y_{R2} - y_{R1}) + x_{R1}^2 + y_{R1}^2 - x_{R2}^2 - y_{R2}^2$$

Finally, solving for x in terms of y yields

$$x = -\frac{y(y_{R2} - y_{R1})}{(x_{R2} - x_{R1})} + \frac{r_1^2 - r_2^2 - x_{R1}^2 - y_{R1}^2 + x_{R2}^2 + y_{R2}^2}{2(x_{R2} - x_{R1})}$$

By defining

$$A = -\frac{(y_{R2} - y_{R1})}{(x_{R2} - x_{R1})}$$

$$B = \frac{r_1^2 - r_2^2 - x_{R1}^2 - y_{R1}^2 + x_{R2}^2 + y_{R2}^2}{2(x_{R2} - x_{R1})}$$

we achieve simplification and obtain

$$x = Ay + B$$

Substituting the preceding solution into the square of the first range equation yields

$$r_1^2 = x_{R1}^2 - 2x_{R1}(Ay + B) + (Ay + B)^2 + y_{R1}^2 - 2y_{R1}y + y^2$$

It is obvious that the preceding equation is a quadratic in terms of y, which can be rewritten as

$$0 = y^2(1 + A^2) + y(-2Ax_{R1} + 2AB - 2y_{R1}) + x_{R1}^2 - 2x_{R1}B + y_{R1}^2 - r_1^2$$

Therefore, we can considerably simplify the preceding quadratic equation by defining

$$a = 1 + A^2$$

$$b = -2Ax_{R1} + 2AB - 2y_{R1}$$

$$c = x_{R1}^2 - 2x_{R1}B + y_{R1}^2 - r_1^2$$

Using the preceding definitions yields the simplified quadratic equation in y as

$$0 = ay^2 + by + c$$

It is well known that the solution to the preceding equation has two roots, indicating that there are two possible receiver locations based only on range information. Using common sense or a priori information (i.e., receiver is not below ground or in outer space), we can throw away the extraneous root to obtain

$$y = \frac{-b - \sqrt{b^2 - 4ac}}{2a}$$

Once we have solved for y we can then solve for x as

$$x = Ay + B$$

In practice the solution to the two preceding equations yields the coordinates of the estimated location of the receiver. Only in the case in which the satellite location is known precisely and the range measurements from the satellite to the receiver are perfect should the preceding equations yield exact solutions. We will see later that satellite geometry will also be an important factor in determining the solutions.

To help us better understand the satellite geometry, we will use the angle θ, shown in Fig. 11.1, between the two range measurements as a way of specifying the geometry. We can see from Fig. 11.1 that the angle θ can be calculated from the vector dot product of the two ranges or

$$\theta = \cos^{-1} \frac{\overline{r}_1 \cdot \overline{r}_2}{|\overline{r}_1||\overline{r}_2|}$$

Figure 11.1 shows that the range vectors can be expressed as

$$\overline{r}_1 = (x_{R1} - x)\overline{i} + (y_{R1} - y)\overline{j}$$

$$\overline{r}_2 = (x_{R2} - x)\overline{i} + (y_{R2} - y)\overline{j}$$

whereas the range magnitudes are simply

$$|\overline{r}_1| = r_1$$

$$|\overline{r}_2| = r_2$$

Therefore, substitution yields the solution for the angle θ as

$$\theta = \cos^{-1} \frac{(x_{R1} - x)(x_{R2} - x) + (y_{R1} - y)(y_{R2} - y)}{r_1 r_2}$$

To check the preceding equations, a case was set up in which the satellites were considered to be in an orbit similar to the GPS satellites. In the planar flat Earth model of Fig. 11.1, this means that the satellites travel in a straight line at a fixed altitude. We are assuming that for this example the receiver is located at the origin (i.e., 0 ft downrange and 0 ft in altitude) of Fig. 11.1. Both satellites are at a 20,000 km altitude and are traveling at a speed of 14,600 ft/s (to the left).[1] Initially, let us assume that the first satellite is 1,000,000 ft downrange of the receiver, while the second satellite is 500,000 ft downrange of the receiver. Estimates of the receiver location are calculated every second for 100 s. In the simulation we integrate satellite velocity to get the satellite location. Because in this example the satellite velocity is constant, we could have integrated the satellite velocity, obtaining the location of the satellites as

$$x_{R1} = \dot{x}_{R1}t + x_{R1}(0)$$

$$x_{R2} = \dot{x}_{R2}t + x_{R2}(0)$$

$$y_{R1} = y_{R1}(0)$$

$$y_{R2} = y_{R2}(0)$$

For this example the initial satellite locations are given by

$$x_{R1}(0) = 1,000,000 \, \text{ft}$$

$$x_{R2}(0) = 500,000 \, \text{ft}$$

$$y_{R1}(0) = 20,000 * 3280 \, \text{ft}$$

$$y_{R2}(0) = 20,000 * 3280 \, \text{ft}$$

and the satellite velocities are

$$\dot{x}_{R1} = -14,600 \, \text{ft/s}$$

$$\dot{x}_{R2} = -14,600 \, \text{ft/s}$$

Because the satellites are moving, the angle θ will change with time.

The preceding equations for the motion of the two satellites, the angle between the two range vectors, and, most importantly, the solution for the receiver location were programmed and appear in Listing 11.1. Although it was not necessary, we can see from the listing that the satellite velocities are numerically integrated to get the satellite position. Every second, the actual and estimated receiver location is printed out along with the angle between the two range vectors.

The nominal case of Listing 11.1 was run, and the estimated location of the receiver (i.e., XH and YH) matched the actual location of the receiver (i.e., X and Y), indicating the formulas that were just derived are correct. Note that the calculations are repeated every second because the satellites are moving. However, all calculations yield the same location for the receiver, which is at the origin of our coordinate system.

Estimation Without Filtering

To get a feeling for how well we can estimate the location of the stationary receiver in the presence of noisy range measurements from both satellites, the simulation of Listing 11.1 was slightly modified to account for the fact that the range measurements might not be perfect. The resultant simulation with the noisy range measurements appears in Listing 11.2. We can see from the listing that zero-mean, Guassian noise with a standard deviation 300 ft was added to each range measurement every second (i.e., $T_s = 1$). The formulas of the preceding section were used to estimate the receiver location from the noisy measurements, as shown in Listing 11.2. Changes to the original simulation of Listing 11.1 to account for the noisy range measurements are highlighted in bold in Listing 11.2.

The nominal case of Listing 11.2 was run. Figures 11.2 and 11.3 compare the actual and estimated downrange and altitude receiver locations based on the noisy range measurements. We can see that a 300-ft 1 σ noise error on range yields extremely large downrange location errors for this particular geometry. In this

Listing 11.1 Simulation to see if receiver location can be determined from two perfect range measurements

```
IMPLICIT REAL*8(A-H)
IMPLICIT REAL*8(O-Z)
X=0.
Y=0.
XR1=1000000.
YR1=20000.*3280.
XR2=500000.
YR2=20000.*3280.
OPEN(1,STATUS='UNKNOWN',FILE='DATFIL')
TS=1.
TF=100.
T=0.
S=0.
H=.01
WHILE(T<=TF)
        XR1OLD=XR1
        XR2OLD=XR2
        XR1D=-14600.
        XR2D=-14600.
        XR1=XR1+H*XR1D
        XR2=XR2+H*XR2D
        T=T+H
        XR1D=-14600.
        XR2D=-14600.
        XR1=.5*(XR1OLD+XR1+H*XR1D)
        XR2=.5*(XR2OLD+XR2+H*XR2D)
        S=S+H
        IF(S>=(TS-.00001))THEN
                S=0.
                R1=SQRT((XR1-X)**2+(YR1-Y)**2)
                Rs=SQRT((XR2-X)**2+(YR2-Y)**2)
                A=(YR1-YR2)/(XR2-XR1)
                B1=(R1**2-R2**2-XR1**2-YR1**2+XR2**2+YR2**2)/
1                   (2.*(XR2-XR1))
                A=1.+A1**2
                B=2.*A1*B1-2.*A1*XR1-2.*YR1
                C=XR1**2-2.*XR1*B1+YR1**2-R1**2
                YH=(-B-SQRT(B**2-4.*A*C))/(2.*A)
                XH=A1*YH+B1
                THET=ACOS(((XR1-X)*(XR2-X)+(YR1-Y)*(YR2-Y))/
1                   (R1*R2))
                WRITE(9,*)T,X,XH,Y,YH,57.3*THET
                WRITE(1,*)T,X,XH,Y,YH,57.3*THET
        ENDIF
END DO
PAUSE
CLOSE(1)
END
```

Listing 11.2   Simulation to see if receiver location can be determined from two noisy range measurements

```
C THE FIRST THREE STATEMENTS INVOKE THE ABSOFT RANDOM
  NUMBER GENERATOR ON THE MACINTOSH
  GLOBAL DEFINE
          INCLUDE 'quickdraw.inc'
  END
  IMPLICIT REAL*8(A-H)
  IMPLICIT REAL*8(O-Z)
  SIGNOISE=300.
  X=0.
  Y=0.
  XR1=1000000.
  YR1=20000.*3280.
  YR2=500000.
  YR2=20000.*3280.
  OPEN(1,STATUS='UNKNOWN',FILE='DATFIL')
  TS=1.
  TF=100.
  T=0.
  S=0.
  H=.01
  WHILE(T < =TF)
          XR1OLD=XR1
          XR2OLD=XR2
          XR1D=-14600.
          XR2D=-14600.
          XR1=XR1+H*XR1D
          XR2=XR2+H*XR2D
          T=T+H
          XR1D=-14600.
          XR2D=-14600.
          XR1=.5*(XR1OLD+XR1+H*XR1D)
          XR2=.5*(XR2OLD+XR2+H*XR2D)
          S=S+H
          IF(S>=(TS-.00001))THEN
                  S=0.
                  CALL GAUSS(R1NOISE,SIGNOISE)
                  CALL GAUSS(R2NOISE,SIGNOISE)
                  R1=SQRT((XR1-X)**2+(YR1-Y)**2)
                  R2=SQRT((XR2-X)**2+(YR2-Y)**2)
                  R1S=R1+R1NOISE
                  R2S=R2+R2NOISE
                  A1=(YR1-YR2)/(XR2-XR1)
                  B1=(R1S**2-R2S**2-XR1**2-YR1**2+XR2**2+YR2**2)/
1                    (2.*(XR2-XR1))
                  A=1.+A1**2
                  B=2.*A1*B1-2.*A1*XR1-2.*YR1
                  C=XR1**2-2.*XR1*B1+YR1**2-R1S**2
                  YH=(-B-SQRT(B**2-4.*A*C))/(2.*A)
                  XH=A1*YH+B1
```

(*continued*)

Listing 11.2   (*Continued*)

1	THET=ACOS(((XR1-X)*(XR2-X)+(YR1-Y)*(YR2-Y))/ (R1*R2))

```
            THET=ACOS(((XR1-X)*(XR2-X)+(YR1-Y)*(YR2-Y))/
  1             (R1*R2))
            WRITE(9,*)T,X,XH,Y,YH,57.3*THET,R1,R2
            WRITE(1,*)T,X,XH,Y,YH,57.3*THET,R1,R2
        ENDIF
    END DO
    PAUSE
    CLOSE(1)
    END
C SUBROUTINE GAUSS IS SHOWN IN LISTING 1.8
```

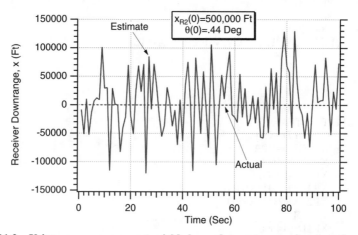

Fig. 11.2   Using raw measurements yields large downrange receiver location errors.

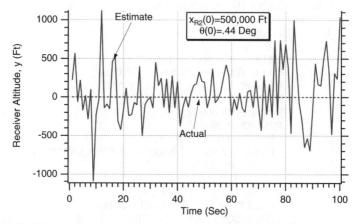

Fig. 11.3   Using raw range measurements yields large altitude receiver location errors.

example the error in estimating the downrange location of the receiver was often in excess of 50,000 ft, whereas the error in estimating the altitude of the receiver was often in excess of 300 ft. The reason for the large downrange location errors is caused by poor geometry. In this example the initial angle between the two range vectors is only 0.44 deg [i.e., $\theta(0) = 0.44$ deg]. In other words, the angle between the range vectors is so small that it appears to the receiver that both satellites are nearly on top of each other (i.e., nearly only one satellite), making a solution to this problem difficult. We can see from Fig. 11.3 that the receiver altitude location errors are much smaller than the downrange errors. Although the altitude errors can be in excess of 300 ft, this is not unreasonable because the standard deviation of the range errors is also 300 ft.

To get a more favorable geometry, the initial downrange location of the second satellite was increased by an order of magnitude from 500,000 to 5,000,000 ft [i.e., $x_{R2}(0) = 5,000,000$]. In this case the initial angle between the two range vectors has increased from 0.44 deg to 3.5 deg. Figure 11.4 shows that the error in estimating the downrange location of the receiver has also been reduced by an order of magnitude due to the increase in the initial geometrical angle between the two satellites. Rather than the downrange errors being in excess of 50,000 ft, they are now only occasionally in excess of 5,000 ft. Figure 11.5 shows that this more favorable geometry only slightly influences the altitude errors. The error in estimating the altitude of the receiver is only slightly less than 300 ft.

To get an even more favorable geometry, the initial downrange location of the second satellite was increased by two orders of magnitude from 500,000 to 50,000,000 ft (i.e., $x_{R2} = 50,000,000$). In this case the initial angle between the two range vectors has increased from 0.44 to 36.9 deg. Figure 11.6 shows that the error in estimating the downrange location of the receiver has also been reduced by two orders of magnitude when the initial geometrical angle between the two satellites is increased by two orders of magnitude. Rather than the downrange errors being in excess of 50,000 ft, they are now only occasionally in excess of

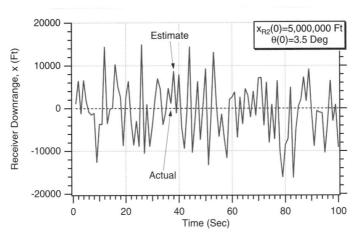

Fig. 11.4   Receiver downrange errors decrease by an order of magnitude when geometrical angle between satellites is increased.

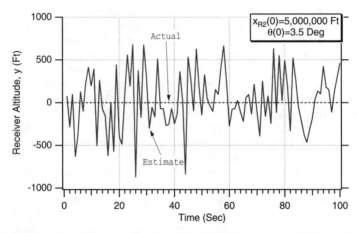

Fig. 11.5 Receiver altitude errors decrease slightly when geometrical angle between satellites is increased.

500 ft. Figure 11.7 shows again that this more favorable geometry only slightly influences the altitude errors. The error in estimating the altitude of the receiver is only slightly less than 300 ft.

Therefore, we have seen that without filtering we could only get the receiver location errors down to the level of the range errors if the geometry is favorable. A favorable geometry is one in which the angle between the two range vectors is large. Theoretically, the most favorable geometry will occur when that angle between the two range vectors approaches 90 deg. The least favorable geometry will be one in which the range angle is zero because it will appear to the receiver that there is only one satellite.

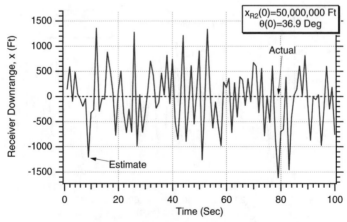

Fig. 11.6 Downrange errors decrease by two orders of magnitude when geometrical angle is increased by two orders of magnitude.

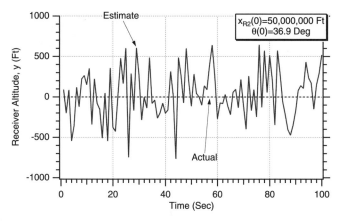

Fig. 11.7 Altitude errors decrease slightly when geometrical angle is increased by two orders of magnitude.

Linear Filtering of Range

It seems reasonable to ask if filtering could have further reduced the receiver location errors. The simplest possible approach would be to filter the noisy range measurements before they are used to calculate the receiver location. Before we can determine the type and order of the filter to use, we must see how the range from each of the satellites to the receiver varies with time. For the geometry in which the initial angle between the two range vectors is 36.9 deg [i.e., $x_{R2}(0) = 50,000,000$], the ranges from each of the satellites to the receiver is displayed in Figs. 11.8 and 11.9. For 100 s of satellite flight, we can see that the range from the first satellite to the receiver is a parabolic function of time, while the range from the second satellite to the receiver is a linear function of time. Therefore, to play it safe, a second-order filter is required to prevent truncation

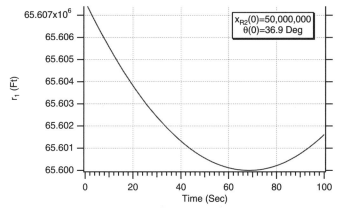

Fig. 11.8 Range from receiver to first satellite is parabolic.

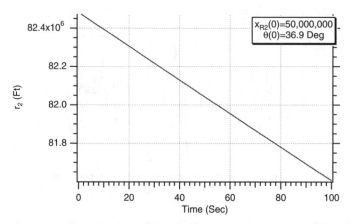

Fig. 11.9   Range from receiver to second satellite is a straight line.

error or filter divergence. For simplicity we will consider using a linear second-order recursive least-squares filter. If the flight times were longer, we might have to use a second-order linear polynomial Kalman filter with process noise.

We will now proceed to develop the second-order recursive least-squares filter equations for this application. Recall in Chapter 3 that we showed that the gains of a second-order recursive least-squares filter were given by

$$K_{1_k} = \frac{3(3k^2 - 3k + 2)}{k(k+1)(k+2)} \qquad k = 1, 2, \ldots, n$$

$$K_{2_k} = \frac{18(2k-1)}{k(k+1)(k+2)T_s}$$

$$K_{3_k} = \frac{60}{k(k+1)(k+2)T_s^2}$$

Therefore, these gains are only a function of the number of measurements taken k and the time between measurements or sampling interval T_s. We can see from the preceding three equations that all three gains eventually go to zero as more measurements are taken (i.e., k gets larger). If we think of the measurement as range in this application, then the estimates from this second-order filter will be range, range rate, and the second derivative of range. For this second-order recursive least-squares filter a residual must be formed that is the difference between the present range measurement and the projected range estimate based on past estimates of range, range rate, and the second derivative of range or

$$\text{Res}_k = r_k^* - \hat{r}_{k-1} - \hat{\dot{r}}_{k-1}T_S - 0.5\hat{\ddot{r}}_{k-1}T_s^2$$

Using the second-order recursive least-squares filter equations of Chapter 3, the new estimates of range, range rate, and the second derivative of range are a combination of the preceding estimates plus a gain multiplied by the residual or

$$\hat{r}_k = \hat{r}_{k-1} + \hat{\dot{r}}_{k-1}T_s + 0.5\hat{\ddot{r}}_{k-1}T_s^2 + K_{1_k}\text{Res}_k$$

$$\hat{\dot{r}}_k = \hat{\dot{r}}_{k-1} + \hat{\ddot{r}}_{k-1}T_s + K_{2_k}\text{Res}_k$$

$$\hat{\ddot{r}}_k = \hat{\ddot{r}}_{k-1} + K_{3_k}\text{Res}_k$$

The preceding simulation of Listing 11.2 was modified to include two decoupled second-order recursive least-squares filters acting on each range measurement. The necessary filtering equations, which were just reviewed in this section, are highlighted in bold in Listing 11.3. In other words, Listing 11.3 is virtually identical to Listing 11.2 except for the addition of the two decoupled second-order recursive least-squares filters. We can see from Listing 11.3 that all of the initial filter estimates are set to zero. This type of filter does not require any type of a priori information. In fact, the filter's initial estimates could be set to any value, and the same results would be achieved because the performance of a recursive least-squares filter is independent of initialization. Recall that the second-order recursive least-squares filter is equivalent to a linear second-order polynomial Kalman filter with infinite initial covariance matrix and zero process noise.

The nominal case of Listing 11.3, in which the initial angle between to the two range vectors was 36.9 deg, was run. The downrange and altitude estimates of the location of the receiver are displayed in Figs. 11.10 and 11.11. By comparing Figs. 11.10 and 11.11 with Figs. 11.6 and 11.7, we can see that filtering the raw range measurements has definitely reduced the errors in locating the receiver. The downrange errors have been reduced from approximately 500 to 300 ft, whereas the altitude errors have been reduced from approximately 300 to 100 ft. In addition, we can see from Figs 11.10 and 11.11 that the estimated receiver location with filtering is smoother (i.e., it seems to bounce around a great deal less) than it is when filtering is not used.

Now that we can see that the filtering of range measurements reduces errors in locating the receiver, it is worthwhile to ask if we could have done better by an even more systematic filtering approach.

Using Extended Kalman Filtering

To see if we can get even better estimates of the receiver location errors, let us try to apply extended Kalman filtering to this problem. Because we are trying to estimate the location of the receiver, it seems reasonable to make the receiver location states. For now, the receiver is assumed to be stationary or fixed in location, which means that the receiver velocity (i.e., derivative of position) is zero. Therefore, we can say that

$$\dot{x} = 0$$

$$\dot{y} = 0$$

Listing 11.3   Simulation to see if receiver location can be determined from two noisy range measurements that are filtered

```
C THE FIRST THREE STATEMENTS INVOKE THE ABSOFT RANDOM
  NUMBER GENERATOR ON THE MACINTOSH
  GLOBAL DEFINE
            INCLUDE 'quickdraw.inc'
  END
  IMPLICIT REAL*8(A-H)
  IMPLICIT REAL*8(O-Z)
  REAL*8K1,K2,K3
  SIGNOISE=300.
  X=0.
  Y=0.
  XR1=1000000.
  YR1=20000.*3280.
  YR2=50000000.
  YR2=20000.*3280.
  R1H=0.
  R1DH=0.
  R1DDH=0.
  R2H=0.
  R2DH=0.
  R2DDH=0.
  OPEN(1,STATUS='UNKNOWN',FILE='DATFIL')
  TS=1.
  TF=100.
  T=0.
  S=0.
  H=.01
  XN=0.
  WHILE(T < =TF)
        XR1OLD=XR1
        XR2OLD=XR2
        XR1D=-14600.
        XR2D=-14600.
        YR1=XR1+H*XR1D
        XR2=XR2+H*XR2D
        T=T+H
        XR1D=-14600.
        XR2D=-14600.
        XR1=.5*(XR1OLD+XR1+H*XR1D)
        XR2=.5*(XR2OLD+XR2+H*XR2D)
        S=S+H
        IF(S>=(TS-.00001))THEN
                S=0.
                XN=XN+1.
          .     K1=3*(3*XN*XN-3*XN+2)/(XN*(XN+1)*(XN+2))
                K2=18*(2*XN-1)/(XN*(XN+1)*(XN+2)*TS)
                K3=60/(XN*(XN+1)*(XN+2)*TS*TS)
                CALL GAUSS(R1NOISE,SIGNOISE)
                CALL GAUSS(R2NOISE,SIGNOISE)
                R1=SQRT(XR1-X)**2+(YR1-Y)**2)
```

(continued)

Listing 11.3 (*Continued*)

```
                    R2=SQRT((XR2-X)**2+(YR2-Y)**2)
                    R1S=R1+R1NOISE
                    R2S=R2+R2NOISE
                    RES1=R1S-R1H-TS*R1DH-.5*TS*TS*R1DDH
                    R1H=R1H+R1DH*TS+.5*R1DDH*TS*TS*K1*RES1
                    R1DH=R1DH+R1DDH*TS+K2*RES1
                    R1DDH=R1DDH+K3*RES1
                    RES2=R2S-R2H-TS*R2DH-.5*TS*TS*R2DDH
                    R2H=R2H+R2DH*TS+.5*R2DDH*TS*TS+K1*RES2
                    R2DH=R2DH+R2DDH*TS+K2*RES2
                    R2DDH=R2DDH+K3*RES2
                    A1=(YR1-YR2)/(XR2-XR1)
                    B1=(R1H**2-R2H**2-XR1**2-YR1**2+XR2**2+YR2**2)/
1                     (2.*(XR2-XR1))
                    A=1.+A1**2
                    B=2.*A1*B1-2.*A1*XR1-2.*YR1
                    C=XR1**2-2.*XR1*B1+YR1**2-R1H**2
                    YH=(-B-SQRT(B**2-4.*A*C))/(2.*A)
                    XH=A1*YH+B1
                    THET=ACOS(((XR1-X)*(XR2-X)+(YR1-Y)*(YR2-Y))/
1                     (R1*R2))
                    WRITE(9,*)T,X,XH,Y,YH,57.3*THET,R1,R2
                    WRITE(1,*)T,X,XH,Y,YH,57.3*THET,R1,R2
          ENDIF
      END DO
      PAUSE
      CLOSE(1)
      END
C SUBROUTINE GAUSS IS SHOWN IN LISTING 1.8
```

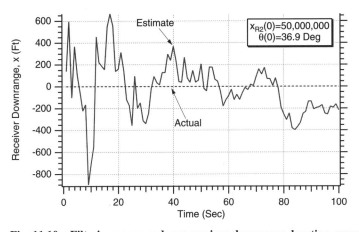

Fig. 11.10   Filtering range reduces receiver downrange location errors.

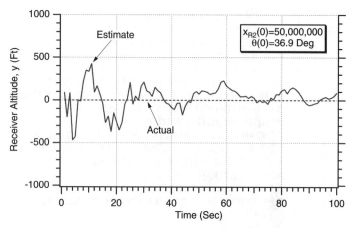

Fig. 11.11   Filtering range reduces receiver altitude location errors.

If we neglect process noise, we can rewrite the preceding equation as two scalar equations in state-space form, yielding

$$\begin{bmatrix} \dot{x} \\ \dot{y} \end{bmatrix} = \begin{bmatrix} 0 & 0 \\ 0 & 0 \end{bmatrix} \begin{bmatrix} x \\ y \end{bmatrix}$$

This means that the systematic dynamics matrix is zero or

$$F = \begin{bmatrix} 0 & 0 \\ 0 & 0 \end{bmatrix}$$

Therefore, only a one-term Taylor series is required to find the fundamental matrix, which means that the fundamental matrix is the identity matrix or

$$\Phi_k = \begin{bmatrix} 1 & 0 \\ 0 & 1 \end{bmatrix}$$

The measurements are of the ranges from each of the satellites to the receiver. We have already shown from Fig. 11.1 that the true or exact ranges from each of the satellites to the receiver are given by

$$r_1 = \sqrt{(x_{R1} - x)^2 + (y_{R1} - y)^2}$$

$$r_2 = \sqrt{(x_{R2} - x)^2 + (y_{R2} - y)^2}$$

Because the actual range equations are nonlinear, we can express the range measurements as linearized functions of the states according to

$$
\begin{bmatrix} \Delta r_1^* \\ \Delta r_2^* \end{bmatrix} = \begin{bmatrix} \dfrac{\partial r_1}{\partial x} & \dfrac{\partial r_1}{\partial y} \\ \dfrac{\partial r_2}{\partial x} & \dfrac{\partial r_2}{\partial y} \end{bmatrix} \begin{bmatrix} \Delta x \\ \Delta y \end{bmatrix} + \begin{bmatrix} v_{r1} \\ v_{r2} \end{bmatrix}
$$

where v_{r1} and v_{r2} are the measurement noises associates with the range measurements. If each of the noise sources is uncorrelated, the discrete measurement noise matrix can be written by inspection of the preceding measurement equation as

$$
\boldsymbol{R}_k = \begin{bmatrix} \sigma_{r1}^2 & 0 \\ 0 & \sigma_{r2}^2 \end{bmatrix}
$$

where σ_{r1}^2 and σ_{r2}^2 are the measurement noise variances. From the measurement equation we can see that the linearized measurement matrix is the matrix of partial differentials given by

$$
\boldsymbol{H}_k = \begin{bmatrix} \dfrac{\partial r_1}{\partial x} & \dfrac{\partial r_1}{\partial y} \\ \dfrac{\partial r_2}{\partial x} & \dfrac{\partial r_2}{\partial y} \end{bmatrix}
$$

where the partial derivatives are evaluated at the projected state estimates. However, because the fundamental matrix is the identity matrix, the projected state estimates and the current state estimates are identical. From the range equation we can evaluate each of the partial derivatives as

$$
\frac{\partial r_1}{\partial x} = 0.5[(x_{R1} - x)^2 + (y_{R1} - y)^2]^{-0.5} 2(x_{R1} - x)(-1) = \frac{-(x_{R1} - x)}{r_1}
$$

$$
\frac{\partial r_1}{\partial y} = 0.5[(x_{R1} - x)^2 + (y_{R1} - y)^2]^{-0.5} 2(y_{R1} - y)(-1) = \frac{-(y_{R1} - y)}{r_1}
$$

$$
\frac{\partial r_2}{\partial x} = 0.5[(x_{R1} - x)^2 + (y_{R1} - y)^2]^{-0.5} 2(x_{R2} - x)(-1) = \frac{-(x_{R2} - x)}{r_2}
$$

$$
\frac{\partial r_2}{\partial y} = 0.5[(x_{R1} - x)^2 + (y_{R1} - y)^2]^{-0.5} 2(y_{R2} - y)(-1) = \frac{-(y_{R2} - y)}{r_2}
$$

and so the linearized measurement matrix is given by

$$
H_k = \begin{bmatrix} \dfrac{-(x_{R1} - x)}{r_1} & \dfrac{-(y_{R1} - y)}{r_1} \\ \dfrac{-(x_{R2} - x)}{r_2} & \dfrac{-(y_{R2} - y)}{r_2} \end{bmatrix}
$$

As was already mentioned, the linearized measurement matrix is evaluated at the current state estimates in this application.

If we assume zero process noise, we now have enough information to solve the Riccati equations. As in other examples in extended Kalman filtering, we saw that we do not have to use the linearized equations for the actual filter. Because the fundamental matrix is exact and equal to the identity matrix, we can say that the projected estimate of the receiver location is simply the last estimate or

$$
\bar{x}_k = \hat{x}_{k-1}
$$

$$
\bar{y}_k = \hat{y}_{k-1}
$$

Therefore, each of the projected ranges from the satellite to the receiver are given by

$$
\bar{r}_{1_k} = \sqrt{(x_{R1_k} - \bar{x}_k)^2 + (y_{R1_k} - \bar{y}_k)^2}
$$

$$
\bar{r}_{2_k} = \sqrt{(x_{R2_k} - \bar{x}_k)^2 + (y_{R2_k} - \bar{y}_k)^2}
$$

The residual does not have to be calculated from the linearized measurement matrix but instead can be calculated as simply the measured range minus the projected range or

$$
\text{Res}_{1_k} = r_{1_k}^* - \bar{r}_{1_k}
$$

$$
\text{Res}_{2_k} = r_{2_k}^* - \bar{r}_{2_k}
$$

The extended Kalman-filtering equations are given by

$$
\hat{x}_k = \bar{x}_k + K_{11_k}\text{Res}_{1_k} + K_{12_k}\text{Res}_{2_k}
$$

$$
\hat{y}_k = \bar{y}_k + K_{21_k}\text{Res}_{1_k} + K_{22_k}\text{Res}_{2_k}
$$

The extended Kalman filter for the satellite tracking problem appears in Listing 11.4. Statements that are specifically required for the extended Kalman filter are highlighted in bold. The initial covariance matrix is chosen so that the off-diagonal elements are zero and the diagonal elements represent the square of our best guess in the errors in locating the receiver. The initial filter estimates (i.e., XH = 1000, YH = 2000) show that the filter is not perfectly initialized because the actual receiver is at the origin (i.e., $x = 0$, $y = 0$). The measurement noise matrix is no longer a scalar as it has been for many of the problems in this text but is a two-dimensional square matrix. Therefore, the inverse of a 2×2 matrix will be required in the Riccati equations. A formula for the exact inverse of a 2×2 matrix was presented in Chapter 1 and appears in this listing.

The nominal case of Listing 11.4 was run in which the initial angle between the two range vectors was 36.9 deg. We can see, by comparing Figs. 11.12 and 11.13 with Figs. 11.10 and 11.11, that extended Kalman filtering dramatically reduces the errors in locating the receiver. When we only filtered the range measurements, the errors in estimating the receiver location were approximately 300 ft in downrange and 100 ft in altitude. Now the errors in estimating the receiver location appear to be significantly smaller in both downrange and altitude. In other words, building an extended Kalman filter that takes into account the dynamics of both the stationary receiver and the moving satellites is much more effective than simply filtering the range measurements directly.

To ensure that the extended Kalman filter was working properly even though there was no process noise, the information of the preceding two figures was displayed in another way. Figures 11.14 and 11.15 display the single-run errors in the estimates of downrange and altitude. Superimposed on each of the figures are the theoretical errors in the estimates, which are obtained by taking the square root of the diagonal elements of the covariance matrix (shown in Listing 11.4). We can see that, because in both figures the single-run estimates appear to lie within the theoretical bounds, the extended Kalman filter appears to be working properly. We can see that after 100 s of filtering the receiver downrange location error has been reduced to approximately 60 ft, while the receiver altitude location error has been reduced to approximately 30 ft. This accuracy was achieved even though we did not know where the receiver was initially (i.e., 1000-ft error in downrange and 2000-ft error in altitude), and the range measurement error standard deviation was 300 ft.

Earlier we saw that the angle between the two range vectors was important in determining the accuracy to which the receiver could be located when no filtering was applied. Experiments with extended Kalman filtering were conduced in which the initial angle between the two range vectors was decreased from its nominal value of 36.9 deg. The square roots of the diagonal elements of the covariance matrix were computed in each case to represent the theoretical errors in the estimates of downrange and altitude. Only single runs were made in each case, and the covariance matrix prediction of the error in the estimate of receiver downrange and altitude was calculated. Figure 11.16 shows that as the initial angle between the range vectors decreases from 36.9 to 0.44 deg the error in the estimate of downrange increases. For example, after 100 s of track we have 60 ft of downrange error when the initial angle is 36.9 deg, 550 ft of downrange error when the initial angle is 3.5 deg, and 950 ft of downrange error when the initial angle is 0.44 deg. On the other hand, Figure 11.17 indicates that as the initial angle between the range vectors decreases from 36.9 to 0.44 deg the error in the estimate of altitude remains approximately the same. In fact, it even decreases

Listing 11.4   Extended Kalman filter for locating receiver based on measurements from two satellites

```
C THE FIRST THREE STATEMENTS INVOKE THE ABSOFT RANDOM
  NUMBER GENERATOR ON THE MACINTOSH
      GLOBAL DEFINE
              INCLUDE 'quickdraw.inc'
      END
      IMPLICIT REAL*8(A-H)
      IMPLICIT REAL*8(O-Z)
      REAL*8 PHI(2,2),P(2,2),M(2,2),PHIP(2,2),PHIPPHIT(2,2),GAIN(2,2)
      REAL*8 Q(2,2),HMAT(2,2),HM(2,2),MHT(2,2)
      REAL*8 PHIT(2,2),RMAT(2,2),HMHTRINV(2,2)
      REAL*8 HMHT(2,2),HT(2,2),KH(2,2),IDN(2,2),IKH(2,2),HMHTR(2,2)
      INTEGER ORDER
      SIGNOISE=300.
      X=0.
      Y=0.
      XH=1000.
      YR=2000.
      XR1=1000000.
      YR1=20000.*3280.
      XR1D=-14600.
      XR2=50000000.
      YR2=20000.*3280.
      XR2D=-14600.
      OPEN(2,STATUS='UNKNOWN',FILE='COVFIL')
      OPEN(1,STATUS='UNKNOWN',FILE='DATFIL')
      ORDER=2
      TS=1.
      TF=100.
      PHIS=0.
      T=0.
      S=0.
      H=.01
      DO 1000 I=1,ORDER
      DO 1000 J=1,ORDER
              PHI(I,J)=0.
              P(I,J)=0.
              Q(I,J)=0.
              IDN(I,J)=0.
1000  CONTINUE
      IDN(1,1)=1.
      IDN(2,2)=1.
      IDN(2,2)=1.
      P(1,1)=1000.**2
      P(2,2)=2000.**2
      RMAT(1,1)=SIGNOISE**2
      RMAT(1,2)=0.
      RMAT(2,1)=0.
      RMAT(2,2)=SIGNOISE**2
```

(*continued*)

Listing 11.4

```
WHILE(T<=TF)
       XR1OLD=XR1
       XR2OLD=XR2
       XR1D=-14600.
       XR2D=-14600.
       XR1=XR1+H*XR1D
       XR2=XR2+H*XR2D
       T=T+H
       XR1D=-14600.
       XR2D=-14600.
       XR1=.5*(XR1OLD+XR1+H*XR1D)
       XR2=.5*(XR2OLD+XR2+H*XR2D)
       S=S+H
       IF(S>=(TS-.00001))THEN
              S=0.
              R1H=SQRT((XR1-XH)**2+(YR1-YH)**2)
              R2H=SQRT((XR2-XH)**2+(YR2-YH)**2)
              HMAT(1,1)=-(XR1-XH)/R1H
              HMAT(1,2)=-(YR1-YH)/R1H
              HMAT(2,1)=-(XR2-XH)/R2H
              HMAT(2,2)=-(YR2-YH)/R2H
              CALL  MATTRN(HMAT,2,2,HT)
              PHI(1,1)=1.
              PHI(2,2)=1.
              CALL  MATTRN(PHI,ORDER,ORDER,PHIT)
              CALL  MATMUL(PHI,ORDER,ORDER,P,ORDER,ORDER,
                  PHIP)
              CALL  MATMUL(PHIP,ORDER,ORDER,PHIT,ORDER,
                  ORDER,PHIPPHIT)
              CALL  MATADD(PHIPPHIT,ORDER,ORDER,Q,M)
              CALL  MATMUL(HMAT,2,ORDER,M,ORDER,
                  ORDER,HM)
              CALL  MATMUL(HM,2,ORDER,HT,ORDER,2,HMHT)
              CALL  MATADD(HMHT,2,2,RMAT,HMHTR)
              DET=HMHTR(1,1)*HMHTR(2,2)-HMHTR(1,2)*HMHTR
                  (2,1)
              HMHTRINV(1,1)=HMHTR(2,2)/DET
              HMHTRINV(1,2)=-HMHTR(1,2)/DET
              HMHTRINV(2,1)=-HMHTR(2,1)/DET
              HMHTRINV(2,2)=HMHTR(1,1)/DET
              CALL  MATMUL(M,ORDER,ORDER,HT,ORDER,2,MHT)
              CALL  MATMUL(MHT,ORDER,2,HMHTRINV,2,2,GAIN)
              CALL  MATMUL(GAIN,ORDER,2,HMAT,2,ORDER,KH)
              CALL  MATSUB(IDN,ORDER,ORDER,KH,IKH)
              CALL  MATMUL(IKH,ORDER,ORDER,M,ORDER,
                  ORDER,P)
              CALL  GAUSS(R1NOISE,SIGNOISE)
              CALL  GAUSS(R2NOISE,SIGNOISE)
              R1=SQRT((XR1-X)**2+(YR1-Y)**2)
```

(continued)

Listing 11.4 (*Continued*)

```
                R2=SQRT((XR2-X)**2+(YR2-Y)**2)
                RES1=R1+R1NOISE-R1H
                RES2=R2+R2NOISE+R2H
                XH=XH+GAIN(1,1)*RES1+GAIN(1,2)*RES2
                YH=YH+GAIN(2,1)*RES1+GAIN(2,2)*RES2
                ERRX=X-XH
                SP11=SQRT(P(1,1))
                ERRY=Y-YH
                SP22=SQRT(P(2,2))
                WRITE(9,*)T,X,XH,Y,YH
                WRITE(1,*)T,X,XH,Y,YH
                WRITE(2,*)T,ERRX,SP11,-SP11,ERRY,SP22,-SP22
        ENDIF
    END DO
    PAUSE
    CLOSE(1)
    CLOSE(2)
    END

C SUBROUTINE GAUSS IS SHOWN IN LISTING 1.8
C SUBROUTINE MATTRN IS SHOWN IN LISTING 1.3
C SUBROUTINE MATMUL IS SHOWN IN LISTING 1.4
C SUBROUTINE MATADD IS SHOWN IN LISTING 1.1
C SUBROUTINE MATSUB IS SHOWN IN LISTING 1.2
```

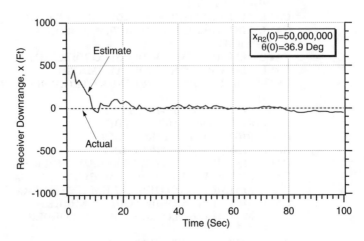

Fig. 11.12 Extended Kalman filtering dramatically reduces receiver downrange location errors.

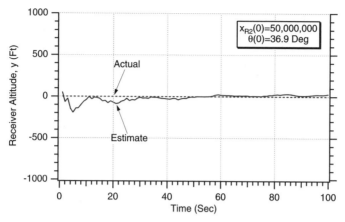

Fig. 11.13   Extended Kalman filtering dramatically reduces receiver altitude location errors.

slightly. For example, after 100 s of track we have 30 ft of altitude error when the angle is 36.9 deg, 28 ft of altitude error when the angle is 3.5 deg, and 22 ft of altitude error when the angle is 0.44 deg.

Using Extended Kalman Filtering with One Satellite

At the beginning of this chapter, we saw that from an algebraic point of view two satellites were necessary in determining the location of a stationary receiver when only range measurements were available. We saw that when the range measurements were perfect we obtained two equations (i.e., for range) with two unknowns (i.e., downrange and altitude of receiver). The quadratic nature of the equations yielded two possible solutions, but we were able to eliminate one by

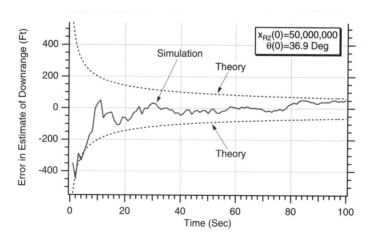

Fig. 11.14   Extended Kalman filter appears to be working correctly in downrange.

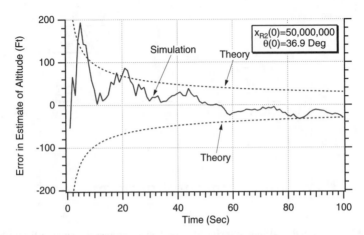

Fig. 11.15   Extended Kalman filter appears to be working correctly in altitude.

using common sense. Because one satellite with one range measurement would yield one equation with two unknowns, it would appear that it would be impossible from an algebraic point of view to estimate the stationary receiver location. Let use see if this is true if we attempt to apply extended Kalman filtering to the problem.

Recall that in this problem the receiver is fixed in location, which means that the receiver velocity (i.e., derivative of position) is zero. Therefore, we can say that

$$\dot{x} = 0$$

$$\dot{y} = 0$$

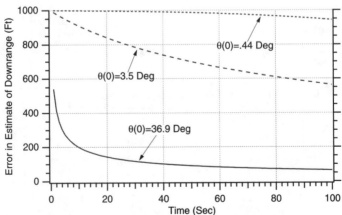

Fig. 11.16   Extended Kalman filter's downrange estimation errors decrease with increasing angle between range vectors.

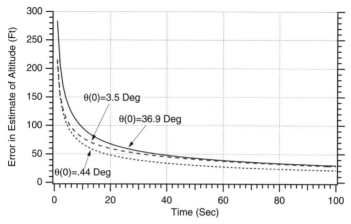

Fig. 11.17 Extended Kalman filter's altitude estimation errors remain approximately constant with increasing angle between range vectors.

If we neglect process noise, we can rewrite the preceding two scalar equations in state-space form, yielding

$$\begin{bmatrix} \dot{x} \\ \dot{y} \end{bmatrix} = \begin{bmatrix} 0 & 0 \\ 0 & 0 \end{bmatrix} \begin{bmatrix} x \\ y \end{bmatrix}$$

From the preceding linear state-space equation we can see that the systems dynamics matrix is zero or

$$F = \begin{bmatrix} 0 & 0 \\ 0 & 0 \end{bmatrix}$$

We showed in the preceding section that for a systems dynamics matrix, which is zero, the fundamental matrix was the identity matrix or

$$\Phi_k = \begin{bmatrix} 1 & 0 \\ 0 & 1 \end{bmatrix}$$

The measurement is now the range from the only satellite to the receiver. We have already shown that the true or exact range from the satellite to the receiver is given by

$$r_1 = \sqrt{(x_{R1} - x)^2 + (y_{R1} - y)^2}$$

Therefore, we can linearize the range measurement as function of the states according to

$$\Delta r_1^* = \begin{bmatrix} \dfrac{\partial r_1}{\partial x} & \dfrac{\partial r_1}{\partial y} \end{bmatrix} \begin{bmatrix} \Delta x \\ \Delta y \end{bmatrix} + v_{r1}$$

where v_{r1} is the measurement noise associated with the range measurement r_1. Because there is only one measurement source, the discrete measurement noise matrix is therefore a scalar given by

$$R_k = \sigma_{r1}^2$$

where σ_{r1}^2 is the measurement noise variance. From the measurement equation we can see that the linearized measurement matrix is given by

$$H_k = \begin{bmatrix} \dfrac{\partial r_1}{\partial x} & \dfrac{\partial r_1}{\partial y} \end{bmatrix}$$

where the partial derivatives are evaluated at the projected state estimates. However, because the fundamental matrix is the identity matrix, the last state estimates and the projected state estimates are identical. From the range equation we can evaluate each of the partial derivatives as

$$\frac{\partial r_1}{\partial x} = 0.5[(x_{R1} - x)^2 + (y_{R1} - y)^2]^{-0.5}2(x_{R1} - x)(-1) = \frac{-(x_{R1} - x)}{r_1}$$

$$\frac{\partial r_1}{\partial y} = 0.5[(x_{R1} - x)^2 + (y_{R1} - y)^2]^{-0.5}2(y_{R1} - y)(-1) = \frac{-(y_{R1} - y)}{r_1}$$

Therefore, the linearized measurement matrix is given by

$$H_k = \begin{bmatrix} \dfrac{-(x_{R1} - x)}{r_1} & \dfrac{-(y_{R1} - y)}{r_1} \end{bmatrix}$$

If we assume zero process noise, we now have enough information to solve the Riccati equations. As in other examples in extended Kalman filtering and in the preceding section, we saw that we did not have to use the linearized equations for the actual filter. Because the fundamental matrix is exact, we can say that the projected estimate of the receiver location is simply the last estimate or

$$\bar{x}_k = \hat{x}_{k-1}$$

$$\bar{y}_k = \hat{y}_{k-1}$$

Therefore, the projected range from the satellite to the receiver is given by

$$\bar{r}_{1_k} = \sqrt{(x_{R1_k} - \bar{x}_k)^2 + (y_{R1_k} - \bar{y}_k)^2}$$

As was the case in the preceding section, we do not have to use the linearized measurement matrix to compute the residual. Instead, the residual is simply the measurement minus the projected value or

$$\text{Res}_{1_k} = r^*_{1_k} - \bar{r}_{1_k}$$

and the extended Kalman-filtering equations are given by

$$\hat{x}_k = \bar{x}_k + K_{11_k}\text{Res}_{1_k}$$

$$\hat{y}_k = \bar{y}_k + K_{21_k}\text{Res}_{1_k}$$

The new extended Kalman filter for the one satellite tracking problem appears in Listing 11.5. Statements that are specifically required for the new extended Kalman filter are highlighted in bold. The initial covariance matrix is still chosen so that the off-diagonal elements are zero. Recall that the diagonal elements represent the square of our best guess in the errors in locating the receiver. The initial filter estimates (i.e., XH = 1000, and YH = 2000) still show that the filter is not perfectly initialized because the receiver is actually located at the origin (i.e., $x = 0$, and $y = 0$). Now the measurement noise matrix is a scalar, and so the inverse required by the Riccati equations can easily be obtained by taking a reciprocal, as indicated in the listing.

The nominal case of Listing 11.5 was run, and the results for the receiver location estimates appear in Figs. 11.18 and 11.19. We can see from Fig. 11.18 that we cannot be quite sure if the extended Kalman filter permits us to estimate the downrange receiver location based on range measurements from only one satellite. For approximately 90 s the downrange estimate does not change significantly from its initial value of 1000 ft, indicating that either the extended Kalman filter is not working or this particular state is not observable because we are only using range measurements from one satellite. However, after 90 s the downrange estimate of the receiver appears to improve. After 100 s of tracking, the downrange error in the receiver location is reduced to approximately 700 ft. Figure 11.19, on the other hand, indicates that we are able to estimate the receiver altitude, based only on range measurements from one satellite. After approximately 40 s the receiver altitude estimates appear to be in error by only 50 ft.

To determine if the preceding results were just luck, or if we really are able to locate the receiver based on range measurements from only one satellite, the run was repeated, but this time the measurement time was increased from 100 to 1000 s. During this extra amount of time, the satellite geometry will change considerably. We can see from Fig. 11.20 that we are now positive that we can track the receiver's downrange location. After 600 s our downrange estimates are in error by less than 100 ft. Of course, these results are not as good as using range measurements from two satellites under favorable geometrical conditions (i.e.,

Listing 11.5   Extended Kalman filter for locating receiver based on measurements from one satellite

```
C THE FIRST THREE STATEMENTS INVOKE THE ABSOFT RANDOM
  NUMBER GENERATOR ON THE MACINTOSH
      GLOBAL DEFINE
              INCLUDE 'quickdraw.inc'
      END
      IMPLICIT REAL*8(A-H)
      IMPLICIT REAL*8(O-Z)
      REAL*8 PHI(2,2),P(2,2),M(2,2),PHIP(2,2),PHIPPHIT(2,2),GAIN(2,1)
      REAL*8 Q(2,2),HMAT(1,2),HM(1,2),MHT(2,1)
      REAL*8 PHIT(2,2),RMAT(1,1),HMHTRINV(1,1)
      REAL*8 HMHT(1,1),HT(2,1),KH(2,2),IDN(2,2),IKH(2,2),HMHTR(1,1)
      INTEGER ORDER
      SIGNOISE=300.
      X=0.
      Y=0.
      XH=1000.
      YH=2000.
      XR1=1000000.
      YR1=20000.*3280.
      OPEN(2,STATUS='UNKNOWN',FILE='COVFIL')
      OPEN(1,STATUS='UNKNOWN',FILE='DATFIL')
      ORDER=2
      TS=1.
      TF=100.
      T=0.
      S=0.
      H=.01
      DO 1000 I=1,ORDER
      DO 1000 J=1,ORDER
              PHI(I,J)=0.
              P(I,J)=0.
              Q(I,J)=0.
              IDN(I,J)=0.
1000      CONTINUE
      IDN(1,1)=1.
      IDN(2,2)=1.
      P(1,1)=1000.**2
      P(2,2)=2000.**2
      RMAT(1,1)=SIGNOISE**2
      WHILE(T < =TF)
          XR1OLD=XR1
              XR1D=-14600.
              XR1=XR1+H*XR1D
              T=T+H
              XR1D=-14600.
              XR1=.5*(XR1OLD+XR1+H*XR1D)
              S=S+H
              IF(S>=(TS-.00001))THEN
                      S=0.
                      R1H=SQRT((XR1-XH)**2+(YR1-YH)**2)
```

(continued)

Listing 11.5 *(Continued)*

```
              HMAT(1,1)=-(XR1-XH)/R1H
              HMAT(1,2)=-(YR1-YH)/R1H
              CALL  MATTRN(HMAT,1,2,HT)
              PHI(1,1)=1.
              PHI(2,2)=1.
              CALL  MATTRN(PHI,ORDER,ORDER,PHIT)
              CALL  MATMUL(PHI,ORDER,ORDER,P,ORDER,ORDER,
                 PHIP)
              CALL  MATMUL(PHIP,ORDER,ORDER,PHIT,ORDER,
   1             ORDER,PHIPPHIT)
              CALL  MATADD(PHIPPHIT,ORDER,ORDER,Q,M)
              CALL  MATMUL(HMAT,1,ORDER,M,ORDER,ORDER,
                 HM)
              CALL  MATMUL(HM,1,ORDER,HT,ORDER,1,HMHT)
              CALL  MATADD(HMHT,1,1,RAMT,HMHTR)
              HMHTRINV(1,1)=1./HMHTR(1,1)
              CALL  MATMUL(M,ORDER,ORDER,HT,ORDER,1,MHT)
              CALL  MATMUL(MHT,ORDER,1,HMTRINV,1,1,GAIN)
              CALL  MATMUL(GAIN,ORDER,1,HMAT,1,ORDER,KH)
              CALL  MATSUB(IDN,ORDER,ORDER,KH,IKH)
              CALL  MATMUL(IKH,ORDER,ORDER,M,ORDER,
                 ORDER,P)
              CALL  GAUSS(R1NOISE,SIGNOISE)
              R1=SQRT((XR1-X)**2+(YR1-Y)**`)
              RES1=R1+R1NOISE-R1H
              XH=XH+GAIN(1,1)*RES1
              YH=YH+GAIN(2,1)*RES1
              ERRX=X-XH
              SP11=SQRT(P(1,1))
              ERRY=Y-YH
              SP22=SQRT(P(2,2))
              WRITE(9,*)T,X,XH,Y,YH
              WRITE(1,*)T,X,XH,Y,YH
              WRITE(2,*)T,ERRX,SP11,-SP11,ERRY,SP22,-SP22
           ENDIF
        END DO
        PAUSE
        CLOSE(1)
        CLOSE(2)
        END

C SUBROUTINE GAUSS IS SHOWN IN LISTING 1.8
C SUBROUTINE MATTRN IS SHOWN IN LISTING 1.3
C SUBROUTINE MATMUL IS SHOWN IN LISTING 1.4
C SUBROUTINE MATADD IS SHOWN IN LISTING 1.1
C SUBROUTINE MATSUB IS SHOWN IN LISTING 1.2
```

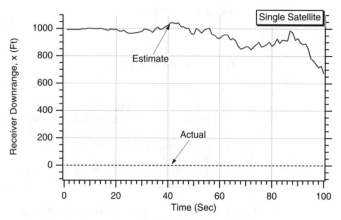

Fig. 11.18   It is not clear if the extended Kalman filter can estimate the receiver's downrange location if only one satellite is used.

large initial angle between range vectors). Similarly, Fig. 11.21 shows that we are able to get fairly good altitude estimates. After approximately 150 s our knowledge of the altitude of the receiver is in error by less than 20 ft. These altitude estimates are comparable to the altitude estimates obtained when we were obtaining range measurements from two satellites. Of course, with two satellites we were able to achieve this accuracy in a fraction of the time. The single satellite case works over time because the filter sees different geometries (i.e., the satellite is moving), and this is the equivalent of having multiple range measurements to a fixed receiver.

To ensure that the extended Kalman filter was operating properly, the errors in the estimate of the downrange and altitude receiver location for 1000 s of

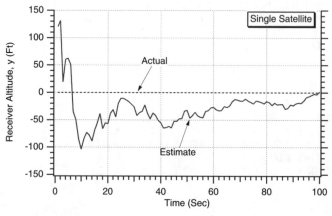

Fig. 11.19   Extended Kalman filter appears to be able to estimate receiver's altitude if only one satellite is used.

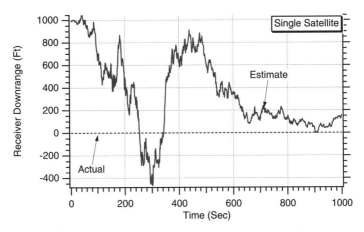

Fig. 11.20   Extended Kalman filter appears to be able to estimate receiver down-range after approximately 600 s if only one satellite is used.

measurements were compared to the theoretical projections of the covariance matrix (i.e., square root of P_{11} and P_{22}, respectively). We can see from both Figs 11.22 and 11.23 that the single-run errors appear to be within the theoretical covariance matrix bounds approximately 68% of the time, indicating that the extended Kalman filter for this problem appears to be operating properly.

Although we might not want to wait for 1000 s to obtain estimates of the receiver location, we can at least see that it is theoretically possible to estimate the receiver location from only one moving satellite. Therefore, it is not always safe to analyze a problem on the basis of equations and unknowns matching each other in order to make a decision on whether or not a Kalman filter will work.

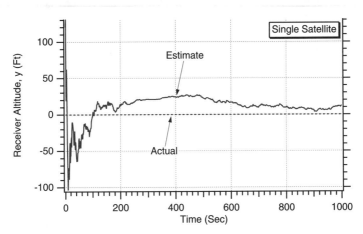

Fig. 11.21   Extended Kalman filter appears to be able to estimate receiver's altitude if only one satellite is used.

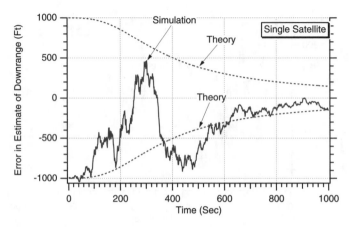

Fig. 11.22   Extended Kalman filter appears to be operating properly in downrange.

Sometimes it is easier just to build the filter and do the analysis on why it works or does not work later.

Using Extended Kalman Filtering with Constant Velocity Receiver

Up until now we have considered only a stationary receiver. Suppose the receiver was moving at a constant velocity. Let us try to apply extended Kalman filtering to the problem of tracking a moving receiver with two moving satellites. Now we are trying to estimate both the location and velocity of the receiver. It seems reasonable to make the receiver location and velocity states. If the receiver travels at constant velocity, that means the receiver acceleration (i.e., derivative of velocity) is zero. However, to protect ourselves in case the velocity is not constant

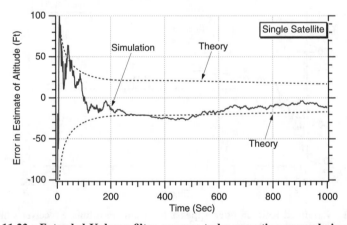

Fig. 11.23   Extended Kalman filter appears to be operating properly in altitude.

we will include some process noise. Therefore, in our model of the real world, we will assume that the acceleration of the receiver is white noise or

$$\ddot{x} = u_s$$

$$\ddot{y} = u_s$$

where the same process noise u_s has been added in both channels. We can rewrite the preceding two scalar equations, representing our new model of the real world, in state-space form, yielding

$$\begin{bmatrix} \dot{x} \\ \ddot{x} \\ \dot{y} \\ \ddot{y} \end{bmatrix} = \begin{bmatrix} 0 & 1 & 0 & 0 \\ 0 & 0 & 0 & 0 \\ 0 & 0 & 0 & 1 \\ 0 & 0 & 0 & 0 \end{bmatrix} \begin{bmatrix} x \\ \dot{x} \\ y \\ \dot{y} \end{bmatrix} + \begin{bmatrix} 0 \\ u_s \\ 0 \\ u_s \end{bmatrix}$$

where u_s is white process noise with spectral density Φ_s. The continuous process-noise matrix can be written by inspection of the linear state-space equation as

$$\boldsymbol{Q} = \begin{bmatrix} 0 & 0 & 0 & 0 \\ 0 & \Phi_s & 0 & 0 \\ 0 & 0 & 0 & 0 \\ 0 & 0 & 0 & \Phi_s \end{bmatrix}$$

In addition, the systems dynamics matrix also can be written by inspection of the linear state-space equation as

$$\boldsymbol{F} = \begin{bmatrix} 0 & 1 & 0 & 0 \\ 0 & 0 & 0 & 0 \\ 0 & 0 & 0 & 1 \\ 0 & 0 & 0 & 0 \end{bmatrix}$$

Because $\mathbf{F}^2$ is zero, a two-term Taylor-series expansion can be used to exactly derive the fundamental matrix or

$$\Phi = \boldsymbol{I} + \boldsymbol{F}t = \begin{bmatrix} 1 & 0 & 0 & 0 \\ 0 & 1 & 0 & 0 \\ 0 & 0 & 1 & 0 \\ 0 & 0 & 0 & 1 \end{bmatrix} + \begin{bmatrix} 0 & 1 & 0 & 0 \\ 0 & 0 & 0 & 0 \\ 0 & 0 & 0 & 1 \\ 0 & 0 & 0 & 0 \end{bmatrix} t = \begin{bmatrix} 1 & t & 0 & 0 \\ 0 & 1 & 0 & 0 \\ 0 & 0 & 1 & t \\ 0 & 0 & 0 & 1 \end{bmatrix}$$

Substituting the sample time for time in the preceding expression yields the discrete fundamental matrix

$$\Phi_k = \begin{bmatrix} 1 & T_s & 0 & 0 \\ 0 & 1 & 0 & 0 \\ 0 & 0 & 1 & T_s \\ 0 & 0 & 0 & 1 \end{bmatrix}$$

The measurements are still the ranges from the satellites to the receiver. We already have shown from Fig. 11.1 that the true or exact ranges from each of the satellites to the receiver are given by

$$r_1 = \sqrt{(x_{R1} - x)^2 + (y_{R1} - y)^2}$$

$$r_2 = \sqrt{(x_{R2} - x)^2 + (y_{R2} - y)^2}$$

However, because the number of states has increased, we must rewrite the linearized measurement equation as

$$\begin{bmatrix} \Delta r_1^* \\ \Delta r_2^* \end{bmatrix} = \begin{bmatrix} \dfrac{\partial r_1}{\partial x} & \dfrac{\partial r_1}{\partial \dot{x}} & \dfrac{\partial r_1}{\partial y} & \dfrac{\partial r_1}{\partial \dot{y}} \\ \dfrac{\partial r_2}{\partial x} & \dfrac{\partial r_2}{\partial \dot{x}} & \dfrac{\partial r_2}{\partial y} & \dfrac{\partial r_1}{\partial \dot{y}} \end{bmatrix} \begin{bmatrix} \Delta x \\ \Delta \dot{x} \\ \Delta y \\ \Delta \dot{y} \end{bmatrix} + \begin{bmatrix} v_{r1} \\ v_{r2} \end{bmatrix}$$

where v_{R1} and v_{R2} are the measurement noises associated with the range measurements r_1 and r_2. The discrete measurement noise matrix is therefore given by

$$R_k = \begin{bmatrix} \sigma_{r1}^2 & 0 \\ 0 & \sigma_{r2}^2 \end{bmatrix}$$

where σ_{r1}^2 and σ_{r2}^2 are the measurement noise variances. From the linearized measurement equation we can see that the linearized measurement matrix is given by

$$H_k = \begin{bmatrix} \dfrac{\partial r_1}{\partial x} & \dfrac{\partial r_1}{\partial \dot{x}} & \dfrac{\partial r_1}{\partial y} & \dfrac{\partial r_1}{\partial \dot{y}} \\ \dfrac{\partial r_2}{\partial x} & \dfrac{\partial r_2}{\partial \dot{x}} & \dfrac{\partial r_2}{\partial y} & \dfrac{\partial r_1}{\partial \dot{y}} \end{bmatrix}$$

where the partial derivatives are evaluated at the projected state estimates. From the range equation we can evaluate each of the partial derivatives as

$$\frac{\partial r_1}{\partial x} = 0.5[(x_{R1} - x)^2 + (y_{R1} - y)^2]^{-0.5} 2(x_{R1} - x)(-1) = \frac{-(x_{R1} - x)}{r_1}$$

$$\frac{\partial r_1}{\partial \dot{x}} = 0$$

$$\frac{\partial r_1}{\partial y} = 0.5[(x_{R1} - x)^2 + (y_{R1} - y)^2]^{-0.5} 2(y_{R1} - y)(-1) = \frac{-(y_{R1} - y)}{r_1}$$

$$\frac{\partial r_1}{\partial \dot{y}} = 0$$

$$\frac{\partial r_2}{\partial x} = 0.5[(x_{R1} - x)^2 + (y_{R1} - y)^2]^{-0.5} 2(x_{R2} - x)(-1) = \frac{-(x_{R2} - x)}{r_2}$$

$$\frac{\partial r_2}{\partial \dot{x}} = 0$$

$$\frac{\partial r_2}{\partial y} = 0.5[(x_{R1} - x)^2 + (y_{R1} - y)^2]^{-0.5} 2(y_{R2} - y)(-1) = \frac{-(y_{R2} - y)}{r_2}$$

$$\frac{\partial r_2}{\partial \dot{y}} = 0$$

Therefore, the linearized measurement matrix is given by

$$H_k = \begin{bmatrix} \dfrac{-(x_{R1} - x)}{r_1} & 0 & \dfrac{-(y_{R1} - y)}{r_1} & 0 \\ \dfrac{-(x_{R2} - x)}{r_2} & 0 & \dfrac{-(y_{R2} - y)}{r_2} & 0 \end{bmatrix}$$

where each of the elements in the matrix are evaluated at their projected values.

Recall that the discrete process-noise matrix can be derived from the continuous process-noise matrix according to

$$Q_k = \int_0^{T_s} \Phi(\tau) Q \Phi^T(\tau) \, d\tau$$

Substitution of the appropriate matrices yields

$$Q_k = \int_0^{T_s} \begin{bmatrix} 1 & \tau & 0 & 0 \\ 0 & 1 & 0 & 0 \\ 0 & 0 & 1 & \tau \\ 0 & 0 & 0 & 1 \end{bmatrix} \begin{bmatrix} 0 & 0 & 0 & 0 \\ 0 & \Phi_s & 0 & 0 \\ 0 & 0 & 0 & 0 \\ 0 & 0 & 0 & \Phi_s \end{bmatrix} \begin{bmatrix} 1 & 0 & 0 & 0 \\ \tau & 1 & 0 & 0 \\ 0 & 0 & 1 & 0 \\ 0 & 0 & \tau & 1 \end{bmatrix} d\tau$$

After multiplication of the three matrices and simple integration, we obtain for the discrete process-noise matrix

$$Q_k = \Phi_S \begin{bmatrix} \dfrac{T_s^3}{3} & \dfrac{T_s^2}{2} & 0 & 0 \\[2ex] \dfrac{T_s^2}{2} & T_s & 0 & 0 \\[2ex] 0 & 0 & \dfrac{T_s^3}{3} & \dfrac{T_s^2}{2} \\[2ex] 0 & 0 & \dfrac{T_s^2}{2} & T_s \end{bmatrix}$$

We now have enough information to solve the Riccati equations. As in other examples in extended Kalman filtering, we saw that we do not have to use the linearized equations for the actual filter. However, because the fundamental matrix is exact we can use it to project the various state estimates forward or

$$\bar{x}_k = \hat{x}_{k-1} + T_s \hat{\dot{x}}_{k-1}$$

$$\bar{\dot{x}}_k = \hat{\dot{x}}_{k-1}$$

$$\bar{y}_k = \hat{y}_{k-1} + T_s \hat{\dot{y}}_{k-1}$$

$$\bar{\dot{y}}_k = \hat{\dot{y}}_{k-1}$$

Therefore, each of the projected ranges from the satellite to the receiver are given by

$$\bar{r}_{1_k} = \sqrt{(x_{R1_k} - \bar{x}_k)^2 + (y_{R1_k} - \bar{y}_k)^2}$$

$$\bar{r}_{2_k} = \sqrt{(x_{R2_k} - \bar{x}_k)^2 + (y_{R2_k} - \bar{y}_k)^2}$$

We have already seen that we do not have to use the linearized measurement matrix to compute the residual. Instead, the residual is simply the range measurement minus the projected range value or

$$\text{Res}_{1_k} = r_{1_k}^* - \bar{r}_{1_k}$$

$$\text{Res}_{2_k} = r_{2_k}^* - \bar{r}_{2_k}$$

Therefore, the extended Kalman-filtering equations are given by

$$\hat{x}_k = \bar{x}_k + K_{11_k}\text{Res}_{1_k} - K_{12_k}\text{Res}_{2_k}$$

$$\hat{\dot{x}}_k = \bar{\dot{x}}_k + K_{21_k}\text{Res}_{1_k} - K_{22_k}\text{Res}_{2_k}$$

$$\hat{y}_k = \bar{y}_k + K_{31_k}\text{Res}_{1_k} - K_{32_k}\text{Res}_{2_k}$$

$$\hat{\dot{y}}_k = \bar{\dot{y}}_k + K_{41_k}\text{Res}_{1_k} + K_{42_k}\text{Res}_{2_k}$$

The extended Kalman filter for estimating the states of a receiver moving at constant velocity appears in Listing 11.6. Statements that are specifically required for the extra two states in the new extended Kalman filter are highlighted in bold. Again, the initial covariance matrix is chosen so that the off-diagonal elements are zero and the diagonal elements represent the square of our best guess in the errors in locating the receiver and errors in our initial estimate of the receiver velocity. The initial filter estimates (i.e., XH = 1000, XDH = 0, YH = 2000, and YDH = 0) show that the filter is not perfectly initialized. The measurement noise matrix is no longer a scalar as it was in the preceding section but is now a two-dimensional square matrix. Therefore, the inverse of a 2×2 matrix will again be required in the Riccati equations. Recall that the formula for the exact inverse of a 2×2 matrix was presented in Chapter 1 and appears in the listing.

The nominal case of Listing 11.6 was run in which the receiver is traveling downrange at a constant velocity but is not traveling in altitude. We can see from Fig. 11.24 that the extended Kalman filter is able to estimate the downrange location of the receiver quite well. We can also see that the filter downrange estimate never diverges from the true location of the receiver. Figure 11.25 shows that we are able to estimate the receiver downrange velocity quite rapidly even though the filter was not initialized correctly. After only 25 s the filter is able to establish a very accurate downrange velocity track of the receiver. Figures 11.26 and 11.27 also indicate that the filter is able to estimate the receiver altitude (which is zero) to within a 100 ft and receiver altitude velocity (which is zero) to within a few feet per second after only 25 s. Thus, we can conclude that the extended Kalman filter appears to be able to track a receiver traveling at constant velocity.

To ensure that the extended Kalman filter is working properly, the single-run simulation errors in all of the state estimates are compared to the theoretical predictions of the covariance matrix (i.e., square root of P_{11}, P_{22}, P_{33}, and P_{44}, respectively) in Figs 11.28–11.31. We can see that because the single-run simulation results appear to be within the theoretical bounds approximately 68% of the time the filter appears to be working properly.

Single Satellite with Constant Velocity Receiver

It is of academic interest to see if we can estimate the constant velocity receiver with only a single satellite. As in the preceding section, it seems reasonable to make the receiver location and velocity states in our model of the real world. If the receiver travels at constant velocity, that means the receiver acceleration (i.e., derivative of velocity) is zero. Again, to protect ourselves in

Listing 11.6   Extended Kalman filter for estimating the states of a receiver moving at constant velocity

```
C THE FIRST THREE STATEMENTS INVOKE THE ABSOFT RANDOM
  NUMBER GENERATOR ON THE MACINTOSH
     GLOBAL DEFINE
              INCLUDE 'quickdraw.inc'
     END
     IMPLICIT REAL*8(A-H)
     IMPLICIT REAL*8(O-Z)
     REAL*8 PHI(4,4),P(4,4),M(4,4),PHIP(4,4),PHIPPHIT(4,4),GAIN(4,2)
     REAL*8 Q(4,4),HMAT(2,4),HM(2,4),MHT(4,2)
     REAL*8 PHIT(4,4),RMAT(2,2),HMHTRINV(2,2)
     REAL8* HMHT(2,2),HT(4,2),KH(4,4),IDN(4,4),IKH(4,4),HMHTR(2,2)
     INTEGER ORDER
     SIGNOISE=300.
     X=0.
     Y=0.
     XH=1000.
     YH+2000.
     XDH=0.
     YDH=0.
     XR1=1000000.
     XR1=20000.*3280.
     XR2=50000000.
     YR2=20000.*3280.
     OPEN(2,STATUS='UNKNOWN',FILE='COVFIL')
     OPEN(1,STATUS='UNKNOWN',FILE='DATFIL')
     ORDER=4
     TS=1.
     TF=200.
     PHIS=0.
     T=0.
     S=0.
     H=.01
     DO 1000 I=1,ORDER
     DO 1000 J=1,ORDER
              PHI(I,J)=0.
              P(I,J)=0.
              Q(I,J)=0.
              IDN(I,J)=0.
1000      CONTINUE
     IDN(1,1)=1.
     IDN(2,2)=1.
     IDN(3,3)=1.
     IDN(4,4)=1.
     P(1,1)=1000.**2
     P(2,2)=100.**2
     P(3,3)=2000.**2
     p(4,4)=100.**2
     RMAT(1,1)=SIGNOISE**2
     RMAT(1,2)=0.
```

(*continued*)

Listing 11.6 (*Continued*)

```
RMAT(2,1)=0.
RMAT(2,2)=SIGNOISE**2
TS2=TS*TS
TS3=TS2*TS
Q(1,1)=PHIS8TS3/3.
Q(1,2)=PHIS*TS2/2.
Q(2,1)=Q(1,2)
Q(2,2)=PHIS*TS
Q(3,3)=PHIS*TS3/3.
Q(3,4)=PHIS*TS2/2.
Q(4,3)=Q(3,4)
Q(4,4)=PHIS*TS
WHILE(T < =TF)
        XR1OLD=XR1
        XR2OLD=XR2
        XOLD=X
        YOLD=Y
        XR1D=-14600.
        XR2D=-14600.
        XD=100.
        YD=0.
        XR1=XR1+H*XR1D
        YR2=XR2+H*XR2D
        X=X+H*XD
        Y=Y+H*YD
        T=T+H
        XR1D=-14600.
        XR2D=-14600.
        XD=100.
        YD=0.
        XR1=.5*(XR1OLD+XR1+H*XR1D)
        XR2=.5*(XR2OLD+XR2+H*XR2D)
        X=.5*(XOLD+X+H*XD)
        Y=.5*(YOLD+Y+H*YD)
        S=S+H
        IF(S>=(TS-.00001))THEN
                S=0.
                XB=XH+XDH*TS
                YB=YH+YDH*TS
                R1B=SQRT((XR1-XB)**2+(YR1-YB)**2)
                R2B=SQRT((XR2-XB)**2+(YR2-YB)**2)
                HMAT(1,1)=-(XR1-XB)/R1B
                HMAT(1,2)=0.
                HMAT(1,3)=-(YR1-YB)/R1B
                HMAT(1,4)=0.
                HMAT(2,1)=-(XR2-XB)/R2B
                HMAT(2,2)=0.
                HMAT(2,3)=-(YR2-YB)/R2B
                HMAT(2,4)=0.
                CALL MATTRN(HMAT,2,4,HT)
```

(*continued*)

Listing 11.6 (*Continued*)

```
                PHI(1,1)=1.
                PHI(1,2)=TS
                PHI(2,2)=1.
                PHI(3,3)=1.
                PHI(3,4)=TS
                PHI(4,4)=1.
                CALL  MATTRN(PHI,ORDER,ORDER,PHIT)
                CALL  MATRMUL(PHI,ORDER,ORDER,P,ORDER,ORDER,
                  PHIP)
                CALL  MATMUL(PHIP,ORDER,ORDER,PHIT,ORDER,
1                 ORDER,PHIPPHIT)
                CALL  MATADD(PHIPPHIT,ORDER,ORDER,Q,M)
                CALL  MATMUL(HMAT,2,ORDER,M,ORDER,ORDER,HM)
                CALL  MATMUL(HM,2,ORDER,HT,ORDER,2,HMHT)
                CALL  MATADD(HMHT,2,2,RMAT,HMHTR)DET=
1                 HMHTR(1,1)*HMHTR(2,2)-HMHTR(1,2)*HMHTR(2,1)
1                 HMHTRINV(1,1)=HMHTR(2,2)/DETHMHTRINV(1,2)=
1                 -HMHTR(1,2)/DETHMHTRINV(2,1)=-HMHTR(2,1)/DET
1                 HMHTRINV(2,2)=HMHTR(1,1)/DET
                CALL  MATMUL(M,ORDER,ORDER,HT,ORDER,2,MHT)
                CALL  MATMUL(MHT,ORDER,2,HMHTRINV,2,2,GAIN)
                GAIN  MATMUL(GAIN,ORDER,2,HMAT,2,ORDER,KH)
                CALL  MATSUB(IDN,ORDER,ORDER,KH,IKH)
                CALL  MATMUL(IKH,ORDER,ORDER,M,ORDER,ORDER,P)
                CALL  GAUSS(R1NOISE,SIGNOISE)
                CALL  GAUSS(R2NOISE,SIGNOISE)
                R1=SQRT((XR1-X)**2+(YR1-Y)**2)
                R2=SQRT((XR2-X)**2+(YR2-Y)**2)
                RES1=R1+R1NOISE-R1B
                RES2=R2+R2NOISE-R2B
                XH=XB+GAIN(1,1)*RES1+GAIN(1,2)*RES2
                XDH=XDH+GAIN(2,1)*RES1+GAIN(2,2)*RES2
                YH=YB+GAIN(3,1)*RES1+GAIN(3,2)*RES2
                YDH=YDH+GAIN(4,1)*RES1+GAIN(4,2)*RES2
                ERRX=X-XH
                SP11=SQRT(P(1,1))
                ERRXD=XD-XDH
                SP22=SQRT(P(2,2))
                ERRY=Y-YH
                SP33=SQRT(P(3,3))
                ERRYD=YD-YDH
                SP44=SQRT(P(4,4))
                WRITE(9,*)T,X,XH,XD,XDH,Y,YH,YD,YDH
                WRITE(1,*)T,X,XH,XD,XDH,Y,YH,YD,YDH
                WRITE(2,*)T,ERRX,SP11,-SP11,ERRXD,SP22,-SP22
1                 ERRY,SP33,-SP33,ERRYD,SP44,-SP44
         ENDIF
    END DO
    PAUSE
```

(continued)

Listing 11.6 (*Continued*)

```
CLOSE(1)
CLOSE(2)
END

C SUBROUTINE GAUSS IS SHOWN IN LISTING 1.8
C SUBROUTINE MATTRN IS SHOWN IN LISTING 1.3
C SUBROUTINE MATMUL IS SHOWN IN LISTING 1.4
C SUBROUTINE MATADD IS SHOWN IN LISTING 1.1
C SUBROUTINE MATSUB IS SHOWN IN LISTING 1.2
```

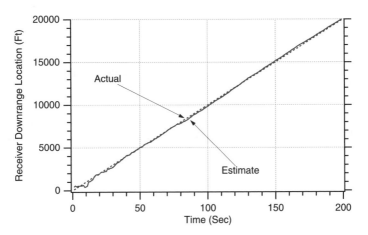

Fig. 11.24   Extended Kalman filter is able to estimate the downrange location of the receiver quite well.

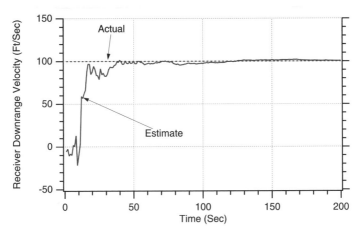

Fig. 11.25   After only 25 s the filter is able to establish a very accurate downrange velocity track of the receiver.

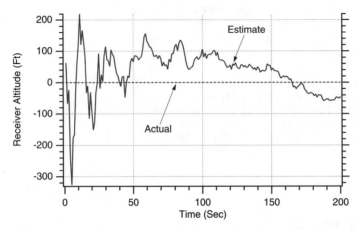

Fig. 11.26 Extended Kalman filter is able to estimate the altitude of the receiver to within 100 ft.

case the velocity is not constant we will include some process noise. Therefore, in our model of the real world, we will assume that the acceleration of the receiver is white noise or

$$\ddot{x} = u_s$$

$$\ddot{y} = u_s$$

where the same process noise u_s has been added in both channels. We can rewrite the preceding two scalar equations, representing our new model of the real world,

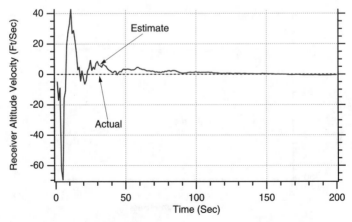

Fig. 11.27 Extended Kalman filter is able to estimate the altitude velocity of the receiver to within a few feet per second.

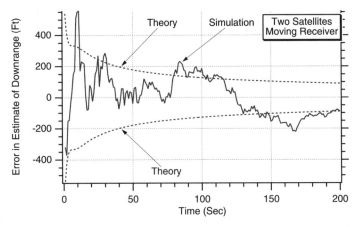

Fig. 11.28 New extended Kalman filter appears to be operating properly in down-range.

in state-space form, yielding

$$\begin{bmatrix} \dot{x} \\ \ddot{x} \\ \dot{y} \\ \ddot{y} \end{bmatrix} = \begin{bmatrix} 0 & 1 & 0 & 0 \\ 0 & 0 & 0 & 0 \\ 0 & 0 & 0 & 1 \\ 0 & 0 & 0 & 0 \end{bmatrix} \begin{bmatrix} x \\ \dot{x} \\ y \\ \dot{y} \end{bmatrix} + \begin{bmatrix} 0 \\ u_s \\ 0 \\ u_s \end{bmatrix}$$

where u_s is white process noise with spectral density Φ_s. The continuous process-noise matrix can be written by inspection of the linear state-space

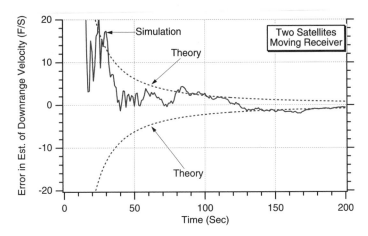

Fig. 11.29 New extended Kalman filter appears to be operating properly in down-range velocity.

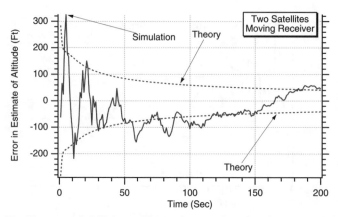

Fig. 11.30   New extended Kalman filter appears to be operating properly in altitude.

equation as

$$Q = \begin{bmatrix} 0 & 0 & 0 & 0 \\ 0 & \Phi_s & 0 & 0 \\ 0 & 0 & 0 & 0 \\ 0 & 0 & 0 & \Phi_s \end{bmatrix}$$

In addition, the systems dynamics matrix can be written by inspection of the linear state-space equation as

$$F = \begin{bmatrix} 0 & 1 & 0 & 0 \\ 0 & 0 & 0 & 0 \\ 0 & 0 & 0 & 1 \\ 0 & 0 & 0 & 0 \end{bmatrix}$$

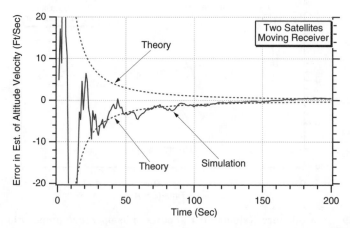

Fig. 11.31   New extended Kalman filter appears to be operating properly in altitude velocity.

Because F^2 is zero, a two-term Taylor-series expansion can be used to derive exactly the fundamental matrix or

$$\Phi = I + Ft = \begin{bmatrix} 1 & 0 & 0 & 0 \\ 0 & 1 & 0 & 0 \\ 0 & 0 & 1 & 0 \\ 0 & 0 & 0 & 1 \end{bmatrix} + \begin{bmatrix} 0 & 1 & 0 & 0 \\ 0 & 0 & 0 & 0 \\ 0 & 0 & 0 & 1 \\ 0 & 0 & 0 & 0 \end{bmatrix} t = \begin{bmatrix} 1 & t & 0 & 0 \\ 0 & 1 & 0 & 0 \\ 0 & 0 & 1 & t \\ 0 & 0 & 0 & 1 \end{bmatrix}$$

Substituting the sampling time for time in the preceding expression yields the discrete fundamental matrix

$$\Phi_k = \begin{bmatrix} 1 & T_s & 0 & 0 \\ 0 & 1 & 0 & 0 \\ 0 & 0 & 1 & T_s \\ 0 & 0 & 0 & 1 \end{bmatrix}$$

The measurement is now the range from a single satellite to the receiver. We have already shown in Fig. 11.1 that the range from a satellite to the receiver is given by

$$r_1 = \sqrt{(x_{R1} - x)^2 + (y_{R1} - y)^2}$$

However, because the number of states has increased, we must rewrite the linearized measurement equation as

$$\Delta r_1^* = \begin{bmatrix} \dfrac{\partial r_1}{\partial x} & \dfrac{\partial r_1}{\partial \dot{x}} & \dfrac{\partial r_1}{\partial y} & \dfrac{\partial r_1}{\partial \dot{y}} \end{bmatrix} \begin{bmatrix} \Delta x \\ \Delta \dot{x} \\ \Delta y \\ \Delta \dot{y} \end{bmatrix} + v_{r1}$$

where v_{r1} is the measurement noise associated with the range measurement r_1. The discrete measurement noise matrix is therefore a scalar given by

$$R_k = \sigma_{r1}^2$$

where σ_{r1}^2 is the measurement noise variance. From the linearized measurement equation we can see that the linearized measurement matrix is given by

$$H_k = \begin{bmatrix} \dfrac{\partial r_1}{\partial x} & \dfrac{\partial r_1}{\partial \dot{x}} & \dfrac{\partial r_1}{\partial y} & \dfrac{\partial r_1}{\partial \dot{y}} \end{bmatrix}$$

where the partial derivatives are evaluated as the projected state estimates. From the range equation we can evaluate each of the partial derivatives as

$$\frac{\partial r_1}{\partial x} = 0.5[(x_{R1} - x)^2 + (y_{R1} - y)^2]^{-0.5}2(x_{R1} - x)(-1) = \frac{-(x_{R1} - x)}{r_1}$$

$$\frac{\partial r_1}{\partial \dot{x}} = 0$$

$$\frac{\partial r_1}{\partial y} = 0.5[(x_{R1} - x)^2 + (y_{R1} - y)^2]^{-0.5}2(y_{R1} - y)(-1) = \frac{-(y_{R1} - y)}{r_1}$$

$$\frac{\partial r_1}{\partial \dot{y}} = 0$$

Therefore, the linearized measurement matrix is given by

$$H_k = \begin{bmatrix} \dfrac{-(x_{R1} - x)}{r_1} & 0 & \dfrac{-(y_{R1} - y)}{r_1} & 0 \end{bmatrix}$$

where, as was just mentioned, each of the elements in the matrix are evaluated at their projected values.

Recall that the discrete process-noise matrix can be derived from the continuous process-noise matrix according to

$$Q_k = \int_0^{T_s} \Phi(\tau) Q \Phi^T(\tau) \, d\tau$$

Substitution of the appropriate matrices yields

$$Q_k = \int_0^{T_s} \begin{bmatrix} 1 & \tau & 0 & 0 \\ 0 & 1 & 0 & 0 \\ 0 & 0 & 1 & \tau \\ 0 & 0 & 0 & 1 \end{bmatrix} \begin{bmatrix} 0 & 0 & 0 & 0 \\ 0 & \Phi_s & 0 & 0 \\ 0 & 0 & 0 & 0 \\ 0 & 0 & 0 & \Phi_s \end{bmatrix} \begin{bmatrix} 1 & 0 & 0 & 0 \\ \tau & 1 & 0 & 0 \\ 0 & 0 & 1 & 0 \\ 0 & 0 & \tau & 1 \end{bmatrix} d\tau$$

After multiplication of the three matrices and simple integration, we obtain for the discrete process-noise matrix

$$Q_k = \Phi_s \begin{bmatrix} \dfrac{T_s^3}{3} & \dfrac{T_s^2}{2} & 0 & 0 \\[2mm] \dfrac{T_s^2}{2} & T_s & 0 & 0 \\[2mm] 0 & 0 & \dfrac{T_s^3}{3} & \dfrac{T_s^2}{2} \\[2mm] 0 & 0 & \dfrac{T_s^2}{2} & T_s \end{bmatrix}$$

We now have enough information to solve the Riccati equations. As in other examples in extended Kalman filtering, we saw that we do not have to use the linearized equations for the actual filter. However, because the fundamental matrix is exact we can use it to project the various state estimates forward or

$$\bar{x}_k = \hat{x}_{k-1} + T_s\hat{\dot{x}}_{k-1}$$

$$\bar{\dot{x}}_k = \hat{\dot{x}}_{k-1}$$

$$\bar{y}_k = \hat{y}_{k-1} + T_s\hat{\dot{y}}_{k-1}$$

$$\bar{\dot{y}}_k = \hat{\dot{y}}_{k-1}$$

Therefore, the projected range from the satellite to the receiver is given by

$$\bar{r}_{1_k} = \sqrt{(x_{R1_k} - \bar{x}_k)^2 + (y_{R1_k} - \bar{y}_k)^2}$$

The residual does not have to be computed from the linearized measurement matrix but instead can be computed as simply the range measurement minus the projected range value or

$$\mathrm{Res}_{1_k} = r^*_{1_k} - \bar{r}_{1_k}$$

Therefore, the extended Kalman-filtering equations are given by

$$\hat{x}_k = \bar{x}_k + K_{11_k}\mathrm{Res}_{1_k}$$

$$\hat{\dot{x}}_k = \bar{\dot{x}}_k + K_{21_k}\mathrm{Res}_{1_k}$$

$$\hat{y}_k = \bar{y}_k + K_{31_k}\mathrm{Res}_{1_k}$$

$$\hat{\dot{y}}_k = \bar{\dot{y}}_k + K_{41_k}\mathrm{Res}_{1_k}$$

The extended Kalman filter for estimating the states of a receiver moving at constant velocity based on single satellite measurements appears in Listing 11.7. As before, the initial covariance matrix is chosen so that the off-diagonal elements are zero and the diagonal elements represent the square of our best guess in the errors in locating the receiver and errors in our initial estimate of the receiver velocity. The initial filter estimates (i.e., XH = 1000, XDH = 0, YH = 2000, and YDH = 0) show that the filter is not perfectly initialized.

The nominal case of Listing 11.7 was run, and the errors in the estimates of the various states are displayed in Figs. 11.32–11.35. First, we can see that because the single-run simulation results appear to be within the theoretical bounds approximately 68% of the time, the filter appears to be working properly. We can see by comparing these figures with Figs. 11.28–11.31 that, in general, the errors in the estimates are larger for the one-satellite case than for the two-satellite case.

Listing 11.7   Extended Kalman filter for single satellite estimating the states of a receiver moving at constant velocity

```
C THE FIRST THREE STATEMENTS INVOKE THE ABSOFT RANDOM
  NUMBER GENERATOR ON THE MACINTOSH
        GLOBAL DEFINE
                INCLUDE 'quickdraw.inc'
        END
        IMPLICIT REAL*8(A-H)
        IMPLICIT REAL*8(O-Z)
        REAL*8 PHI(4,4),P(4,4),M(4,4),PHIP(4,4),PHIPPHIT(4,4),GAIN(4,1)
        REAL*8 Q(4,4),HMAT(1,4),HM(1,4),MHT(4,1)
        REAL*8 PHIT(4,4),RMAT(1,1),HMHTRINV(1,1)
        REAL*8 HMHT(1,1),HT(4,1),KH(4,4),IDN(4,4),IKH(4,4),HMHTR(1,1)
        INTEGER ORDER
        SIGNOISE=300.
        PHIS=0.
        X=0.
        Y=0.
        XH=1000.
        YH=2000.
        XH=0.
        YH=0.
        XR1=1000000.
        YR1=20000.*3280.
        OPEN(2,STATUS='UNKNOWN',FILE='COVFIL')
        OPEN(1,STATUS='UNKNOWN',FILE='DATFIL')
        ORDER=4
        TS=1.
        TF=1000.
        T=0.
        S=0.
        H=.01
        DO 1000 I=1,ORDER
        DO 1000 J=1,ORDER
                PHI(I,J)=0.
                P(I,J)=0.
                Q(I,J)=0.
                IDN(I,J)=0.
1000        CONTINUE
        IDN(1,1)=1.
        IDN(2,2)=1.
        IDN(3,3)=1.
        IDN(4,4)=1.
        P(1,1)=1000.**2
        P(2,2)=100.**2
        P(3,3)=2000.**2
        P(4,4)=100.**2
        RMAT(1,1)=SIGNOISE**2
        TS2=TS*TS
        TS3=TS2*TS
        Q(1,1)=PHIS*TS3/3.
```

(continued)

Listing 11.7 *(Continued)*

```
Q(1,2)=PHIS*TS2/2.
Q(2,1)=Q(1,2)
Q(2,2)=PHIS*TS
Q(3,3)=PHIS*TS3/3.
Q(3,4)=PHIS*TS2/2.
Q(4,3)=Q(3,4)
Q(4,4)=PHIS*TS
WHILE(T<=TF)
        XR1OLD=XR1
        XOLD=X
        YOLD=Y
        XR1D=-14600.
        XD=100.
        YD=0.
        YR1=XR1+H*XR1D
        X=X+H*XD
        Y=Y+H*TD
        T=T+H
        XR1D=-14600.
        XD=100.
        YD=0.
        XR1=.5*(XR1OLD+XR1+H*XR1D)
        X=.5*(XOLD+X+H*XD)
        Y=.5*(YOLD+Y+H*YD)
        S=S+H
        IF(S>=(TS-.00001))THEN
                S=0.
                XB=XH+XDH*TS
                YB=YH+YDH*TS
                R1B=SQRT((XR1-XB)**2+(YR1-YB)**2)
                HMAT(1,1)=-(XR1-XB)/R1B
                HMAT(1,2)=0.
                HMAT(1,3)=-(YR1-YB)/R1B
                HMAT(1,4)=0.
                CALL MATTRN(HMAT,1,ORDER,HT)
                PHI(1,1)=1.
                PHI(1,2)=TS
                PHI(2,2)=1.
                PHI(3,3)=1.
                PHI(3,4)=TS
                PHI(4,4)=1.
                CALL MATTRN(PHI,ORDER,ORDER,PHIT)
                CALL MATMUL(PHI,ORDER,ORDER,P,ORDER,ORDER,
                    PHIP)
                CALL MATMUL(PHIP,ORDER,ORDER,PHIT,ORDER,
                    ORDER,PHIPPHIT)
                CALL MATADD(PHIPPHIT,ORDER,ORDER,Q,M)
                CALL MATMUL(HMAT,1,ORDER,M,ORDER,ORDER,HM)
                CALL MATMUL(HM,1,ORDER,HT,ORDER,1,HMHT)
```

(continued)

Listing 11.7 (*Continued*)

```
          CALL  MATADD(HMHT,1,1,RMAT,HMHTR)HMHTRINV
            (1,1)=1./HMHTR(1,1)
          CALL  MATMUL(M,ORDER,ORDER,HT,ORDER,1,MHT)
          CALL  MATMUL(MHT,ORDER,1,HMHTRINV,1,1,GAIN)
          CALL  MATMUL(GAIN,ORDER,1,HMAT,1,ORDER,KH)
          CALL  MATSUB(IDN,ORDER,ORDER,KH,IKH)
          CALL  MATMUL(IKH,ORDER,ORDER,M,ORDER,ORDER,P)
          CALL  GAUSS(R1NOISE,SIGNOISE)
          R1=SQRT((XR1-X)**2+(YR1-Y)**2)
          RES1=R1+R1NOISE-R1B
          XH=XB+GAIN(1,1)*RES1
          XDH=XDH+GAIN(2,1)*RES1
          YH=YB+GAIN(3,1)*RES1
          YDH=YDH+GAIN(4,1)*RES1
          ERRX=X-XH
          SP11=SQRT(P(1,1))
          ERRXD=XD-XDH
          SP22=SQRT(P(2,2))
          ERRY=Y-YH
          SP33=SQRT(P(3,3))
          ERRYD=YD-YDH
          SP44=SQRT(P(4,4))
          WRITE(9,*)T,X,XH,XD,XDH,Y,YH,YD,YDH
          WRITE(1,*)T,X,XH,XD,XDH,Y,YH,YD,YDH
          WRITE(2,*)T,ERRX,SP11,-SP11,ERRXD,SP22,-SP22,
  1           ERRY,SP33,-SP33,ERRYD,SP44,-SP44
        ENDIF
      END DO
      PAUSE
      CLOSE(1)
      CLOSE(2)
      END

C SUBROUTINE GAUSS IS SHOWN IN LISTING 1.8
C SUBROUTINE MATTRN IS SHOWN IN LISTING 1.3
C SUBROUTINE MATMUL IS SHOWN IN LISTING 1.4
C SUBROUTINE MATADD IS SHOWN IN LISTING 1.1
C SUBROUTINE MATSUB IS SHOWN IN LISTING 1.2
```

For example, by comparing Figs. 11.28 and 11.32 we can seen that the error in the estimate of downrange has increased from approximately 100 ft in the two-satellite case to 1000 ft in the one-satellite case. By comparing Figs. 11.29 and 11.33, we can see that the error in the estimate of downrange velocity is approximately the same at 1 ft/s. By comparing Figs. 11.30 and 11.34, we can see that the error in the estimate of altitude has increased from 40 ft in the two-satellite case to 250 ft in the one-satellite case. By comparing Figs. 11.31 and 11.35, we can see that the error in the estimate of altitude velocity is approximately the same for both the one- and two-satellite cases. As we have seen before,

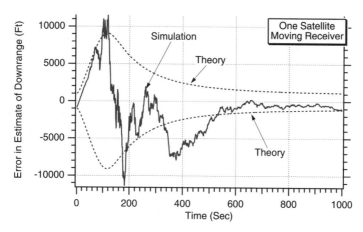

Fig. 11.32 New extended Kalman filter appears to be operating properly in down-range.

in all cases it takes much longer to reduce the errors in the estimates using only one satellite than it does with two satellites.

Thus, it appears that only one satellite is required in the two-dimensional case for tracking a constant velocity receiver. Of course, two satellites will yield smaller position errors and will converge on good estimates much faster than when only using one satellite.

Using Extended Kalman Filtering with Variable Velocity Receiver

We have seen that the extended Kalman filter, using range measurements from two satellites, was able to track a receiver moving at constant velocity. The

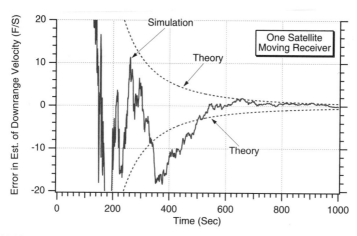

Fig. 11.33 New extended Kalman filter appears to be operating properly in down-range velocity.

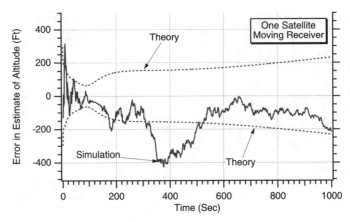

Fig. 11.34 New extended Kalman filter appears to be operating properly in altitude.

extended Kalman filter that was used did not even require process noise to keep track of the moving receiver. In this section we will see how the same extended Kalman filter behaves when the receiver is moving erratically.

For our real world model we will still assume that the receiver is still traveling at constant altitude, but the velocity in the downrange direction will be variable. We will use a Gauss–Markov model to represent the variable velocity. This is the same model (i.e., white noise through a low-pass filter) we simulated in Chapter 1. A block diagram of the model in the downrange direction appears in Fig. 11.36. Here we can see that white noise u_{s1} with spectral density Φ_{s1} enters a low-pass filter with time constant T to form the random velocity. The time constant

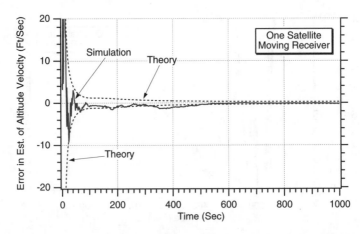

Fig. 11.35 New extended Kalman filter appears to be operating properly in altitude velocity.

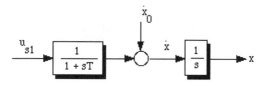

Fig. 11.36 Gauss–Markov model for downrange velocity and location of the receiver.

will determine the amount of correlation in the resulting noise. Larger values of the time constant will result in more correlation, and the resultant signal will look less noisy. The average value of the velocity is $\dot{x}_0$, which is added to the random component of the velocity to form the total velocity $\dot{x}$. The total velocity is integrated once to yield the new receiver downrange location x.

There is a steady-state relationship between the spectral density of the white noise input to the low-pass filter and the variance of the output that is given by[2]

$$\sigma^2 = \frac{\Phi_s}{2T}$$

This relationship was demonstrated via simulation in Chapter 1. Suppose that the receiver was on a sports car that had an average speed of 100 ft/s. We desire a Gauss–Markov noise with a standard deviation of 30 ft/s and a correlation time constant of 5 s to modify the average speed of the sports car. According to the preceding calculation, the spectral density Φ_{s1} of the white noise going into the low-pass filter would be given by

$$\Phi_{s1} = 2T\sigma^2 = 2*5*30^2 = 9000$$

To model white noise in a digital simulation, we simply have to generate Gaussian random noise σ_{Noise} every integration interval. The standard deviation of the Gaussian noise was shown in Chapter 1 and in Ref. 2 to be given by

$$\sigma_{\text{Noise}} = \sqrt{\frac{\Phi_{s1}}{H}}$$

where H is the integration interval used in the simulation to integrate numerically the differential equations.

Listing 11.8 models the variable velocity sports car (i.e., receiver) model of Fig. 11.36. Second-order Runge–Kutta numerical integration is used to integrate the appropriate equations. We can show that Gaussian noise is called into the simulation every integration interval in order to model the white noise using the technique already described.

The nominal case of Listing 11.8 was run, and the resultant receiver downrange velocity appears in Fig. 11.37. We can see that although the average downrange receiver velocity is approximately 100 ft/s the velocity can change quite a bit with the Gauss–Markov model used. The reason for this is that the

Listing 11.8   Using a Gauss–Markov model to represent receiver velocity

```
C THE FIRST THREE STATEMENTS INVOKE THE ABSOFT RANDOM
  NUMBER GENERATOR ON THE MACINTOSH
      GLOBAL DEFINE
              INCLUDE 'quickdraw.inc'
      END
      IMPLICIT REAL*8(A-H)
      IMPLICIT REAL*8(O-Z0
      TAU=5.
      PHI=9000.
      OPEN(1,STATUS='UNKNOWN',FILE='DATFIL')
      T=0.
      S=0.
      H=.01
      SIG=SQRT(PHI/H)
      X=0.
      Y1=0.
      XDP=100.
      TS=1.
      TF=200.
      WHILE(T<=TF)
              CALL GAUSS(X1,SIG)
              XOLD=X
              Y1OLD=Y1
              Y1D=(X1-Y1)/TAU
              XD=XDP+Y1
              X=X+H*XD
              Y1=Y1+H*Y1D
              T=T+H
              Y1D=(X1-Y1)/TAU
              D=XDP+Y1
              X=.5*(XOLD+X+H*XD)
              Y1=.5*(Y1OLD+Y1+H*Y1D)
              S=S+H
              IF(S>=(TS-.00001))THEN
                      S=0.
                      WRITE(9,*)T,XD,X
                      WRITE(1,*)T,XD,X
              ENDIF
      END DO
      PAUSE
      CLOSE(1)
      END

C SUBROUTINE GAUSS IS SHOWN IN LISTING 1.8
```

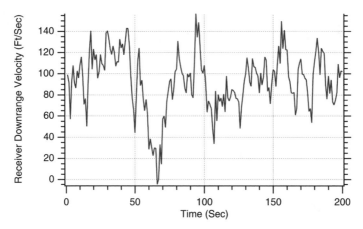

Fig. 11.37   Downrange velocity of receiver varies quite a bit.

standard deviation of the velocity noise is 30 ft/s, which means 68% of the time the velocity should be between 70 and 130 ft/s. Although Fig. 11.37 might not be the most realistic model of a moving receiver in that the velocities are not smoother, it is good enough for our purposes. Integrating the velocity will yield the receiver location. We can see from Fig. 11.38 that although the velocity profile of the receiver is quite erratic the downrange location of the receiver as a function of time is nearly a straight line.

Combining the extended Kalman filter for the two-satellite problem (i.e., Listing 11.6) and the variable velocity model of this section, we obtain Listing 11.9. The differences between this listing and Listing 11.6 are highlighted in bold.

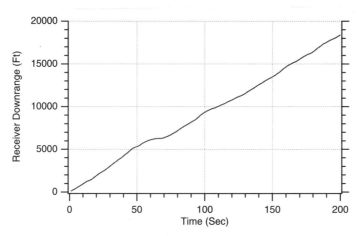

Fig. 11.38   Downrange location of the receiver as a function of time is nearly a straight line.

Listing 11.9   Influence of variable velocity receiver on extended Kalman filter

```
C THE FIRST THREE STATEMENTS INVOKE THE ABSOFT RANDOM
  NUMBER GENERATOR ON THE MACINTOSH
      GLOBAL DEFINE
              INCLUDE 'quickdraw.inc'
      END
      IMPLICIT REAL*8(A-H)
      IMPLICIT REAL*8(O-Z)
      REAL*8 PHI(4,4),P(4,4),M(4,4),PHIP(4,4),PHIPPHIT(4,4),GAIN(4,2)
      REAL*8 Q(4,4),HMAT(2,4),HM(2,4),MHT(4,2)
      REAL*8 PHIT(4,4),RMAT(2,2),HMHTRINV(2,2)
      REAL*8 HMHT(2,2),HT(4,2),KH(4,4),IDN(4,4),IKH(4,4),HMHTR(2,2)
      INTEGER ORDER
      SIGNOISE=300.
      PHIREAL=9000.
      TAU=5.
      PHIS=0.
      X=0.
      Y=0.
      Y1=0.
      XDP=100.
      YD=0.
      XH=1000.
      YH=2000.
      XDH=0.
      YDH=0.
      XR1=1000000.
      YR1=20000.*3280.
      XR1D=-14600.
      XR2=50000000.
      YR2=20000.*3280.
      XR2D=-14600.
      OPEN(2,STATUS='UNKNOWN',FILE='COVFIL')
      OPEN(1,STATUS='UNKNOWN',FILE='DATFIL')
      ORDER=4
      RS=1.
      TF=200.
      T=0.
      S=0.
      H=.01
      SIG=SQRT(PHIREAL/H)
      DO 1000 I=1,ORDER
      DO 1000 J=1,ORDER
              PHI(I,J)=0.
              P(I,J)=0.
              Q(I,J)=0.
              IDN(I,J)=0.
1000      CONTINUE
      IDN(1,1)=1.
      IDN(2,2)=1.
      IDN(3,3)=1.
```

(*continued*)

Listing 11.9 *(Continued)*

```
IDN(4,4)=1.
P(1,1)=1000.**2
P(2,2)=100.**2
P(3,3)=2000.**2
P(4,4)=100.**2
TS2=TS*TS
TS3=TS2*TS
Q(1,1)=PHIS*TS3/3.
Q(1,2)=PHIS*TS2/2.
Q(2,1)=Q(1,2)
Q(2,2)=PHIS*TS
Q(3,3)=PHIS*TS3/3.
Q(3,4)=PHIS*TS2/2.
Q(4,3)=Q(3,4)
Q(4,4)=PHIS*TS
RMAT(1,1)=SIGNOISE**2
RMAT(1,2)=0.
RMAT(2,1)=0.
RMAT(2,2)=SIGNOISE**2
WHILE(T < =TF)
        CALL  GAUSS(X1,SIG)
        XR1OLD=XR1
        XR2OLD=XR2
        XOLD=X
        YOLD=Y
        Y1OLD=Y1
        XR1D=-14600.
        XR2D=-14600.
        Y1D=(X1-Y1)/TAU
        XD=XDP+Y1
        YD=0.
        XR1=XR1+H*XR1D
        XR2=XR2+H*XR2D
        X=X+H*XD
        Y=Y+H*YD
        Y1=Y1+H*Y1D
        T=T+H
        XR1D=-14600.
        XR2D=-14600.
        Y1D=(X1-Y1)/TAU
        XD=XDP+Y1
        YD=0.
        XR1=.5*(XR1OLD+XR1+H*XR1D)
        XR2=.5*(XR2OLD+XR2+H*HR2D)
        X=.5*(XOLD+X+H*XD)
        Y=.5*(YOLD+Y+H*YD)
        Y1=.5*(Y1OLD+Y1+H*Y1D)
        S=S+H
        IF(S>=(TS-.00001))THEN
```

(continued)

Listing 11.9 (*Continued*)

```
S=0.
XB=XH+XDH*TS
YB=YH+YDH*TS
R1B=SQRT((XR1-XB)**2+(YR1-YB)**2)
R2B=SQRT((XR2-XB)**2+(YR2-YB)**2)
HMAT(1,1)=-(XR1-XB)/R1B
HMAT(1,2)=0.
HMAT(1,3)=-(YR1-YB)/R1B
HMAT(1,4)=0.
HMAT(2,1)=-(XR2-XB)/R2B
HMAT(2,2)=0.
HMAT(2,3)=-(YR2-YB)/R2B
HMAT(2,4)=0.
CALL MATTRN(HMAT,2,4,HT)
PHI(1,1)=1.
PHI(1,2)=TS
PHI(2,2)=1.
PHI(3,3)=1.
PHI(3,4)=TS
PHI(4,4)=1.
CALL MATTRN(PHI,ORDER,ORDER,PHIT)
CALL MATMUL(PHI,ORDER,ORDER,P,ORDER,ORDER,
     PHIP)
CALL MATMUL(PHIP,ORDER,ORDER,PHIT,ORDER,
     ORDER,PHIPPHIT)
CALL MATADD(PHIPPHIT, ORDER,ORDER,Q,M)
CALL MATMUL(HMAT,2,ORDER,M,ORDER,ORDER,HM)
CALL MATMUL(HM,2,ORDER,HT,ORDER,2,HMHT)
CALL MATADD(HMHT,2,2,RMAT,HMHTR)DET=
     HMHTR(1,1)*HMHTR(2,2)-HMHTR(1,2)*HMHTR(2,1)
     HMHTRINV(1,1)=HMHTR(2,2)/DETHMHTRINV(1,2)=
     -HMHTR(1,2)/DETHMHTRINV(2,1)=-HMHTR(2,1)/DET
     HMHTRINV(2,2)=HMHTR(1,1)/DET
CALL MATMUL(M,ORDER,ORDER,HT,ORDER,2,MHT)
CALL MATMUL(MHT,ORDER,2,HMHTRINV,2,2,GAIN)
CALL MATMUL(GAIN,ORDER,2,HMAT,2,ORDER,KH)
CALL MATSUB(IDN,ORDER,ORDER,KH,IKH)
CALL MATMUL(IKH,ORDER,ORDER,M,ORDER,ORDER,P)
CALL GAUSS(R1NOISE,SIGNOISE)
CALL GAUSS(R2NOISE,SIGNOISE)
R1=SQRT((XR1-X)**2+(YR1-Y)**2)
R2=SQRT((XR2-X)**2(YR2-Y)**2)
RES1=R1+R1NOISE-R1B
RES2=R2+R2NOISE-R2B
XH=XB+GAIN(1,1)*RES1+GAIN(1,2)*RES2
XDH=XDH+GAIN(2,1)*RES1+GAIN(2,2)*RES2
YH=YB+GAIN(3,1)*RES1+GAIN(3,2)*RES2
YDH=YDH+GAIN(4,1)*RES1+GAIN(4,2)*RES2
ERRX=X-XH
SP11=SQRT(P(1,1))
ERRXD=XD-XDH
```

(*continued*)

Listing 11.9   (*Continued*)

```
                SP22=SQRT(P(2,2))
                ERRY=Y-YH
                SP33=SQRT(P(3,3))
                ERRYD=YD-YDH
                SP44=SQRT(P(4,4))
                WRITE(9,*)T,X,XH,XD,XDH,Y,YH,YD,YDH
                WRITE(1,*)T,X,XH,XD,XDH,Y,YH,YD,YDH
                WRITE(2,*)T,ERRX,SP11,-SP11,ERRXD,SP22,-SP22,
 1              ERRY,SP33,-SP33,ERRYD,SP44,-SP44
          ENDIF
       END DO
       PAUSE
       CLOSE(1)
       CLOSE(2)
       END

C SUBROUTINE GAUSS IS SHOWN IN LISTING 1.8
C SUBROUTINE MATTRN IS SHOWN IN LISTING 1.3
C SUBROUTINE MATMUL IS SHOWN IN LISTING 1.4
C SUBROUTINE MATADD IS SHOWN IN LISTING 1.1
C SUBROUTINE MATSUB IS SHOWN IN LISTING 1.2
```

The nominal case of Listing 11.9 does not include process noise for the extended Kalman filter (i.e., $\Phi_s = 0$). This means that the filter is expecting a constant velocity receiver even though the real receiver velocity is changing erratically. We can see from Fig. 11.39 that having zero process noise does not cause the filter's estimate of the receiver location to severely diverge from the actual receiver location. However, we can see from Fig. 11.40 that the filter is not

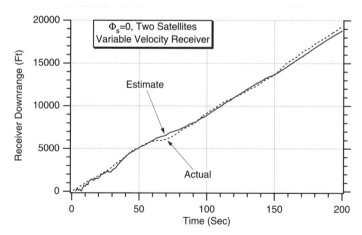

Fig. 11.39   Filter is able to track receiver downrange location when filter does not have process noise.

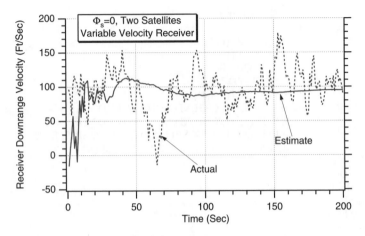

Fig. 11.40 Filter is unable to follow receiver downrange velocity variations when filter does not have process noise.

able to track the receiver velocity accurately. The filter is able to track the average velocity of the receiver, but it is not able to keep up with the velocity variations.

Figure 11.41 indicates that we can track the receiver altitude fairly accurately. A comparison of this figure with Fig. 11.26, in which the receiver downrange velocity was a constant, indicates that the quality of the altitude estimates are approximately the same. Similarly, Fig. 11.42 indicates that the quality of the receiver altitude velocity estimate is quite good. A comparison of Fig. 11.42 with Fig. 11.27 indicates that the quality of the receiver altitude velocity estimates is approximately the same.

The simulation of Listing 11.9 was rerun with the Gauss–Markov model for the downrange receiver velocity, but this time process noise was used in the extended Kalman filter (i.e., $\Phi_s = 100$). By comparing Fig. 11.43 with Fig.

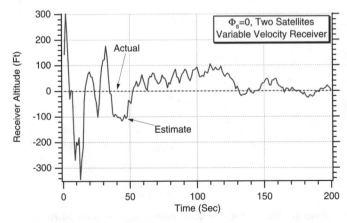

Fig. 11.41 Receiver altitude estimate is good when downrange velocity varies.

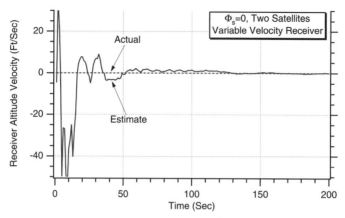

Fig. 11.42 **Receiver altitude velocity estimate is excellent when downrange velocity varies.**

11.39, we can see that the receiver downrange estimate is slightly noisier when process noise is used. The addition of process noise appears to have widened the bandwidth of the extended Kalman filter. By comparing Fig. 11.44 with Fig. 11.40, we can see that the addition of process noise has a significant effect on the receiver downrange velocity estimate. The wider bandwidth of the extended Kalman filter with process noise enables it to better follow the actual velocity variations of the receiver.

Figures 11.45 and 11.46 show that the receiver altitude and altitude velocity errors increase when process noise is added. By comparing Fig. 11.45 with Fig. 11.41, we can see that the receiver altitude error increased from approximately 50 to 150 ft with the addition of process noise. Similarly, by comparing Fig. 11.46

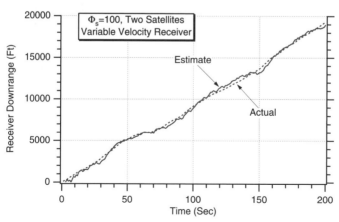

Fig. 11.43 **Process noise makes extended Kalman filter's receiver downrange estimate slightly noisier.**

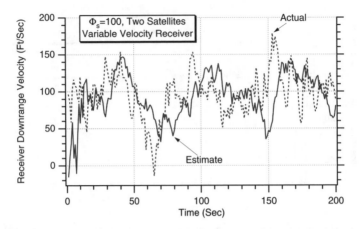

Fig. 11.44 Process noise enables extended Kalman filter to better track receiver downrange velocity.

with Fig. 11.42, we can see that the receiver altitude velocity error increased from a few feet per second to approximately 15 ft/s with the addition of process noise. Therefore, we can see that the addition of process noise was a good thing in the channel in which the receiver velocity was changing, but a bad thing in the channel in which the receiver velocity was zero.

Finally, to check that the errors in the estimates of the states of the extended Kalman filter are within the theoretical tolerances, four more sets of curves were generated. Figures 11.47–11.50 display the errors in the estimates of receiver downrange, downrange velocity, altitude, and altitude velocity, respectively. Superimposed on these plots are the theoretical predictions from the Riccati

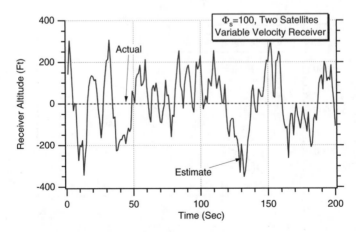

Fig. 11.45 Errors in estimating altitude increase when process noise is added to extended Kalman filter.

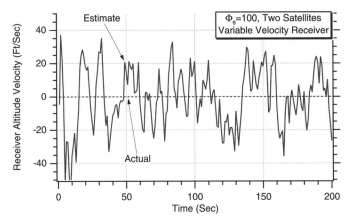

Fig. 11.46 Errors in estimating altitude velocity increase when process noise is added to extended Kalman filter.

equation covariance matrix (i.e., square root of P_{11}, P_{22}, P_{33}, and P_{44}, respectively). We can see that the solid curves, which represent the actual errors in the estimates for a single simulation run, fall within the theoretical bounds approximately 68% of the time, indicating that the filter is working properly.

Variable Velocity Receiver and Single Satellite

Because we were successful in tracking the variable velocity receiver with an extended Kalman filter and two satellites, it is of considerable interest to see if we can be as successful with only one satellite. We have already demonstrated that the extended Kalman filter of Listing 11.8 was able to track a constant velocity

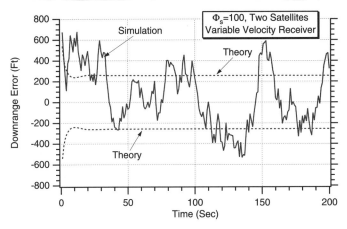

Fig. 11.47 Error in estimate of receiver downrange is within theoretical bounds when downrange velocity varies.

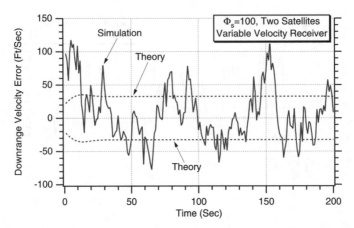

Fig. 11.48   Error in estimate of receiver downrange velocity is within theoretical bounds when downrange velocity varies.

receiver using only one satellite. Combining the same extended Kalman filter for the one-satellite problem (i.e., Listing 11.8) and the variable velocity receiver, we obtain Listing 11.10. The differences between this listing and Listing 11.8 (i.e., variable velocity receiver model) are highlighted in bold.

The nominal case of Listing 11.10 does not include process noise for the extended Kalman filter (i.e., $\Phi_s = 0$). This means that the filter is expecting a constant velocity receiver. We can see from Fig. 11.51 that having zero process noise does not cause the filter's estimate of the receiver location to diverge severely from the actual receiver location. However, we can see from Fig. 11.52 that the filter is not able to track the receiver velocity accurately. The filter is able

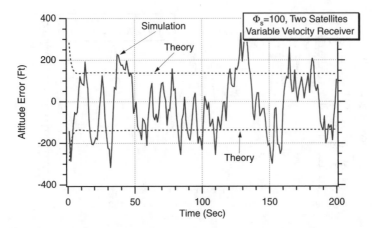

Fig. 11.49   Error in estimate of receiver altitude is within theoretical bounds when downrange velocity varies.

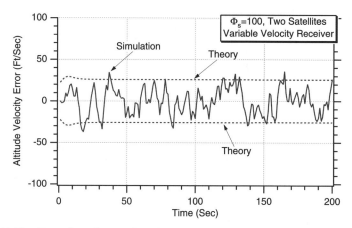

Fig. 11.50 Error in estimate of receiver altitude velocity is within theoretical bounds when downrange velocity varies.

to track the average velocity of the receiver, but it is not able to keep up with the velocity variations. These results are similar to those obtained with two satellites. However, when we only have one satellite, it takes much longer to get satisfactory estimates.

When we added process noise to the extended Kalman filter that used measurements from two satellites, we were able to follow more closely the velocity variations of the receiver. The same amount of process noise (i.e., PHIS = 100) was added to the extended Kalman filter, which used measurements from only one satellite. We can see from Fig. 11.53 that the addition of process noise actually makes the estimates coming from the filter worthless. We already saw that the addition of process noise widened the bandwidth of the Kalman

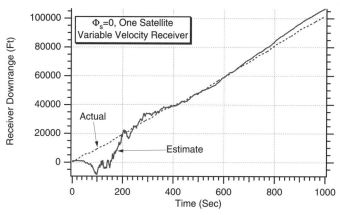

Fig. 11.51 After a while the filter appears to be able to track receiver downrange location when filter does not have process noise.

**Listing 11.10   Influence of variable velocity receiver on one-satellite extended
Kalman filter**

```
C THE FIRST THREE STATEMENTS INVOKE THE ABSOFT RANDOM
  NUMBER GENERATOR ON THE MACINTOSH
      GLOBAL DEFINE
              INCLUDE 'quickdraw.inc'
      END
      IMPLICIT REAL*8(A-H)
      IMPLICIT REAL*8(O-Z)
      REAL*8 PHI(4,4),P(4,4),M(4,4),PHIP(4,4),PHIPPHIT(4,4),GAIN(4,1)
      REAL*8 Q(4,4),HMAT(1,4),HM(1,4),MHT(4,1)
      REAL*8 PHIT(4,4),RMAT(1,1),HMHTRINV(1,1)
      REAL*8 HMHT(1,1),HT(4,1),KH(4,4),IDN(4,4),IKH(4,4),HMHTR(1,1)
      INTEGER ORDER
      SIGNOISE=300.
      PHIREAL=9000.
      TAU=5.
      PHIS=0.
      X=0.
      Y=0.
      Y1=0.
      XDP=100.
      YD=0.
      XH=1000.
      YH=2000.
      XDH=0.
      YDH=0.
      XR1=1000000.
      YR1=20000.*3280.
      XR1D=-14600.
      OPEN(2,STATUS='UNKNOWN',FILE='COVIL')
      OPEN(1,STATUS='UNKNOWN',FILE='DATFIL')
      ORDER=4
      TS=1.
      TF=1000.
      T=0.
      S=0.
      H=.01
      SIG=SQRT(PHIREAL/H)
      DO 1000 I=1,ORDER
      DO 1000 J=1,ORDER
              PHI(I,J)=0.
              P(I,J)=0.
              Q(I,J)=0.
              IDN(I,J)=0.
 1000 CONTINUE
      IDN(1,1)=1.
      IDN(2,2)=1.
      IDN(3,3)=1.
      IDN(4,4)=1.
      P(1,1)=1000.**2
```

(*continued*)

Listing 11.10 (*Continued*)

```
P(2,2)=100.**2
P(3,3)=2000.**2
P(4,4)=100.**2
RMAT(1,1)=SIGNOISE**2
TS2=TS*TS
TS3=TS2*TS
Q(1,1)=PHIS*TS3/3.
Q(1,2)=PHIS*TS2/2.
Q(2,1)=Q(1,2)
Q(2,2)=PHIS*TS
Q(3,3)=PHIS*TS3/3.
Q(3,4)=PHIS*TS2/2.
Q(4,3)=Q(3,4)
Q(4,4)=PHIS*TS
WHILE(T < =TF)
        CALL  GAUSS(X1,SIG)
        XR1OLD=XR1
        XOLD=X
        YOLD=Y
        Y1OLD=Y1
        XR1D=-14600.
        Y1D=(X1-Y1)/TAU
        XD=XDP+Y1
        YD=0.
        XR1=XR1+H*XR1D
        X=X+H*XD
        Y=Y+H*YD
        Y1=Y1+H*Y1D
        T=T+H
        XR1D=-14600.
        Y1D=(X1-Y1)/TAU
        XD=XDP+Y1
        YD=0.
        XR1=.5*(XR1OLD+XR1+H*XR1D)
        X=.5*(XOLD+X+H*XD)
        Y=.5*(YOLD+Y+H*YD)
        Y1=.5*(Y1OLD+Y1+H*Y1D)
        S=S+H
        IF(S>=(TS-.00001))THEN
                S=0.
                XB=XH+XDH*TS
                YB=YH+YDH*TS
                R1B=SQRT((XR1-XB)**2(YR1-YB)**2)
                HMAT(1,1)=-(XR1-XB)/R1B
                HMAT(1,2)=0.
                HMAT(1,3)=-(YR1-YB)/R1B
                HMAT(1,4)=0.
                CALL  MATTRN(HMAT,1,ORDER,HT)
                PHI(1,1)=1.
                PHI(1,2)=TS
```

(*continued*)

Listing 11.10 (*Continued*)

```
                    PHI(2,2)=1.
                    PHI(3,3)=1.
                    PHI(3,4)=TS
                    PHI(4,4)=1.
                    CALL MATTRN(PHI,ORDER,ORDER,PHIT)
                    CALL MATMUL(PHI,ORDER,ORDER,P,ORDER,ORDER,
                        PHIP)
                    CALL MATMUL(PHIP,ORDER,ORDER,PHIT,ORDER,
      1                 ORDER,PHIPPHIT)
                    CALL MATADD(PHIPPHIT,ORDER,ORDER,Q,M)
                    CALL MATMUL(HMAT,1,ORDER,M,ORDER,ORDER,HM)
                    CALL MATMUL(HM,1,ORDER,HT,ORDER,1,HMHT)
                    CALL MATADD(HMHT,1,1,RMAT,HMHR)
                        HMHTRINV(1,1)=1./HMHTR(1,1)
                    CALL MATMUL(M,ORDER,ORDER,HT,ORDER,1,MHT)
                    CALL MATMUL(MHT,ORDER,1,HMHTRINV,1,1,GAIN)
                    CALL MATMUL(GAIN,ORDER,1,HMAT,1,ORDER,KH)
                    CALL MATSUB(IDN,ORDER,ORDER,KH,IKH)
                    CALL MATMUL(IKH,ORDER,ORDER,M,ORDER,ORDER,P)
                    CALL GAUSS(R1NOISE,SIGNOISE)
                    R1=SQRT((XR1-X)**2+(YR1-Y)**2)
                    RES1=R1+R1NOISE-R1B
                    XH=XB+GAIN(1,1)*RES1
                    XDH=XDH+GAIN(2,1)*RES1
                    YH=YB+GAIN(3,1)*RES1
                    YDH=YDH+GAIN(4,1)*RES1
                    ERRX=X-XH
                    SP11=SQRT(P(1,1))
                    ERRXD=XD-XDH
                    SP22=SQRT(P(2,2))
                    ERRY=Y-YH
                    SP33=SQRT(P(3,3))
                    ERRYD=YD-YDH
                    SP44=SQRT(P(4,4))
                    WRITE(9,*)T,X,XH,XD,XDH,Y,YH,YD,YDH
                    WRITE(1,*)T,X,XH,XD,XDH,Y,YH,YD,TDH
                    WRITE(2,*)T,ERRX,SP11,-SP11,ERRXD,SP22,-SP22,
      1                 ERRY,SP33,-SP33,ERRYD,SP44,-SP44
            ENDIF
        END DO
        PAUSE
        CLOSE(1)
        CLOSE(2)
        END

C SUBROUTINE GAUSS IS SHOWN IN LISTING 1.8
C SUBROUTINE MATTRN IS SHOWN IN LISTING 1.3
C SUBROUTINE MATMUL IS SHOWN IN LISTING 1.4
C SUBROUTINE MATADD IS SHOWN IN LISTING 1.1
C SUBROUTINE MATSUB IS SHOWN IN LISTING 1.2
```

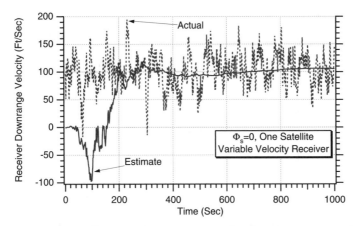

Fig. 11.52   Filter is unable to follow receiver downrange velocity variations when filter does not have process noise.

filter, which usually improved the filter's tracking ability at the expense of more noise transmission. However, we can see from Fig. 11.53 that the addition of process noise has actually ruined the tracking ability of the filter.

Because we know that we can track the receiver with zero process noise and we know that $\Phi_s = 100$ is too much process noise, we dramatically reduced the process noise in the experiment (i.e., PHIS = 0.1). We can see from Fig. 11.54 that reducing the process noise has enabled us to track the receiver but with a very significant lag. These tracking results are still much worse than that of the case in which there was no process noise.

At this point it is of interest to investigate the error in the estimate of the downrange location of the receiver and compare it to the theoretical estimate of

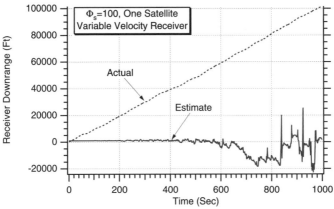

Fig. 11.53   Addition of process noise has ruined the tracking ability of the extended Kalman filter when only one satellite is available.

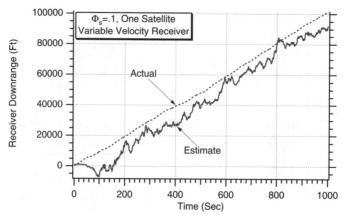

Fig. 11.54   Reducing the process noise has enabled the extended Kalman filter to track the receiver with a very significant lag when only one satellite is available.

the filter's covariance matrix. We can see from Fig. 11.55 that the simulation results are always inside the theoretical bounds of the covariance matrix, which means that the filter appears to be working properly. However, we can also see that the theoretical errors in the estimate of receiver downrange are very large and are increasing with time. Therefore, these results are indicating that it is not possible to track the variable velocity receiver with only a single satellite. We saw that it takes hundreds of seconds for the filter to converge when a single satellite is used for a stationary receiver. Because the time constant of the velocity variations is 5 s (see Fig. 11.36), we can see that the bandwidth of the velocity model is outside the bandwidth of the filter, thus making the task impossible.

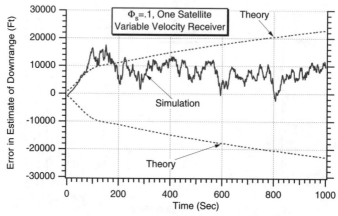

Fig. 11.55   It is not possible to track the variable velocity receiver with only a single satellite.

Summary

In this chapter we have investigated various filtering options, including no filtering at all, for determining the location of a receiver based on noisy range measurements from a satellite pair. The importance of satellite geometry was demonstrated, and it was shown why an extended Kalman filter is better than a simple second-order recursive least-squares filter for this problem. The value of adding process noise to the extended Kalman filter in tracking a receiver traveling along an erratic path also was demonstrated. It also was shown that an extended Kalman filter could be built that could track a stationary or constant velocity receiver based on one satellite only if enough time were available. However, if the receiver velocity was varying, one satellite was not sufficient for maintaining track.

References

[1]Parkinson, B. W., Spilker, J. J., Axelrad, P., and Enge, P., *Global Positioning System Theory and Applications*, Vol. 1, Progress in Astronautics and Aeronautics, AIAA, Reston, VA, 1996, pp. 29–54.

[2]Zarchan, P., *Tactical and Strategic Missile Guidance*, 3rd ed. Progress in Astronautics and Aeronautics, AIAA, Reston, VA, 1998, pp. 61–64.

Chapter 12

Biases

Introduction

I N THE previous chapter we saw how a receiver could be located based on extended Kalman filtering of range measurements to the receiver from either one or two satellites. In those examples of Chapter 11, it was assumed that the satellite location was known perfectly and that the range measurements were only corrupted by zero-mean Gaussian noise. In this chapter we will first see that estimation accuracy is degraded as a result of a possible bias in the range measurement. Two cases will be considered. First, the range biases will be assumed to be caused solely by the satellites themselves. In this case each range measurement from the satellite is assumed to have a bias. Second, the range bias will be assumed to be caused solely by the receiver (i.e., because of a timing error). Various methods for improving performance by estimating the bias will be explored.

Influence of Bias

In the preceding chapter, one of the extended Kalman filters that was developed used range measurements from two satellites to a stationary receiver. In that formulation, which is based on Fig. 12.1, it was assumed that the satellite locations were known perfectly and that the range measurements were only corrupted by zero-mean uncorrelated Gaussian noise.

Listing 12.1 presents the same extended Kalman filter of Chapter 11 in the environment of Fig. 12.1. However, this time the noisy range measurements also have a 10,000-ft bias on them. Highlighted in bold in Listing 12.1 are the changes that had to be made to the preceding extended Kalman-filtering simulation to model the range measurement bias. We can see from Listing 12.1 that the extended Kalman filter is ignorant of the bias on the range measurement. The nominal case, which has range measurement noise with a standard deviation of 300 ft, is set up with favorable satellite geometry in which the initial angle θ between the two range vectors is 36.9 deg.

The nominal case of Listing 12.1 was run in which there was zero-mean uncorrelated Gaussian-range measurement noise with a standard deviation of 300 and 10,000 ft of bias on each range measurement. Figures 12.2 and 12.3 display the actual and estimated receiver downrange and altitude location. We can see the downrange receiver estimation errors are more than 3,000 ft, whereas the altitude

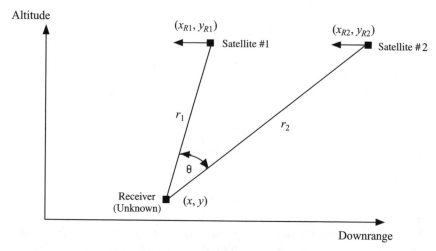

Fig. 12.1 Satellite geometry for filtering experiment.

Listing 12.1 Simulation of extended Kalman filter using biased range measurements from two satellites

```
C THE FIRST THREE STATEMENTS INVOKE THE ABSOFT RANDOM
  NUMBER GENERATOR ON THE MACINTOSH
      GLOBAL DEFINE
              INCLUDE 'quickdraw.inc'
      END
      IMPLICIT REAL*8 (A-H)
      IMPLICIT REAL*8 (O-Z)
      REAL*8 PHI(2,2),P(2,2),M(2,2),PHIP(2,2)PHIPPHIT(2,2),GAIN(2,2)
      REAL*8 Q(2,2),HMAT(2,2)HM(2,2)MHT(2,2)
      REAL*8 PHIT(2,2),RMAT(2,2),HMHTRINV(2,2)
      REAL*8 HMHT(2,2),HT(2,2),KH(2,2),IDN(2,2),IKH(2,2),HMHTR(2,2)
      INTEGER ORDER
      SIGNOISE=300.
      BIAS=10000.
      X=0.
      Y=0.
      XH=1000.
      YH=2000.
      XR1=1000000.
      YR1=20000.*3280.
      XR2=50000000.
      YR2=20000.*3280.
      OPEN(2,STATUS='UNKNOWN',FILE='COVFIL')
      OPEN(1,STATUS='UNKNOWN',FILE='DATFIL')
      ORDER=2
      TS=1.
      TF=100.
      PHIS=0.
```

(continued)

Listing 12.1   (*Continued*)

```
        T=0.
        S=0.
        H=.01
        DO  1000  I=1,ORDER
        DO  1000  J=1,ORDER
                PHI(I,J)=0.
                P(I,J)=0.
                Q(I,J)=0.
                IDN(I,J)=0.
1000 CONTINUE
        IDN(1,1)=1.
        IDN(2,2)=1.
        P(1,1)=1000.**2
        P(2,2)=2000.**2
        RMAT(1,1)=SIGNOISE**2
        RMAT(1,2)=0.
        RMAT(2,1)=0.
        RMAT(2,2)=SIGNOISE**2
        WHILE(T<=TF)
                XR1OLD=XR1
                XR2OLD=XR2
                XR1D=-14600.
                XR2D=-14600.
                XR1=XR1+H*XR1D
                XR2=XR2+H*XR2D
                T=T+H
                XR1D=-14600.
                XR2D=-14600.
                XR1=.5*(XR1OLD+XR1+H*XR1D)
                XR2=.5*(XR2OLD+XR2+H*XR2D)
                S=S+H
                IF(S>=(TS-.00001))THEN
                        S=0.
                        R1H=SQRT((XR1-XH)**2+(YR1-YH)**2)
                        R2H=SORT((XR2-XH)**2+(YR2-YH)**2)
                        HMAT(1,1)=-(XR1-XH)/R1H
                        HMAT(1,2)=-(YR1-YH)/R1H
                        HMAT(2,1)=-(XR2-XH)/R2H
                        HMAT(2,2)=-(YR2-YH)/R2H
                        CALL  MATTRN(HMAT,2,2,HT)
                        PHI(1,1)=1.
                        PHI(2,2)=1.
                        CALL  MATTRN(PHI,ORDER,ORDER,PHIT)
                        CALL  MATMUL(PHI,ORDER,ORDER,P,ORDER,ORDER,
                          PHIP)
                        CALL  MATMUL(PHIP,ORDER,ORDER,PHIT,ORDER,
                          ORDER,PHIPPHIT)
                        CALL  MATADD(PHIPPHIT,ORDER,ORDER,Q,M)
                        CALL  MATMUL(HMAT,2,ORDER,M,ORDER,ORDER,HM)
```

1

                                                                        (*continued*)

Listing 12.1 *(Continued)*

```
                    CALL  MATMUL(HM,2,ORDER,HT,ORDER,2,HMT)
                    CALL  MATADD(HMHT,2,2,RMAT,HMHTR)
                    DET=HMHTR(1,1)*HMHTR(2,2)-HMHTR(1,2)*HMHTR(2,1)
                    HMHTRINV(1,1)=HMHTR(2,2)/DET
                    HMHTRINV(1,2)=-HMHTR(1,2)/DET
                    HMHTRINV(2,1)=-HMHTR(2,1)/DET
                    HMHTRINV(2,2)=HMHTR(1,1)/DET
                    CALL  MATMUL(M,ORDER,ORDER,HT,ORDER,2,MHT)
                    CALL  MATMUL(MHT,ORDER,2,HMHTRINV,2,2,GAIN)
                    CALL  MATMUL(GAIN,ORDER,2,HMAT,2,ORDER,KH)
                    CALL  MATSUB(IDN,ORDER,ORDER,KH,IKH)
                    CALL  MATMUL(IKH,ORDER,ORDER,M,ORDER,
                      ORDER,P)
                    CALL  GAUSS(R1NOISE,SIGNOISE)
                    CALL  GAUSS(R2NOISE,SIGNOISE)
                    R1=SQRT((XR1-X)**2+(YR1-Y)**2)+BIAS
                    R2=SQRT((XR2-X)**2+(YR2-Y)**2)+BIAS
                    RES1=R1+R1NOISE-R1H
                    RES2=R2+R2NOISE-R2H
                    XH=XH+GAIN(1,1)*RES1+GAIN(1,2)*RES2
                    YH=YH+GAIN(2,1)*RES1+GAIN(2,2)*RES2
                    ERRX=X-XH
                    SP11=SQRT(P(1,1))
                    ERRY=Y-YH
                    SP22=SQRT(P(2,2))
                    R1T=SQRT((XR1-X)**2+(YR1-Y)**2)
                    R2T=SQRT((XR2-X)**2+(YR2-Y)**2)
                    TEMP=(XR1-X)*(XR2-X)+(YR1-Y)*(YR2-Y)
                    THET=57.3*ACOS(TEMP/(R1T*R2T))
                    WRITE(9,*)T,X,XH,Y,TH,THET
                    WRITE(1,*)T,X,XH,Y,YH,THET
                    WRITE(2,*)T,ERRX,SP11,-SP11,ERRY,SP22,-SP22
          ENDIF
      END DO
      PAUSE
      CLOSE(1)
      CLOSE(2)
      END
C SUBROUTINE GAUSS IS SHOWN IN LISTING 1.8
C SUBROUTINE MATTRN IS SHOWN IN LISTING 1.3
C SUBROUTINE MATMUL IS SHOWN IN LISTING 1.4
C SUBROUTINE MATADD IS SHOWN IN LISTING 1.1
C SUBROUTINE MATSUB IS SHOWN IN LISTING 1.2
```

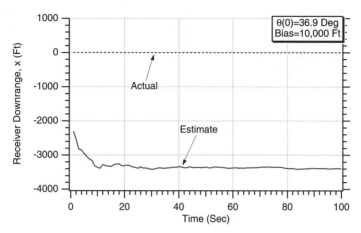

Fig. 12.2   Filter is unable to estimate receiver downrange when range bias is present.

receiver estimation errors are approximately 10,000 ft. These errors are more than an order of magnitude larger than they were in Chapter 11. In other words, the range bias error cannot be handled by the extended Kalman filter.

Estimating Satellite Bias with Known Receiver Location

We saw in the preceding section that a bias in our range measurements can degrade the extended Kalman filter's ability to estimate the downrange and altitude location of a stationary receiver. In this problem the receiver location is, of course, unknown. If the range bias errors are caused by a satellite bias error (either intentional or unintentional),[1] then we can attempt to estimate the bias if the receiver location is known (i.e., surveyed). In other words, if we know the

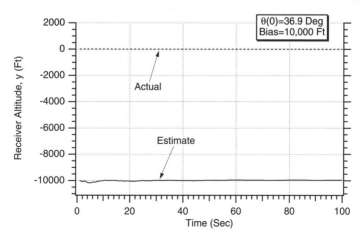

Fig. 12.3   Filter is unable to estimate receiver altitude when range bias is present.

satellite location, we would like to take range measurements to estimate the satellite bias on the noisy range measurements. If the scheme works, then the satellite bias information can be broadcast and taken into account by other receivers whose location is unknown.

Figure 12.4 depicts the scenario in which there is one satellite with coordinates x_{R1}, y_{R1} and one receiver with fixed location x, y. The satellite is moving, but it is assumed that the satellite coordinates are always known. We measure the range from the satellite to the receiver r_1. The range measurements are corrupted by zero-mean uncorrelated Gaussian noise plus the bias that we are attempting to estimate.

From Fig. 12.4 we can see that the actual range from the satellite to the receiver is always given by

$$r_1 = \sqrt{(x_{R1} - x)^2 + (y_{R1} - y)^2}$$

However, the measured range from the satellite to the receiver is

$$r_1^* = \sqrt{(x_{R1} - x)^2 + (y_{R1} - y)^2} + \text{BIAS} + v_{r1}$$

where v_{r1} is the zero-mean uncorrelated Gaussian measurement noise associated with the satellite range measurement and BIAS is the unknown bias caused by the satellite on the range measurement. We would like to estimate the range bias. For this one-satellite problem the discrete noise matrix is a scalar, which is given by

$$\boldsymbol{R}_k = \sigma_{r1}^2$$

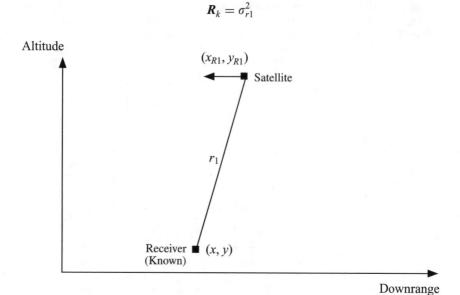

Fig. 12.4   Geometry for attempting to estimate range bias.

where σ_{r1}^2 is the measurement noise variance. In this system the only quantity that must be estimated is the range bias because the satellite and receiver locations are already known. Because we are assuming that the bias is a constant, its derivative must be zero. We can add process noise to the derivative of the bias to play it safe in case the bias varies with time. Therefore, the scalar state-space equation for this problem is simply

$$\dot{\text{BIAS}} = u_s$$

From the preceding equation we can tell that the systems dynamics matrix is zero (i.e., $\mathbf{F} = 0$). Therefore, the fundamental matrix is given by a one-term Taylor-series expansion, which means that the fundamental matrix is the identity matrix. For this scalar problem that means the fundamental matrix is simply unity or

$$\Phi_k = 1$$

The continuous process-noise matrix is also a scalar and can be written by inspection of the state-space equation as

$$Q = \Phi_s$$

where Φ_s is the spectral density of the white process noise u_s.

Recall that the discrete process-noise matrix can be derived from the continuous process-noise matrix according to

$$Q_k = \int_0^{T_s} \Phi(\tau) Q \Phi^T(\tau) \, d\tau$$

Substitution of the appropriate scalars yields

$$Q_k = \int_0^{T_s} 1 * \Phi_s^* 1 \, d\tau = \Phi_s T_s$$

Therefore, we can see that the discrete process noise is simply

$$Q_k = \Phi_s T_s$$

Because the measurement is given by

$$r_1^* = \sqrt{(x_{R1} - x)^2 + (y_{R1} - y)^2} + \text{BIAS} + v_{r1}$$

and BIAS is a state, the linearized measurement equation can be written as

$$\Delta r_1^* = \frac{\partial r_1^*}{\partial \text{BIAS}} \Delta \text{BIAS} + v$$

where the partial derivative can be easily evaluated as

$$\frac{\partial r_1^*}{\partial \text{BIAS}} = 1$$

Therefore, the linearized measurement matrix is also a scalar given by

$$\boldsymbol{H}_k = 1$$

We now have enough information to solve the Riccati equations. As in other examples in extended Kalman filtering, we saw that we do not have to use the linearized equations for the actual filter. Because the fundamental matrix is exact and equal to unity, we can say that the projected estimate of the bias is simply the last estimate of the bias or

$$\overline{\text{BIAS}}_k = \widehat{\text{BIAS}}_{k-1}$$

Therefore, the projected range from the satellite to the receiver is given by

$$\bar{r}_{1_k} = \sqrt{(x_{R1_k} - x_k)^2 + (y_{R1_k} - y_k)^2} + \overline{\text{BIAS}}_k$$

We do not have to use the linearized measurement matrix to calculate the filter residual. Instead, we can say that the residual is simply the measured range minus the projected value of range or

$$\text{Res}_k = r_{1_k}^* - \bar{r}_{1_k}$$

Therefore, the scalar extended Kalman-filtering equation is given by

$$\widehat{\text{BIAS}}_k = \overline{\text{BIAS}}_k + K_{1_k} \text{Res}_k$$

The scalar extended Kalman filter for estimating bias appears in Listing 12.2. The true bias is set to 10,000 ft, and the initial estimate of that bias is set to zero. The initial covariance matrix is chosen to reflect the fact that the initial error in the estimate of the bias is 10,000 ft (i.e., $P = 10{,}000{**}2$). The process noise is nominally set to zero (i.e., PHIS = 0). Because the measurement noise matrix and all of the other equations are scalar, no matrix inverse is required in the Riccati equations.

The nominal case of Listing 12.2 was run in which there was no process noise. Figure 12.5 shows that we can in fact estimate the bias rather quickly if the receiver location is known. Figure 12.6 shows that the single-term errors in the estimate of the bias falls within the theoretical approximately 68% of the time, indicating that the extended Kalman filter appears to be working correctly. We are able to reduce 10,000 ft of uncertainty in the bias (at time zero) to less than 100 ft of bias error in approximately 20 s.

Listing 12.2   Extended Kalman filter to estimate range bias if receiver location is known

```
C THE FIRST THREE STATEMENTS INVOKE THE ABSOFT RANDOM
  NUMBER GENERATOR ON THE MACINTOSH
      GLOBAL DEFINE
              INCLUDE 'quickdraw.inc'
      END
      IMPLICIT REAL*8 (A-H)
      IMPLICIT REAL*8 (O-Z)
      REAL*8 M,MHT,KH,IKH
      INTEGER ORDER
      SIGNOISE=300.
      BIAS=10000.
      BIASH=0.
      X=0.
      Y=0.
      XR1=1000000.
      YR1=20000.*3280.
      OPEN(2,STATUS='UNKNOWN',FILE='COVFIL')
      OPEN(1,STATUS='UNKNOWN',FILE='DATFIL')
      ORDER=1
      TS=1.
      TF=100.
      PHIS=0.
      Q=PHIS*TS
      T=0.
      S=0.
      H=.01
      P=10000.**2
      RMAT=SIGNOISE**2
      WHILE(T<=TF)
              XR1OLD=XR1
              XR1D=-14600.
              XR1=XR1+H*XR1D
              T=T+H
              XR1D=-14600.
              XR1=.5*(XR1OLD+XR1+H*XR1D)
              S=S+H
              IF(S>=(TS-.00001))THEN
                      S=0.
                      BIASB=BIASH
                      R1B=SQRT((XR1-X)**2+(YR1-Y)**2)+BIASB
                      R1H=SQRT((XR1-X)**2+(YR1-Y)**2)+BIASH
                      HMAT=1.
                      M=P+Q
                      HMHT=HMAT*HMAT*M
                      HMHTR=HMHT+RMAT
                      HMHTRINV=1./HMHTR
                      MHT=M*HMAT
                      GAIN=MHT*HMHTRINV
```

(*continued*)

Listing 12.2 (*Continued*)

```
                    KH=GAIN*HMAT
                    IKH=1.-KH
                    P=IKH*M
                    CALL GAUSS(R1NOISE,SIGNOISE)
                    R1=SQRT((XR1-X)**2+(YR1-Y)**2)+BIAS
                    RES1=R1+R1NOISE-R1H
                    BIASH=BIASB+GAIN*RES1
                    ERRBIAS=BIAS-BIASH
                    SP11=SQRT(P)
                    WRITE(9,*)T,BIAS,BIASH
                    WRITE(1,*)T,BIAS,BIASH
                    WRITE(2,*)T,ERRBIAS,SP11,-SP11
         ENDIF
      END DO
      PAUSE
      CLOSE(1)
      CLOSE(2)
      END

C SUBROUTINE GAUSS IS SHOWN IN LISTING 1.8
```

The preceding case was run without process noise, and we saw that the estimates were quite good. Figures 12.7 and 12.8 show that adding process noise in this example simply increases the errors in estimating the bias. Figure 12.8 shows that when the filter has process noise, the errors in the estimate no longer reduce as more measurements are taken. As was the case in many of our examples from other chapters, the addition of process noise causes the errors in the estimate of the bias to approach a steady state.

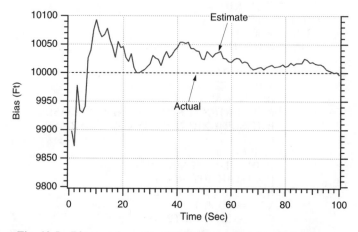

Fig. 12.5   Bias can be estimated if the receiver location is known.

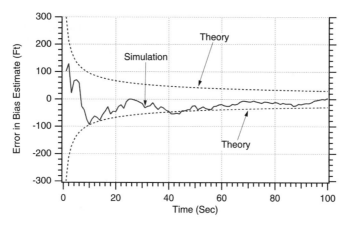

Fig. 12.6 Extended Kalman filter appears to be working properly.

Estimating Receiver Bias with Unknown Receiver Location and Two Satellites

We saw in the preceding section that the range bias caused by the satellite could be estimated with only one satellite if the receiver location were known in advance. In this section we will see if an extended Kalman filter can be used to estimate a range bias caused by the receiver. In this example, we will consider two satellites whose locations are always considered to be known and a receiver whose location is unknown. The purpose of the extended Kalman filter in this case is to not only estimate the bias but also to estimate the receiver location. The geometry for this example is depicted in Fig. 12.9.

Because we are trying to estimate the location of the receiver and the bias caused by the receiver, it seems reasonable to include both the receiver location and the bias states in the proposed extended Kalman filter. For this problem the

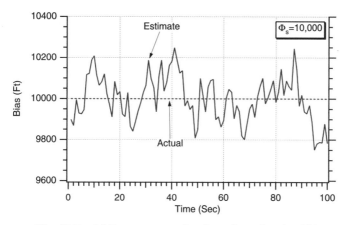

Fig. 12.7 Adding process noise degrades estimate of bias.

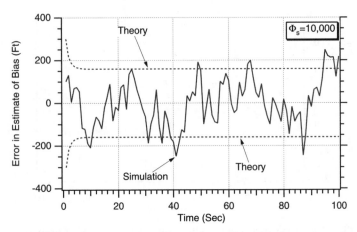

Fig. 12.8 Adding process noise causes errors in estimate of bias to approach a steady state.

receiver is fixed in location. Because the receiver is not moving, the receiver velocity (i.e., derivative of position) must be zero. Because we are assuming that the bias is a constant, its derivative also must be zero. We can add process noise to the derivative of bias to play it safe in case the bias varies with time. Therefore, the equations for this problem, upon which the extended Kalman filter will be based, are given by

$$\dot{x} = 0$$

$$\dot{y} = 0$$

$$\dot{BIAS} = u_s$$

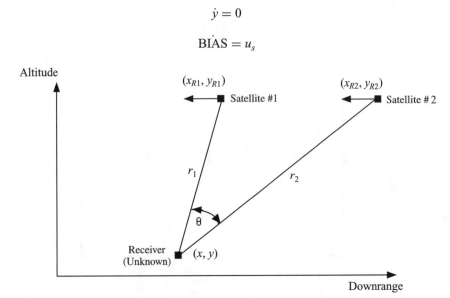

Fig. 12.9 Geometry for estimating receiver location and range bias caused by the receiver based on range measurements from two satellites.

where u_s is white noise with spectral density Φ_s. We can rewrite the preceding three scalar equations in state-space form, yielding

$$\begin{bmatrix} \dot{x} \\ \dot{y} \\ \dot{\text{BIAS}} \end{bmatrix} = \begin{bmatrix} 0 & 0 & 0 \\ 0 & 0 & 0 \\ 0 & 0 & 0 \end{bmatrix} \begin{bmatrix} x \\ y \\ \text{BIAS} \end{bmatrix} + \begin{bmatrix} 0 \\ 0 \\ u_s \end{bmatrix}$$

This means that the systems dynamics matrix is zero or

$$F = \begin{bmatrix} 0 & 0 & 0 \\ 0 & 0 & 0 \\ 0 & 0 & 0 \end{bmatrix}$$

Therefore, only a one-term Taylor series is required to find the fundamental matrix, which means that the fundamental matrix is the identity matrix or

$$\Phi_k = \begin{bmatrix} 1 & 0 & 0 \\ 0 & 1 & 0 \\ 0 & 0 & 1 \end{bmatrix}$$

The continuous process-noise matrix can be written by inspection of the state-space equation as

$$Q = \begin{bmatrix} 0 & 0 & 0 \\ 0 & 0 & 0 \\ 0 & 0 & \Phi_s \end{bmatrix}$$

As was the case in the preceding section, because the fundamental matrix is the identity matrix the discrete process-noise matrix is simply the continuous process-noise matrix multiplied by the sampling time or

$$Q_k = \begin{bmatrix} 0 & 0 & 0 \\ 0 & 0 & 0 \\ 0 & 0 & \Phi_s T_s \end{bmatrix}$$

The measurements for this example are still the ranges from the satellites to the receiver. We can see from Fig. 12.7 that the true or exact ranges from each of the satellites to the receiver are given by

$$r_1 = \sqrt{(x_{R1} - x)^2 + (y_{R1} - y)^2}$$

$$r_2 = \sqrt{(x_{R2} - x)^2 + (y_{R2} - y)^2}$$

whereas the measured ranges from each of the satellites to the receiver are given by

$$r_1^* = \sqrt{(x_{R1} - x)^2 + (y_{R1} - y)^2} + \text{BIAS} + v_{r1}$$

$$r_2^* = \sqrt{(x_{R2} - x)^2 + (y_{R2} - y)^2} + \text{BIAS} + v_{r2}$$

where v_{r1} and v_{r2} are zero-mean, uncorrelated Gaussian measurement noises associated with the range measurements r_1 and r_2 and BIAS is the unknown bias on each of the range measurements. Therefore, we can express the linearized range measurements as functions of the proposed states according to

$$\begin{bmatrix} \Delta r_1^* \\ \Delta r_2^* \end{bmatrix} = \begin{bmatrix} \dfrac{\partial r_1^*}{\partial x} & \dfrac{\partial r_1^*}{\partial y} & \dfrac{\partial r_1^*}{\partial \text{BIAS}} \\ \dfrac{\partial r_2^*}{\partial x} & \dfrac{\partial r_2^*}{\partial y} & \dfrac{\partial r_2^*}{\partial \text{BIAS}} \end{bmatrix} \begin{bmatrix} \Delta x \\ \Delta y \\ \Delta \text{BIAS} \end{bmatrix} + \begin{bmatrix} v_{r1} \\ v_{r2} \end{bmatrix}$$

The discrete measurement noise matrix can be obtained directly from the preceding equation and is given by

$$R_k = \begin{bmatrix} \sigma_{r1}^2 & 0 \\ 0 & \sigma_{r2}^2 \end{bmatrix}$$

where σ_{r1}^2 and σ_{r2}^2 are the measurement noise variances for each of the range measurements. From the measurement equation we can see that the linearized measurement matrix is given by

$$H_k = \begin{bmatrix} \dfrac{\partial r_1^*}{\partial x} & \dfrac{\partial r_1^*}{\partial y} & \dfrac{\partial r_1^*}{\partial \text{BIAS}} \\ \dfrac{\partial r_2^*}{\partial x} & \dfrac{\partial r_2^*}{\partial y} & \dfrac{\partial r_2^*}{\partial \text{BIAS}} \end{bmatrix}$$

where the partial derivatives are evaluated at the projected state estimates. Because the fundamental matrix is the identity matrix, the projected state estimates are simply the last state estimates. From the two range equations we can evaluate each of the partial derivatives as

$$\frac{\partial r_1^*}{\partial x} = 0.5[(x_{R1} - x)^2 + (y_{R1} - y)^2]^{-0.5}2(x_{R1} - x)(-1) = \frac{-(x_{R1} - x)}{r_1}$$

$$\frac{\partial r_1^*}{\partial y} = 0.5[(x_{R1} - x)^2 + (y_{R1} - y)^2]^{-0.5}2(y_{R1} - y)(-1) = \frac{-(y_{R1} - y)}{r_1}$$

$$\frac{\partial r_1^*}{\partial \text{BIAS}} = 1$$

$$\frac{\partial r_2^*}{\partial x} = 0.5[(x_{R1} - x)^2 + (y_{R1} - y)^2]^{-0.5}2(x_{R2} - x)(-1) = \frac{-(x_{R2} - x)}{r_2}$$

$$\frac{\partial r_2^*}{\partial y} = 0.5[(x_{R1} - x)^2 + (y_{R1} - y)^2]^{-0.5}2(y_{R2} - y)(-1) = \frac{-(y_{R2} - y)}{r_2}$$

$$\frac{\partial r_2^*}{\partial \text{BIAS}} = 1$$

and so the linearized measurement matrix, which was already shown to be a 2×3 matrix of partial derivatives, is given by

$$H_k = \begin{bmatrix} \dfrac{-(x_{R1} - x)}{r_1} & \dfrac{-(y_{R1} - y)}{r_1} & 1 \\ \dfrac{-(x_{R2} - x)}{r_2} & \dfrac{-(y_{R2} - y)}{r_2} & 1 \end{bmatrix}$$

We now have enough information to solve the Riccati equations. Again, as in other examples in extended Kalman filtering, we saw that we do not have to use the linearized equations for the actual filter. Because the fundamental matrix is exact and equal to the identity matrix, we can say that the projected estimate of the receiver location and the bias is simply the last estimate or

$$\bar{x}_k = \hat{x}_{k-1}$$

$$\bar{y}_k = \hat{y}_{k-1}$$

$$\overline{\text{BIAS}}_k = \widehat{\text{BIAS}}_{k-1}$$

Therefore, each of the projected ranges from the satellite to the receiver are given by

$$\bar{r}_{1_k} = \sqrt{(x_{R1_k} - \bar{x}_k)^2 + (y_{R1_k} - \bar{y}_k)^2} + \overline{\text{BIAS}}_k$$

$$\bar{r}_{2_k} = \sqrt{(x_{R2_k} - \bar{x}_k)^2 + (y_{R2_k} - \bar{y}_k)^2} + \overline{\text{BIAS}}_k$$

As was already mentioned, we do not have to use the linearized measurement matrix to compute the filter residual. Instead, we can say that each of the residuals is simply the range measurement minus the projected range or

$$\text{Res}_{1_k} = r_{1_k}^* - \bar{r}_{1_k}$$

$$\text{Res}_{2_k} = r_{2_k}^* - \bar{r}_{2_k}$$

Now we can write the extended Kalman-filtering equations as

$$\hat{x}_k = \bar{x}_k + K_{11_k}\text{Res}_{1_k} + K_{12_k}\text{Res}_{2_k}$$

$$\hat{y}_k = \bar{y}_k + K_{21_k}\text{Res}_{1_k} + K_{22_k}\text{Res}_{2_k}$$

$$\widehat{\text{BIAS}}_k = \overline{\text{BIAS}}_k + K_{31_k}\text{Res}_{1_k} + K_{32_k}\text{Res}_{2_k}$$

The scalar extended Kalman filter for estimating bias appears in Listing 12.3. The true bias is set to 10,000 ft, and the initial estimate of that bias is set to zero. The initial covariance matrix is chosen to reflect the fact that in the error the estimate of bias is 10,000 [i.e., $P(3, 3) = 10,000**2$]. The process noise is nominally set to zero (i.e., PHIS = 0). The measurement noise standard deviation is considered to be 300 ft. Each of the range measurements is assumed to be uncorrelated, and so the measurement noise matrix is a two-dimensional square matrix with the off-diagonal terms set to zero and each of the diagonal terms set to $300.^2$

The nominal case of Listing 12.3, in which there was no process noise, was run. We can see from Fig. 12.10 that the new extended Kalman filter is unable to estimate the bias. The actual bias is 10,000 ft, and the filter estimates that the bias is 12,000 ft (up from its initial estimate of zero). The filter estimate of the bias does not change as more measurements are taken. As a consequence of the filter's inability to estimate bias, Figs. 12.11 and 12.12 show that the presence of the bias still prevents the filter from estimating the receiver downrange and altitude. The near-flat line plots for all of the estimates indicate that perhaps bias is unobservable (for more on the subject, see Chapter 14) from range measurements from only two satellites.

Another experiment was conducted in which the process noise was increased from its value of zero to a value of 10,000. We can see from Fig. 12.13 that although adding process noise slightly makes the value of the estimated bias less constant we cannot say that the filter is working to our expectations. It appears

Listing 12.3 **Extended Kalman filter for estimating range bias if receiver location is unknown and range measurements are available from two satellites**

```
C THE FIRST THREE STATEMENTS INVOKE THE ABSOFT RANDOM
  NUMBER GENERATOR ON THE MACINTOSH
    GLOBAL DEFINE
            INCLUDE 'quickdraw.inc'
    END
    IMPLICIT REAL*8 (A-H)
    IMPLICIT REAL*8 (O-Z)
    REAL*8 PHI(3,3),P(3,3),M(3,3),PHIP(3,3),PHIPPHIT(3,3),GAIN(3,2)
    REAL*8 Q(3,3),HMAT(2,3),HM(2,3),MHT(3,2)
    REAL*8 PHIT(3,3),RMAT(2,2),HMHTRINV(2,2)
    REAL*8 HMHT(2,2),HT(3,2)KH(3,3),IDN(3,3),IKH(3,3),HMHTR(2,2)
    INTEGER ORDER
    SIGNOISE=300.
    TF=100.
    X=0.
    Y=0.
    XH=1000.
    YH=2000.
    XR1=1000000.
    YR1=20000.*3280.
    XR2=50000000.
    YR2=20000.*3280.
    BIAS=10000.
    BIASH=0.
    PHIS=0.
C   PHIS=(1.*(BIAS-BIASH)**2)/TF
    OPEN(2,STATUS='UNKNOWN',FILE='COVFIL')
    OPE(1,STATUS='UNKNOWN',FILE='DATFIL')
    ORDER=3
    TS=1.
    T=0.
    S=0.
    H=.01
    DO 1000 I=1,ORDER
    DO 1000 J=1,ORDER
            PHI(I,J)=0.
            P(I,J)=0.
            Q(I,J)=0.
            IDN(I,J)=0.
1000 CONTINUE
    Q(3,3)=PHIS*TS
    IDN(1,1)=1.
    IDN(2,2)=1.
    IDN(3,3)=1.
    P(1,1)=1000.**2
    P(2,2)=2000.**2
    P(3,3)=10000.**2
    RMAT(1,1)=SIGNOISE**2
```

(continued)

Listing 12.3 (*Continued*)

```
       RMAT(1,2)=0.
       RMAT(2,1)=0.
       RMAT(2,2)=SIGNOISE**2
       WHILE(T<=TF)
              XR1OLD=XR1
              XR2OLD=XR2
              XR1D=-14600.
              XR2D=-14600.
              XR1=XR1+H*XR1D
              XR2=XR2+H*XR2D
              T=T+H
              XR1D=-14600.
              XR2D=-14600.
              XR1=.5*(XR1OLD+XR1+H*XR1D)
              XR2=.5*(XR2OLD+XR2+H*XR2D)
              S=S+H
              IF(S>=(TS-.00001))THEN
                     S=0.
                     R1H=SQRT((XR1-XH)**2+(YR1-YH)**2)+BIASH
                     R2H=SQRT((XR2-XH)**2+(YR2-YH)**2)+BIASH
                     HMAT(1,1)=-(XR1-XH)/SQRT((XR1-XH)**2+
1                       (YR1-YH)**2)
                     HMAT(1,2)=-(YR1-YH)/SQRT((XR1-XH)**2+
1                       (YR1-YH)**2)
                     HMAT(1,3)=1.
                     HMAT(2,1)=-(XR2-XH)/SQRT((XR2-XH)**2+
1                       (YR2-YH)**2)
                     HMAT(2,2)=-(YR2-YH)/SQRT((XR2-XH)**2+
1                       (YR2-YH)**2)
                     HMAT(2,3)=1.
                     CALL  MATTRN(HMAT,2,3,HT)
                     PHI(1,1)=1.
                     PHI(2,2)=1.
                     PHI(3,3)=1.
                     CALL  MATTRN(PHI,ORDER,ORDER,PHIT)
                     CALL  MATMUL(PHI,ORDER,ORDER,P,ORDER,ORDER,
                       PHIP)
                     CALL  MATMUL(PHIP,ORDER,ORDER,PHIT,ORDER,
1                       ORDER,PHIPPHIT)
                     CALL  MATADD(PHIPPHIT,ORDER,ORDER,Q,M)
                     CALL  MATMUL(HMAT,2,ORDER,M,ORDER,ORDER,HM)
                     CALL  MATMUL(HM,2,ORDER,HT,ORDER,2,HMHT)
                     CALL  MATADD9HMHT,2,2,RMAT,HMHTR)
                     DET=HMHTR(1,1)*HMHTR(2,2)-HMHTR(1,2)*HMHTR(2,1)
                       HMHTRINV(1,1)=HMHTR(2,2)/DET
                       HMHTRINV(1,2)=-HMHTR(1,2)/DET
                       HMHTRINV(2,1)=-HMHTR(2,1)/DET
                       HMHTRINV(2,2)=HMHTR(1,1)/DET
                     CALL  MATMUL(M,ORDER,ORDER,HT,ORDER,2,MHT)
                     CALL  MATMUL(MHT,ORDER,2,HMHTRINV,2,2,GAIN)
```
(*continued*)

Listing 12.3   (*Continued*)

```
                    CALL  MATMUL(GAIN,ORDER,2,HMAT,2,ORDER,KH)
                    CALL  MATSUB(IDN,ORDER,ORDER,KH,IKH)
                    CALL  MATMUL(IKH,ORDER,ORDER,M,ORDER,
                      ORDER,P)
                    CALL  GAUSS(R1NOISE,SIGNOISE)
                    CALL  GAUSS(R2NOISE,SIGNOISE)
                    R1=SQRT((XR1-X)**2+YR1-Y)**2)
                    R2=SQRT((XR2-X)**2+(YR2-Y)**2)
                    RES1=R1+R1NOISE+BIAS-R1H
                    RES2=R2+R2NOISE+BIAS-R2H
                    XH=XH+GAIN(1,1)*RES1+GAIN(1,2)*RES2
                    YH=YH+GAIN(2,1)*RES1+GAIN(2,2)*RES2
                    BIASH=BIASH+GAIN(3,1)*RES1+GAIN(3,2)*RES2
                    ERRX=X-XH
                    SP11=SQRT(P(1,1))
                    ERRY=Y-YH
                    SP22=SQRT(P(2,2))
                    ERRBIAS=BIAS-BIASH
                    SP33=SQRT(P(3,3))
                    R1T=SQRT((XR1-X)**2+(YR1-Y)**2)
                    R2T=SQRT((XR2-X)**2+(YR2-Y)**2)
                    TEMP=(XR1-X)*(XR2-X)+(YR1-Y)*(YR2-Y)
                    THET=57.3*ACOS(TEMP/(R1T*R2T))
                    WRITE(9,*)T,X,XH,Y,YH,BIAS,BIASH,THET
                    WRITE(1,*)T,X,XH,Y,YH,BIAS,BIASH,THET
                    WRITE(2,*)T,ERRX,SP11,-SP11,ERRY,SP22,-SP22,
1                     ERRBIAS,SP33,-SP33
          ENDIF
      END DO
      PAUSE
      CLOSE(1)
      CLOSE(2)
      END


C SUBROUTINE GAUSS IS SHOWN IN LISTING 1.8
C SUBROUTINE MATTRN IS SHOWN IN LISTING 1.3
C SUBROUTINE MATMUL IS SHOWN IN LISTING 1.4
C SUBROUTINE MATADD IS SHOWN IN LISTING 1.1
C SUBROUTINE MATSUB IS SHOWN IN LISTING 1.2
```

that the extended Kalman filter with range measurements from two satellites is not suitable for estimating the range bias when the receiver location is unknown.

Estimating Receiver Bias with Unknown Receiver Location and Three Satellites

We saw in the preceding section that the receiver range bias could not be estimated with only two satellites if the receiver location were unknown. In this section we will see if an extended Kalman filter can be used with three satellites

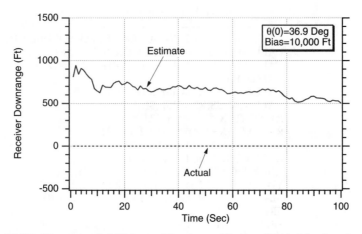

Fig. 12.10 **New extended Kalman filter is unable to estimate bias based on range measurements from two satellites.**

and a receiver whose location is unknown. As was the case in the preceding section, the purpose of the extended Kalman filter is not only to estimate the bias but also to estimate the receiver location. The geometry for the three satellite examples is depicted in Fig. 12.14.

Because we are trying to estimate the location of the receiver and the bias, it still seems reasonable to make both the receiver location and the bias states in the proposed extended Kalman filter. In addition, because for this problem the receiver is fixed in location, the receiver velocity (i.e., derivative of position) must be zero. Also, because we are assuming that the bias is a constant, its

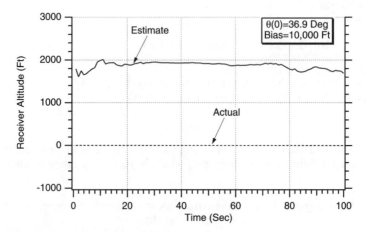

Fig. 12.11 **New extended Kalman filter is unable to estimate receiver downrange based on range measurements from two satellites.**

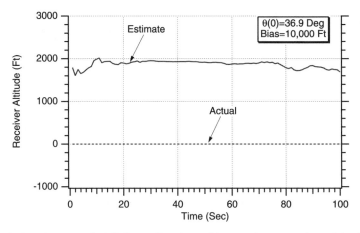

Fig. 12.12 New extended Kalman filter is unable to estimate receiver altitude based on range measurements from two satellites.

derivative must also be zero. We can add process noise to the derivative of the bias to play it safe in case the bias varies with time. Therefore, the equations describing the real world for this problem, upon which the extended Kalman filter will be based, are still the same as those for the preceding section and are given by

$$\dot{x} = 0$$

$$\dot{y} = 0$$

$$\dot{\text{BIAS}} = u_s$$

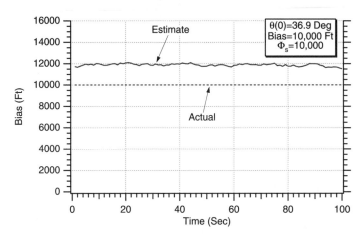

Fig. 12.13 New extended Kalman filter is unable to estimate bias based on range measurements from two satellites.

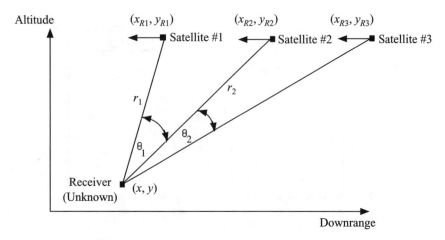

Fig. 12.14 Geometry for estimating receiver location and range bias based on range measurements from three satellites.

where u_s is white noise with spectral density Φ_s. We can rewrite the preceding three scalar equations in state-space form, yielding

$$
\begin{bmatrix} \dot{x} \\ \dot{y} \\ \text{BIAS} \end{bmatrix} = \begin{bmatrix} 0 & 0 & 0 \\ 0 & 0 & 0 \\ 0 & 0 & 0 \end{bmatrix} \begin{bmatrix} x \\ y \\ \text{BIAS} \end{bmatrix} + \begin{bmatrix} 0 \\ 0 \\ u_s \end{bmatrix}
$$

As was the case in the preceding section, this means that the systems dynamics matrix is zero or

$$
F = \begin{bmatrix} 0 & 0 & 0 \\ 0 & 0 & 0 \\ 0 & 0 & 0 \end{bmatrix}
$$

Therefore, only a one-term Taylor series is required to find the fundamental matrix, which means that the fundamental matrix is the identity matrix or

$$
\Phi_k = \begin{bmatrix} 1 & 0 & 0 \\ 0 & 1 & 0 \\ 0 & 0 & 1 \end{bmatrix}
$$

The continuous process-noise matrix can be written by inspection of the state-space equation as

$$
Q = \begin{bmatrix} 0 & 0 & 0 \\ 0 & 0 & 0 \\ 0 & 0 & \Phi_s \end{bmatrix}
$$

Because the fundamental matrix is the identity matrix, the discrete process-noise matrix is simply the continuous process-noise matrix multiplied by the sampling time or

$$Q_k = \begin{bmatrix} 0 & 0 & 0 \\ 0 & 0 & 0 \\ 0 & 0 & \Phi_s T_s \end{bmatrix}$$

The measurements for this example are still the ranges from the satellites to the receiver. However, this time, because of the extra satellite, we have a third range r_3. We can see from Fig. 12.7 that the true or exact ranges from each of the satellites to the receiver are given by

$$r_1 = \sqrt{(x_{R1} - x)^2 + (y_{R1} - y)^2}$$

$$r_2 = \sqrt{(x_{R2} - x)^2 + (y_{R2} - y)^2}$$

$$r_3 = \sqrt{(x_{R3} - x)^2 + (y_{R3} - y)^2}$$

However, the measured ranges from each of the satellites to the receiver are

$$r_1^* = \sqrt{(x_{R1} - x)^2 + (y_{R1} - y)^2} + \text{BIAS} + v_{r1}$$

$$r_2^* = \sqrt{(x_{R2} - x)^2 + (y_{R2} - y)^2} + \text{BIAS} + v_{r2}$$

$$r_3^* = \sqrt{(x_{R3} - x)^2 + (y_{R3} - y)^2} + \text{BIAS} + v_{r3}$$

where v_{r1}, v_{r2}, and v_{r3} are all zero-mean, uncorrelated Gaussian measurement noises associated with the range measurements r_1^*, r_2^*, and r_3^*, whereas BIAS is the unknown bias on each of the three range measurements. Therefore, we can develop a linearized range measurement equation by expressing the range measurements as functions of the proposed states according to

$$\begin{bmatrix} \Delta r_1^* \\ \Delta r_2^* \\ \Delta r_3^* \end{bmatrix} = \begin{bmatrix} \dfrac{\partial r_1^*}{\partial x} & \dfrac{\partial r_1^*}{\partial y} & \dfrac{\partial r_1^*}{\partial \text{BIAS}} \\ \dfrac{\partial r_2^*}{\partial x} & \dfrac{\partial r_2^*}{\partial y} & \dfrac{\partial r_2^*}{\partial \text{BIAS}} \\ \dfrac{\partial r_3^*}{\partial x} & \dfrac{\partial r_3^*}{\partial y} & \dfrac{\partial r_3^*}{\partial \text{BIAS}} \end{bmatrix} \begin{bmatrix} \Delta x \\ \Delta y \\ \Delta \text{BIAS} \end{bmatrix} + \begin{bmatrix} v_{r1} \\ v_{r2} \\ v_{r3} \end{bmatrix}$$

The discrete measurement noise matrix can be obtained directly from the preceding equation and is given by

$$R_k = \begin{bmatrix} \sigma_{r1}^2 & 0 & 0 \\ 0 & \sigma_{r2}^2 & 0 \\ 0 & 0 & \sigma_{r3}^2 \end{bmatrix}$$

where σ_{r1}^2, σ_{r2}^2, and σ_{r3}^2 are the measurement noise variances for each of the range measurements. From the linearized measurement equation we can see that the linearized measurement matrix is given by

$$H_k = \begin{bmatrix} \dfrac{\partial r_1^*}{\partial x} & \dfrac{\partial r_1^*}{\partial y} & \dfrac{\partial r_1^*}{\partial \text{BIAS}} \\[2mm] \dfrac{\partial r_2^*}{\partial x} & \dfrac{\partial r_2^*}{\partial y} & \dfrac{\partial r_2^*}{\partial \text{BIAS}} \\[2mm] \dfrac{\partial r_3^*}{\partial x} & \dfrac{\partial r_3^*}{\partial y} & \dfrac{\partial r_3^*}{\partial \text{BIAS}} \end{bmatrix}$$

where the partial derivatives are evaluated at the projected state estimates. Because the fundamental matrix is exact and equal to the identity matrix, we can say that the state projections are simply the last state estimates. From the three range equations we can evaluate each of the partial derivatives as

$$\frac{\partial r_1^*}{\partial x} = 0.5[(x_{R1} - x)^2 + (y_{R1} - y)^2]^{-0.5}2(x_{R1} - x)(-1) = \frac{-(x_{R1} - x)}{r_1}$$

$$\frac{\partial r_1^*}{\partial y} = 0.5[(x_{R1} - x)^2 + (y_{R1} - y)^2]^{-0.5}2(y_{R1} - y)(-1) = \frac{-(y_{R1} - y)}{r_1}$$

$$\frac{\partial r_1^*}{\partial \text{BIAS}} = 1$$

$$\frac{\partial r_2^*}{\partial x} = 0.5[(x_{R2} - x)^2 + (y_{R2} - y)^2]^{-0.5}2(x_{R2} - x)(-1) = \frac{-(x_{R2} - x)}{r_2}$$

$$\frac{\partial r_2^*}{\partial y} = 0.5[(x_{R2} - x)^2 + (y_{R2} - y)^2]^{-0.5}2(y_{R2} - y)(-1) = \frac{-(y_{R2} - y)}{r_2}$$

$$\frac{\partial r_2^*}{\partial \text{BIAS}} = 1$$

$$\frac{\partial r_3^*}{\partial x} = 0.5[(x_{R3} - x)^2 + (y_{R3} - y)^2]^{-0.5}2(x_{R3} - x)(-1) = \frac{-(x_{R3} - x)}{r_3}$$

$$\frac{\partial r_3^*}{\partial y} = 0.5[(x_{R3} - x)^2 + (y_{R3} - y)^2]^{-0.5}2(y_{R3} - y)(-1) = \frac{-(y_{R3} - y)}{r_3}$$

$$\frac{\partial r_3^*}{\partial \text{BIAS}} = 1$$

Therefore, the linearized measurement matrix is now a 3×3 matrix of partial derivatives and is given by

$$
H_k = \begin{bmatrix}
\dfrac{-(x_{R1} - x)}{r_1} & \dfrac{-(y_{R1} - y)}{r_1} & 1 \\[2ex]
\dfrac{-(x_{R2} - x)}{r_2} & \dfrac{-(y_{R2} - y)}{r_2} & 1 \\[2ex]
\dfrac{-(x_{R3} - x)}{r_3} & \dfrac{-(y_{R2} - y)}{r_2} & 1
\end{bmatrix}
$$

We now have enough information to solve the Riccati equations. As in other examples in extended Kalman filtering, we saw that we do not have to use the linearized equations for the actual filter. Because the fundamental matrix is exact and equal to the identity matrix, we can say that the projected estimate of the receiver location and bias is simply the last estimate or

$$
\bar{x}_k = \hat{x}_{k-1}
$$

$$
\bar{y}_k = \hat{y}_{k-1}
$$

$$
\overline{\text{BIAS}}_k = \widehat{\text{BIAS}}_{k-1}
$$

Therefore, each of the projected ranges from the satellite to the receiver is given by

$$
\bar{r}_{1_k} = \sqrt{(x_{R1_k} - \bar{x}_k)^2 + (y_{R1_k} - \bar{y}_k)^2} + \overline{\text{BIAS}}_k
$$

$$
\bar{r}_{2_k} = \sqrt{(x_{R2_k} - \bar{x}_k)^2 + (y_{R2_k} - \bar{y}_k)^2} + \overline{\text{BIAS}}_k
$$

$$
\bar{r}_{3_k} = \sqrt{(x_{R3_k} - \bar{x}_k)^2 + (y_{R3_k} - \bar{y}_k)^2} + \overline{\text{BIAS}}_k
$$

We do not have to use the linearized measurement matrix to compute the filter residuals. Instead, each of the residuals is simply the range measurements minus the projected range or

$$
\text{Res}_{1_k} = r_{1_k}^* - \bar{r}_{1_k}
$$

$$
\text{Res}_{2_k} = r_{2_k}^* - \bar{r}_{2_k}
$$

$$
\text{Res}_{3_k} = r_{3_k}^* - \bar{r}_{3_k}
$$

and, therefore, the extended Kalman-filtering equations are given by

$$\hat{x}_k = \bar{x}_k + K_{11_k}\text{Res}_{1_k} + K_{12_k}\text{Res}_{2_k} + K_{13_k}\text{Res}_{3_k}$$

$$\hat{y}_k = \bar{y}_k + K_{21_k}\text{Res}_{1_k} + K_{22_k}\text{Res}_{2_k} + K_{23_k}\text{Res}_{3_k}$$

$$\widehat{\text{BIAS}}_k = \overline{\text{BIAS}}_k + K_{31_k}\text{Res}_{1_k} + K_{32_k}\text{Res}_{2_k} + K_{33_k}\text{Res}_{3_k}$$

The scalar extended Kalman filter for estimating bias based on range measurements from three satellites appears in Listing 12.4. As was the case in the preceding sections, the true bias is set to 10,000 ft, and the initial estimate of that bias is set to zero. The third diagonal element of the initial covariance matrix was chosen to reflect the fact that the error in the estimate of bias is 10,000 [i.e., $P(3,3) = 10,000**2$], whereas the other two diagonal elements reflect anticipated errors in the estimates of the initial receiver downrange and altitude. The process noise is nominally set to zero (i.e., PHIS = 0). The measurement noise standard deviation for each of the range measurements from the three satellites is considered to be 300 ft. Each of the range measurements is assumed to be uncorrelated, and so the measurement noise matrix is a three-dimensional square matrix with the off-diagonal terms set to zero and each of the diagonal terms set to 300^2. In the Riccati equations we will now have to take the inverse of a three-dimensional square matrix. There is an exact formula for this inverse that was presented in Chapter 1, and that formula appears in the subroutine MTINV in Listing 12.4.

The nominal case of Listing 12.4 was run in which the first two satellites were still separated by an initial angle of 36.9 deg, whereas the third satellite was separated initially from the second satellite by 49.8 deg. We can see from Fig. 12.15 that we are now able to estimate the bias with the new extended Kalman filter using range measurements from all three satellites. Figures 12.16 and 12.17 show that the accurate estimate of bias leads to accurate estimates of the downrange and altitude location of the receiver.

To ensure that the filter is working properly, errors in the estimates of downrange, altitude, and bias, along with the theoretical predictions from the filter's covariance matrix (i.e., square root of appropriate diagonal element), are displayed in Figs. 12.18–12.20. Because the single-run errors in the estimates are within the theoretical bounds, we can say that the filter appears to be working properly. However, if we compare the errors in the estimates of downrange and altitude of Figs. 12.18 and 12.19 to the case in which there was no bias at all (i.e., Figs. 11.16 and 11.17), we can see that the process of estimating the bias led to somewhat larger estimation errors in downrange and altitude. Of course, these errors are much smaller than if we had not estimated the bias at all (see Figs. 12.2 and 12.3).

In the preceding example we had good estimates because the initial satellite-receiver geometry was good [i.e., large angles for $\theta_1(0)$ and $\theta_2(0)$]. As $\theta_2(0)$ approaches zero, we essentially have a two-satellite problem, and we already know that the bias cannot be estimated successfully. Figure 12.21 shows that when θ_2 reduces from 49.8 (XR3 = 1,000,000,000) to 9.8 deg

Listing 12.4. **New extended Kalman filter for estimating receiver location and range bias based on range measurements from three satellites**

```
C THE FIRST THREE STATEMENTS INVOKE THE ABSOFT RANDOM
  NUMBER GENERATOR ON THE MACINTOSH
      GLOBAL DEFINE
              INCLUDE 'quickdraw.inc'
      END
      IMPLICIT REAL*8 (A-H)
      IMPLICIT REAL*8 (O-Z)
      REAL*8 PHI(3,3),P(3,3),M(3,3),PHIP(3,3),PHIPPHIT(3,3),GAIN(3,3)
      REAL*8 Q(3,3),HMAT(3,3),HM(3,3),MHT(3,3)
      REAL*8 PHIT(3,3),RMAT(3,3),HMHTRINV(3,3)
      REAL*8 HMHT(3,3),HT(3,3)KH(3,3),IDN(3,3),IKH(3,3),HMHTR(3,3)
      INTEGER ORDER
      TF=100.
      SIGNOISE=300.
      X=0.
      Y=0.
      XH=1000.
      YH=2000.
      XR1=1000000.
      YR1=20000.*3280.
      XR2=50000000.
      YR2=20000.*3280.
      XR3=1000000000.
      YR3=20000.*3280.
      BIAS=10000.
      BIASH=0.
      PHIS=0.
      OPEN(2,STATUS='UNKNOWN',FILE='COVFIL')
      OPEN(1,STATUS='UNKNOWN',FILE='DATFIL')
      ORDER=3
      TS=1.
      T=0.
      S=0.
      H=.01
      DO 1000 I=1,ORDER
      DO 1000 J=1,ORDER
              PHI(I,J)=0.
              P(I,J)=0.
              Q(I,J)=0.
              IDN(I,J)=0.
1000      CONTINUE
      DO 1001 I=1,3
      DO 1001 J=1,3
              RMAT(I,J0=0.
1001  CONTINUE
      Q(3,3)=PHIS*TS
      IDN(1,1)=1.
      IDN(2,2)=1.
```

(*continued*)

Listing 12.4 (*Continued*)

```
        IDN(3,3)=1.
        IDN(3,3)=1.
        P(1,1)=1000.**2
        P(2,2)=2000.**2
        P(3,3)=10000.**2
        RMAT(1,1)=SIGNOISE**2
        RMAT(2,2)=SIGNOISE**2
        RMAT(3,3)=SIGNOISE**2
        WHILE(T<=TF)
                XR1OLD=XR1
                XR2OLD=XR2
                XR3OLD=XR3
                XR1D=-14600.
                XR2D=-14600.
                XR3D=-14600.
                XR1=XR1+H*XR1D
                XR2=XR2+H*XR2D
                XR3=XR3+H*XR3D
                T+T+H
                XR1D=-14600.
                XR2D=-14600
                XR3D=-14600.
                XR1=.5*(XR1OLD+XR1+H*XR1D)
                XR2=.5*(XR2OLD+XR2+H*XR2D)
                XR3-.5*(XR3OLD+XR3+H*XR3D)
                S=S+H
                IF(S>=(TS-.00001))THEN
                        S=0.
                        R1H=SQRT((XR1-XH)**2+(YR1-YH)**2)+BIASH
                        R2H=SQRT((XR2-XH)**2+(YR2-YH)**2)+BIASH
                        R3H+SQRT((XR3-XH)**2+(YR3-YH)**2)+BIASH
                        HMAT(1,1)+-(XR1-XH)/SQRT((XR1-XH)**2+
1                           (YR1-YH)**2)
                        HMAT(1,2)=-(YR1-YH)/SQRT((XR1-XH)**2+
1                           (YR1-YH)**2)
                        HMAT(1,3)=1.
                        HMAT(2,1)=-(XR2-XH)SQRT((XR2-XH)**2+
1                           (YR2-YH)**2)
                        HMAT(2,2)=-(YR2-YH)/SQRT((XR2-XH)**2+
1                           (YR2-YH)**2)
                        HMAT(2,3)=1.
                        HMAT(3,1)=-(XR3-XH)/SQRT((XR3-XH)**2+
1                           (YR3-YH)**2)
                        HMAT(3,2)=-(YR3-YH)/SQRT((XR3-XH)**2+
1                           (YR3-YH)**2)
                        HMAT(3,3)=1.
                        CALL  MATTRN(HMAT,3,3,HT)
                        PHI(1,1)=1.
                        PHI(2,2)=1.
                        PHI(3,3)=1.
```

(*continued*)

Listing 12.4 (*Continued*)

```
         CALL  MATTRN(PHI,ORDER,ORDER,PHIT)
         CALL  MATMUL(PHI,ORDER,ORDER,P,ORDER,ORDER,
         PHIP)
         CALL  MATMUL(PHIP,ORDER,ORDER,PHIT,ORDER,
1        ORDER,PHIPPHIT)
         CALL  MATADD(PHIPPHIT,ORDER,ORDER,Q,M)
         CALL  MATMUL(HMAT,3,ORDER,M,ORDER,ORDER,HM)
         CALL  MATMUL(HM,3,ORDER,HT,ORDER,3,HMHT)
         CALL  MATADD(HMHT,3,3,RMAT,HMHTR)
         CALL  MATINV(HMHTR,3,HMHTRINV)
         CALL  MATMUL(M,ORDER,ORDER,HT,ORDER,3,MHT)
         CALL  MATMUL(MHT,ORDER,3,HMHTRINV,3,3,GAIN)
         CALL  MATMUL(GAIN,ORDER,3,HMAT,3,ORDER,KH)
         CALL  MATSUB(IDN,ORDER,ORDER,KH,IKH)
         CALL  MATMUL(IKH,ORDER,ORDER,M,ORDER,
         ORDER,P)
         CALL  GAUSS(R1NOISE,SIGNOISE)
         CALL  GAUSS(R2NOISE,SIGNOISE)
         CALL  GAUSS(R3NOISE,SIGNOISE)
         R1=SQRT((XR1-X)**2+(YR1-Y)**2)
         R2=SQRT((XR2-X)2+(YR2-Y)**2)
         R3=SQRT((XR3-X)**2+(YR3-Y)**2)
         RES1=R1+R1NOISE+BIAS-R1H
         RES2=R2+R2NOISE+BIAS-R2H
         RES3=R3+R3NOISE+BIAS-R3H
         XH=XH+GAIN(1,1)*RES1+GAIN(1,2)*RES2+GAIN(1,3)
1        *RES3
         YH=YH+GAIN(2,1)*RES1+GAIN(2,2)*RES2+GAIN(2,3)
1        *RES3
         BIASH=BIASH+GAIN(3,1)*RES1+GAIN(3,2)*RES2
1        +GAIN(3,3)*RES3
         ERRX+X-XH
         SP11=SQRT(P(1,1))
         ERRY=Y-YH
         SP22=SQRT(P(2,2))
         ERRBIAS+BIAS-BIASH
         SP33=SQRT(P(3,3))
         R1T=SQRT((XR1-X)**2+(YR1-Y)**2)
         R2T=SQRT((XR2-X)**2+(YR2-Y)**2)
         TEMP=(XR1-X)*(XR2-X)+(YR1-Y)*(YR2-Y)
         THET1=57.3*ACOS(TEMP/(R1T*RT2))
         R3T=SQRT((XR3-X)**2+(YR3-Y)**2)
         TEMP3=(XR2-X)*(XR3-X)+(YR2-Y)*(YR3-Y)
         THET2=57.3*ACOS(TEMP3/(RT2*R3T))
         WRITE(9,*)T,X,XH,Y,YH,BIAS,BIASH,THET1,THET2
         WRITE(1,*)T,X,XH,Y,YH,BIAS,BIASH,THET1,THET2
         WRITE(2,*)T,ERRX,SP11,-SP11,ERRY,SP22,-SP22,
1        ERRBIAS,SP33,-SP33
      ENDIF
```

(*continued*)

Listing 12.4 *(Continued)*

```
END  DO
PAUSE
CLOSE(1)
CLOSE(2)
END
SUBROUTINE  MTINV(Q,N,AINV)
IMPLICIT  REAL*8      (A-H,O-Z)
REAL*8  I,Q(3,3),AINV(3,3)
A=Q(1,1)
B=Q(1,2)
C=Q(1,3)
D=Q(2,1)
E=Q(2,2)
F=Q(2,3)
G=Q(3,1)
H=Q(3,2)
I=Q(3,3)
DET=A*E*I+B*F*G+C*D*H-C*E*G-B*D*I-A*F*H
AINV(1,1)=(E*I-F*H)/DET
AINV(1,2)=(C*H-B*I)/DET
AINV(1,3)=(B*F-E*C)/DET
AINV(2,1)=(G*F-D*I)/DET
AINV(2,2)=(A*I-G*C)/DET
AINV(2,3)=(D*C-A*F)/DET
AINV(3,1)=(D*H-G*E)/DET
AINV(3,2)=(G*B-A*H)/DET
AINV(3,3)=(A*E-B*D)/DET
RETURN
END


C SUBROUTINE GAUSS IS SHOWN IN LISTING 1.8
C SUBROUTINE MATTRN IS SHOWN IN LISTING 1.3
C SUBROUTINE MATMUL IS SHOWN IN LISTING 1.4
C SUBROUTINE MATADD IS SHOWN IN LISTING 1.1
C SUBROUTINE MATSUB IS SHOWN IN LISTING 1.2
```

(XR3 = 70,000,000) the error in the estimate of bias increases from 200 to 1000 ft after 100 s of filtering.

Summary

In this chapter we have shown two ways of estimating the range bias for a two-dimensional problem involving satellites and a receiver. We showed that if the satellite were causing the bias in the range measurement we could estimate the bias by taking range measurements from one satellite to a receiver whose location was known precisely. A simple one-state extended Kalman filter yielded excellent bias estimates in this application. We also showed that if the receiver were causing the bias and the receiver location were unknown, three satellites were required in

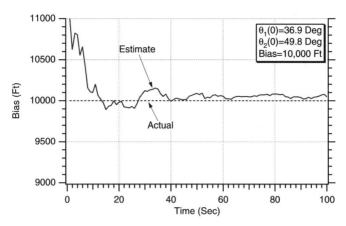

Fig. 12.15   Three satellite range measurements enable new extended Kalman filter to estimate bias.

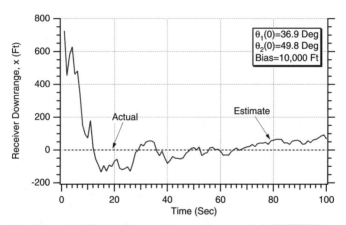

Fig. 12.16   Three satellite range measurements enable new extended Kalman filter to estimate receiver downrange.

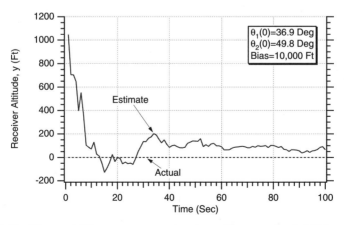

Fig. 12.17   Three satellite range measurements enable new extended Kalman filter to estimate receiver altitude.

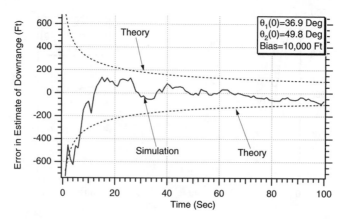

Fig. 12.18 New extended Kalman filter appears to be yielding correct estimates of receiver downrange.

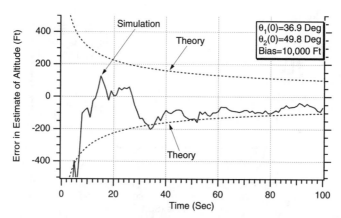

Fig. 12.19 New extended Kalman filter appears to be yielding correct estimates of receiver altitude.

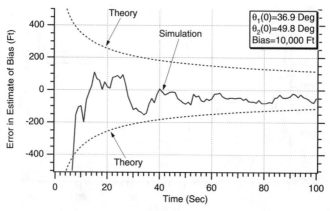

Fig. 12.20 New extended Kalman filter appears to be yielding correct estimates of bias.

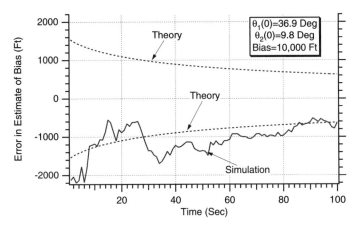

Fig. 12.21   Errors in the estimate of bias get larger as the angle between the second and third satellite decreases.

the two-dimensional problem for an extended Kalman filter to estimate the bias and receiver location. A three-state extended Kalman filter yielded excellent bias and receiver location estimates in this application. However, if only two satellites were available in the two-dimensional problem, neither good bias estimates nor good receiver location estimates could be obtained.

Reference

[1]Parkinson, B. W., Spilker, J. J., Axelrad, P., and Enge, P., *Global Positioning System: Theory and Applications*, Vol. 1, Progress in Astronautics and Aeronautics, AIAA, Reston, VA, 1996, pp. 29–54.

Chapter 13

Linearized Kalman Filtering

Introduction

SO FAR we have seen how linear Kalman filters are designed and how they perform when the real world can be described by linear differential equations expressed in state-space form and when the measurements are linear functions of the states. In addition, we have seen how extended Kalman filtering can be used when the real world is either described by nonlinear differential equations or measurements are not linear functions of the filter states. With both the linear and extended Kalman filters, no a priori assumption was made concerning additional information that might be available. The only information required to start the extended Kalman filter was the initial state estimates and an initial covariance matrix.

In some problems, such as those involving satellite tracking, a great deal of additional a priori information exists. For example, in satellite tracking we know in advance the approximate orbit of the satellite. In principle, this type of information should make the overall filtering problem easier. In this chapter two examples will be presented showing how a linearized Kalman filter can be built if nominal trajectory information exists for a nonlinear problem. In addition, we will also see what happens to the performance of the linearized Kalman filter if the nominal trajectory information is not accurate.

Theoretical Equations[1]

To apply linearized Kalman filtering techniques, it is first necessary to describe the real world by a set of nonlinear differential equations. We have already seen from the chapters on extended Kalman filtering that these equations also can be expressed in nonlinear state-space form as a set of first-order nonlinear differential equations or

$$\dot{x} = f(x) + w$$

where x is a vector of the system states, $f(x)$ is a nonlinear function of those states, and w is a random zero-mean process. The process-noise matrix describing the random process w for the preceding model is giving by

$$Q = E(ww^T)$$

The measurement equation, required for the application of a linearized Kalman filter, is considered to be a nonlinear function of the states according to

$$z = h(x) + v$$

where v is the measurement noise, which is considered to be a zero-mean uncorrelated random process described by the measurement noise matrix R, which is defined as

$$R = E(vv^T)$$

If a nominal trajectory for this nonlinear problem exists, then we can say that there is also an additional set of nonlinear differential equations. We can write the nonlinear differential equations describing the trajectory in compact form as

$$\dot{x}_{NOM} = f(x_{NOM})$$

where x_{NOM} represents the states of the nominal trajectory. We can now develop an error equation that is the difference between the states of the actual trajectory x and the states of the nominal trajectory x_{NOM} and is given by

$$\Delta x = x - x_{NOM}$$

where Δx represent error states. We can then write an approximate linearized differential equation for the error states (i.e., difference between actual states and nominal trajectory states) as

$$\Delta \dot{x} = \left. \frac{\partial f(x)}{\partial x} \right|_{x=x_{NOM}} \Delta x + w$$

The preceding equation is of the same form as the extended Kalman-filtering state-space equation, except we are now working with error states and the partial derivative matrix is evaluated on the nominal trajectory, rather than at the current state estimate.

In a similar way we can also develop an error equation for the measurement. The error measurement is the difference between the actual measurement and the nominal measurement or

$$\Delta z = z - z_{NOM}$$

Therefore, the error measurement equation is given by

$$\Delta z = \left. \frac{\partial h(x)}{\partial x} \right|_{x=x_{NOM}} \Delta x + v$$

Again, as was the case before, the preceding linearized error measurement equation is of the same form as the extended Kalman-filtering linearized measurement equation, except we are now working with error states and the partial derivative matrix is evaluated on the nominal trajectory, rather than at the projected state estimate.

From the preceding error equations we can define the systems dynamics matrix and the measurement matrix to be matrices of partial derivatives or

$$F = \frac{\partial f(x)}{\partial x}\bigg|_{x=x_{\text{NOM}}}$$

$$H = \frac{\partial h(x)}{\partial x}\bigg|_{x=x_{\text{NOM}}}$$

For systems in which the measurements are discrete, we can rewrite the measurement equation as

$$\Delta z_k = H \Delta x_k + v_k$$

The fundamental matrix, which is required for discrete systems, can be obtained directly from the systems dynamics matrix. We have already shown in preceding chapters that the fundamental matrix can be approximated by the Taylor-series expansion for $\exp(F T_s)$ and is given by

$$\Phi_k = I + F T_s + \frac{F^2 T_s^2}{2!} + \frac{F^3 T_s^3}{3!} + \cdots$$

where T_s is the sample time and I is the identity matrix. As was shown in the chapters pertaining to extended Kalman filtering, the Taylor series is often approximated by only the first two terms or

$$\Phi_k \approx I + F T_s$$

The measurement and fundamental matrices of the linearized Kalman filter may be considered to be linear functions of time because they can be computed offline based on information from the nominal trajectory. The Riccati equations, required for the computation of the Kalman gains, are identical to those of the linear and extended Kalman filters and are repeated here for convenience.

$$M_k = \Phi_k P_{k-1} \Phi_k^T + Q_k$$
$$K_k = M_k H^T (H M_k H^T + R_k)^{-1}$$
$$P_k = (I - K_k H) M_k$$

where P_k is a covariance matrix representing errors in the state estimates after an update and M_k is the covariance matrix representing errors in the state estimates before an update. As was just mentioned, because Φ_k and H can be considered linear functions of time, the Kalman gains can also be computed offline, as is possible with a linear Kalman filter. The discrete process-noise matrix Q_k can still be found from the continuous process-noise matrix, according to

$$Q_k = \int_0^{T_s} \Phi(\tau) Q \Phi^T(\tau) \, d\tau$$

Finally, the equation for the linearized Kalman filter is given by

$$\Delta \hat{x}_k = \Phi \Delta \hat{x}_{k-1} + K_k (\Delta z_k - H \Phi \Delta \hat{x}_{k-1})$$

We have just seen that the linearized Kalman filter gives estimates of the error states. Therefore, in order to find estimates of the actual states we must combine nominal trajectory information with the error state estimates or

$$\hat{x}_k = \Delta \hat{x}_k + x_{\text{NOM}}$$

where x_{NOM} are the states obtained from the nominal trajectory. The best way to understand the equations for the linearized Kalman filter is by working an example.

Falling Object Revisited

Let us again consider the one-dimensional example of an object falling on a tracking radar, as was shown in Fig. 8.1 but now redrawn for convenience in Fig. 13.1. In this problem we are assuming that the drag is unknown, which means we have to estimate the ballistic coefficient in order to get a good estimate of the position and velocity of the object. The ballistic coefficient of the object in this example is $500 \, \text{lb/ft}^2$. As was the case in Chapter 8, the object was initially at 200,000 ft above the radar and had a velocity of 6000 ft/s toward the radar that

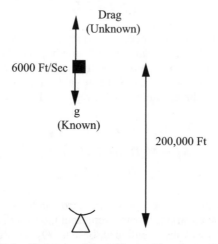

Fig. 13.1   Radar tracking object with unknown drag.

is located on the surface of the Earth. The radar measures the range from the radar to the object (i.e., altitude of the object) with a 25-ft standard deviation measurement accuracy. The radar takes measurements 10 times a second until ground impact. Let us assume that we have nominal trajectory information for a 6000-ft/s falling object with a ballistic coefficient of $500\,\text{lb/ft}^2$ starting at an altitude of 200,000 ft. We would like to build a linearized Kalman filter to estimate the altitude, velocity, and ballistic coefficient of the object based on both the nominal trajectory information and the radar measurements. *The following is an academic exercise because if this type of information were really available it might not be necessary to build the estimator.*

As was the case in Chapter 8, if x is the distance from the radar to the object (i.e., altitude), then the acceleration in the altitude direction acting on the object consists of gravity and drag or

$$\ddot{x} = \text{Drag} - g = \frac{Q_p g}{\beta} - g$$

where β is the ballistic coefficient of the object. The ballistic coefficient, which is considered to be a constant in this problem, is a term describing the amount of drag acting on the object. The dynamic pressure Q_p in the preceding equation is given by

$$Q_p = 0.5\rho\dot{x}^2$$

where ρ is the air density and $\dot{x}$ is the velocity of the object in the altitude direction. As was the case in Chapter 8, we can still assume that the air density is an exponential function of altitude[2] given by

$$\rho = 0.0034e^{-x/22,000}$$

Therefore, the second-order differential equation describing the acceleration on the object can be expressed as

$$\ddot{x} = \frac{Q_p g}{\beta} - g = \frac{0.5g\rho\dot{x}^2}{\beta} - g = \frac{0.0034ge^{-x/22,000}\dot{x}^2}{2\beta} - g$$

Before we proceed any further, it is of interest to see how accurate the nominal trajectory is for our problem. Listing 13.1 presents a simulation in which both the actual and nominal equations of motion of the object are both integrated. We can see from Listing 13.1 that the initial conditions for the nominal and actual trajectories of the object are identical. The actual ballistic coefficient is $500\,\text{lb/ft}^2$, and the nominal ballistic coefficient is also $500\,\text{lb/ft}^2$. We will keep the actual ballistic coefficient fixed but vary the nominal ballistic coefficient from 500 to $1100\,\text{lb/ft}^2$ (i.e., a change of more than 100%) in steps of $300\,\text{lb/ft}^2$. Every 0.1 s we print out the difference between the actual and nominal altitudes, actual and nominal velocities, and actual and nominal ballistic coefficients. In addition, we also print out the actual and nominal altitudes and velocities of the object.

Listing 13.1 Simulation for actual and nominal trajectories of falling object

```
IMPLICIT REAL*8 (A-H)
IMPLICIT REAL*8 (O-Z)
X=200000.
XD=-6000.
BETA=500.
XNOM=200000.
XDNOM=-6000.
BETANOM=800.
OPEN(1,STATUS='UNKNOWN',FILE='DATFIL')
TS=.1
TS=0.
S=0.
H=.001
WHILE(X>=0.)
      XOLD=X
      XDOLD=XD
      XNOMOLD=XNOM
      XDNOMOLD=XDNOM
      XDD=.0034*32.2*XD*XD*EXP(-X/22000.)/(2.*BETA)-32.2
      XDDNOM=.0034*32.2*XDNOM*XDNOM*EXP(-XNOM/22000.)/
1         (2.*BETANOM)-32.2
      X=X+H*XD
      XD=XD+H*XDD
      XNOM=XNOM+H*XDNOM
      XDNOM=XDNOM+H*XDDNOM
      T=T+H
      XDD=.0034*32.2*XD*XD*EXP(-X/22000.)/(2.*BETA)-32.2
      XDDNOM=.0034*32.2*XDNOM*XDNOM*EXP(-XNOM/22000.)/
1         (2.*BETANOM)-32.2
      X=.5*(XOLD+X+H*XD)
      XD=.5*(XDOLD+XD+H*XDD)
      XNOM=.5*(XNOMOLD+XNOM+H*XDNOM)
      XDNOM=.5*(XDNOMOLD+XDNOM+H*XDDNOM)
      S=S+H
      IF(S>=(TS-.00001))THEN
            S=0.
            DELXH=X-XNOM
            DELXDH=XD-XDNOM
            DELBETAH=BETA-BETANOM
            XH=XNOM+DELXH
            XDH=XDNOM+DELXDH
            BETAH=BETANOM+DELBETAH
            WRITE(9,*)T,DELXH,DELXDH,DELBETAH,X,XH,
1               XD,XDH
            WRITE(1,*)T,DELXH,DELXDH,DELBETAH,X,XH,
1               XD,XDH
      ENDIF
END DO
PAUSE
CLOSE(1)
END
```

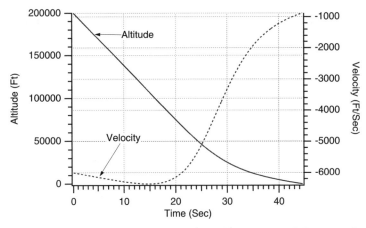

Fig. 13.2   Object slows up significantly as it races toward the ground.

The nominal case of Listing 13.1 was run in which the actual and nominal trajectories were identical. Figure 13.2 presents the altitude and velocity profiles for the falling object. We can see that it takes nearly 45 s for the falling object to reach the ground from its initial altitude of 200,000 ft. The velocity of the object decreases from 6000 ft/s at 200,000 ft altitude to approximately 1000 ft/s near ground impact.

Figures 13.3 and 13.4 show how the differences between the actual and nominal altitude and velocity occur when the nominal value of the ballistic coefficient differs from the actual value of the ballistic coefficient. We can see from Fig. 13.3 that a 100 lb/ft^2 difference in the ballistic coefficient (i.e., $\beta_{ACT} = 500$ lb/ft^2 and $\beta_{NOM} = 600$ lb/ft^2 or a 20% difference) can yield a 4000-ft altitude error, whereas we can see from Fig. 13.4 that the same error in ballistic coefficient yields a 50-ft/s velocity error by the time the object hits the

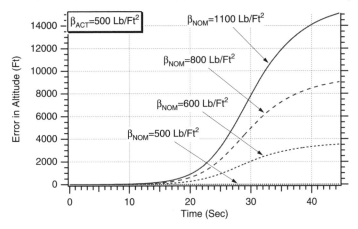

Fig. 13.3   Differences between actual and nominal ballistic coefficient can yield significant differences between actual and nominal altitude of object.

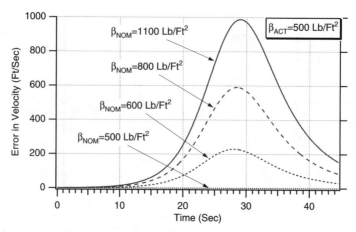

Fig. 13.4 Differences between actual and nominal ballistic coefficient can yield significant differences between actual and nominal velocity of object.

ground. A 300-lb/ft^2 difference in the ballistic coefficient (i.e., $\beta_{ACT} = 500$ lb/ft^2 and $\beta_{NOM} = 800$ lb/ft^2 or a 60% difference) can yield double the altitude and velocity errors (i.e., 8000-ft altitude error and 100-ft/s velocity error). We can also see from Fig. 13.3 that a 600-lb/ft^2 difference in the ballistic coefficient (i.e., $\beta_{ACT} = 500$ lb/ft^2 and $\beta_{NOM} = 1100$ lb/ft^2 or a 120% difference) can yield a 14,000-ft altitude error, and from Fig. 13.4 we can see that the same ballistic coefficient error yields a 200-ft/s velocity error by the time the object hits the ground.

Developing a Linearized Kalman Filter

In Chapter 8 the states of the extended Kalman filter were altitude, velocity, and ballistic coefficient. Therefore, it makes sense that the error states for the linearized Kalman filter used to track the falling object should be

$$\Delta x = \begin{bmatrix} \Delta x \\ \Delta \dot{x} \\ \Delta \beta \end{bmatrix} = \begin{bmatrix} x - x_{NOM} \\ \dot{x} - \dot{x}_{NOM} \\ \beta - \beta_{NOM} \end{bmatrix}$$

Recall that the acceleration acting on the object is a nonlinear function of the states and was shown to be

$$\ddot{x} = \frac{0.0034 g e^{-x/22,000} \dot{x}^2}{2\beta} - g$$

In addition, if we believe that the ballistic coefficient is constant, that means its derivative must be zero or

$$\dot{\beta} = 0$$

For safety we can add process noise u_s to this equation to account for the fact that the ballistic coefficient may be varying. In that case the new equation for the derivative of the ballistic coefficient becomes

$$\dot{\beta} = u_s$$

Therefore, our model of the real world is given by

$$\dot{x} = f(x) + w$$

and we want the error equations to be of the form

$$\Delta\dot{x} = \left.\frac{\partial f(x)}{\partial x}\right|_{x=x_{\text{NOM}}} \Delta x + w$$

Because we have chosen the appropriate states for this tracking problem, substitution yields our error equation model of the real world. As was already mentioned, in this equation the partial derivatives are evaluated along the nominal trajectory

$$\begin{bmatrix} \Delta\dot{x} \\ \Delta\ddot{x} \\ \Delta\dot{\beta} \end{bmatrix} = \begin{bmatrix} \dfrac{\partial\dot{x}}{\partial x} & \dfrac{\partial\dot{x}}{\partial\dot{x}} & \dfrac{\partial\dot{x}}{\partial\beta} \\[2mm] \dfrac{\partial\ddot{x}}{\partial x} & \dfrac{\partial\ddot{x}}{\partial\dot{x}} & \dfrac{\partial\ddot{x}}{\partial\beta} \\[2mm] \dfrac{\partial\dot{\beta}}{\partial x} & \dfrac{\partial\dot{\beta}}{\partial\dot{x}} & \dfrac{\partial\dot{\beta}}{\partial\beta} \end{bmatrix}_{x=x_{\text{NOM}}} \begin{bmatrix} \Delta x \\ \Delta\dot{x} \\ \Delta\beta \end{bmatrix} + \begin{bmatrix} 0 \\ 0 \\ u_s \end{bmatrix}$$

We can see from the preceding linearized state-space differential equation that the systems dynamic matrix is the matrix of partial derivatives evaluated along the nominal trajectory or

$$F = \begin{bmatrix} \dfrac{\partial\dot{x}}{\partial x} & \dfrac{\partial\dot{x}}{\partial\dot{x}} & \dfrac{\partial\dot{x}}{\partial\beta} \\[2mm] \dfrac{\partial\ddot{x}}{\partial x} & \dfrac{\partial\ddot{x}}{\partial\dot{x}} & \dfrac{\partial\ddot{x}}{\partial\beta} \\[2mm] \dfrac{\partial\dot{\beta}}{\partial x} & \dfrac{\partial\dot{\beta}}{\partial\dot{x}} & \dfrac{\partial\dot{\beta}}{\partial\beta} \end{bmatrix}_{x=x_{\text{NOM}}}$$

We can also see from the linearized state-space equation that the continuous process-noise matrix is given by

$$Q = E(ww^T)$$

Therefore, we can say that for this example the continuous process-noise matrix is

$$\boldsymbol{Q}(t) = \begin{bmatrix} 0 & 0 & 0 \\ 0 & 0 & 0 \\ 0 & 0 & \Phi_s \end{bmatrix}$$

where Φ_s is the spectral density of the white noise source assumed to be on the derivative of the error equation for the ballistic coefficient. To evaluate the systems dynamics matrix, we must take the necessary partial derivatives as indicated by the systems dynamics matrix $\boldsymbol{F}$. The first row of partial derivatives yields

$$\frac{\partial \dot{x}}{\partial x} = 0$$

$$\frac{\partial \dot{x}}{\partial \dot{x}} = 1$$

$$\frac{\partial \dot{x}}{\partial \beta} = 0$$

whereas the second row of partial derivatives yields

$$\frac{\partial \ddot{x}}{\partial x} = \frac{-0.0034 e^{-x/22,000} \dot{x}^2 g}{2\beta(22,000)} = \frac{-\rho g \dot{x}^2}{44,000\beta}$$

$$\frac{\partial \ddot{x}}{\partial \dot{x}} = \frac{2*0.0034 e^{-x/22,000} \dot{x} g}{2\beta} = \frac{\rho g \dot{x}}{\beta}$$

$$\frac{\partial \ddot{x}}{\partial \beta} = \frac{-0.0034 e^{-x/22,000} \dot{x}^2 g}{2\beta^2} = \frac{-\rho g \dot{x}^2}{2\beta^2}$$

Finally, the last row of partial derivatives yields

$$\frac{\partial \dot{\beta}}{\partial x} = 0$$

$$\frac{\partial \dot{\beta}}{\partial \dot{x}} = 0$$

$$\frac{\partial \dot{\beta}}{\partial \beta} = 0$$

Therefore, the systems dynamics matrix turns out to be

$$
F = \begin{bmatrix}
0 & 1 & 0 \\[2mm]
\dfrac{-\rho_{\text{NOM}}g\dot{x}^2_{\text{NOM}}}{44,000\beta_{\text{NOM}}} & \dfrac{\rho_{\text{NOM}}g\dot{x}_{\text{NOM}}}{\beta_{\text{NOM}}} & \dfrac{-\rho_{\text{NOM}}g\dot{x}^2_{\text{NOM}}}{2\beta^2_{\text{NOM}}} \\[3mm]
0 & 0 & 0
\end{bmatrix}
$$

where gravity g is assumed to be known perfectly and the other quantities are obtained from the nominal trajectory. The nominal air density in the preceding expression is given by

$$
\rho_{\text{NOM}} = 0.0034e^{-x_{\text{NOM}}/22,000}
$$

The fundamental matrix can be found by a two-term Taylor-series approximation, yielding

$$
\Phi(t) \approx I + Ft
$$

From the systems dynamics matrix we see that it make sense to define

$$
f_{21} = \frac{-\rho_{\text{NOM}}g\dot{x}^2_{\text{NOM}}}{44,000\beta_{\text{NOM}}}
$$

$$
f_{22} = \frac{\rho_{\text{NOM}}\dot{x}_{\text{NOM}}g}{\beta_{\text{NOM}}}
$$

$$
f_{23} = \frac{-\rho_{\text{NOM}}g\dot{x}^2_{\text{NOM}}}{2\beta^2_{\text{NOM}}}
$$

Now the systems dynamics matrix can be written more compactly as

$$
F(t) = \begin{bmatrix}
0 & 1 & 0 \\
f_{21} & f_{22} & f_{23} \\
0 & 0 & 0
\end{bmatrix}
$$

If we only use the first two terms of the Taylor-series expansion for the fundamental matrix, we obtain

$$
\Phi(t) \approx I + Ft = \begin{bmatrix}
1 & 0 & 0 \\
0 & 1 & 0 \\
0 & 0 & 1
\end{bmatrix} + \begin{bmatrix}
0 & 1 & 0 \\
f_{21} & f_{22} & f_{23} \\
0 & 0 & 0
\end{bmatrix}t = \begin{bmatrix}
1 & t & 0 \\
f_{21}t & 1+f_{22}t & f_{23}t \\
0 & 0 & 1
\end{bmatrix}
$$

or more simply

$$\Phi(t) \approx \begin{bmatrix} 1 & t & 0 \\ f_{21}t & 1+f_{22}t & f_{23}t \\ 0 & 0 & 1 \end{bmatrix}$$

Therefore, the discrete fundamental matrix can be found by substituting T_s for t or

$$\Phi_k \approx \begin{bmatrix} 1 & T_s & 0 \\ f_{21}T_s & 1+f_{22}T_s & f_{23}T_s \\ 0 & 0 & 1 \end{bmatrix}$$

In this example the error altitude measurement is a linear function of the states and is given by

$$\Delta x^* = \begin{bmatrix} 1 & 0 & 0 \end{bmatrix} \begin{bmatrix} \Delta x \\ \Delta \dot{x} \\ \Delta \beta \end{bmatrix} + v$$

where v is the measurement noise. We can see from the preceding equation that the measurement noise matrix can be written by inspection and is given by

$$H = \begin{bmatrix} 1 & 0 & 0 \end{bmatrix}$$

Because the discrete measurement noise matrix is given by

$$R_k = E(v_k v_k^T)$$

we can say that for this problem the discrete measurement noise matrix turns out to be a scalar and is given by

$$R_k = \sigma_v^2$$

where σ_v^2 is the variance of the altitude noise. We have already stated that the continuous process-noise matrix is given by

$$Q = E(ww^T)$$

Because the process noise is on the derivative of the ballistic coefficient, the continuous process-noise matrix for this example is given by

$$Q(t) = \begin{bmatrix} 0 & 0 & 0 \\ 0 & 0 & 0 \\ 0 & 0 & \Phi_s \end{bmatrix}$$

where Φ_s is the spectral density of the white noise source assumed to be on the derivative of the third state. The discrete process-noise matrix can be derived from the continuous process-noise matrix according to

$$Q_k = \int_0^{T_s} \Phi(\tau) Q \Phi^T(\tau) \, d\tau$$

Therefore, substitution yields

$$Q_k = \Phi_s \int_0^{T_s} \begin{bmatrix} 1 & \tau & 0 \\ f_{21}\tau & 1+f_{22}\tau & f_{23}\tau \\ 0 & 0 & 1 \end{bmatrix} \begin{bmatrix} 0 & 0 & 0 \\ 0 & 0 & 0 \\ 0 & 0 & 1 \end{bmatrix} \begin{bmatrix} 1 & f_{21}\tau & 0 \\ \tau & 1+f_{22}\tau & 0 \\ 0 & f_{23}\tau & 0 \end{bmatrix} d\tau$$

After some algebra we get

$$Q_k = \Phi_s \int_0^{T_s} \begin{bmatrix} 0 & 0 & 0 \\ 0 & f_{23}^2\tau^2 & f_{23}\tau \\ 0 & f_{23}\tau & 1 \end{bmatrix} d\tau$$

Finally, after integration we obtain the final expression for the discrete process-noise matrix as

$$Q_k = \Phi_s \begin{bmatrix} 0 & 0 & 0 \\ 0 & f_{23}^2 \dfrac{T_s^3}{3} & f_{23} \dfrac{T_s^2}{2} \\ 0 & f_{23} \dfrac{T_s^2}{2} & T_s \end{bmatrix}$$

We now have defined all of the matrices required to solve the Riccati equations. The equation for the linearized Kalman filter is given by

$$\begin{bmatrix} \Delta\hat{x}_k \\ \Delta\hat{\dot{x}}_k \\ \Delta\hat{\beta}_k \end{bmatrix} = \Phi \begin{bmatrix} \Delta\hat{x}_{k-1} \\ \Delta\hat{\dot{x}}_{k-1} \\ \Delta\hat{\beta}_{k-1} \end{bmatrix} + \begin{bmatrix} K_{1_k} \\ K_{2_k} \\ K_{3_k} \end{bmatrix} \left(\Delta x_k^* - H\Phi \begin{bmatrix} \Delta\hat{x}_{k-1} \\ \Delta\hat{\dot{x}}_{k-1} \\ \Delta\hat{\beta}_{k-1} \end{bmatrix} \right)$$

where Δx_k^* is the noisy measurement of the difference between the actual and nominal altitudes, and the measurement and fundamental matrices have been defined already.

The preceding equations for the Kalman filter and Riccati equations were programmed and are shown in Listing 13.2, along with a simulation of the real world and a simulation of the nominal trajectory. From Listing 13.2 we can see that initially the nominal and actual ballistic coefficients are identical and are set to 500 lb/ft^2. Therefore, initially we have a perfect nominal trajectory because it is identical to the actual nominal trajectory. We can see that initially there is

Listing 13.2   Simulation of linearized Kalman filter and falling object

```
C THE FIRST THREE STATEMENTS INVOKE THE ABSOFT RANDOM
  NUMBER GENERATOR ON THE MACINTOSH
      GLOBAL DEFINE
              INCLUDE 'quickdraw.inc'
      END
      IMPLICIT REAL*8 (A-H)
      IMPLICIT REAL*8 (O-Z)
      REAL*8 PHI(3,3),P(3,3),PHIP(3,3),PHIPPHIT(3,3),K(3,1)
      REAL*8 Q(3,3),HMAT(1,3),HM(1,3),MHT(3,1),XH(3,1)
      REAL*8 PHIT(3,3)
      REAL*8 HMHT(1,1),HT(3,1),KH(3,3),IDN(3,3),IKH(3,3)
      REAL*8 MEAS(1,1),PHIXH(3,1),HPHIXH(1,1),RES(1,1),KRES(3,1)
      INTEGER ORDER
      X=200000.
      XD=-6000.
      BETA=500.
      XNOM=200000.
      XDNOM=-6000.
      BETANOM=500.
      OPEN(1,STATUS='UNKNOWN',FILE='DATFIL')
      ORDER=3
      TS=.1
      TF=30.
      Q33=300.*300./TF
      T=0.
      S=0.
      H=.001
      SIGNOISE=25.
      XH(1,1)=X-XNOM
      XH(2,1)=XD-XDNOM
      XH(3,1)=-300.
      DO 1000 I=1,ORDER
      DO 1000 J=1,ORDER
              PHI(I,J)=0.
              P(I,J)=0.
              Q(I,J)=0.
              IDN(I,J)=0.
 1000 CONTINUE
      IDN(1,1)=1.
      IDN(2,2)=1.
      IDN(3,3)=1.
      P(1,1)=SIGNOISE*SIGNOISE
      P(2,2)=20000.
      P(3,3)=300.**2
      DO 1100 I=1,ORDER
              HMAT(1,I)=0.
              HT(I,1)=0.
 1100 CONTINUE
      HMAT(1,1)=1.
      HT(1,1)=1.
```

(*continued*)

Listing 13.2   (*Continued*)

```
WHILE(X>=0.)
      XOLD=X
      XDOLD=XD
      XNOMOLD=XNOM
      XDD=.0034*32.2*XD*XD*EXP(-X/22000.)/(2.*BETA)-32.2
      XDDNOM=.0034*32.2*XDNOM*XDNOM*EXP(-XNOM/22000.)/
1        (2.*BETANOM)-32.2
      X=X+H*XDD
      XD=XD+H*XD
      XNOM=XNOM+H*XDNOM
      XDNOM=XDNOM+H*XDDNOM
      T=T+H
      XDD=.0034*32.2*XD*XD*EXP(-X/22000.)/(2.*BETA)-32.2
      XDDNOM=.0034*32.2*XDNOM*XDNOM*EXP(-XNOM/22000.)/
1     (2.*BETANOM)-32.2
      X=.5*(XOLD+X+H*XD)
      XD=.5*(XDOLD+XD+H*XDD)
      XNOM=.5*(XNOMOLD+XNOM+H*XDNOM)
      XDNOM=.5*(XDNOMOLD+XDNOM+H*XDDNOM)
      S=S+H
      IF(S>=(TS-.00001))THEN
            S=0.
            RHONOM=.0034*EXP(-XNOM/22000.)
            F21=-32.2*RHONOM*XDNOM*XDNOM/(2.*22000.
1           *BETANOM)
            F22=RHONOM*32.2*XDNOM/BETANOM
            F23=-RHONOM*32.2*XDNOM*XDNOM/(2.*BETANOM*
1           BETANOM)
            PHI(1,1)=1.
            PHI(1,2)=TS
            PHI(2,1)=F21*TS
            PHI(2,2)=1.+F22*TS
            PHI(2,3)=F23*TS
            PHI(3,3)=1.
            Q(2,2)=F23*F23*Q33*TS*TS*TS/3.
            Q(2,3)=F23*Q33*TS*TS/2.
            Q(3,2)=Q(2,3)
            Q(3,3)=Q33*TS
            CALL  MATTRN(PHI,ORDER,ORDER,PHIT)
            CALL  MATMUL(PHI,ORDER,ORDER,P,ORDER,ORDER,PHIP)
            CALL  MATMUL(PHIP,ORDER,ORDER,PHIT,ORDER,ORDER.
1           PHIPPHIT)
            CALL  MATADD(PHIPPHIT,ORDER,ORDER,Q,M)
            CALL  MATMUL(HMAT,1,ORDER,M,ORDER,ORDER,HM)
            CALL  MATMUL(HM,1,ORDER,HT,ORDER,1,HMHT)
            HMHTR=HMHT(1,1)+SIGNOISE*SIGNOISE
            HMHTRINV=1./HMHTR
            CALL  MATMUL(M,ORDER,ORDER,HT,ORDER,1,MHT)
            DO 150 I=1,ORDER
            K(I,1)=MHT(I,1)*HMHTRINV
```

(*continued*)

Listing 13.2 (*Continued*)

```
150       CONTINUE
          CALL  MATMUL(K,ORDER,1,HMAT,1,ORDER,KH)
          CALL  MATSUB(IDN,ORDER,ORDER,KH,IKH)
          CALL  MATMUL(IKH,ORDER,ORDER,M,ORDER,ORDER,P)
          CALL  GAUSS(XNOISE,SIGNOISE)
          DELX=X-XNOM
          MEAS(1,1)=DELX+XNOISE
          CALL  MATMUL(PHI,ORDER,ORDER,XH,ORDER,1,PHIXH)
          CALL  MATMUL(HMAT,1,ORDER,PHIXH,ORDER,1,HPHIXH)
          CALL  MATSUB(MEAS,1,1,HPHIXH,RES)
          CALL  MATMUL(K,ORDER,1,RES,1,1,KRES)
          CALL  MATADD(PHIXH,ORDER,1,KRES,XH)
          DELXH=XH(1,1)
          DELXDH=XH(2,1)
          DELBETAH=XH(3,1)
          XHH=XNOM+DELXH
          XDH=XDNOM+DELXDH
          BETAH=BETANOM+DELBETAH
          ERRX=X-XHH
          SP11=SQRT(P(1,1))
                  ERRXD=XD-XDH
                  SP22=SQRT(P(2,2))
                  ERRBETA=BETA-BETAH
                  SP33=SQRT(P(3,3))
                  WRITE(9,*)T,BETA,BETAH,ERRBETA,SP33,-SP33
                  WRITE(1,*)T,BETA,BETAH,ERRBETA,SP33,-SP33
          ENDIF
      END DO
      PAUSE
      CLOSE(1)
      END

C SUBROUTINE GAUSS IS SHOWN IN LISTING 1.8
C SUBROUTINE MATTRN IS SHOWN IN LISTING 1.3
C SUBROUTINE MATMUL IS SHOWN IN LISTING 1.4
C SUBROUTINE MATADD IS SHOWN IN LISTING 1.1
C SUBROUTINE MATSUB IS SHOWN IN LISTING 1.2
```

---

process noise and that the third diagonal element of the initial covariance matrix has been set to 300^2 to reflect what we believe the uncertainty to be in our first guess of the estimated ballistic coefficient. This, of course, is simulation fiction because we really know the initial ballistic coefficient perfectly by virtue of the fact that the ballistic coefficient of the nominal trajectory corresponds to truth. We have initialized the error states estimates of the filter to zero, except for the ballistic coefficient error state, which we have set to $-300\,\text{lb/ft}^2$. The standard deviation of the measurement noise on our error altitude measurement is 25 ft.

The nominal case was run, and the ballistic coefficient estimate profile is displayed in Fig. 13.5. We can see that the linearized Kalman filter is able to

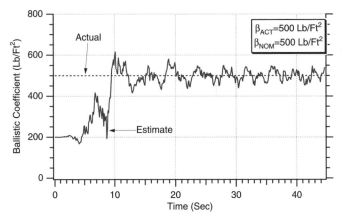

Fig. 13.5 Linearized Kalman filter is able to accurately estimate the ballistic coefficient after only 10 s.

accurately track the ballistic coefficient after only 10 s. The estimates could be made less noisy by simply reducing the value of the process noise. Figure 13.6 shows that the linearized Kalman filter is working correctly because the singly run simulation errors in the estimates of ballistic coefficient lie within the theoretical bounds.

Another case was run with Listing 13.2 in which the nominal value of the ballistic coefficient was increased by 20% to 600 lb/ft^2. This means that the nominal trajectory, upon which the linearized Kalman filter is based, will be in slight error. All other parameters in Listing 13.2 were left alone. Figure 13.7 shows that the ballistic coefficient estimate is still excellent, whereas Fig. 13.8

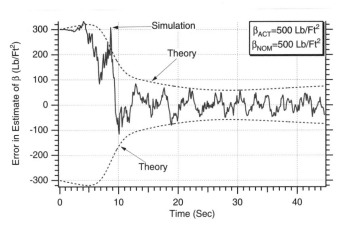

Fig. 13.6 Linearized Kalman filter appears to be working correctly because the single-run errors in the estimates of the ballistic coefficient lie within the theoretical bounds.

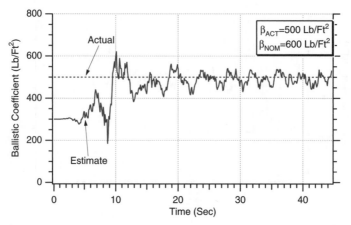

Fig. 13.7 Linearized Kalman filter is able to estimate accurately the ballistic coefficient when there is a slight error in the nominal trajectory.

indicates that the errors in the estimate of the ballistic coefficient are still within the theoretical bounds. Again, this also holds true after many simulation trials. We still can say that the linearized Kalman filter is performing according to theory.

Another case was run with Listing 13.2 in which the nominal value of the ballistic coefficient was increased by 60% to 800 lb/ft². This means that the nominal trajectory, upon which the linearized Kalman filter is based, will be in greater error than the preceding case. Figure 13.9 shows that the ballistic coefficient estimate is always less than the actual ballistic coefficient. This appears to be true even after many repeated simulation trials. Figure 13.10 indicates that now the errors in the estimate of the ballistic coefficient are no

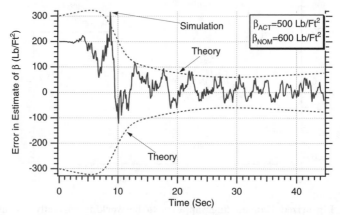

Fig. 13.8 Error in estimate of the ballistic coefficient is between theoretical error bounds when there is a slight error in the nominal trajectory.

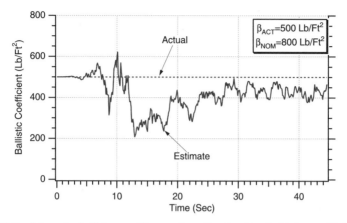

Fig. 13.9 Linearized Kalman filter underestimates the ballistic coefficient when there is 60% error in the nominal ballistic coefficient.

longer within the theoretical bounds but slightly on top of the upper bound. Again, this also holds true after many simulation trials. In this case we cannot say that the linearized Kalman filter is performing according to theory.

Finally, another case was run with Listing 13.2 in which the nominal value of the ballistic coefficient was increased 120% to 1100 lb/ft^2. This means that the nominal trajectory, upon which the linearized Kalman filter is based, was in even greater error than in the preceding case. Figure 13.11 shows that the ballistic coefficient estimate is not only always less than the actual ballistic coefficient but also has a significant steady-state error. This also appears to be true after many repeated simulation trials. Figure 13.12 indicates that now the errors in the

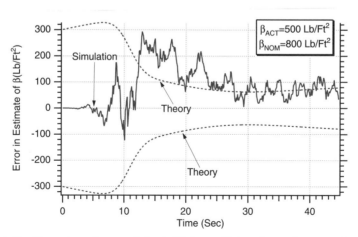

Fig. 13.10 Error in estimate of the ballistic coefficient is not between theoretical error bounds when there is a 60% error in the nominal ballistic coefficient.

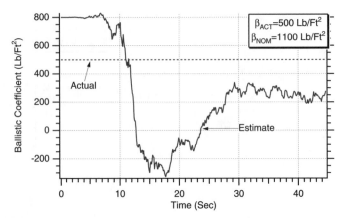

Fig. 13.11   Significant steady-state error in the linearized Kalman filter's estimate of the ballistic coefficient exists when there is a 120% error in the nominal ballistic coefficient.

estimate of the ballistic coefficient are completely outside the theoretical error bounds. This also holds true after many simulation trials. In this case we can definitely say that the linearized Kalman filter is not performing according to theory.

Experiments also were conducted in which the nominal ballistic coefficient was decreased from 500 to 100 lb/ft^2 in steps of 100 lb/ft^2. The linearized Kalman filter gave excellent estimates when the nominal ballistic coefficient was reduced 20% to 400 lb/ft^2. The estimates deteriorated somewhat when the nominal ballistic coefficient was reduced 40% to 300 lb/ft^2. In this case, the errors in the estimates were on the upper theoretical bound. However, when the nominal

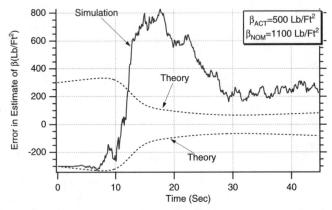

Fig. 13.12   When there is a 120% error in the nominal ballistic coefficient, the errors in the estimates from the linearized Kalman filter are completely outside the theoretical error bounds.

ballistic coefficient was reduced 60% to 200 lb/ft^2, the errors in the estimates were always outside the theoretical error bounds.

Thus, for this example, we can say that the linearized Kalman filter only performed well when the nominal trajectory was close to the actual trajectory (i.e., nominal ballistic coefficient differed from actual ballistic coefficient by no more than 20%). When the nominal trajectory was significantly different from the actual trajectory, the linearized Kalman filter's estimates diverged from the actual values of the ballistic coefficient.

Cannon-Launched Projectile Revisited

Let us revisit the cannon-launched projectile problem of Chapter 9. In this simple example a radar tracks a cannon-launched projectile traveling in a two-dimensional, drag-free environment, as shown in Fig. 13.13. After the projectile is launched at an initial velocity, only gravity acts on the projectile. The tracking radar that is located at coordinates x_R, y_R of Fig. 9.1 measures the range r and angle θ from the radar to the projectile.

In this example the projectile is launched at an initial velocity of 3000 ft/s at a 45-deg angle with respect to the horizontal axis in a zero-drag environment (i.e., atmosphere is neglected). The radar is located on the ground and is 100,000 ft downrange from where the projectile is initially launched. The radar is considered to have an angular measurement accuracy of 0.01 rad and a range measurement accuracy of 100 ft.

Before we can proceed with the development of a filter for this application, it is important to get a numerical and analytical feel for all aspects of the problem. Because only gravity acts on the projectile, we can say that in the downrange or x direction there is no acceleration, whereas in the altitude or y direction there is the acceleration of gravity g or

$$\ddot{x}_T = 0$$
$$\ddot{y}_T = -g$$

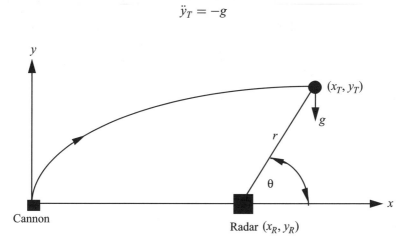

Fig. 13.13   Radar tracking cannon-launched projectile.

where g is the acceleration of gravity and has the value of 32.2 ft/s^2 in the English system of units. Using trigonometry, the angle and range from the radar to the projectile can be obtained by inspection of Fig. 13.13 to be

$$\theta = \tan^{-1}\left(\frac{y_T - y_R}{x_T - x_R}\right)$$

$$r = \sqrt{(x_T - x_R)^2 + (y_T - y_R)^2}$$

Before we proceed any further it is of interest to see how accurate the nominal trajectory is for our problem. Listing 13.3 presents a simulation in which the actual and nominal equations of motion of the object are both integrated. We can see from the listing that the initial conditions for the nominal and actual trajectories of the object are identical. The projectile velocity at launch is 3000 ft/s, whereas the nominal projectile velocity is also 3000 ft/s. We will keep the actual initial projectile velocity fixed but vary the nominal projectile velocity from 3050 to 3200 ft/s. Every second we print out the difference between the actual and nominal downranges, actual and nominal radar angles, and actual and nominal radar ranges.

Figures 13.14–13.16 show how the differences between the actual and nominal downranges, radar angles, and radar ranges vary when the nominal value of the initial projectile velocity differs from the actual initial value of the projectile velocity. We can see from Fig. 13.14 that a 50-ft/s difference or 1.7% increase in the initial projectile velocity can yield a 3000-ft downrange error. We can also see that a 100-ft/s difference or 3.3% increase in the initial projectile velocity can yield a 9000-ft downrange error. Finally, we can see that a 200-ft/s difference or 6.7% increase in the initial projectile velocity can yield an 18,000-ft downrange error. Thus, we can see that small initial velocity errors can yield substantial errors in the downrange component of the nominal trajectory. Figure 13.15 indicates that a 50-ft/s difference or 1.7% increase in the initial projectile velocity can yield a 1-deg radar angle error. The figure also indicates that a 100-ft/s difference or 3.3% increase in the initial projectile velocity can yield a 3-deg radar angle error, whereas a 200-ft/s difference or 6.7% increase in the initial projectile velocity can yield a 5-deg radar angle error. Similarly, Fig. 13.16 indicates that a 50-ft/s difference or 1.7% increase in the initial projectile velocity can yield a 4000-ft radar range error, whereas a 100-ft/s difference or 3.3% increase in the initial projectile velocity can yield an 8000-ft radar range error. Figure 13.16 also shows that a 200-ft/s difference or 6.7% increase in the initial projectile velocity can yield a 20,000-ft radar range error.

Linearized Cartesian Kalman Filter

In Chapter 9 the states of the extended Kalman filter were downrange, downrange velocity, altitude, and altitude velocity. Therefore, it makes sense

Listing 13.3 Simulation for actual and nominal projectile trajectories

```
IMPLICIT  REAL*8(A-H,O-Z)
TS=1.
VT=3000.
VTERR=0.
GAMDEG=45.
GAMDEGERR=0.
VTNOM=VT+VTERR
GAMDEGNOM=GAMDEG+GAMDEGERR
G=32.2
XT=0.
YT=0.
XTD=VT*COS(GAMDEG/57.3)
YTD=VT*SIN(GAMDEG/57.3)
XTNOM=XT
YTNOM=YT
XTDNOM=VTNOM*COS(GAMDEGNOM/57.3)
YTDNOM=VTNOM*SIN(GAMDEGNOM/57.3)
XR=100000.
YR=0.
OPEN(1,STATUS='UNKNOWN',FILE='DATFIL')
T=0.
S=0.
H=.001
WHILE(YT>=0.)
       XTOLD=XT
       XTDOLD=XTD
       YTOLD=YT
       YTDOLD=YTD
       XTNOMOLD=XTNOM
       XTDNOMOLD=XTDNOM
       YTNOMOLD=YTNOM
       YTDNOMOLD=YTDNOM
       XTDD=0.
       YTDD=-G
       XTDDNOM=0.
       YTDDNOM=-G
       XT=XT+H*XTD
       XTD=XTD+H*XTDD
       YT=YT+H*YTD
       YTD=YTD+H*YTDD
       XTNOM=XTNOM+H*XTDNOM
       XTDNOM=XTDNOM+H*XTDDNOM
       YTNOM=YTNOM+H*YTDNOM
       YTDNOM=YTDNOM+H*YTDDNOM
       T=T+H
       XTDD=0.
       YTDD=-G
       XTDDNOM=0.
       YTDDNOM=-G
```

(*continued*)

Listing 13.3 (*Continued*)

```
        XT=.5*(XTOLD+XT+H*XTD)
        XTD=.5*(XTDOLD+XTD+H*XTDD)
        YT=.5*(YTOLD+YT+H*YTD)
        YTD=.5*(YTDOLD+YTD+H*YTDD)
        XTNOM=.5*(XTNOMOLD+XTNOM+H*XTDNOM)
        XTDNOM=.5*(XTDNOMOLD+XTDNOM+H*XTDDNOM)
        YTNOM=.5*(YTNOMOLD+YTNOM+H*YTDNOM)
        YTDNOM=.5*(YTDNOMOLD+YTDNOM+H*YTDDNOM)
        S=S+H
        IF(S>=(TS-.00001))THEN
            S=0.
            RTNOM=SQRT((XTNOM-XR)**2+(YTNOM-YR)***2)
            THET=ATAN2((YT-YR),(XT-XR))
            RT=SQRT((XT-XR)**2+(YT-YR)**2)
            THETNOM=ATAN2((YTNOM-YR),(XTNOM-XR))
            DELTHET=57.3*(THET-THETNOM)
            DELRT=RT-RTNOM
            DELXT=XT-XTNOM
            DELXTD=XTD-XTDNOM
            DELYT=YT-YTNOM
            DELYTD=YTD-YTDNOM
            WRITE(9,*)T,DELXT,DELXTD,DELYT,DELYTD,DELTHET,
1               DELRT
            WRITE(1,*)T,DELXT,DELXTD,DELYT,DELYTD,DELTHET,
1               DELRT
        ENDIF
    END DO
    PAUSE
    CLOSE(1)
    END
```

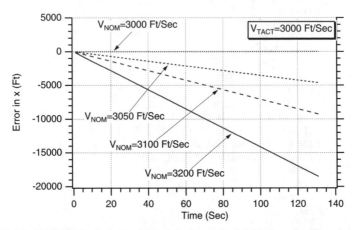

Fig. 13.14 Differences between actual and nominal initial projectile velocities can yield significant differences between actual and nominal downrange of the projectile.

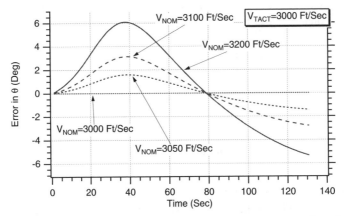

Fig. 13.15 Differences between actual and nominal initial projectile velocities can yield significant differences between actual and nominal angles from the radar to the projectile.

that the error states for the linearized Kalman filter used to track the projectile should be

$$\Delta x = \begin{bmatrix} \Delta x_T \\ \Delta \dot{x}_T \\ \Delta y_T \\ \Delta \dot{y}_T \end{bmatrix} = \begin{bmatrix} x_T - x_{TNOM} \\ \dot{x}_T - \dot{x}_{TNOM} \\ y_T - y_{TNOM} \\ \dot{y}_T - \dot{y}_{TNOM} \end{bmatrix}$$

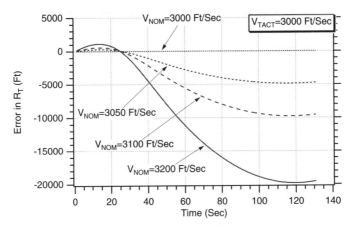

Fig. 13.16 Differences between actual and nominal initial projectile velocities can yield significant differences between actual and nominal ranges from the radar to the projectile.

When the preceding Cartesian error states are chosen, the state-space error differential equation describing projectile motion is linear and becomes

$$\begin{bmatrix} \Delta \dot{x}_T \\ \Delta \ddot{x}_T \\ \Delta \dot{y}_T \\ \Delta \ddot{y}_T \end{bmatrix} = \begin{bmatrix} 0 & 1 & 0 & 0 \\ 0 & 0 & 0 & 0 \\ 0 & 0 & 1 & 0 \\ 0 & 0 & 0 & 0 \end{bmatrix} \begin{bmatrix} \Delta x_T \\ \Delta \dot{x}_T \\ \Delta y_T \\ \Delta \dot{y}_T \end{bmatrix} + \begin{bmatrix} 0 \\ u_s \\ 0 \\ u_s \end{bmatrix}$$

In the preceding error equation gravity g is missing because it is assumed to be known exactly and therefore cannot cause any errors. We also have added process noise u_s to the acceleration portion of the error equations as protection for effects that may not be considered by the Kalman filter. However, in our initial experiments we will assume the process noise to be zero because our state-space equations upon which the Kalman filter is based will initially be a perfect match to the real world.

From the preceding state-space equation we can see that systems dynamics matrix is given by

$$F = \begin{bmatrix} 0 & 1 & 0 & 0 \\ 0 & 0 & 0 & 0 \\ 0 & 0 & 0 & 1 \\ 0 & 0 & 0 & 0 \end{bmatrix}$$

The fundamental matrix can be found from the Taylor-series approximation, which is given by

$$\Phi(t) = I + Ft + \frac{F^2 t^2}{2!} + \frac{F^3 t^3}{3!} + \cdots$$

Because in this example

$$F^2 = \begin{bmatrix} 0 & 1 & 0 & 0 \\ 0 & 0 & 0 & 0 \\ 0 & 0 & 0 & 1 \\ 0 & 0 & 0 & 0 \end{bmatrix} \begin{bmatrix} 0 & 1 & 0 & 0 \\ 0 & 0 & 0 & 0 \\ 0 & 0 & 0 & 1 \\ 0 & 0 & 0 & 0 \end{bmatrix} = \begin{bmatrix} 0 & 0 & 0 & 0 \\ 0 & 0 & 0 & 0 \\ 0 & 0 & 0 & 0 \\ 0 & 0 & 0 & 0 \end{bmatrix}$$

all of the higher-order terms must be zero, and the fundamental matrix becomes

$$\Phi(t) = I + Ft = \begin{bmatrix} 1 & 0 & 0 & 0 \\ 0 & 1 & 0 & 0 \\ 0 & 0 & 1 & 0 \\ 0 & 0 & 0 & 1 \end{bmatrix} + \begin{bmatrix} 0 & 1 & 0 & 0 \\ 0 & 0 & 0 & 0 \\ 0 & 0 & 0 & 1 \\ 0 & 0 & 0 & 0 \end{bmatrix} t$$

or more simply

$$\Phi(t) = \begin{bmatrix} 0 & t & 0 & 0 \\ 0 & 1 & 0 & 0 \\ 0 & 0 & 1 & t \\ 0 & 0 & 0 & 1 \end{bmatrix}$$

Therefore, the discrete fundamental matrix can be found by substituting the sample time T_s for time t and is given by

$$\Phi_k = \begin{bmatrix} 0 & T_s & 0 & 0 \\ 0 & 1 & 0 & 0 \\ 0 & 0 & 1 & T_s \\ 0 & 0 & 0 & 1 \end{bmatrix}$$

Because our states have been chosen to be Cartesian, the radar error measurements Δr and $\Delta \theta$ will automatically be nonlinear functions of those states. Therefore, we must write the measurement equation as

$$\begin{bmatrix} \Delta \theta^* \\ \Delta r^* \end{bmatrix} = \begin{bmatrix} \dfrac{\partial \theta}{\partial x_T} & \dfrac{\partial \theta}{\partial \dot{x}_T} & \dfrac{\partial \theta}{\partial y_T} & \dfrac{\partial \theta}{\partial \dot{y}_T} \\ \dfrac{\partial r}{\partial x_T} & \dfrac{\partial r}{\partial \dot{x}_T} & \dfrac{\partial r}{\partial y_T} & \dfrac{\partial r}{\partial \dot{y}_T} \end{bmatrix} \begin{bmatrix} \Delta x_T \\ \Delta \dot{x}_T \\ \Delta y_T \\ \Delta \dot{y}_T \end{bmatrix} + \begin{bmatrix} v_\theta \\ v_r \end{bmatrix}$$

where v_θ and v_r represent the measurement noise on the error angle and error range, respectively. Because the angle from the radar to the projectile is given by

$$\theta = \tan^{-1}\left(\frac{y_T - y_R}{x_T - x_R} \right)$$

the four partial derivatives of the angle with respect to each of the states can be computed as

$$\frac{\partial \theta}{\partial x_T} = \frac{1}{1 + (y_T - y_R)^2/(x_T - x_R)^2} \frac{(x_T - x_R)*0 - (y_T - y_R)*1}{(x_T - x_R)^2} = \frac{-(y_T - y_R)}{r^2}$$

$$\frac{\partial \theta}{\partial \dot{x}_T} = 0$$

$$\frac{\partial \theta}{\partial y_T} = \frac{1}{1 + (y_T - y_R)^2/(x_T - x_R)^2} \frac{(x_T - x_R)*1 - (y_T - y_R)*0}{(x_T - x_R)^2} = \frac{(x_T - x_R)}{r^2}$$

$$\frac{\partial \theta}{\partial \dot{y}_T} = 0$$

Similarly, because the range from the radar to the projectile is given by

$$r = \sqrt{(x_T - x_R)^2 + (y_T - y_R)^2}$$

the four partial derivatives of the range with respect to each of the states can be computed as

$$\frac{\partial r}{\partial x_T} = \frac{1}{2}[(x_T - x_R)^2 + (y_T - y_R)^2]^{-1/2} 2(x_T - x_R) = \frac{(x_T - x_R)}{r}$$

$$\frac{\partial r}{\partial \dot{x}_T} = 0$$

$$\frac{\partial r}{\partial y_T} = \frac{1}{2}[(x_T - x_R)^2 + (y_T - y_R)^2]^{-1/2} 2(y_T - y_R) = \frac{(y_T - y_R)}{r}$$

$$\frac{\partial r}{\partial \dot{y}_T} = 0$$

Because the measurement matrix is given by

$$H = \begin{bmatrix} \dfrac{\partial \theta}{\partial x_T} & \dfrac{\partial \theta}{\partial \dot{x}_T} & \dfrac{\partial \theta}{\partial y_T} & \dfrac{\partial \theta}{\partial \dot{y}_T} \\ \dfrac{\partial r}{\partial x_T} & \dfrac{\partial r}{\partial \dot{x}_T} & \dfrac{\partial r}{\partial y_T} & \dfrac{\partial r}{\partial \dot{y}_T} \end{bmatrix}$$

we can use substitution of the partial derivatives already derived to yield the measurement matrix for this example and obtain

$$H = \begin{bmatrix} \dfrac{-(y_T - y_R)}{r^2} & 0 & \dfrac{x_T - x_R}{r^2} & 0 \\ \dfrac{x_T - x_R}{r} & 0 & \dfrac{y_T - y_R}{r} & 0 \end{bmatrix}$$

For this problem it is assumed that we know where the radar is so that x_R and y_R are known and do not have to be estimated. The discrete measurement matrix will be based on the nominal trajectory states or

$$H_k = \begin{bmatrix} \dfrac{-(y_{T\text{NOM}} - y_R)}{r^2_{\text{NOM}}} & 0 & \dfrac{x_{T\text{NOM}} - x_R}{r^2_{\text{NOM}}} & 0 \\ \dfrac{x_{T\text{NOM}} - x_R}{r_{\text{NOM}}} & 0 & \dfrac{y_{T\text{NOM}} - y_R}{r_{\text{NOM}}} & 0 \end{bmatrix}$$

Because the discrete measurement noise matrix is given by

$$R_k = E(v_k v_k^T)$$

we can say that for this problem the discrete measurement noise matrix is

$$R_k = \begin{bmatrix} \sigma_\theta^2 & 0 \\ 0 & \sigma_r^2 \end{bmatrix}$$

where σ_θ^2 and σ_r^2 are the variances of the angle noise and range noise measurements, respectively. Similarly because the continuous process-noise matrix is given by

$$Q = E(ww^T)$$

we can easily show from the linear state-space equation for this example that the continuous process-noise matrix is

$$Q(t) = \begin{bmatrix} 0 & 0 & 0 & 0 \\ 0 & \Phi_s & 0 & 0 \\ 0 & 0 & 0 & 0 \\ 0 & 0 & 0 & \Phi_s \end{bmatrix}$$

where Φ_s is the spectral density of the white noise sources assumed to be on the downrange and altitude accelerations acting on the projectile. The discrete process-noise matrix can be derived from the continuous process-noise matrix according to

$$Q_k = \int_0^{T_s} \Phi(\tau)Q\Phi^T(\tau)\, dt$$

Therefore, substitution of the appropriate matrices into the preceding expression yields

$$Q_k = \int_0^{T_s} \begin{bmatrix} 1 & \tau & 0 & 0 \\ 0 & 1 & 0 & 0 \\ 0 & 0 & 1 & \tau \\ 0 & 0 & 0 & 1 \end{bmatrix} \begin{bmatrix} 0 & 0 & 0 & 0 \\ 0 & \Phi_s & 0 & 0 \\ 0 & 0 & 0 & 0 \\ 0 & 0 & 0 & \Phi_s \end{bmatrix} \begin{bmatrix} 1 & 0 & 0 & 0 \\ \tau & 1 & 0 & 0 \\ 0 & 0 & 1 & 0 \\ 0 & 0 & \tau & 1 \end{bmatrix} d\tau$$

After some algebra we get

$$Q_k = \int_0^{T_s} \begin{bmatrix} \tau^2\Phi_s & \tau\Phi_s & 0 & 0 \\ \tau\Phi_s & \Phi_s & 0 & 0 \\ 0 & 0 & \tau^2\Phi_s & \tau\Phi_s \\ 0 & 0 & \tau\Phi_s & \Phi_s \end{bmatrix} d\tau$$

Finally, after integration we obtain the final expression for the discrete process-noise matrix to be

$$
Q_k = \begin{bmatrix}
\dfrac{T_s^3 \Phi_s}{3} & \dfrac{T_s^2 \Phi_s}{2} & 0 & 0 \\[2ex]
\dfrac{T_s^2 \Phi_s}{2} & T_s \Phi_s & 0 & 0 \\[2ex]
0 & 0 & \dfrac{T_s^3 \Phi_s}{3} & \dfrac{T_s^2 \Phi_s}{2} \\[2ex]
0 & 0 & \dfrac{T_s^2 \Phi_s}{2} & T_s \Phi_s
\end{bmatrix}
$$

We now have defined all of the matrices required to solve the Riccati equations. The equation for the linearized Kalman filter is given by

$$
\begin{bmatrix}
\Delta \hat{x}_{T_k} \\
\Delta \hat{\dot{x}}_{T_k} \\
\Delta \hat{y}_{T_k} \\
\Delta \hat{\dot{y}}_{T_k}
\end{bmatrix}
= \Phi
\begin{bmatrix}
\Delta \hat{x}_{T_{k-1}} \\
\Delta \hat{\dot{x}}_{T_{k-1}} \\
\Delta \hat{y}_{T_{k-1}} \\
\Delta \hat{\dot{y}}_{T_{k-1}}
\end{bmatrix}
+
\begin{bmatrix}
K_{11_k} & K_{12_k} \\
K_{21_k} & K_{22_k} \\
K_{31_k} & K_{32_k} \\
K_{41_k} & K_{42_k}
\end{bmatrix}
\left(
\begin{bmatrix}
\Delta \theta^* \\
\Delta r^*
\end{bmatrix}
- H \Phi
\begin{bmatrix}
\Delta \hat{x}_{T_{k-1}} \\
\Delta \hat{\dot{x}}_{T_{k-1}} \\
\Delta \hat{y}_{T_{k-1}} \\
\Delta \hat{\dot{y}}_{T_{k-1}}
\end{bmatrix}
\right)
$$

where the measurement H and fundamental Φ matrices have been defined already.

The preceding equations for the Kalman filter and Riccati equations were programmed and are shown in Listing 13.4, along with a simulation of the real world and nominal trajectory. We can see that the process-noise matrix is set to zero in this example (i.e., $\Phi_s = 0$). In addition, we have initially zero nominal errors (i.e., VTERR = 0, GAMDEGERR = 0), which means that the nominal trajectory is a perfect match to the actual trajectory. The initial covariance matrix has the same errors as the one in Chapter 9.

The nominal case of Listing 13.4 was run, and the errors in the estimates of the downrange, downrange velocity, altitude, and altitude velocity are displayed in Figs 13.17–13.20. We can see that because the single-run results fall within the theoretical error bounds approximately 68% of the time then we can say that the linearized Kalman filter appears to be working correctly.

Three other cases were run in which an error was introduced to the nominal velocity. We can see from Fig. 13.21 that if the nominal velocity error is only 50 ft/s or 1.7% the single-run results lie within the theoretical error bounds approximately 68% of the time, indicating that the linearized Kalman filter still appears to be working correctly. However, if the velocity error is increased to 100 ft/s or 3.3%, Fig. 13.22 indicates that the single-run error in the estimate of downrange starts to drift outside the error bounds near the end of the run. This means that the linearized filter is beginning to have problems. Finally, Fig. 13.23 shows that when the nominal velocity is in error by 200 ft/s or 6.7% there is total disagreement between the single-run errors in the estimates of downrange and the

Listing 13.4   Linearized Kalman filter for tracking cannon-launched projectile

```
C THE FIRST THREE STATEMENTS INVOKE THE ABSOFT RANDOM
  NUMBER GENERATOR ON THE MACINTOSH
     GLOBAL DEFINE
           INCLUDE 'quickdraw.inc'
     END
     IMPLICIT REAL*8(A-H,O-Z)
     REAL*8 P(4,4),Q(4,4),M(4,4),PHI(4,4),HMAT(2,4),HT(4,2),PHIT(4,4)
     REAL*8 RMAT(2,2),IDN(4,4),PHIP(4,4),PHIPPHIT(4,4),HM(2,4)
     REAL*8 HMHT(2,2),HMHTR(2,2),HMHTRINV(2,2),MHT(4,2),K(4,2)
     REAL*8 KH(4,4),IKH(4,4),XH(4,1),MEAS(2,1),PHIXH(4,1),HPHIXH(2,1)
     REAL*8 RES(2,1),KRES(4,1)
     INTEGER ORDER
     TS=1.
     ORDER=4
     PHIS=0.
     SIGTH=.01
     SIGR=100.
     VT=3000.
     VTERR=0.
     GAMDEG=45.
     GAMDEGERR=0.
     VTNOM=VT+VTERR
     GAMDEGNOM=GAMDEG+GAMDEGERR
     G=32.2
     XT=0.
     YT=0.
     XTD=VT*COS(GAMDEG/57.3)
     YTD=VT*SIN(GAMDEG/57.3)
     XTNOM=XT
     YTNOM=YT
     XTDNOM=VTNOM*COS(GAMDEGNOM/57.3)
     YTDNOM=VTNOM*SIN(GAMDEGNOM/57.3)
     XR=100000.
     YR=0.
     OPEN(1,STATUS='UNKNOWN',FILE='DATFIL')
     OPEN(2,STATUS='UNKNOWN',FILE='COVFIL')
     T=0.
     S=0.
     H=.001
     XH(1,1)=XT-XTNOM
     XH(2,1)=XTD-XTDNOM
     XH(3,1)=YT-YTNOM
     XH(4,1)=YTD-YTDNOM
     DO 14 I=1,ORDER
     DO 14 J=1,ORDER
     PHI(I,J)=0.
     P(I,J)=0.
     Q(I,J)=0.
     IDN(I,J)=0.
```

(*continued*)

Listing 13.3 (*Continued*)

```
14 CONTINUE
   TS2=TS*TS
   TS3=TS2*TS
   Q(1,1)=PHIS*TS3/3.
   Q(1,2)=PHIS*TS2/2.
   Q(2,1)=Q(1,2)
   Q(2,2)=PHIS*TS
   Q(3,3)=Q(1,1)
   Q(3,4)=Q(1,2)
   Q(4,3)=Q(3,4)
   Q(4,4)=Q(2,2)
   PHI(1,1)=1.
   PHI(1,2)=TS
   PHI(2,2)=1.
   PHI(3,3)=1.
   PHI(3,4)=TS
   PHI(4,4)=1.
   CALL MATTRN(PHI,ORDER,ORDER,PHIT)
   RMAT(1,1)=SIGTH**2
   RMAT(1,2)=0.
   RMAT(2,1)=0.
   RMAT(2,2)=SIGR**2
   IDN(1,1)=1.
   IDN(2,2)=1.
   IDN(3,3)=1.
   IDN(4,4)=1.
   P(1,1)=1000.**2
   P(2,2)=100.**2
   P(3,3)=1000.**2
   P(4,4)=100.**2
   WHILE(YT>=0.)
        XTOLD=XT
        XTDOLD=XTD
        YTOLD=YT
        YTDOLD=YTD
        XTNOMOLD=XTNOM
        XTDNOMOLD=XTDNOM
        YTNOMOLD=YTNOM
        YTDNOMOLD=YTDNOM
        XTDD=0.
        YTDD=-G
        XTDDNOM=0.
        YTDDNOM=-G
        XT=XT+H*XTD
        XTD=XTD+H*XTDD
        YT=YT+H*YTD
        YTD=YTD+H*YTDD
        XTNOM=XTNOM+H+XTDNOM
        XTDNOM=XTDNOM+H*XTDDNOM
        YTNOM=YTNOM+H*YTDNOM
```

(*continued*)

Listing 13.3   (*Continued*)

```
YTDNOM=YTDNOM+H*YTDDNOM
T=T+H
XTDD=0.
YTDD=-G
XTDDNOM=0.
YTDDNOM=-G
XT=.5*(XTOLD+XT+H*XTD)
XTD=.5*(XTDOLD+XTD+H*XTDD)
YT=.5*(YTOLD+YT+H*YTD)
YTD=.5*(YTDOLD+YTD+H*YTDD)
XTNOM=.5*(XTNOMOLD+XTNOM+h*XTDNOM)
XTDNOM=.5*(XTDNOMOLD+XTDNOM+H*XTDDNOM)
XTNOM=.5*(YTNOMOLD+YTNOM+H*YTDNOM)
YTDNOM=.5*(YTDNOMOLD+YTDNOM+H*YTDDNOM)
S=S+H
IF(S>=(TS-.00001))THEN
        S=0.
        RTNOM=SQRT((XTNOM-XR)**22+(YTNOM-YR)**2)
        HMAT(1,1)=-(YTNOM-YR)/RTNOM**2
        HMAT(1,2)=0.
        HMAT(1,3)=(XTNOM-YR)/RTNOM**2
        HMAT(1,4)=0.
        HMAT(2,1)=(XTNOM-XR)/RTNOM
        HMAT(2,2)=0.
        HMAT(2,3)=(YTNOM-YR)/RTNOM
        HMAT(2,4)=0.
        CALL  MATTRN(HMAT,2,ORDER,HT)
        CALL  MATMUL(PHI,ORDER,ORDER,P,ORDER,ORDER,
          PHIP)
        CALL  MATMUL(PHIP,ORDER,ORDER,PHIT,ORDER,
          ORDER,PHIPPHIT)
        CALL  MATADD(PHIPPHIT,ORDER,ORDER,Q,M)
        CALL  MATMUL(HMAT,2,ORDER,M,ORDER,ORDER,HM)
        CALL  MATMUL(HM,2,ORDER,HT,ORDER,2,HMHT)
        CALL  MATADD(HMHT,ORDER,ORDER,RMAT,HMHTR)
        DET=HMHTR(1,1)*HMHTR(2,2)-HMHTR(1,2)*HMHTR(2,1)
        HMHTRINV(1,1)=HMHTR(2,2)/DET
        HMHTRINV(1,2)=-HMHTR(1,2)/DET
        HMHTRINV(2,1)=-HMHTR(2,1)/DET
        HMHTRINV(2,2)=HMHTR(1,1)/DET
        CALL  MATMUL(M,ORDER,ORDER,HT,ORDER,2,MHT)
        CALL  MATMUL(MHT,ORDER,2,HMHTRINV,2,2,K)
        CALL  MATMUL(K,ORDER,2,HMAT,2,ORDER,KH)
        CALL  MATSUB(IDN,ORDER,ORDER,KH,IKH)
        CALL  MATMUL(IKH,ORDER,ORDER,M,ORDER,
          ORDER,P)
        CALL  GAUSS(THETNOISE,SIGTH)
        CALL  GAUSS(RTNOISE,SIGR)
        THET=ATAN2((YT-YR),(XT-XR))
```

1

(*continued*)

Listing 13.3 *(Continued)*

```
                RT=SQRT((XT-XR)**2+(YT-YR)**2)
                THETNOM=ATAN2((YTNOM-YR),(XTNOM-XR))
                RTNOM=SQRT((XTNOM-XR)**2+(YTNOM-YR)**2)
                DELTHET=THET-THETNOM
                DELRT=RT-RTNOM
                MEAS(1,1)=DELTHET+THETNOISE
                MEAS(2,1)=DELRT+RTNOISE
                CALL  MATMUL(PHI,ORDER,ORDER,XH,ORDER,1,
                      PHIXH)
                CALL  MATMUL(HMAT,2,ORDER,PHIXH,ORDER,1,
                      HPHIXH)
                CALL  MATSUB(MEAS,2,1,HPHIXH,RES)
                CALL  MATMUL(K,ORDER,2,RES,2,1,KRES)
                CALL  MATADD(PHIXH,ORDER,1,KRES,XH)
                XTH=XTNOM+XH(1,1)
                XTDH=XTDNOM+XH(2,1)
                YTH=YTNOM+XH(3,1)
                YTDH=YTDNOM+XH(4,1)
                ERRX=XT-XTH
                SP11=SQRT(P(1,1))
                ERRXD=XTD-XTDH
                SP22=SQRT(P(2,2))
                ERRY=YT-YTH
                SP33=SQRT(P(3,3))
                ERRYD=YTD-YTDH
                SP44=SQRT(P(4,4))
                WRITE(9,*)T,XT,YT,THET*57.3,RT
                WRITE(1,*)T,XT,XTH,XTD,XTDH,YT,YTH,YTD,YTDH
                WRITE(2,*)T,ERRX,SP11,-SP11,ERRXD,SP22,-SP22,
     1                ERRY,SP33,-SP33,ERRYD,SP44,-SP44
            ENDIF
        END DO
        PAUSE
        CLOSE(1)
        CLOSE(2)
        END

C SUBROUTINE GAUSS IS SHOWN IN LISTING 1.8
C SUBROUTINE MATTRN IS SHOWN IN LISTING 1.3
C SUBROUTINE MATMUL IS SHOWN IN LISTING 1.4
C SUBROUTINE MATADD IS SHOWN IN LISTING 1.1
C SUBROUTINE MATSUB IS SHOWN IN LISTING 1.2
```

theoretical predictions of the covariance matrix. This type of behavior indicates that the linearized Kalman filter cannot tolerate this large of a discrepancy between the nominal and actual initial projectile velocities. A small error in the nominal initial velocity causes a large trajectory error and is thus able to ruin the effectiveness of the extended Kalman filter. Similar results were obtained when the nominal initial velocity was decreased by 50, 100, and 200 ft/s.

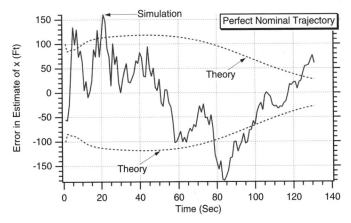

Fig. 13.17 Downrange errors in the estimates are within the theoretical bounds when the nominal trajectory is perfect.

Summary

The messages from the two examples of this chapter are quite clear. When the nominal trajectory, upon which the linearized Kalman filter is based, is accurate, then wonderful estimates can be obtained from the linearized Kalman filter. Of course, if the nominal trajectory were accurate, estimates would not be required because all information could be obtained directly from the nominal trajectory. We also have observed, via simulation experiments, that when the nominal trajectory is inaccurate the estimates from the linearized Kalman filter deteriorate. In fact, significant errors in the nominal trajectory can render the linearized Kalman filter worthless.

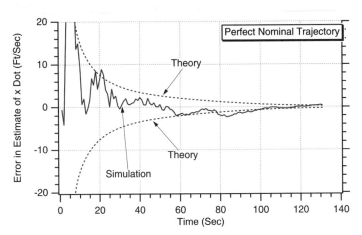

Fig. 13.18 Downrange velocity errors in the estimates are within the theoretical bounds when the nominal trajectory is perfect.

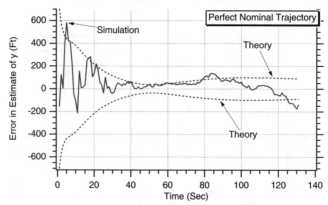

Fig. 13.19 Altitude errors in the estimates are within the theoretical bounds when the nominal trajectory is perfect.

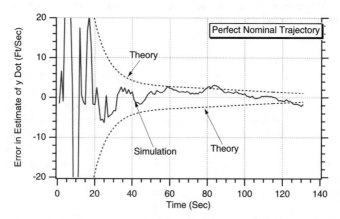

Fig. 13.20 Altitude velocity errors in the estimates are within the theoretical bounds when the nominal trajectory is perfect.

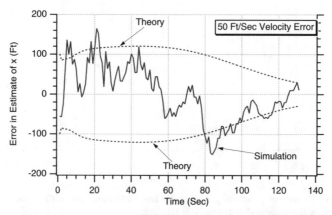

Fig. 13.21 Error in estimate of projectile downrange is between theoretical error bounds when there is a 1.7% error in the nominal velocity.

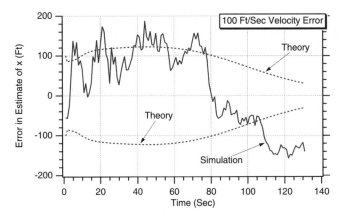

Fig. 13.22   Error in estimate of projectile downrange starts to drift outside theoretical error bounds when there is a 3.3% error in the nominal velocity.

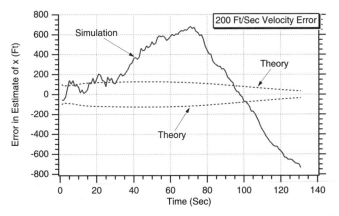

Fig. 13.23   Error in estimate of projectile downrange has no relationship to theoretical error bounds when there is a 6.7% error in the nominal velocity.

References

[1]Gelb, A., *Applied Optimal Estimation*, Massachusetts Inst. of Technology Press, Cambridge, MA, 1974, pp. 180–190.

[2]Zarchan, P., *Tactical and Strategic Missile Guidance*, 3rd ed., Progress in Astronautics and Aeronautics, AIAA, Reston, VA, 1998, pp. 210–211.

Chapter 14

Miscellaneous Topics

Introduction

IN THIS chapter we will discuss a few important topics that have not been addressed in preceding chapters. Although a Kalman filter can be built that gives the correct theoretical answers, we often want to see if those answers are useful for a particular application. The first section will address this issue by investigating how filter performance can degrade if there is too much measurement noise. In preceding chapters of this text, most of the examples were for cases in which there were hundreds of measurements. In the second section we will investigate how the performance of a polynomial Kalman filter behaves as a function of filter order (i.e., number of states in filter) when only a few measurements are available. Filter divergence is fairly easy to recognize in a simulation where we have access to the true states in our model of the real world. We will demonstrate in the third section that the filter residual and its theoretical bounds can also be used as an indication of whether or not the filter is working properly outside the world of simulation (i.e., the world in which the truth is not available). In the fourth section we will show that a Kalman filter can be built even if some states are unobservable. Because the filter will appear to work, we will then demonstrate some practical tests that can be used to detect that the estimates obtained from the unobservable states are worthless. Finally, in the last section we will discuss aiding. We will show an example in which there are two sensors, both of which do not work satisfactorily by themselves but when combined work well under adverse circumstances.

Sinusoidal Kalman Filter and Signal-to-Noise Ratio

In Chapter 5 we designed a linear Kalman filter that was optimized for a measurement that consisted of a sinusoidal signal corrupted by measurement noise. We showed that a linear two-state special purpose or sinusoidal Kalman filter could be designed, which yielded superior estimates when compared to other more general Kalman filters for the same problem. The fundamental and measurement matrices for the linear sinusoidal two-state Kalman filter were shown to be

$$\Phi_k = \begin{bmatrix} \cos \omega T_s & \dfrac{\sin \omega T_s}{\omega} \\ -\omega \sin \omega T_s & \cos \omega T_s \end{bmatrix}$$

$$H = \begin{bmatrix} 1 & 0 \end{bmatrix}$$

where ω is the frequency of the sinusoid, which is assumed to be known. For that system we showed in Chapter 5 that the Kalman-filtering equations were given by

$$\hat{x}_k = \cos \omega T_s \hat{x}_{k-1} + \frac{\sin \omega T_s}{\omega} \hat{\dot{x}}_{k-1} + K_{1_k} \text{Res}_k$$

$$\hat{\dot{x}}_k = -\omega \sin \omega T_s \hat{x}_{k-1} + \cos \omega T_s \hat{\dot{x}}_{k-1} + K_{2_k} \text{Res}_k$$

where the residual was defined as

$$\text{Res}_k = x_k^* - \cos \omega T_s \hat{x}_{k-1} - \frac{\sin \omega T_s}{\omega} \hat{\dot{x}}_{k-1}$$

Listing 14.1, which is identical to Listing 5.3 of Chapter 5, programs the preceding Kalman-filtering equations, which make use of the fact that the signal is sinusoidal. The noisy sinusoidal signal is an input to the Kalman filter. In this listing the nominal inputs of Chapter 5 are repeated. For example, the frequency of the sinusoid is 1 rad/s, whereas the amplitude is unity. The measurement noise has zero mean and unity standard deviation. Therefore, we can see that for this nominal case the signal-to-noise ratio is unity (i.e., amplitude of sinusoid divided by standard deviation of noise equals one or, in the notation of Listing 14.1, A/SIGNOISE = 1). We also notice from Listing 14.1 that the Kalman filter is set up without process noise, and the diagonal terms of the initial covariance matrix are set to infinity. Because the initial covariance matrix is infinite, we are really saying that we have no a priori information on how the states of the filter (i.e., estimate of x and its derivative) should be initialized. Therefore, for lack of better choices, we simply initialize the two states of the filter to zero.

The nominal case of Fig. 14.1 was run. We can see from Fig. 14.1 that when the signal-to-noise ratio is unity (i.e., in the notation of the simulation, Signal to Noise Ratio = A/SIGNOISE = 1/1 = 1) the measurement appears to be a very noisy sinusoid. Figure 14.2 shows how the sinusoidal Kalman-filter estimate of the first state compares to the actual signal. We can see that after a brief transient period the filter estimate and the actual signal are virtually identical. Therefore, we can say that the sinusoidal Kalman filter gives excellent estimates of the signal when the signal-to-noise ratio is unity.

Another case was run with Listing 14.1 in which the magnitude of the measurement noise was increased by a factor of 10. The standard deviation of the noise is 10 or, in Listing 14.1, SIGNOISE = 10. This means that the signal-to-noise ratio is now 0.1 (i.e., in the notation of the simulation the Signal to Noise Ratio = A/SIGNOISE = 1/10 = 0.1). Figure 14.3 compares the filter measurement with the actual signal. We can no longer tell that the actual measurement is a sinusoid plus noise. In fact, the actual measurement now appears to be correlated noise because it does not appear to be totally random. Figure 14.4 shows that under the new condition of a low signal-to-noise ratio the filter estimate of the signal deteriorates. However, after a brief transient period the Kalman filter estimate of x is sinusoidal in shape and appears to have the same frequency as the signal. We can also see from Fig. 14.4 that the filter estimate of x tracks the signal

Listing 14.1   Sinusoidal Kalman filter with noisy sinusoidal measurement

```
C THE FIRST THREE STATEMENTS INVOKE THE ABSOFT RANDOM
  NUMBER GENERATOR ON THE MACINTOSH
  GLOBAL DEFINE
        INCLUDE 'quickdraw.inc'
  END
  IMPLICIT REAL*8(A-H,O-Z)
  REAL*8 P(2,2),Q(2,2),M(2,2),PHI(2,2),HMAT(1,2),HT(2,1),PHIT(2,2)
  REAL*8 RMAT(1,1),IDN(2,2),PHIP(2,2),PHIPPHIT(2,2),HM(1,2)
  REAL*8 HMHT(1,1),HMHTR(1,1),HMHTRINV(1,1),MHT(2,1),K(2,1)
  REAL*8 KH(2,2),IKH(2,2)
  INTEGER ORDER
  OPEN(1,STATUS='UNKNOWN',FILE='DATFIL')
  ORDER=2
  PHIS=0.
  W=1
  A=1
  TS=.1
  XH=0.
  XDH=0.
  SIGNOISE=1.
  DO 14 I=1,ORDER
  DO 14 J=1,ORDER
  PHI(I,J)=0.
  P(I,J)=0.
  Q(I,J)=0.
  IDN(I,J)=0.
14 CONTINUE
  RMAT(1,1)=SIGNOISE**2
  IDN(1,1)=1.
  IDN(2,2)=1.
  P(1,1)=99999999999.
  P(2,2)=99999999999.
  PHI(1,1)=COS(W*TS)
  PHI(1,2)=SIN(W*TS)/W
  PHI(2,1)=-W*SIN(W*TS)
  PHI(2,2)=COS(W*TS)
  HMAT(1,1)=1.
  HMAT(1,2)=0.
  CALL MATTRN(PHI,ORDER,ORDER,PHIT)
  CALL MATTRN(HMAT,1,ORDER,HT)
  Q(1,1)=PHIS*TS**3/3
  Q(1,2)=PHIS*TS*TS/2
  Q(2,1)=Q(1,2)
  Q(2,2)=PHIS*TS
  DO 10 T=0.,20.,TS
        CALL MATMUL(PHI,ORDER,ORDER,P,ORDER,ORDER,PHIP)
        CALL MATMUL(PHIP,ORDER,ORDER,PHIT,ORDER,ORDER,
        PHIPPHIT)
```

(continued)

Listing 14.1   (*Continued*)

```
        CALL  MATADD(PHIPPHIT,ORDER,ORDER,Q,M)
        CALL  MATMUL(HMAT,1,ORDER,M,ORDER,ORDER,HM)
        CALL  MATMUL(HM,1,ORDER,HT,ORDER,1,HMHT)
        CALL  MATADD(HMHT,ORDER,ORDER,RMAT,HMHTR)
          HMHTRINV(1,1)=1./HMHTR(1,1)
        CALL  MATMUL(M,ORDER,ORDER,HT,ORDER,1,MHT)
        CALL  MATMUL(MHT,ORDER,1,HMHTRINV,1,1,K)
        CALL  MATMUL(K,ORDER,1,HMAT,1,ORDER,KH)
        CALL  MATSUB(IDN,ORDER,ORDER,KH,IKH)
        CALL  MATMUL(IKH,ORDER,ORDER,M,ORDER,ORDER,P)
        CALL  GAUSS(XNOISE,SIGNOISE)
        X=A*SIN(W*T)
        XD=A*W*COS(W*T)
        XS=X+XNOISE
        XHOLD=XH
        RES=XS-XH*COS(W*TS)-SIN(W*TS)*XDH/W
        XH=COS(W*TS)*XH+XDH*SIN(W*TS)/W+K(1,1)*RES
        XDH=-W*SIN(W*TS)*XHOLD+XDH*COS(W*TS)+K(2,1)*RES
        ERRX=X-XH
        SP11=SQRT(P(1,1))
        WRITE(9,*)T,X,XS,XH,ERRX,SP11,-SP11
        WRITE(1,*)T,X,XS,XH,ERRX,SP11,-SP11
   10 CONTINUE
        PAUSE
        CLOSE(1)
        END


C SUBROUTINE GAUSS  IS SHOWN IN LISTING 1.8
C SUBROUTINE MATTRN IS SHOWN IN LISTING 1.3
C SUBROUTINE MATMUL IS SHOWN IN LISTING 1.4
C SUBROUTINE MATADD IS SHOWN IN LISTING 1.1
C SUBROUTINE MATSUB IS SHOWN IN LISTING 1.2
```

with a delay. The delay indicates that when the signal-to-noise ration is 0.1 the Kalman filter has a lower bandwidth than it did when the signal-to-noise ratio was unity.

Another case was run with Listing 14.1 in which the magnitude of the measurement noise was increased by a factor of 100 from the nominal case. This means that the standard deviation of the noise is 100 or, in Listing 14.1, SIGNOISE = 100. The signal-to-noise ratio is now 0.01 because in the notation of the simulation A/SIGNOISE = 1/100 = 0.01. Figure 14.5 compares the filter measurement with the actual signal. The measurement still appears to be correlated noise and certainly bears no resemblance to the original sinusoidal signal. We can see from Fig. 14.6 that although the sinusoidal Kalman filter's estimate of x is sinusoidal in shape its frequency and amplitude is not that of the

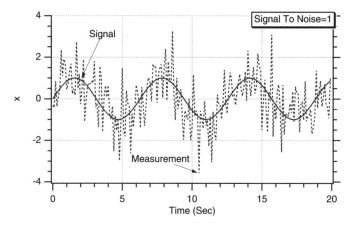

Fig. 14.1 **Measurement appears to be a noisy sinusoid when the signal-to-noise ratio is unity.**

actual signal. Thus, we can conclude that the sinusoidal Kalman filter no longer provides useful estimates if the signal-to-noise ratio is too low.

Even though it was apparent from the estimation results of Fig. 14.6 that the filter was not able to track the actual signal when the signal-to-noise ratio was too low, it is illustrative to look at another Kalman filter measure of performance. For the case in which the signal-to-noise ratio is 0.01, Fig. 14.7 presents the error in the estimate of x as a function of time along with the theoretical predictions of the filter's covariance matrix (i.e., square root of first diagonal element of covariance matrix). We can see from Fig. 14.7 that even though we know that the sinusoidal Kalman filter's estimates are virtually worthless when the signal-to-noise ratio is 0.01 the filter appears to be working correctly because the error in the estimate of

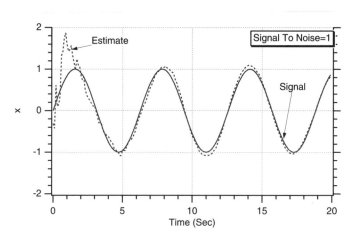

Fig. 14.2 **Sinusoidal Kalman filter yields excellent estimates of the signal when the signal-to-noise ratio is unity.**

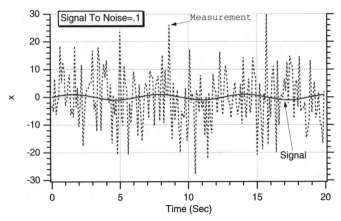

Fig. 14.3   Measurement appears to be correlated noise when signal-to-noise ratio is 0.1.

the first state is within the theoretical bounds of the covariance matrix. However, it is not immediately obvious from Fig. 14.7 that the error in the estimate is too large to be of any value in tracking a signal. If we examine Fig. 14.7 more carefully, we can see that after 20 s of filtering the error in the estimate of x is approximately 10. If we realize that the magnitude of the signal is unity, then it is apparent that an uncertainty of 10 is much too large (i.e., error in estimate is 10 times larger than signal). Therefore, it is always important to look at both the estimation and the error in the estimate results in order to get a clearer picture of the filter's adequacy for a particular application.

In all of the results presented so far, the filter was incorrectly initialized. An experiment was conducted to see how the filter performed when the signal-to-

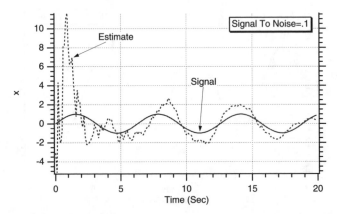

Fig. 14.4   Filter estimate deteriorates and lags signal when signal-to-noise ratio is 0.1.

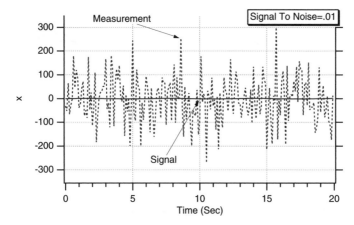

Fig. 14.5   Measurement appears to be correlated noise when signal-to-noise ratio is 0.01.

noise ratio was 0.01 but the filter states were correctly initialized (i.e., XH = 0, XDH = 1). In this example the initial covariance matrix was still infinite. Essentially, an infinite covariance matrix in this example means that the filter does not realize it is correctly initialized. Figure 14.8 shows that the estimation results are virtually unchanged with perfect initialization when the signal-to-noise ratio is 0.01. This means that if we accidently initialize the Kalman filter correctly when the initial covariance matrix is infinite it is not helpful in improving tracking performance (i.e., reducing the estimation errors) because the Kalman filter is assuming no a priori information.

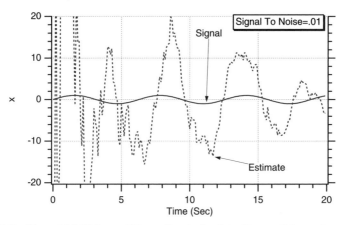

Fig. 14.6   Sinusoidal Kalman filter estimate is virtually worthless when signal-to-noise ratio is 0.01.

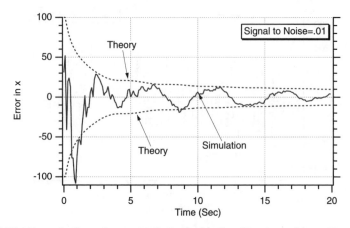

Fig. 14.7　Error in the estimate results indicates that filter is working correctly when signal-to-noise ratio is 0.01.

A final experiment was conducted in which the filter was again correctly initialized (i.e., XH = 0, and XDH = 1), but this time we told the filter about the perfect initial estimates (i.e., zero initial covariance matrix). Figure 14.9 shows that the filter estimates are now virtually perfect even though the signal-to-noise ratio is only 0.01.

Thus, we can see that under conditions of low signal-to-noise ratio the initialization of the filter states and initial covariance matrix can be very important in reducing estimation errors and improving track accuracy. In some applications providing good initial estimates to the Kalman filter is known as "warm starting" the filter.

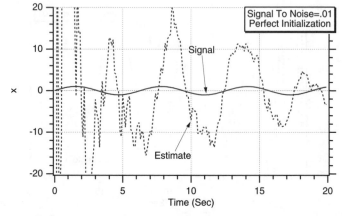

Fig. 14.8　Filter estimate is still virtually worthless if states are perfectly initialized when signal-to-noise ratio is 0.01.

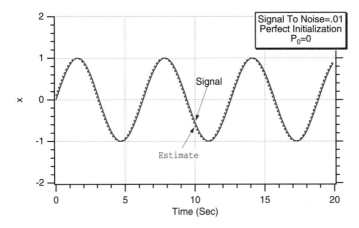

Fig. 14.9   Filter estimate is near perfect if states are perfectly initialized, and initial covariance matrix is zero when signal-to-noise ratio is 0.01.

When Only a Few Measurements Are Available

In all of the examples presented so far in the text, we were in a measurement-rich environment. In those cases often hundreds of measurements were available for the Kalman filter to process. In some applications only a few measurements are available, and the Kalman filter must make the most of them. In addition, many designers attempt to incorporate everything (including the proverbial kitchen sink) into their Kalman-filter design. The belief is that more complex filters are better (or at least to some they are more impressive). In such a case the resultant filter has many states and in fact may be overdesigned. In this section we will see what happens when a high-order Kalman filter operates in an environment in which only a few measurements are available.

In Chapter 4 we designed and evaluated different-order linear polynominal Kalman filters. For example, a first-order or two-state polynomial Kalman filter assumed that the measurement signal was a ramp (i.e., first-order signal) plus noise, whereas a second-order or three-state polynomial Kalman filter assumed that the measurement signal was a parabola (i.e., second-order signal) plus noise. For the polynomial Kalman filter the structure of the fundamental, measurement, and measurement noise matrices had patterns that were identified in Table 4.1. We can develop another table for even higher-order polynomial Kalman filters using the techniques of Chapter 4 or simply by engineering induction. The resultant fundamental, measurement, and noise matrices (scalar) for a three-, four-, and five-state Kalman filter are presented in Table 14.1. Recall that the model used in the derivation of those matrices assumed that the measurement was a polynomial plus noise. In the notation of Table 14.1, a Kalman filter with three states is described by a second-order polynomial, whereas a Kalman filter with five states is described by a fourth-order polynomial. We can see from Table 14.1 that the rows of the fundamental matrices are identical to the terms of a Taylor-series expansion for an exponential.

Table 14.1   Important matrices for different-order linear polynomial Kalman filters

States	Fundamental matrix	Measurement matrix	Measurement noise matrix
3	$\Phi_k = \begin{bmatrix} 1 & T_s & 0.5T_s^2 \\ 0 & 1 & T_s \\ 0 & 0 & 1 \end{bmatrix}$	$H = [1 \quad 0 \quad 0]$	$R_k = \sigma_n^2$
4	$\Phi_k = \begin{bmatrix} 1 & T_s & 0.5T_s^2 & \dfrac{T_s^3}{6} \\ 0 & 1 & T_s & 0.5T_s^2 \\ 0 & 0 & 1 & T_s \\ 0 & 0 & 0 & 1 \end{bmatrix}$	$H = [1 \quad 0 \quad 0 \quad 0]$	$R_k = \sigma_n^2$
5	$\Phi_k = \begin{bmatrix} 1 & T_s & 0.5T_s^2 & \dfrac{T_s^3}{6} & \dfrac{T_s^4}{24} \\ 0 & 1 & T_s & 0.5T_s^2 & \dfrac{T_s^3}{6} \\ 0 & 0 & 1 & T_s & 0.5T_s^2 \\ 0 & 0 & 0 & 1 & T_s \\ 0 & 0 & 0 & 0 & 1 \end{bmatrix}$	$H = [1 \quad 0 \quad 0 \quad 0 \quad 0]$	$R_k = \sigma_n^2$

We also showed in Chapter 4 that the various linear polynomial Kalman filters were all described by the matrix difference equation

$$\hat{x}_k = \Phi_k \hat{x}_{k-1} + K_k(z_k - H\Phi_k \hat{x}_{k-1})$$

where z_k is the scalar measurement (i.e., polynomial signal plus noise) and the other matrices are defined in Table 14.1. Recall that the Kalman gains K_k are computed from the discrete set of matrix Riccati equations given by

$$M_k = \Phi_k P_{k-1} \Phi_k^T + Q_k$$

$$K_k = M_k H^T (H M_k H^T + R_k)^{-1}$$

$$P_k = (I - K_k H) M_k$$

For the polynomial Kalman filters of Chapter 4, it was assumed that the diagonal elements of the initial covariance matrix were each infinite and that the off-diagonal elements were zero. In the work that follows we will also make the same assumptions and, in addition, also assume that the process-noise matrix Q_k is zero.

Listing 14.2 presents a simulation of a pure polynomial signal (i.e., no noise added) entering a linear polynomial Kalman filter. By changing the variable STATE, we can define the type of polynomial Kalman filter that will be used. For

Listing 14.2 **Different-order polynomial Kalman filters with a noise-free polynomial measurement**

```
C THE FIRST THREE STATEMENTS INVOKE THE ABSOFT RANDOM
  NUMBER GENERATOR ON THE MACINTOSH
     GLOBAL DEFINE
          INCLUDE 'quickdraw.inc'
     END
     INTEGER STATE
     PARAMETER(STATE=3)
     IMPLICIT REAL*8(A-H,O-Z)
     REAL*8 M(STATE,STATE),P(STATE,STATE),K(STATE,1),PHI(STATE,STATE)
     REAL*8 H(1,STATE),R(1,1),PHIT(STATE,STATE)
     REAL*8 PHIP(STATE,STATE),HT(STATE,1),KH(STATE,STATE)
     REAL*8 IKH(STATE,STATE)
     REAL*8 MHT(STATE,1),HMHT(1,1),HMHTR(1,1),HMHTRINV(1,1)
     REAL*8 IDN(STATE,STATE)
     REAL*8 Q(STATE,STATE),PHIPPHIT(STATE,STATE),XH(STATE,1)
     REAL*8 PHIX(STATE,1),HPHI(1,STATE)
     REAL*8 HPHIX(1,1)
     REAL*8 KRES(STATE,1)
     OPEN(1,STATUS='UNKNOWN',FILE='DATFIL')
     NPTS=5
     PHIS=0.
     TS=1.
     SIGNOISE=1.
     DO 1000 I=1,STATE
     DO 1000 J=1,STATE
          PHI(I,J)=0.
          P(I,J)=0.
          IDN(I,J)=0.
          Q(I,J)=0.
1000 CONTINUE
     DO 1100 I=1,STATE
          H(1,I)=0.
1100 CONTINUE
     H(1,1)=1
     CALL MATTRN(H,1,STATE,HT)
     R(1,1)=SIGNOISE**2
     IF(STATE.EQ.1)THEN
          XH(1,1)=0
          IDN(1,1)=1.
          P(1,1)=99999999999999.
          PHI(1,1)=1
     ELSEIF(STATE.EQ.2)THEN
          XH(1,1)=0
          XH(2,1)=0
          IDN(1,1)=1.
          IDN(2,2)=1.
          P(1,1)=99999999999999.
```

(*continued*)

Listing 14.2   (*Continued*)

```
           P(2,1)=99999999999999.
           PHI(1,1)=1
           PHI(1,2)=TS
           PHI(2,2)=1
     ELSEIF(STATE.EQ.3)THEN
           XH(1,1)=0
           XH(2,1)=0
           XH(3,1)=0
           IDN(1,1)=1.
           IDN(2,2)=1.
           IDN(3,3)=1.
           P(1,1)=99999999999999.
           P(2,2)=99999999999999.
           P(3,3)=99999999999999.
           PHI(1,1)=1
           PHI(1,2)=TS
           PHI(1,3)=.5*TS*TS
           PHI(2,2)=1
           PHI(2,3)=TS
           PHI(3,3)=1
     ELSEIF(STATE.EQ.4)THEN
           XH(1,1)=0
           XH(2,1)=0
           XH(3,1)=0
           XH(4,1)=0
           IDN(1,1)=1.
           IDN(2,2)=1.
           IDN(3,3)=1.
           IDN(4,4)=1.
           P(1,1)=99999999999999.
           P(2,2)=99999999999999.
           P(3,3)=99999999999999.
           P(4,4)=99999999999999.
           PHI(1,1)=1
           PHI(1,2)=TS
           PHI(1,3)=.5*TS*TS
           PHI(1,4)=TS*TS*TS/6.
           PHI(2,2)=1
           PHI(2,3)=TS
           PHI(2,4)=.5*TS*TS
           PHI(3,3)=1.
           PHI(3,4)=TS
           PHI(4,4)=1.
     ELSE
           XH(1,1)=0
           XH(2,1)=0
           XH(3,1)=0
           XH(4,1)=0
           XH(5,1)=0
           IDN(1,1)=1.
```

(*continued*)

Listing 14.2 (*Continued*)

```
            IDN(2,2)=1.
            IDN(3,3)=1.
            IDN(4,4)=1.
            IDN(5,5)=1.
            P(1,1)=99999999999999.
            P(2,2)=99999999999999.
            P(3,3)=99999999999999.
            P(4,4)=99999999999999.
            P(5,5)=99999999999999.
            PHI(1,1)=1
            PHI(1,2)=TS
            PHI(1,3)=.5*TS*TS
            PHI(1,4)=TS*TS*TS/6.
            PHI(1,5)=TS*TS*TS*TS/24.
            PHI(2,2)=1
            PHI(2,3)=TS
            PHI(2,4)=.5*TS*TS
            PHI(2,5)=TS*TS*TS/6
            PHI(3,3)=1.
            PHI(3,4)=TS
            PHI(3,5)=.5*TS*TS
            PHI(4,4)=1.
            PHI(4,5)=TS.
            PHI(5,5)=1.
ENDIF
CALL MATTRN (PHI,STATE,STATE,PHIT)
DO 10 I=1,NPTS
            T=(I-1)*TS
            CALL MATMUL(PHI,STATE,STATE,P,STATE,STATE,PHIP)
            CALL MATMUL(PHIP,STATE,STATE,PHIT,STATE,STATE,PHIPPHIT)
            CALL MATADD(PHIPPHIT,STATE,STATE,Q,M)
            CALL MATMUL(M,STATE,STATE,HT,STATE,1,MHT)
            CALL MATMUL(H,1,STATE,MHT,STATE,1,HMHT)
              HMHTR(1,1)=HMHT(1,1)+R(1,1)HMHTRINV(1,1)=1./HMHTR(1,1)
            CALL MATMUL(MHT,STATE,1,HMHTRINV,1,1,K)
            CALL MATMUL(K,STATE,1,H,1,STATE,KH)
            CALL MATSUB(IDN,STATE,STATE,KH,IKH)
            CALL MATMUL(IKH,STATE,STATE,M,STATE,STATE,P)
            CALL GAUSS(XNOISE,SIGNOISE)
            XNOISE=0.
            X=T**2+15*T-3.
            XD=2.*T+15.
            XDD=2.
            XDDD=0
            XDDDD=0.
            XS=X+XNOISE
            CALL MATMUL(PHI,STATE,STATE,XH,STATE,1,PHIX)
            CALL MATMUL(H,1,STATE,PHI,STATE,STATE,HPHI)
```

(*continued*)

Listing 14.2 (*Continued*)

```
            CALL  MATMUL(HPHI,1,STATE,XH,STATE,1,HPHIX)
              RES=XS-HPHIX(1,1)
            CALL  MATSCA(K,STATE,1,RES,KRES)
            CALL  MATADD(PHIX,STATE,1,KRES,XH)
            IF(STATE.EQ.1)THEN
                    WRITE(9,*)I,X,XH(1,1)
                    WRITE(1,*)I,X,XH(1,1)
            ELSEIF(STATE.EQ.2)THEN
                    WRITE(9,*)I,X,XH(1,1),XD,XH(2,1)
                    WRITE(1,*)I,X,XH(1,1),XD,XH(2,1)
            ELSEIF(STATE.EQ.3)THEN
                    WRITE(9,*)I,X,XH(1,1),XD,XH(2,1),XDD,XH(3,1)
                    WRITE(1,*)I,X,XH(1,1),XD,XH(2,1),XDD,XH(3,1)
            ELSEIF(STATE.EQ.4)THEN
                    WRITE(9,*)I,X,XH(1,1),XD,XH(2,1),XDD,XH(3,1),
1                       XDDD,XH(4,1)
                    WRITE(1,*)I,X,XH(1,1),XD,XH(2,1),XDD,XH(3,1),
1                       XDDD,XH(4,1)
            ELSE
                    WRITE(9,*)I,X,XH(1,1),XD,XH(2,1),XDD,XH(3,1),
1                       XDDD,XH(4,1),XDDDD,XH(5,1)
                    WRITE(1,*)I,X,XH(1,1),XD,XH(2,1),XDD,XH(3,1),
1                       XDDD,XH(4,1),XDDDD,XH(5,1)
            ENDIF
10      CONTINUE
        CLOSE(1)
        PAUSE
        END


C SUBROUTINE GAUSS IS SHOWN IN LISTING 1.8
C SUBROUTINE MATTRN IS SHOWN IN LISTING 1.3
C SUBROUTINE MATMUL IS SHOWN IN LISTING 1.4
C SUBROUTINE MATADD IS SHOWN IN LISTING 1.1
C SUBROUTINE MATSUB IS SHOWN IN LISTING 1.2
```

example, setting STATE $= 3$ will yield a three-state polynomial Kalman filter, and setting STATE $= 5$ will yield a five-state polynomial Kalman filter. For the nominal case considered the actual signal is a parabolic function of time given by

$$x = t^2 + 15t - 3$$

Therefore, the derivatives of the signal can easily be obtained using calculus and are given by

$$\dot{x} = 2t + 15$$

$$\ddot{x} = 2$$

We can see from the preceding equation that all higher-order derivatives for the parabolic signal are zero. In Listing 14.2 we are only considering the case where a few measurements are available. The number of measurements available is denoted by the variable NPTS, and the time between measurements is the sampling time TS, which is considered to be 1 s. Although the parabolic measurement signal is not corrupted by noise in this listing, we must lie to the Riccati equations and tell them there is a small amount of measurement noise (i.e., the standard deviation of the measurement noise is unity or in the notation of the simulation SIGNOISE = 1). Telling the Riccati equations that the measurement noise is zero would cause a division by zero, making the equations useless. Therefore, we always have to be gentle with the Riccati equations because, as with some individuals, they cannot always handle the truth.

The nominal case of Listing 14.2 was run in which five measurements were taken (i.e., NPTS = 5) using a three-state polynomial Kalman filter (i.e., STATE = 3) for the parabolic signal. We can see from Fig. 14.10 that the three-state filter exactly estimates the signal after only one measurement. This should not be surprising because the measurement is noise free. Therefore, the first estimate of the signal x is simply the measurement itself. After the first measurement the filter continues to estimate the signal perfectly.

We can see from Figs. 14.11 and 14.12 that three measurements are required with the three-state filter to get accurate estimates of the first and second derivatives of the signal. Although it seems reasonable that three measurements are required to estimate the second derivative of the signal, common sense indicates that only two points are required to estimate exactly the first derivative of the signal if the measurements are noise free (i.e., take the difference of the two measurements and divide by the sampling time to get first derivative). However, in order to estimate the first derivative of the signal with fewer measurements a two-state polynomial Kalman filter is required.

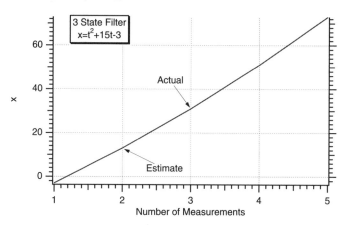

Fig. 14.10   Three-state polynomial Kalman filter estimates signal perfectly after only one measurement.

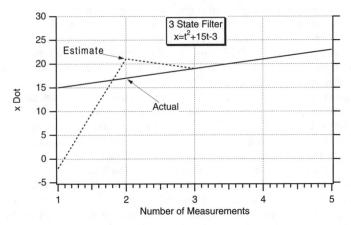

Fig. 14.11 Three-state polynomial Kalman filter requires three measurements to estimate first derivative of the signal.

Another experiment was conducted using the same parabolic, noise-free parabolic measurement signal. In this case a four-state polynomial Kalman filter (i.e., STATE = 4 in Listing 14.2) was used to estimate the various derivatives of the signal. We can see from Fig. 14.13 that the higher order polynomial Kalman filter is also able to estimate the signal after only one measurement.

However, we can also see from Fig. 14.14 that an extra measurement is now required to estimate the first derivative of the signal when using the four-state Kalman filter. Figure 14.15 indicates that the same four measurements also can be used to estimate the second derivative of the signal exactly. In addition, by

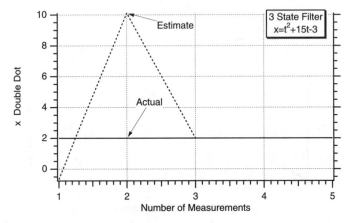

Fig. 14.12 Three-state polynomial Kalman filter requires three measurements to estimate second derivative of the signal.

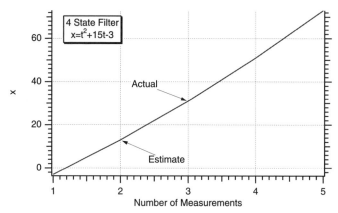

Fig. 14.13   Four-state polynomial Kalman filter estimates signal perfectly after only one measurement.

comparing Fig. 14.14 with Fig. 14.11 and Fig. 14.15 with Fig. 14.12, we can see that the excursions in the estimates of the various derivatives are greater with the higher-order filter. *The higher-order polynomial Kalman filter is less accurate than the lower-order filter in estimating the lower-order derivatives of the signal when only a few measurements are available.*

Of course, the advantage of a higher-order polynomial Kalman filter is that we are able to estimate higher-order derivatives of the signal. In other words, a higher-order filter is able to estimate more derivatives than a lower-order filter. We can see from Fig. 14.16 that four measurements are also required to estimate the third derivative of the signal.

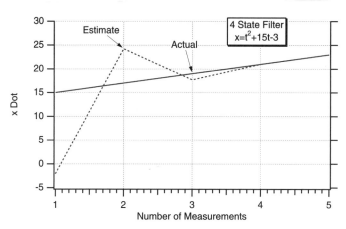

Fig. 14.14   Four-state polynomial Kalman filter requires four measurements to estimate first derivative of the signal.

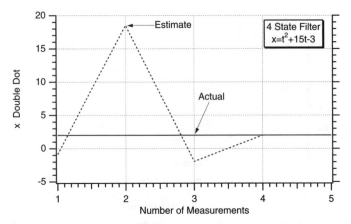

Fig. 14.15 Four-state polynomial Kalman filter requires four measurements to estimate second derivative of the signal.

Another experiment was conducted using the same parabolic, noise-free measurement signal. In this case a five-state polynomial Kalman filter (i.e., STATE = 5 in Listing 14.2) was used to estimate the various derivatives of the signal. We can see from Figs. 14.17–14.19 that the five-state polynomial Kalman filter requires five measurements to estimate the first, second, and third derivatives of measurement signal. Recall that the four-state polynomial Kalman filter only required four measurements to estimate the same states perfectly. In addition, by comparing Fig. 14.17 with Fig. 14.14 we can see that five-state filter is less accurate than the four-state polynomial Kalman filter when only a few (i.e., less than five) measurements are available. Similarly, by comparing Fig. 14.18 with

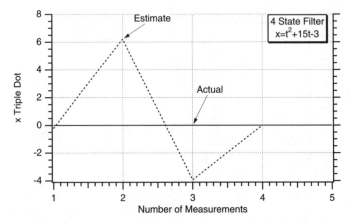

Fig. 14.16 Four-state polynomial Kalman filter requires four measurements to estimate third derivative of the signal.

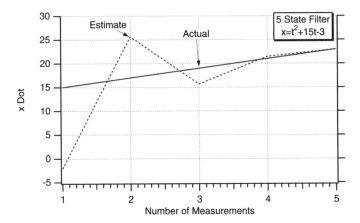

Fig. 14.17 Five-state polynomial Kalman filter requires five measurements to estimate first derivative of the signal.

Fig. 14.15 and by comparing Fig. 14.19 with Fig. 14.16, we can see that the five-state polynomial Kalman filter is less accurate than the four-state polynomial Kalman filter when estimating the second and third derivatives of the signal.

Again, the advantage of the higher-order polynomial Kalman filter is that it is able to estimate more derivatives of the signal than a lower-order filter. Figure 14.20 shows that the five-state polynomial Kalman filter is able to estimate the fourth derivative of the noise-free measurement signal perfectly after only five measurements.

The preceding experiments have demonstrated that when only a few measurements are available the lower-order polynomial Kalman filter is more accurate than the higher-order filter in estimating the lower-order derivatives of the

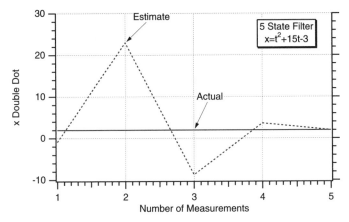

Fig. 14.18 Five-state polynomial Kalman filter requires five measurements to estimate second derivative of the signal.

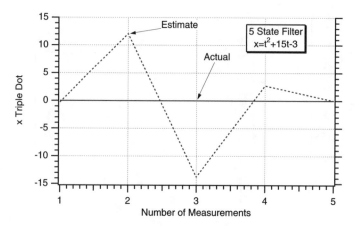

Fig. 14.19 Five-state polynomial Kalman filter requires five measurements to estimate third derivative of the signal.

measurement signal. In the noise-free case a three-state filter required three measurements to get perfect estimates of the first and second derivatives of a parabolic signal; a four-state filter required four measurements to get perfect estimates of the first, second, and third derivatives of the signal; and the five-state polynomial Kalman filter required five measurement to get estimates of the first, second, third, and fourth derivatives of the signal.

Detecting Filter Divergence in the Real World

In all of the work we have done so far in determining if the Kalman filter was working properly, we compared the errors in the estimates obtained from

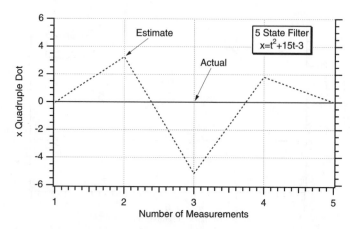

Fig. 14.20 Five-state polynomial Kalman filter requires five measurements to estimate fourth derivative of the signal.

simulation to the theoretical predictions of the covariance matrix. We considered the filter to be working properly when the simulated errors in the estimates were within the theoretical bounds approximately 68% of the time. When the filter was not working, the errors in the estimates usually rapidly departed from the theoretical bounds. Unfortunately this type of experiment can only be conducted in a simulation because the error in the estimate is the difference between the actual state and the estimated state. In actual filter operation we never know the true states of the system (i.e., if we knew the true states of the system we would not need a filter). We only have access to the estimated states provided by the filter. However, we always have access to the filter residual. The residual, which potentially contains important information, is the difference between the measurement and projected state estimates or, more simply, the quantity that multiplies the Kalman gains. In this section we will see if we can compare the measured residual to any theoretical predictions in order to determine if the filter is working properly. If a theoretical prediction of the residual can be derived, we can say that the filter is working properly if the measured residual falls within the theoretical bounds 68% of the time. If the residual continually exceeds the theoretical bounds by a large factor (i.e., three), we can conclude that the filter is diverging and that some action is required to prevent complete catastrophe (i.e., add more process noise or switch to higher-order filter). If the residual instantly exceeds the theoretical threshold bounds by a large value, the data might be bad, and we might want to ignore that measurement. The science of ignoring some measurements is also known as data editing.

Recall that the equation for the Kalman filter is given by

$$\hat{x}_k = \Phi_k \hat{x}_{k-1} + K_k(z_k - H\Phi_k \hat{x}_{k-1})$$

Therefore, the formula for the residual can be seen from the preceding expression to be the quantity in parenthesis or

$$\mathbf{Res} = z_k - H\Phi_k \hat{x}_{k-1}$$

The measurement, which is a linear combination of the states plus noise, can be rewritten as

$$z_k = Hx_k + v_k$$

In the preceding equation the new state x_k can be obtained from the old value of the state according to

$$x_k = \Phi_k x_{k-1} + w_k$$

Using substitution, we can now rewrite the equation for the residual as

$$\mathbf{Res} = H\Phi_k x_{k-1} + Hw_k + v_k - H\Phi_k \hat{x}_{k-1}$$

Combining like terms yields

$$\mathbf{Res} = H\Phi_k(x_{k-1} - \hat{x}_{k-1}) + Hw_k + v_k$$

By definition we can find the covariance of the residual by multiplying the preceding equation by its transpose and taking expectations of both sides of the equation. If we assume that the process and measurement noise are not correlated and that neither noise is correlated with the state or its estimate, we obtain, after some algebra,

$$E(\mathbf{Res*Res}^T) = H\Phi_k E[(x_{k-1} - \hat{x}_{k-1})(x_{k-1} - \hat{x}_{k-1})^T]\Phi_k^T H^T$$
$$+ HE(w_k w_k^T)H^T + E(v_k v_k^T)$$

Because

$$P_k = E[(x_{k-1} - \hat{x}_{k-1})(x_{k-1} - \hat{x}_{k-1})^T]$$

$$Q_k = E(w_k w_k^T)$$

$$R_k = E(v_k v_k^T)$$

The covariance of the residual can be reduced to

$$E(\mathbf{Res*Res}^T) = H\Phi_k P_k \Phi_k^T H^T + HQ_k H^T + R_k$$

or more simply

$$E(\mathbf{Res*Res}^T) = H(\Phi_k P_k \Phi_k^T + Q_k)H^T + R_k$$

From the first Riccati equation we know that

$$M_k = \Phi_k P_k \Phi_k^T + Q_k$$

Therefore, the formula for the covariance in the residual can be further simplified to[1,2]

$$E(\mathbf{Res*Res}^T) = HM_k H^T + R_k$$

To see if the preceding formula is accurate and useful, the three-state polynomial Kalman filter, obtained from Chapter 4, is used where an object is accelerating away from a tracking radar. This time the object travels with an initial velocity of 1000 ft/s and has an acceleration of 6 g (i.e., 193 ft/s²) away from the radar, as shown in Fig. 14.21. Initially, the object is located at the radar. It is assumed in this example that the radar can measure the distance to the object with a standard deviation measurement accuracy of 10 ft.

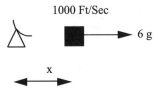

Fig. 14.21 Accelerating object traveling away from radar.

Using basic physics the downrange location of the object x (i.e., x is the distance from the radar to the object) at any time can easily be shown to be

$$x = 1000t + \frac{1}{2} \, 193.2t^2$$

To find the velocity of the object at any time, we take the first derivative of the preceding expression, yielding

$$\dot{x} = 1000 + 193.2t$$

The initial velocity of the object can be obtained from the preceding velocity formula by setting $t = 0$ and obtaining $1000 \, \text{ft/s}$, which matches our initial problem setup. To find the acceleration of the object at any time, we take the derivative of the velocity formula and obtain

$$\ddot{x} = 193.2$$

which again matches the problem set up (i.e., $193.2 \, \text{ft/s}^2 = 6 \, \text{g}$).

We can use the linear three-state polynomial Kalman filter that was derived in Chapter 4 to track the object. The simulation of the accelerating object and Kalman filter appears in Listing 14.3. Essentially, this listing is identical to Listing 4.3 of Chapter 4, except that the coefficients of the polynomial signal and the radar measurement accuracy have been changed for our new problem. In addition, in Listing 14.3 we now calculate the theoretical standard deviation of the residual and compare it to the actual measured residual. The changes from the original simulation of Listing 4.3 have been highlighted in bold in Listing 14.3.

The nominal case of Listing 14.3, in which the linear three-state polynomial Kalman filter had no process noise, was run, and the errors in the estimates of position, velocity, and acceleration appear in Figs. 14.22–14.24. We can see from the figures that the various errors in the estimates from the single run appear to be within the theoretical bounds, obtained from the Riccati equations covariance matrix, approximately 68% of the time, indicating that the Kalman filter appears to be working properly.

The residual from the same single-run results, along with the theoretical estimate of the standard deviation of the residual (derived in this section), is presented in Fig. 14.25. We can see that the formula for the standard deviation of the residual appears to be correct because the simulated results, for the filter we know to be working, lie between the theoretical bounds approximately 68% of the time. Thus, we can say that if we were monitoring the residual in a real world

Listing 14.3   Tracking accelerating object with linear three-state polynomial Kalman filter

```
C THE FIRST THREE STATEMENTS INVOKE THE ABSOFT RANDOM
NUMBER GENERATOR ON THE MACINTOSH
      GLOBAL DEFINE
            INCLUDE 'quickdraw.inc'
      END
      IMPLICIT REAL*8(A-H,O-Z)
      REAL*8 M(3,3),P(3,3),K(3,1),PHI(3,3),H(1,3),R(1,1),PHIT(3,3)
      REAL*8 PHIP(3,3),HT(3,1),KH(3,3)IKH(3,3)
      REAL*8 MHT(3,1),HMHT(1,1),HMHTR(1,1),HMHTRINV(1,1),IDN(3,3)
      REAL*8 Q(3,3),PHIPPHIT(3,3)
      INTEGER STATE
      OPEN(1,STATUS='UNKNOWN',FILE='DATFIL')
      OPEN(2,STATUS='UNKNOWN',FILE='COVFIL')
      ORDER=3
      PHIS=0.
      TS=1.
      A0=0.
      A1=1000.
      A2=193.2/2.
      XH=0
      XDH=0
      XDDH=0
      SIGNOISE=10.
      DO 1000 I=1,ORDER
      DO 1000 J=1,ORDER
            PHI(I,J)=0.
            P(I,J)=0.
            IDN(I,J)=0.
            Q(I,J)=0.
1000  CONTINUE
      IDN(1,1)=1.
      IDN(2,2)=1.
      IDN(3,3)=1.
      P(1,1)=99999999999999.
      P(2,2)=99999999999999.
      P(3,3)=99999999999999.
      PHI(1,1)=1
      PHI(1,2)=TS
      PHI(1,3)=.5*TS*TS
      PHI(2,2)=1
      PHI(2,3)=TS
      PHI(3,3)=1
      DO 1100 I=1,ORDER
            H(1,I)=0.
1100  CONTINUE
      H(1,1)=1
      CALL MATTRN(H,1,ORDER,HT)
      R(1,1)=SIGNOISE**2
```

(continued)

Listing 14.3 (*Continued*)

```
CALL MATTRN(PHI,ORDER,ORDER,PHIT)
Q(1,1)=PHIS*TS**5/20
Q(1,2)=PHIS*TS**4/8
Q(1,3)=PHIS*TS**3/6
Q(1,2)=Q(1,2)
Q(2,2)=PHIS*TS**3/3
Q(2,3)=PHIS*TS*TS/2
Q(3,1)=Q(1,3)
Q(3,2)=Q(2,3)
Q(3,3)=PHIS*TS
DO 10 T=0.,10.,TS
       CALL MATMUL(PHI,ORDER,ORDER,P,ORDER,ORDER,PHIP)
       CALL MATMUL(PHIP,ORDER,ORDER,PHIT,ORDER,ORDER,
       PHIPPHIT)
       CALL MATADD(PHIPPHIT,ORDER,ORDER,Q,M)
       CALL MATMUL(M,ORDER,ORDER,HT,ORDER,1,MHT)
       CALL MATMUL(H,1,ORDER,MHT,ORDER,1,HMHT)HMHTR(1,1)=
       HMHT(1,1)+R(1,1)HMHTRINV(1,1)=1./HMHTR(1,1)
       CALL MATMUL(MHT,ORDER,1,HMHTRINV,1,1,K)
       CALL MATMUL(K,ORDER,1,H,1,ORDER,KH)
       CALL MATSUB(IDN,ORDER,ORDER,KH,IKH)
       CALL MATMUL(IKH,ORDER,ORDER,M,ORDER,ORDER,P)
       CALL GAUSS(XNOISE,SIGNOISE)
       X=A0+A1*T+A2*T*T
       XD=A1+2*A2*T
       XDD=2*A2
       XS=X+XNOISE
       RES=XS-XH-TS*XDH-.5*TS*TS*XDDH
       XH=XH+XDH*TS+.5*TS*TS*XDDH+K(1,1)*RES
       XDH=XDH+XDDH*TS+K(2,1)*RES
       XDDH=XDDH+K(3,1)*RES
       SP11=SQRT(P(1,1))
       SP22=SQRT(P(2,2))
       SP33=SQRT(P(3,3))
       SP44=SQRT(HMHTR(1,1))
       XHERR=X-XH
       XDHERR=XD-XDH
       XDDHERR=XDD-XDDH
       WRITE(9,*)T,X,XH,XD,XDH,XDD,XDDH
       WRITE(1,*)T,X,XH,XD,XDH,XDD,XDDH
       WRITE(2,*)T,XHERR,SP11,-SP11,XDHERR,SP22,-SP22,
     1           XDDHERR,SP33,-SP33,RES,SP44,-SP44
10 CONTINUE
   CLOSE(1)
   CLOSE(2)
   PAUSE
   END
```

(*continued*)

Listing 14.3 (*Continued*)

```
C SUBROUTINE GAUSS IS SHOWN IN LISTING 1.8
C SUBROUTINE MATTRN IS SHOWN IN LISTING 1.3
C SUBROUTINE MATMUL IS SHOWN IN LISTING 1.4
C SUBROUTINE MATADD IS SHOWN IN LISTING 1.1
C SUBROUTINE MATSUB IS SHOWN IN LISTING 1.2
```

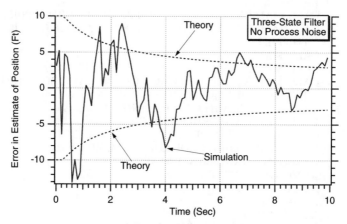

Fig. 14.22   Error in the estimate of position is within the theoretical bounds.

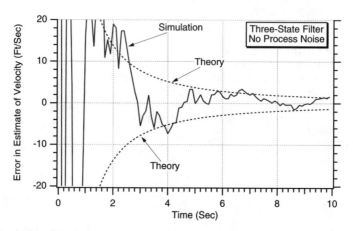

Fig. 14.23   Error in the estimate of velocity is within the theoretical bounds.

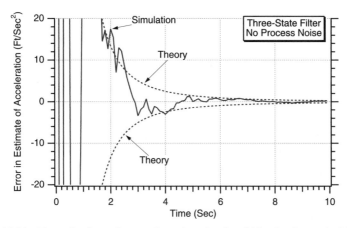

Fig. 14.24 Error in the estimate of acceleration is within the theoretical bounds.

application of the Kalman filter the results of Fig. 14.25 would indicate that the filter was working properly.

The preceding example showed that when the Kalman filter was working properly the residual and its theoretical bounds also indicated that the filter was working. Now we will consider an example in which the filter is not working. We know from Chapter 4 that if we use a linear two-state polynomial Kalman filter without process noise on the same problem the filter should not work because a two-state filter (or position-velocity filter) cannot track an accelerating object. Listing 14.4 presents a two-state liner polynomial Kalman filter and the accelerating object. The listing is virtually identical to Listing 4.4, where the two-state linear polynomial Kalman filter was first derived, except that the polynomial signal coefficients are different and the radar noise standard deviation has been

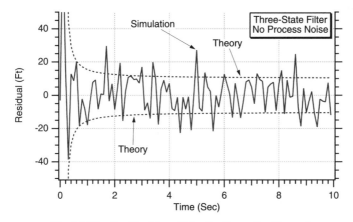

Fig. 14.25 Residual is within the theoretical bounds.

Listing 14.4   Tracking accelerating object with linear two-state polynomial Kalman filter.

```
C THE FIRST THREE STATEMENTS INVOKE THE ABSOFT RANDOM
  NUMBER GENERATOR ON THE MACINTOSH
     GLOBAL DEFINE
            INCLUDE 'quickdraw.inc'
     END
     IMPLICIT REAL*8(A-H,O-Z)
     REAL*8 P(2,2),Q(2,2),M(2,2)PHI(2,2)HMAT(1,2),HT(2,1),PHIT(2,2)
     REAL*8 RMAT(1,1),IDN(2,2),PHIP(2,2),PHIPPHIT(2,2),HM(1,2)
     REAL*8 HMHT(1,1),HMHTR(1,1),HMHTRINV(1,1),MHT(2,1),K(2,1)
     REAL*8 KH(2,2),IKH(2,2)
     INTEGER STATE
     TS=1.
     PHIS=10000.
     A0=0.
     A1=1000.
     A2=193.2/2.
     XH=0.
     XDH=0.
     SIGNOISE=10.
     ORDER=2
     OPEN(1,STATUS='UNKNOWN',FILE='DATFIL')
     OPEN(2,STATUS='UNKNOWN',FILE='COVFIL')
     T=0.
     S=0.
     H=.001
     DO 14 I=1,ORDER
     DO 14 J=1,ORDER
            PHI(I,J)=0.
            P(I,J)=0.
            Q(I,J)=0.
            IDN(I,J)=0.
14   CONTINUE
     RMAT(1,1)=SIGNOISE**2
     IDN(1,1)=1.
     IDN(2,2)=1.
     P(1,1)=99999999999999.
     P(2,2)=99999999999999.
     PHI(1,1)=1.
     PHI(1,2)=TS
     PHI(2,2)=1.
     Q(1,1)=TS*TS*TS*PHIS/3.
     Q(1,2)=.5*TS*TS*PHIS
     Q(2,1)=Q(1,2)
     Q(2,2)=PHIS*TS
     HMAT(1,1)=1.
     HMAT(1,2)=0.
     DO 10 T=0.,10.,TS
     CALL MATTRN(PHI,ORDER,ORDER, PHIT)
```

(*continued*)

Listing 14.4 (*Continued*)

```
      CALL  MATTRN(HMAT,1,ORDER,HT)
      CALL  MATMUL(PHI,ORDER,ORDER,P,ORDER,ORDER,PHIP)
      CALL  MATMUL(PHIP,ORDER,ORDER,PHIT,ORDER,ORDER,PHIPPHIT)
      CALL  MATADD(PHIPPHIT,ORDER,ORDER,Q,M)
      CALL  MATMUL(HMAT,1,ORDER,M,ORDER,ORDER,HM)
      CALL  MATMUL(HM,1,ORDER,HT,ORDER,1,HMHT)
      CALL  MATADD(HMHT,ORDER,ORDER,RMAT,HMHTR)
      HMHTRINV(1,1)=1./HMHTR(1,1)
      CALL  MATMUL(M,ORDER,ORDER,HT,ORDER,1,MHT)
      CALL  MATMUL(MHT,ORDER,1,HMHTRINV,1,1,K)
      CALL  MATMUL(K,ORDER,1,HMAT,1,ORDER,KH)
      CALL  MATSUB(IDN,ORDER,ORDER,KH,IKH)
      CALL  MATMUL(IKH,ORDER,ORDER,M,ORDER,ORDER,P)
      CALL  GAUSS(XNOISE,SIGNOISE)
      X=A0+A1*T+A2*T*T
      XD=A1+2*A2*T
      XS=X+XNOISE
      RES=XS-XH-TS*XDH
      XH=XH+XDH*TS+K(1,1)*RES
      XDH=XDH+K(2,1)*RES
      SP11=SQRT(P(1,1))
      SP22=SQRT(P(2,2))
      SP44=SQRT(HMHTR(1,1))
      XHERR=X-XH
      XDHERR=XD-XDH
      WRITE(9,*)T,XD,XDH,K(1,1),K(2,1)
      WRITE(1,*)T,X,XH,XD,XDH
      WRITE(2,*)T,XHERR,SP11,-SP11,XDHERR,SP22,-SP22,RES,SP44,-SP44
   14 CONTINUE
      PAUSE
      CLOSE(1)
      END


C SUBROUTINE GAUSS  IS SHOWN IN LISTING 1.8
C SUBROUTINE MATTRN IS SHOWN IN LISTING 1.3
C SUBROUTINE MATMUL IS SHOWN IN LISTING 1.4
C SUBROUTINE MATADD IS SHOWN IN LISTING 1.1
C SUBROUTINE MATSUB IS SHOWN IN LISTING 1.2
```

reduced to 10 ft. In addition, the linear polynomial Kalman-filter equations are slightly different because in this example we are not compensating for gravity. Initially, the process-noise parameter PHIS has been set to zero, indicating that there is no process noise. All of the changes from the original listing are highlighted in bold in Listing 14.4.

The nominal case of Listing 14.4, in which the filter had no process noise, was run, and the results are displayed in Fig. 14.26. We can see that the simulated error in the estimate of position for the two-state filter is outside the theoretical

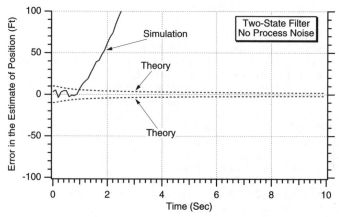

Fig. 14.26   Error in the estimate of position for two-state filter without process noise is diverging.

error bounds, indicating that the filter is not working properly. In fact, the simulated error in the estimate of position is diverging from the error bounds, indicating that the two-state filter is unable to track the accelerating target.

The residual of the two-state linear polynomial Kalman filter which is available in the real world, can also be seen to be diverging from the theoretical predictions in Fig. 14.27. Thus, if we were monitoring the residual, we could tell within 2 s that the two-state filter was clearly not working properly and that some action must be taken.

Another case was run with Listing 14.4. This time process noise was added to the two-state linear polynomial Kalman filter (i.e., PHIS = 10000). Figures 14.28 and 14.29 now indicate that the filter appears to be working because the errors in

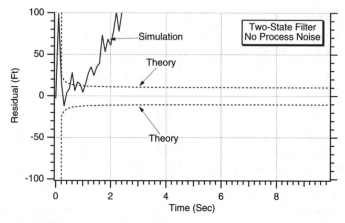

Fig. 14.27   Residual also indicates that two-state filter without process noise can not track accelerating object.

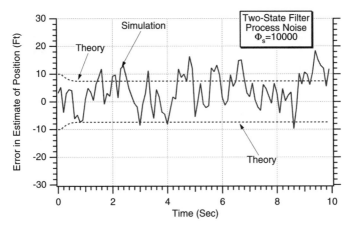

Fig. 14.28 Addition of process noise enables errors in estimate of position to lie within theoretical bounds.

the estimates of position and velocity are within the theoretical bounds. However, if we compare the two-state filter results of Figs. 14.28 and 14.29 with the three-state filter results of Figs. 14.22 and 14.23, we can see that the three-state filter is working better than the two-state filter because the errors in the estimates of position and velocity are smaller.

The residual results of Fig. 14.30 also indicate that the two-state filter with process noise appears to be working properly because the residual is within the theoretical bounds approximately 68% of the time.

Thus, we have demonstrated that the residual and its theoretical bounds also can be used as an indication of whether or not the filter is working properly. The

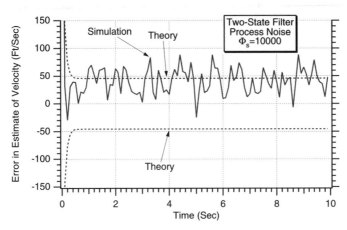

Fig. 14.29 Addition of process noise enables errors in estimate of velocity to lie within theoretical bounds.

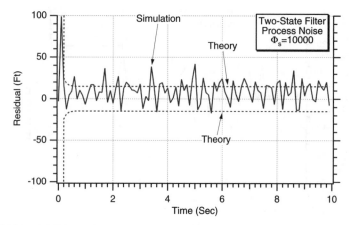

Fig. 14.30   Addition of process noise enables residual to lie within theoretical bounds.

residual test not only appears to be a useful test that can be used in the actual operation of the filter to see if it is performing according to theory, but it also can be used to determine if corrective actions to the filter are required.

Observability Example

Theoretically, there are tests one must conduct in order to see if the states chosen for the Kalman filter are observable. In many cases the tests are too complicated to use and are either avoided or simply forgotten. The purpose of this section is to show that it is possible to build a Kalman filter with an unobservable state. The filter will appear to operate, but it is up to the designer to figure out the appropriate tests to ensure that the filter state estimates are meaningful and observable. If it is determined that a state is not observable, then the state can be eliminated from the filter. The elimination of unnecessary states will reduce the computational complexity of the filter without compromising filter performance.

In Listing 14.4 we had a two-state linear polynomial Kalman filter acting on a parabolic signal (i.e., accelerating object) corrupted by noise. The filter attempted to estimate the signal and its derivative (i.e., x and $\dot{x}$). Let us now use the same program for a simpler example in which the signal is a ramp corrupted by noise. In this case the filter is of high enough order to track the signal without building up truncation error (i.e. the filter will not diverge), and therefore process noise is not required. We will use the same signal and noise as was used in Chapter 3 (see Listing 3.3), where the appropriate parameters are given by

$$\text{SIGNOISE} = 5.$$
$$\text{A0} = 3.$$
$$\text{A1} = 1.$$
$$\text{A2} = 0.$$

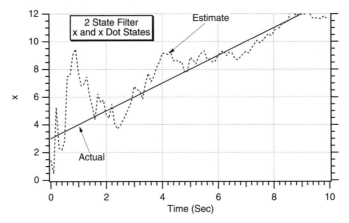

Fig. 14.31   After a brief transient period the two-state filter is able to estimate the signal.

The nominal case of Listing 14.4 was run with the preceding parameter changes, and the actual signal and derivative along with their estimates are presented in Figs. 14.31–14.32. We can see that the two-state linear polynomial Kalman filter is able to track the signal and its derivative accurately after a brief transient period.

The two-state linear polynomial Kalman filter can be formulated in another way so that rather than estimating the signal and its derivative we could alternately estimate the coefficients of a first-order polynomial describing the signal. If the measured signal is given by

$$x^* = a_0 + a_1 t + v$$

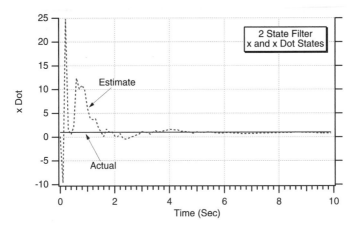

Fig. 14.32   After a brief transient period the two-state filter is able to estimate the derivative of the signal.

where t is time and v is the measurement noise, we desire to estimate the coefficients a_0 and a_1. Because the coefficients of the preceding polynomial are constants, their derivatives must be zero. Therefore, it we choose the coefficients as states, then the resulting state-space equation for our model of the real world becomes

$$\begin{bmatrix} \dot{a}_0 \\ \dot{a}_1 \end{bmatrix} = \begin{bmatrix} 0 & 0 \\ 0 & 0 \end{bmatrix} \begin{bmatrix} a_0 \\ a_1 \end{bmatrix}$$

Because the systems dynamics matrix of the preceding equation is zero, the fundamental matrix must be the identity matrix. The measurement equation can be expressed in state-space form as

$$x^* = \begin{bmatrix} 1 & t \end{bmatrix} \begin{bmatrix} a_0 \\ a_1 \end{bmatrix} + v$$

It can easily be shown from the preceding two equations that the fundamental and measurement matrices are given by

$$\Phi_k \begin{bmatrix} 1 & 0 \\ 0 & 1 \end{bmatrix} \quad \text{and} \quad H = \begin{bmatrix} 1 & t \end{bmatrix}$$

The other matrices are identical to those of Listing 14.4, and we now have enough information to program and solve the Riccati equations. Because the Kalman-filtering equation is given by

$$\hat{x}_k = \Phi_k \hat{x}_{k-1} + K_k(z_k - H\Phi_k\hat{x}_{k-1})$$

substitution of the appropriate matrices yields

$$\begin{bmatrix} \hat{a}_{0_k} \\ \hat{a}_{1_k} \end{bmatrix} = \begin{bmatrix} 1 & 0 \\ 0 & 1 \end{bmatrix} \begin{bmatrix} \hat{a}_{0_{k-1}} \\ \hat{a}_{1_{k-1}} \end{bmatrix} + \begin{bmatrix} K_{1_k} \\ K_{2_k} \end{bmatrix} \left(x^* - \begin{bmatrix} 1 & t \end{bmatrix} \begin{bmatrix} 1 & 0 \\ 0 & 1 \end{bmatrix} \begin{bmatrix} \hat{a}_{0_{k-1}} \\ \hat{a}_{1_{k-1}} \end{bmatrix} \right)$$

After multiplying out the terms of the preceding equation, we get the two scalar Kalman-filtering equations for estimating the polynomial coefficients:

$$\hat{a}_{0_k} = \hat{a}_{0_{k-1}} + K_{1_k}(x_k^* - \hat{a}_{0_{k-1}} - \hat{a}_{1_{k-1}}t)$$

$$\hat{a}_{1_k} = \hat{a}_{1_{k-1}} + K_{2_k}(x_k^* - \hat{a}_{0_{k-1}} - \hat{a}_{1_{k-1}}t)$$

The preceding Kalman-filtering equations, along with the matrices required for the Riccati equations, were programmed and appear in Listing 14.5. The differences between this simulation and that of Listing 14.4 are highlighted in bold.

The nominal case of Listing 14.5 was run, and the actual and estimated ramp coefficients are displayed in Figs. 14.33 and 14.34. We can see from Fig. 14.33 that the Kalman filter approximately estimates the first coefficient (i.e., to within

Listing 14.5   Using coefficients as states for linear two-state Kalman filter tracking noisy ramp

```
C THE FIRST THREE STATEMENTS INVOKE THE ABSOFT RANDOM
  NUMBER GENERATOR ON THE MACINTOSH GLOBAL DEFINE
      GLOBAL DEFINE
              INCLUDE 'quickdraw.inc'
      END
      IMPLICIT REAL*8(A-H,O-Z)
      REAL*8  P(2,2),Q(2,2),M(2,2),PHI(2,2),HMAT(1,2),HT(2,1),PHIT(2,2)
      REAL*8  RMAT(1,1),IDN(2,2),PHIP(2,2),PHIPPHIT(2,2),HM(1,2)
      REAL*8  HMHT(1,1),HMHTR(1,1),HMHTRINV(1,1),MHT(2,1),K(2,1)
      REAL*8  KH(2,2),IKH(2,2)
      INTEGER ORDER
      TS=.1
      PHIS=0.
      A0=3.
      A1=1.
      A2=0.
      A0H=0.
      A1H=0.
      SIGNOISE=5.
      ORDER=2
      OPEN(1,STATUS='UNKNOWN',FILE='DATFIL')
      OPEN(2,STATUS='UNKNOWN',FILE='COVFIL')
      T=0.
      S=0.
      H=.001
      DO 14 I=1,ORDER
      DO 14 J=1,ORDER
      PHI(I,J)=0.
      P(I,J)=0.
      Q(I,J)=0.
      IDN(I,J)=0.
  14  CONTINUE
      RMAT(1,1)=SIGNOISE**2
      IDN(1,1)=1.
      IDN(2,2)=1.
      PHI(1,1)=1.
      PHI(2,2)=1.
      P(1,1)=99999999999.
      P(2,2)=99999999999.
      DO 10 T=0.,10.,TS
      HMAT(1,1)=1.
      HMAT(1,2)=T.
      CALL  MATTRN(PHI,ORDER,ORDER,PHIT)
      CALL  MATTRN(HMAT,1,ORDER,HT)
      CALL  MATMUL(PHI,ORDER,ORDER,P,ORDER,ORDER,PHIP)
      CALL  MATMUL(PHIP,ORDER,ORDER,PHIT,ORDER,ORDER,PHIPPHIT)
      CALL  MATADD(PHIPPHIT,ORDER,ORDER,Q,M)
      CALL  MATMUL(HMAT,1,ORDER,M,ORDER,ORDER,HM)
```

(continued)

Listing 14.5   *(Continued)*

```
      CALL  MATMUL(HM,1,ORDER,HT,ORDER,1,HMHT)
      CALL  MATADD(HMHT,ORDER,ORDER,RMAT,HMHTR)
        HMHTRINV(1,1)=1./HMHTR(1,1)
      CALL  MATMUL(M,ORDER,ORDER,HT,ORDER,1,MHT)
      CALL  MATMUL(MHT,ORDER,1,HMHTRINV,1,1,K)
      CALL  MATMUL(K,ORDER,1,HMAT,1,ORDER,KH)
      CALL  MATSUB(IDN,ORDER,ORDER,KH,IKH)
      CALL  MATMUL(IKH,ORDER,ORDER,M,ORDER,ORDER,P)
      CALL  GAUSS(XNOISE,SIGNOISE)
      X=A0+A1*T+A2*T*T
      XD=A1+2*A2*T
      XS=X+XNOISE
      RES=XS-A0H-A1H*T
      A0H=A0H+K(1,1)*RES
      A1H=A1H+K(2,1)*RES
      SP11=SQRT(P(1,1))
      SP22=SQRT(P(2,2))
      A0HERR=A0-A0H
      A1HERR=A1-A1H
      XH=A0H+A1H*T
      XDH=A1H
      WRITE(9,*)T,A0,A0H,A1,A1H,X,XH,XD,XDH
      WRITE(1,*)T,A0,A0H,A1,A1H,X,XH,XD,XDH
      WRITE(2,*)T,A0HERR,SP11,-SP11,A1HERR,SP22,-SP22
10    CONTINUE
      PAUSE
      CLOSE(1)
      END


C SUBROUTINE GAUSS IS SHOWN IN LISTING 1.8
C SUBROUTINE MATTRN IS SHOWN IN LISTING 1.3
C SUBROUTINE MATMUL IS SHOWN IN LISTING 1.4
C SUBROUTINE MATADD IS SHOWN IN LISTING 1.1
C SUBROUTINE MATSUB IS SHOWN IN LISTING 1.2
```

33%) after a brief transient period. Figure 14.34 shows that the filter does a much better job in estimating the second coefficient. Note that the initial estimates for both coefficients were zero in this example.

From the polynomial coefficients we can reconstruct the estimated signal and its derivative according to

$$\hat{x}_k = \hat{a}_{0_k} + \hat{a}_{1_k} t$$
$$\hat{\dot{x}}_k = \hat{a}_{1_k}$$

Figures 14.35 and 14.36 show that the estimates of the signal and its derivative, based on the alternative Kalman filter's estimates of the polynomial coefficients, are comparable in quality to the estimates of the preceding Kalman

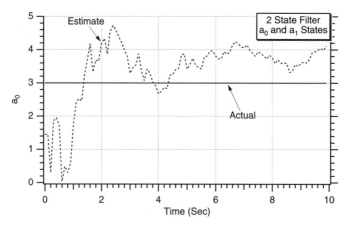

Fig. 14.33 **Alternative Kalman filter estimates first polynomial coefficient to within 33%.**

filter (see Figs. 14.31 and 14.32). Therefore, we can conclude that the alternative Kalman filter for obtaining polynomial coefficients (which in turn can be used to obtain the derivatives of a polynomial in time) appears to be working properly.

To see if the alternative Kalman filter was robust, the first state was severely misinitialized from its nominal value of zero to 100. If the state were perfectly initialized in this example, the correct value would be 3. We can see from Fig. 14.37 that after a brief transient period the filter is able to recover and accurately estimate a_0. *Usually a linear Kalman filter has no problem in recovering from poor initialization.*

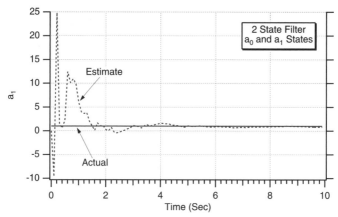

Fig. 14.34 **Alternative Kalman filter does very well in estimating second polynomial coefficient.**

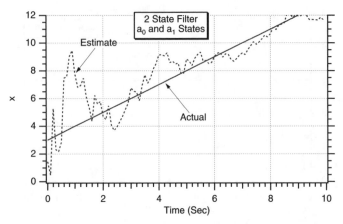

Fig. 14.35   Using alterative Kalman filter estimates yields good estimates of signal.

The original measurement equation can also be written in less efficient form as

$$x^* = a_0 + a_1 t + a_2 t + v$$

Note that the measurement is still a ramp (i.e., the preceding equation is still linear in time) corrupted by noise. In this example the sum of the coefficients a_1 and a_2 will be identical to the a_1 coefficient of the preceding example. We will now attempt to build a Kalman filter that will estimate all three coefficients of the ramp. Because the coefficients of the preceding polynomial are constants, their derivatives must also be zero. Therefore, if we choose the three coefficients as

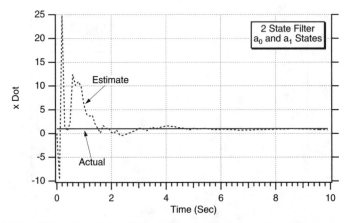

Fig. 14.36   Using alternative Kalman filter estimates yields good estimates of the derivative of the signal.

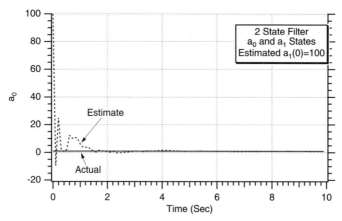

Fig. 14.37   Alternative Kalman filter is able to recover from poor initialization.

states, then the resulting state-space equation for our model of the real world becomes

$$\begin{bmatrix} \dot{a}_0 \\ \dot{a}_1 \\ \dot{a}_2 \end{bmatrix} = \begin{bmatrix} 0 & 0 & 0 \\ 0 & 0 & 0 \\ 0 & 0 & 0 \end{bmatrix} \begin{bmatrix} a_0 \\ a_1 \\ a_2 \end{bmatrix}$$

Because the systems dynamics matrix of the preceding equation is zero, the fundamental matrix must be the identity matrix. The measurement equation can now be expressed in terms of the states as

$$x^* = \begin{bmatrix} 1 & t & t \end{bmatrix} \begin{bmatrix} a_0 \\ a_1 \\ a_2 \end{bmatrix} + v$$

Therefore, the fundamental and measurement matrices are given by

$$\Phi_k = \begin{bmatrix} 1 & 0 & 0 \\ 0 & 1 & 0 \\ 0 & 0 & 1 \end{bmatrix} \quad \text{and} \quad H = \begin{bmatrix} 1 & t & t \end{bmatrix}$$

The other matrices are identical to those of Listing 14.5, and we now have enough information required to program the Riccati equations. Recall that the Kalman-filtering equation is given by

$$\hat{x}_k = \Phi_k \hat{x}_{k-1} + K_k(z_k - H\Phi_k \hat{x}_{k-1})$$

Substitution of the appropriate matrices yields

$$
\begin{bmatrix} \hat{a}_{0_k} \\ \hat{a}_{1_k} \\ \hat{a}_{2_k} \end{bmatrix} = \begin{bmatrix} 1 & 0 & 0 \\ 0 & 1 & 0 \\ 0 & 0 & 1 \end{bmatrix} \begin{bmatrix} \hat{a}_{0_{k-1}} \\ \hat{a}_{1_{k-1}} \\ \hat{a}_{2_{k-1}} \end{bmatrix} + \begin{bmatrix} K_{1_k} \\ K_{2_k} \\ K_{3_k} \end{bmatrix} \left(x^* - \begin{bmatrix} 1 & t & t \end{bmatrix} \begin{bmatrix} 1 & 0 & 0 \\ 0 & 1 & 0 \\ 0 & 0 & 1 \end{bmatrix} \begin{bmatrix} \hat{a}_{0_{k-1}} \\ \hat{a}_{1_{k-1}} \\ \hat{a}_{2_{k-1}} \end{bmatrix} \right)
$$

After multiplying out the terms of the preceding equation, we get the three scalar Kalman-filtering equations for estimating the polynomial coefficients

$$
\hat{a}_{0_k} = \hat{a}_{0_{k-1}} + K_{1_k}(x_k^* - \hat{a}_{0_{k-1}} - \hat{a}_{1_{k-1}}t - \hat{a}_{2_{k-1}}t)
$$

$$
\hat{a}_{1_k} = \hat{a}_{1_{k-1}} + K_{2_k}(x_k^* - \hat{a}_{0_{k-1}} - \hat{a}_{1_{k-1}}t - \hat{a}_{2_{k-1}}t)
$$

$$
\hat{a}_{2_k} = \hat{a}_{2_{k-1}} + K_{3_k}(x_k^* - \hat{a}_{0_{k-1}} - \hat{a}_{1_{k-1}}t - \hat{a}_{2_{k-1}}t)
$$

The preceding Kalman-filtering equations, along with the matrices required for the Riccati equations, were programmed and appear in Listing 14.6. The differences between this simulation and that of Listing 14.5 are highlighted in bold. Notice that the sum of the coefficients multiplying time is unity (i.e., $a_1 + a_2 = 0.1 + 0.9 = 1$), which is the same as the coefficient multiplying the time term in Listing 14.5.

The nominal case of Listing 14.6 was run. We can see from Fig. 14.38 that the estimate of a_1 asymptotically approaches a constant that is slightly on the high side. From Fig. 14.39 we can see that the estimate of a_2 also asymptotically approaches a constant. However, the estimate of a_2 is slightly on the low side.

Although the estimates of the polynomial coefficients approaches asymptotic constants that were slightly wrong, Figs. 14.40 and 14.41 show that the estimates of the signal x and its derivative $\dot{x}$ are of the same quality as the two preceding Kalman filters (see Figs. 14.35 and 14.36 and Figs. 14.31 and 14.32). Therefore, it appears that although the Kalman filter was not able to estimate a_1 and a_2 individually the filter was able to determine the effect of their combination. In fact, we will demonstrate that the estimates of a_1 and a_2 individually are completely dependent on the initial conditions assumed for these states.

To test the robustness of the three-state Kalman filter, the initial estimate of a_1 was changed for 0 to 100. We can see from Figs. 14.42 and 14.43 that now the Kalman filter is clearly unable to estimate either a_1 or a_2. The filter's estimates of these states are dependent on the initial conditions and do not change after only a few measurements are taken. Normally if there is no process noise, the filter estimates should get better as more measurements are taken. The lack of improvement in the estimates of a_1 and a_2 indicate that these states are not observable individually.

Although the Kalman filter was unable to estimate a_1 and a_2 individually under conditions of poor initialization, Fig. 14.44 shows that the sum of a_1 and a_2 (obtained by adding the estimates from Figs. 14.42 and 14.43) can be estimated near perfectly. Of course, we essentially estimated the sum with the preceding two-state filter with far less computation (i.e., Riccati equations consisted of 2×2 matrices, rather than 3×3 matrices).

Listing 14.6   Using coefficients as states for linear three-state Kalman filter

```
C THE FIRST THREE STATEMENTS INVOKE THE ABSOFT RANDOM
  NUMBER GENERATOR ON THE MACINTOSH
    GLOBAL DEFINE
            INCLUDE 'quickdraw.inc'
    END
    IMPLICIT REAL*8(A-H,O-Z)
    REAL*8 P(3,3),Q(3,3),M(3,3),PHI(3,3),HMAT(1,3),HT(3,1),PHIT(3,3)
    REAL*8 RMAT(1,1),IDN(3,3),PHIP(3,3),PHIPPHIT(3,3),HM(1,3)
    REAL*8 HMHT(1,1),HMHTR(1,1),HMHTRINV(1,1),MHT(3,1),K(3,1)
    REAL*8 KH(3,3),IKH(3,3)
    INTEGER ORDER
    TS=.1
    PHIS=0.
    A0=3.
    A1=.1
    A2=.9
    A0H=0.
    A1H=0.
    A2H=0.
    SIGNOISE=5.
    ORDER=3
    OPEN(1,STATUS='UNKNOWN',FILE='DATFIL')
    OPEN(2,STATUS='UNKNOWN',FILE='COVFIL')
    T=0.
    S=0.
    H=.001
    DO 14 I=1,ORDER
    DO 14 J=1,ORDER
    PHI(I,J)=0.
    P(I,J)=0.
    Q(I,J)=0.
    IDN(I,J)=0.
 14 CONTINUE
    RMAT(1,1)=SIGNOISE**2
    IDN(1,1)=1.
    IDN(2,2)=1.
    IDN(3,3)=1.
    PHI(1,1)=1.
    PHI(2,2)=1.
    PHI(3,3)=1.
    P(1,1)=99999999999.
    P(2,2)=99999999999.
    P(3,3)=99999999999.
    DO 10 T=0.,10.,TS
    HMAT(1,1)=1.
    HMAT(1,2)=T
    HMAT(1,3)=T
    CALL MATTRN(PHI,ORDER,ORDER,PHIT)
    CALL MATTRN(HMAT,1,ORDER,HT)
```

(continued)

Listing 14.6 *(Continued)*

```
CALL  MATMUL(PHI,ORDER,ORDER,P,ORDER,ORDER,PHIP)
CALL  MATMUL(PHIP,ORDER,ORDER,PHIT,ORDER,ORDER,PHIPPHIT)
CALL  MATADD(PHIPPHIT,ORDER,ORDER,Q,M)
CALL  MATMUL(HMAT,1,ORDER,M,ORDER,ORDER,HM)
CALL  MATMUL(HM,1,ORDER,HT,ORDER,1,HMHT)
CALL  MATADD(HMHT,ORDER,ORDER,RMAT,HMHTR)
   HMHTRINV(1,1)=1./HMHTR(1,1)
CALL  MATMUL(M,ORDER,ORDER,HT,ORDER,1,MHT)
CALL  MATMUL(MHT,ORDER,1,HMHTRINV,1,1,K)
CALL  MATMUL(K,ORDER,1,HMAT,1,ORDER,KH)
CALL  MATSUB(IDN,ORDER,ORDER,KH,IKH)
CALL  MATMUL(IKH,ORDER,ORDER,M,ORDER,ORDER,P)
CALL  GAUSS(XNOISE,SIGNOISE)
X=A0+A1*T+A2*T
XD=A1*A2
XS=X+XNOISE
RES=XS-A0H-A1H*T-A2H*T
A0H=A0H+K(1,1)*RES
A1H=A1H+K(2,1)*RES
A2H=A2H+K(3,1)*RES
SP11=SQRT(P(1,1))
SP22=SQRT(P(2,2))
SP33=SQRT(P(3,3))
A0HERR=A0-A0H
A1HERR=A1-A1H
A2HERR=A2-A2H
XH=A0H+A1H*T+A2H*T
XDH=A1H+A2H
WRITE(9,*)T,A0,A0H,A1,A1H,A2,A2H,X,XH,XD,XDH
WRITE(1,*)T,A0,A0H,A1,A1H,A2,A2H,X,XH,XD,XDH
WRITE(2,*)T,A0HERR,SP11,-SP11,A1HERR,SP22,-SP22
1  A2HERR,SP33,-SP33
10 CONTINUE
PAUSE
CLOSE(1)
END


C SUBROUTINE GAUSS IS SHOWN IN LISTING 1.8
C SUBROUTINE MATTRN IS SHOWN IN LISTING 1.3
C SUBROUTINE MATMUL IS SHOWN IN LISTING 1.4
C SUBROUTINE MATADD IS SHOWN IN LISTING 1.1
C SUBROUTINE MATSUB IS SHOWN IN LISTING 1.2
```

The experiments of this section showed that even if states are not observable a Kalman filter can be designed and built, which might appear to work if the designer is not careful. We showed that when filter states were not observable the filter state estimates were a function of the initial conditions and the state estimates did not improve as more measurements were taken. If the initial

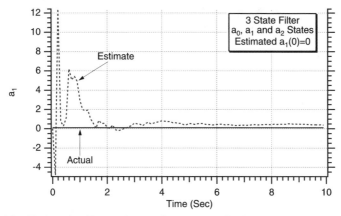

Fig. 14.38 Three-state filter estimate of a_1 asymptotically approaches a constant that is on the high side.

conditions were accurate, the state estimates of the unobservable states would be accurate. However, if the initial conditions were inaccurate the state estimates of the unobservable states would also be inaccurate. In the simple example presented here the nonobservability of a_1 and a_2 was fairly easy to determine. However, in complex filters with hundreds of states the nonobservability of some states may not be as obvious.

Aiding

In some applications separate sensor measurements can sometimes be used to help when one sensor is failing.[3] For example, consider a case in which a wildly accelerating vehicle with a global positioning system (GPS) receiver is traveling

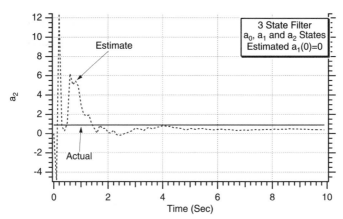

Fig. 14.39 Three-state filter estimate of a_2 asymptotically approaches a constant that is on the low side.

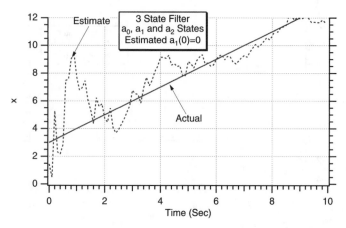

Fig. 14.40   Using three-state Kalman filter yields good estimates of signal.

down a road. In this simplified example we will assume that the GPS receiver measures the vehicle's location (rather than the distance from a satellite to the receiver) to within a certain accuracy. As was shown in Chapter 11, if GPS information is available continuously we can determine the location of the vehicle crudely at any time by using the raw measurements directly or more precisely by the use of filtering. However, sometimes GPS information is either lost (i.e., vehicle is going through a tunnel) or is denied (i.e., GPS is jammed). Under these circumstances we would determine the vehicle's location by coasting our filter (i.e., predicting ahead using the fundamental matrix and the last good state

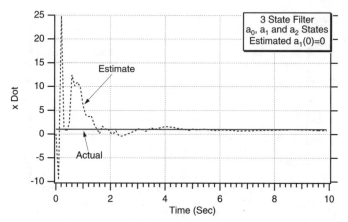

Fig. 14.41   Using three-state Kalman filter yields good estimates of the derivative of the signal.

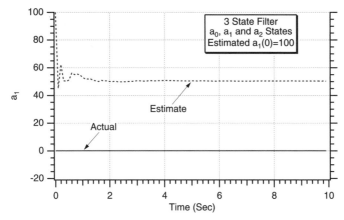

Fig. 14.42 **With bad initial conditions the three-state filter is unable to estimate** a_1.

estimates) based on our last estimates of the position, velocity, and acceleration of the vehicle. However, if our model of the real world were not accurate, coasting the filter would fail, and we eventually would not be able to tell where the vehicle is.

We will demonstrate in this section that by using an inertial navigation system (INS) in conjunction with GPS we can locate the vehicle when GPS information is either lost or denied. In this case the INS, although not as accurate as the GPS, can provide sufficient accuracy to enable quick restarting of the GPS estimates when GPS measurements begin again. This resultant process, in which two sensors help each other, is sometimes called aiding.

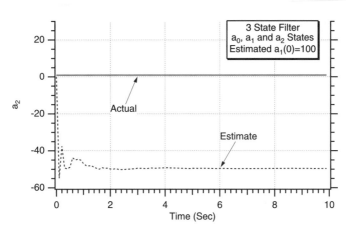

Fig. 14.43 **With bad initial conditions the three-state filter is unable to estimate** a_2.

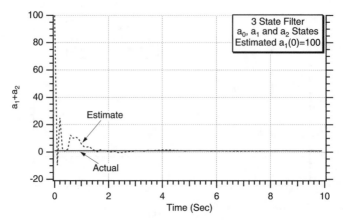

Fig. 14.44 **Sum of a_1 and a_2 can be estimated with three-state filter when the initialization is poor.**

Let us assume that a vehicle travels down a straight road at an average velocity of 100 ft/s with a 20-ft/s sinusoidal velocity excursion. With such a wildly maneuvering model the vehicle velocity is given by

$$\dot{x} = 100 + 20 \sin \omega t$$

If we assume that initially the vehicle is at the origin [i.e., $x(0) = 0$], then integrating the preceding expression yields the location of the vehicle as

$$x = 100t - \frac{20 \cos \omega t}{\omega} + \frac{20}{\omega}$$

where $20/\omega$ is simply the appropriate constant of integration to yield zero initial condition on position. Because the vehicle's velocity is not constant, we can find the acceleration of the vehicle by differentiating the velocity formula and obtaining

$$\ddot{x} = 20\omega \cos \omega t$$

We can see from the preceding expression that if the frequency of the sinusoid is 0.1 rad/s the maximum vehicle acceleration or deceleration is 2 ft/s^2.

In this example we are assuming that the sinusoidal motion of the vehicle is unknown to us, and we would first like to build a Kalman filter to track the object given GPS position measurements every second. It is assumed that GPS provides vehicle position measurements with a 1-σ measurement accuracy of 50 ft (i.e., the standard deviation of the measurement noise on position is 50 ft). Because the motion of the object is unknown to us (i.e., we do not know it is sinusoidal), for safety and robustness we will build a three-state polynomial linear Kalman filter to track the object. We have already shown in Chapter 4 that having less than three filter states (i.e., position, velocity, and acceleration) might result in filter divergence. The appropriate matrices for the linear three-state polynomial Kalman filter were already presented in Table 4.1 of Chapter 4.

Listing 14.7 is a simulation of the one-dimensional moving vehicle and the linear three-state polynomial Kalman filter used to track it. We can see from the simulation that we are assuming in this example that GPS is providing the vehicle location corrupted by zero-mean Gaussian noise with a standard deviation of 50 ft (SIGNOISE = 50). The three-state linear polynomial Kalman filter provides estimates of the vehicle's position, velocity, and acceleration (i.e., XH, XDH, and XDDH). It is assumed that initially the filter is ignorant of the states of the vehicle, and so all of the initial filter estimates are set to zero and the diagonal elements of the initial covariance matrix are set to infinity. The process noise level, used to set the discrete process-noise matrix in the Riccati equations, was selected by experiment (PHIS = 1). There is a parameter XLOSE that allows us to simulate the effects of lost data. When time is greater than XLOSE, the measurement noise variance in the Riccati equations is set to a large number, which is turn causes the Kalman gains to go to zero. Under these circumstances the filter coasts because the residual multiplied by the gains is zero and no more measurements will be processed.

The nominal case of Listing 14.7 was run in which it was assumed that 200 s of GPS measurement data were available. Figures 14.45 and 14.46 display the actual and estimated vehicle position and velocity for 200 s. We can see from Fig. 14.45 that the estimate of the vehicle's position appears to be excellent. Figure 14.46 indicates that we are also able to track the vehicle velocity quite well and capture its sinusoidal motion.

It is difficult to tell from Figs. 14.45 and 14.46 exactly how well the polynomial Kalman filter is working on the GPS data. For more precise work we require error in the estimate information. Figure 14.47 displays the errors in the estimates of the vehicle's position and compares them to the predictions of the covariance matrix. The performance predictions obtained from the covariance matrix are not exact because the covariance matrix believes that the real vehicle motion is parabolic, whereas it is actually sinusoidal. Nonetheless, from the figure we can tell that the Kalman filter appears to be working properly because the single-run simulation results appear to lie within the theoretical bounds approximately 68% of the time. We can also tell that on the average we know the vehicle's location to within 30 ft even though the vehicle is wildly accelerating. Of course, without any filtering at all we could locate the target to within 50 ft (1σ) because that is the measurement accuracy of the data.

Another case was run in which we had GPS data for the first 100 s and then lost the remaining data (i.e., XLOSE = 100 in Listing 14.7). We can see from Fig. 14.48 that under these new circumstances the error in the estimate of the vehicle's location is very small for the first 100 s (i.e., approximately 30 ft) and then starts to grow as soon as the measurements are lost. After being denied GPS data for nearly 100 s, we are not able to determine where the vehicle is to within nearly 2 miles. The Kalman filter is not able to coast successfully when the data are lost because the Kalman filter model of the vehicle motion is parabolic, whereas the actual vehicle motion is sinusoidal.

Listing 14.7   Simulation of moving vehicle and three-state polynomial Kalman filter using GPS information

```
C THE FIRST THREE STATEMENTS INVOKE THE ABSOFT RANDOM
  NUMBER GENERATOR ON THE MACINTOSH
      GLOBAL DEFINE
              INCLUDE 'quickdraw.inc'
      END
      IMPLICIT REAL*8(A-H,O-Z)
      REAL*8 M(3,3),P(3,3),K(3,1),PHI(3,3),H(1,3),R(1,1),PHIT(3,3)
      REAL*8 PHIP(3,3),HT(3,1),KH(3,3),IKH(3,3)
      REAL*8 MHT(3,1),HMHT(1,1),HMHTR(1,1),HMHTRINV(1,1),IDN(3,3)
      REAL*8 Q(3,3),PHIPPHIT(3,3)
      INTEGER ORDER
      OPEN(1,STATUS='UNKNOWN',FILE='DATFIL')
      OPEN(2,STATUS='UNKNOWN',FILE='COVFIL')
      ORDER=3
      XLOSE=99999.
      W=.1
      PHIS=1.
      TS=1.
      XH=0
      XDH=0
      XDDH=0
      SIGNOISE=50.
      DO 1000 I=1,ORDER
      DO 1000 J=1,ORDER
              PHI(I,J)=0.
              P(I,J)=0.
              IDN(I,J)=0.
              Q(I,J)=0.
 1000 CONTINUE
      IDN(1,1)=1.
      IDN(2,2)=1.
      IDN(3,3)=1.
      P(1,1)=99999999999999.
      P(2,2)=99999999999999.
      P(3,3)=99999999999999.
      PHI(1,1)=1
      PHI(1,2)=TS
      PHI(1,3)=5*TS*TS
      PHI(2,2)=1
      PHI(2,3)=TS
      PHI(3,3)=1
      DO 1100 I=1,ORDER
              H(1,1)=0.
 1100 CONTINUE
      H(1,1)=1
      CALL MATTRN(H,1,ORDER,HT)
      R(1,1)=SIGNOISE**2
      CALL MATTRN(PHI,ORDER,ORDER,PHIT)
      Q(1,1)=PHIS*TS**5/20
      Q(1,2)=PHIS*TS**4/8
```

(continued)

Listing 14.7   (*Continued*)

```
Q(1,3)=PHIS*TS**3/6
Q(2,1)=Q(1,2)
Q(2,2)=PHIS*TS**3/3
Q(2,3)=PHIS*TS*TS/2
Q(3,1)=Q(1,3)
Q(3,2)=Q(2,3)
Q(3,3)=PHIS*TS
DO 10 T=0.,200.,TS
        IF(T>XLOSE)THEN
                R(1,1)=999999999999999.
        ENDIF
        CALL MATMUL(PHI,ORDER,ORDER,P,ORDER,ORDER,PHIP)
        CALL MATMUL(PHIP,ORDER,ORDER,PHIT,ORDER,ORDER,
            PHIPPHIT)
        CALL MATADD(PHIPPHIT,ORDER,ORDER,Q,M)
        CALL MATMUL(M,ORDER,ORDER,HT,ORDER,1,MHT)
        CALL MATMUL(H,1,ORDER,MHT,ORDER,1,HMHT)HMHTR(1,1)=
            HMHT(1,1)+R(1,1)HMHTRINV(1,1)=1./HMHTR(1,1)
        CALL MATMUL(MHT,ORDER,1,HMHTRINV,1,1,K)
        CALL MATMUL(K,ORDER,1,H,1,ORDER,KH)
        CALL MATSUB(IDN,ORDER,ORDER,KH,IKH)
        CALL MATMUL(IKH,ORDER,ORDER,M,ORDER,ORDER,P)
        CALL GAUSS(XNOISE,SIGNOISE)
        X=100.*T-20.*COS(W*T)/W+20./W
        XD=100.+20.*SIN(W*T)
        XDD=20.*W*COS(W*T)
        XS=X+XNOISE
        RES=XS-XH-TS*XDH-.5*TS*TS*XDDH
        XH=XH+XDH*TS+.5*TS*TS*XDDH+K(1,1)*RES
        XDH=XDH+XDDH*TS+K(2,1)*RES
        XDDH=XDDH+K(3,1)*RES
        SP11=SQRT(P(1,1))
        SP22=SQRT(P(2,2))
        SP33=SQRT(P(3,3))
        XHERR=X-XH
        XDHERR=XD-XDH
        XDDHERR=XDD-XDDH
        WRITE(9,*)T,X,XH,XD,XDH,XDD,XDDH
        WRITE(1,*)T,X,XH,XD,XDH,XDD,XDDH
        WRITE(2,*)T,XHERR,SP11,-SP11,XDHERR,SP22,-SP22,XDDHERR,
1           SP33,-SP33
10 CONTINUE
    CLOSE(1)
    CLOSE(2)
    PAUSE
    END

C SUBROUTINE GAUSS IS SHOWN IN LISTING 1.8
C SUBROUTINE MATTRN IS SHOWN IN LISTING 1.3
C SUBROUTINE MATMUL IS SHOWN IN LISTING 1.4
C SUBROUTINE MATADD IS SHOWN IN LISTING 1.1
C SUBROUTINE MATSUB IS SHOWN IN LISTING 1.2
```

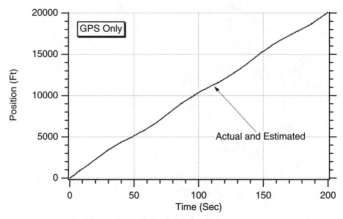

Fig. 14.45   With all of the GPS data, we are able to track the vehicle's position extremely well.

To locate the vehicle when data are denied, an INS can be used. For simplicity let us assume that an inertial system, assumed to be located on the vehicle along with the GPS receiver, can provide noise-free vehicle position information at the same rate as GPS. However, let us also assume that the INS information is contaminated with an acceleration bias of 1 ft/s². If we could correctly initialize the inertial system with the vehicle's location (i.e., we are ignoring the details on how this is accomplished) and only use the INS, we would still have a location error of $0.5bt^2$ where b is the INS acceleration bias. After 200 s the vehicle location uncertainty would be 20,000 ft (i.e., error = $0.5*1*200^2 = 20,000$) or approximately a 4-mile uncertainty in the vehicle location, which is certainly no improvement of the GPS case when data are denied for 100 s.

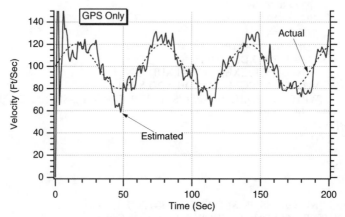

Fig. 14.46   With all of the GPS data, we are able to follow the vehicle's sinusoidal motion.

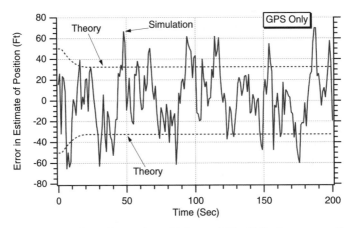

Fig. 14.47   We are able to locate the vehicle to within 30 ft when we continually get GPS measurements.

The goal will be to see if we can combine the GPS and INS data so that we can estimate the bias and then use that information to aid the GPS system. In this example both GPS and INS are inadequate by themselves. If x represents the vehicle's location, we have assumed that the GPS measurement is given by

$$z_{\text{GPS}} = x + v_{\text{GPS}}$$

and that the INS measurement is corrupted by an acceleration bias or

$$z_{\text{INS}} = x + 0.5bt^2$$

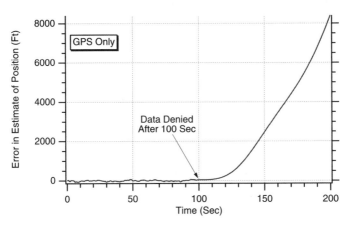

Fig. 14.48   Filter is unable to track vehicle after GPS data are lost.

where v_{GPS} is the GPS measurement noise and b is the INS acceleration bias. By subtracting the GPS measurement from the INS measurement, we form a difference measurement

$$\Delta z = z_{INS} - z_{GPS} = x + 0.5bt^2 - x - v_{GPS} = 0.5bt^2 - v_{GPS}$$

Because the combined measurement is linearly related to the bias and corrupted by zero-mean Gaussian measurement noise, we can build a one-state Kalman filter to estimate the bias. Because the bias is a constant, its derivative must be zero. However, because we would like to give our one-state Kalman filter some bandwidth and robustness (see Chapter 6), we will assume that the derivative of the bias is white process noise u_s or

$$\dot{b} = u_s$$

where the spectral density of the process noise is Φ_s.

From the preceding equations we can see that the systems dynamics matrix, which is a scalar, is zero. Therefore, the fundamental matrix is a 1×1 identity matrix or unity. In summary, the fundamental, measurement, process-noise, and measurement noise matrices for the one-state Kalman filter are all scalar (i.e., 1×1 matrices) and are given by

$$\Phi_k = 1$$

$$H = 0.5t^2$$

$$Q_k = \Phi_s T_s$$

$$R_k = \sigma_{GPS}^2$$

The preceding scalar matrices are used to find the Kalman gains from the Riccati equations. The one-state Kalman filter can then be derived from

$$\hat{x}_k = \Phi_k \hat{x}_{k-1} + K_k(z_k - H\Phi_k \hat{x}_{k-1})$$

In scalar form the one-state Kalman filter becomes

$$\hat{b}_k = \hat{b}_{k-1} + K_k(\Delta z_k - 0.5\hat{b}_{k-1}t^2)$$

Listing 14.8 programs the wildly maneuvering vehicle and the preceding one-state Kalman filter for estimating the INS bias. This Kalman filter makes use of both GPS and INS measurements. Also included in the listing is the logic for telling the filter that GPS data are lost. When the GPS data are lost, INS data are still available. The process noise for the one-state filter was selected by experiment (i.e., PHIS = 10).

The nominal case of Listing 14.8 was run. We can see from Fig. 14.49 that when both GPS and INS measurement data are available the one-state Kalman

Listing 14.8 Kalman filter for estimating INS bias by using both GPS and INS measurement

```
C THE FIRST THREE STATEMENTS INVOKE THE ABSOFT RANDOM
  NUMBER GENERATOR ON THE MACINTOSH
     GLOBAL DEFINE
                INCLUDE 'quickdraw.inc'
     END
     IMPLICIT REAL*8(A-H,O-Z)
     REAL*8 M(1,1),P(1,1),K(1,1),PHI(1,1),H(1,1),R(1,1),PHIT(1,1)
     REAL*8 PHIP(1,1),HT(1,1),KH(1,1),IKH(1,14)
     REAL*8 MHT(1,1),HMHT(1,1),HMHTR(1,1),HMHTRINV(1,1),IDN(1,1)
     REAL*8 Q(1,1),PHIPPHIT(1,1)
     INTEGER ORDER
     OPEN(1,STATUS='UNKNOWN',FILE='DATFIL')
     ORDER=1
     XLOSE=99999999999.
     W=.1
     PHIS=10.
     TS=1.
     BIASH=0.
     BIAS=1.
     SIGNOISE=50.
     DO 1000 I=1,ORDER
     DO 1000 J=1,ORDER
                PHI(I,J)=0.
                P(I,J)=0.
                IDN(I,J)=0.
                Q(I,J)=0.
1000 CONTINUE
     IDN(1,1)=1.
     P(1,1)=99999999999999.
     PHI(1,1)=1
     DO 1100 I=1,ORDER
                H(1,1)=0.
1100 CONTINUE
     R(1,1)=SIGNOISE**2
     CALL MATTRN(PHI,ORDER,ORDER,PHIT)
     Q(1,1)=PHIS*TS
     DO 10 T=0.,200.,TS
                IF(T>XLOSE)THEN
                          R(1,1)=999999999999999.
                ENDIF
                H(1,1)=5*T*T
                CALL MATTRN(H,1,ORDER,HT)
                CALL MATMUL(PHIP,ORDER,ORDER,P,ORDER,ORDER,PHIP)
                CALL MATMUL(PHIP,ORDER,ORDER,PHIT,ORDER,ORDER,
                  PHIPPHIT)
                CALL MATADD(PHIPPHIT,ORDER,ORDER,Q,M)
                CALL MATMUL(M,ORDER,ORDER,HT,ORDER,1,MHT)
```

(continued)

Listing 14.8 *(Continued)*

```
          CALL  MATMUL(H,1,ORDER,MHT,ORDER,1,HMHT)HMHTR(1,1)=
            HMHT(1,1)+R(1,1)HMHTRINV(1,1)=1./HMHTR(1,1)
          CALL  MATMUL(MHT,ORDER,1,HMHTRINV,1,1,K)
          CALL  MATMUL(K,ORDER,1,H,1,ORDER,KH)
          CALL  MATSUB(IDN,ORDER,ORDER,KH,IKH)
          CALL  MATMUL(IKH,ORDER,ORDER,M,ORDER,ORDER,P)
          CALL  GAUSS(XNOISE,SIGNOISE)
          X=100.*T-20.*COS(W*T)/W+20./W
          XD=100.+20.*SIN(W*T)
          XDD=20.*W*COS(W*T)
          XGPSS=X+XNOISE
          XINSS=X+.5*BIAS*T*T
          XS=XINSS-XGPSS
          RES=XS-.5*BIASH*T*T
          BIASH=BIASH+K(1,1)*RES
          SP11=SQRT(P(1,1))
          BIASERR=BIAS-BIASH
          XH=XINSS-.5*BIASH*T*T
          WRITE(9,*)T,BIAS,BIASH,BIASERR,SP11,-SP11
          WRITE(1,*)T,BIAS,BIASH,BIASERR,SP11,-SP11
   10  CONTINUE
       CLOSE(1)
       PAUSE

C  SUBROUTINE  GAUSS  IS  SHOWN  IN  LISTING  1.8
C  SUBROUTINE  MATTRN  IS  SHOWN  IN  LISTING  1.3
C  SUBROUTINE  MATMUL  IS  SHOWN  IN  LISTING  1.4
C  SUBROUTINE  MATADD  IS  SHOWN  IN  LISTING  1.1
C  SUBROUTINE  MATSUB  IS  SHOWN  IN  LISTING  1.2
```

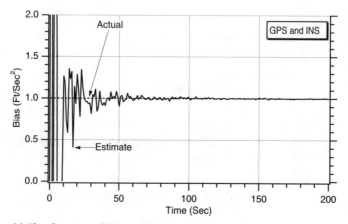

Fig. 14.49   One-state Kalman filter is able to estimate bias in less than 50 s.

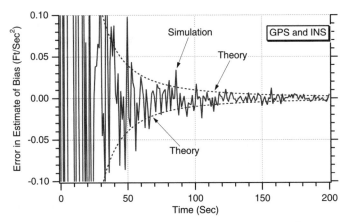

Fig. 14.50   One-state Kalman filter results match theory.

filter is able to estimate the INS bias accurately in less than 50 s. Figure 14.50 shows that the errors in estimating the bias are within the theoretical error bounds of the covariance matrix approximately 68% of the time, indicating that the one-state Kalman filter appears to be working properly.

Another case was run in which the GPS measurements were lost after 100 s. We can see from Fig. 14.51 that we are still able to estimate the INS bias after the GPS measurements are lost. The reason for this is that our one-state filter is able to coast intelligently when data are lost because its model of the real world is correct (i.e., if there was no process noise, we are assuming that the bias is constant). This filter does not care that the vehicle is wildly maneuvering.

We can get the best of both worlds by having the GPS and combined GPS filters running in parallel. When GPS data are available, we simply use the output

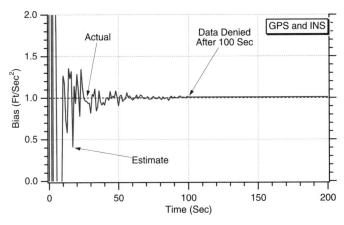

Fig. 14.51   New Kalman filter is able to estimate bias after GPS data are lost.

of the GPS three-state linear polynomial Kalman filter to locate the vehicle. When GPS data are denied, we compute the vehicle location from the INS measurement of the vehicle location $x_{k\text{INS}}$ and the estimated bias from the one-state filter or

$$\hat{x}_k = x_{k\text{INS}} - 0.5\hat{b}_k t^2$$

Listing 14.9 has both the three-state linear polynomial Kalman filter and one-state Kalman filter running in parallel along with the wildly maneuvering vehicle. The simulation is set up so that both GPS and INS measurements are available for the first 100 s. After 100 s only INS data are available. Care has to be taken to set $R(1,1)$ to a large enough number to ensure that the Kalman gains go to zero when measurement data are denied.

The nominal case of Listing 14.9 was run, and we can see from Fig. 14.52 that after the GPS data are lost the error in the estimate to the vehicle's position builds up with time. The initial condition for the drift depends on the last estimate of the bias because the filter is coasting (i.e., the Kalman gains are zero). By comparing Fig. 14.52 with Fig. 14.48, we can see that the error in locating the vehicle has been reduced from slightly less than 2 miles to approximately 200 ft. Thus, we can see that when GPS data are lost using imperfect INS information can be advantageous. However, we can also see that the longer the GPS data are lost the more uncertainty there will be in locating the vehicle.

In practice, the INS and GPS estimates are made with only one Kalman filter, which uses both GPS and INS outputs as measurements. Using INS information can also be advantageous in reinitializing the GPS receiver after jamming or blocking if the GPS signal has been lost.

Summary

In this chapter we investigated how filter performance degraded with decreasing signal-to-noise ratio. We saw that although the filter was performing correctly in the sense that the errors in the estimates matched theoretical expectations the quality of the estimates was too low to be of practical utility. It was also shown that the initialization of the filter states and initial covariance matrix could be very important in reducing estimation errors under conditions of low signal-to-noise ratio. We also investigated how the performance of a polynomial Kalman filter behaved as a function of filter order (i.e., number of states in filter) when only a few measurements were available. We showed that, in general, lower-order filters provided better estimates more rapidly than the higher-order filters. The number of measurements required to get perfect estimates in the noise-free case was shown to be equal to the number of states in the polynomial Kalman filter. We also demonstrated in this chapter that the residual and its theoretical bounds can also be used as an indication of whether or not the filter is working properly. Unlike the normal tests that are conducted with the filter's covariance matrix, which can only be used in simulations, the residual test can be used both in simulations and actual practice. An example was chosen in another section in which a filter was intentionally designed in which some states were not observable. We showed, via numerical examples, some easy-to-use practical tests that could be conducted to determine the nonobservability of the states in

Listing 14.9   GPS Kalman filter and new Kalman filter running in parallel

```
C THE FIRST THREE STATEMENTS INVOKE THE ABSOFT RANDOM
NUMBER GENERATOR ON THE MACINTOSH
      GLOBAL DEFINE
            INCLUDE 'quickdraw.inc'
      END
      IMPLICIT REAL*8(A-H,O-Z)
      REAL*8 M(3,3),P(3,3),K(3,1),PHI(3,3),H(1,3),R(1,1),PHIT(3,3)
      REAL*8 PHIP(3,3),HT(3,1),KH(3,3),IKH(3,3)
      REAL*8 MHT(3,1),HMHT(1,1),HMHTR(1,1),HMHTRINV(1,1),IDN(3,3)
      REAL*8 Q(3,3),PHIPPHIT(3,3)
      REAL*8 MPZ(1,1),PPZ(1,1),KPZ(1,1),PHIPZ(1,1),HPZ(1,1)
      REAL*8 PHIPPZ(1,1),HTPZ(1,1),KHPZ(1,1),IKHPZ(1,1)
      REAL*8 MHTPZ(1,1),HMHTPZ(1,1),HMHTRPZ(1,1),HMHTRINVPZ(1,1)
      REAL*8 QPZ(1,1),PHIPPHITPZ(1,1),IDNPZ(1,1)PHITPZ(1,1)
      INTEGER ORDER,ORDERPZ
      OPEN(1,STATUS='UNKNOWN',FILE='DATFIL')
      OPEN(2,STATUS='UNKNOWN',FILE='COVFIL')
      ORDER=3
      ORDERPZ=1
      XLOSE=100.
      W=.1
      PHIS=1.
      PHISPZ=10.
      TS=1.
      XH=0
      XDH=0
      XDDH=0
      BIASH=0.
      BIAS=1.
      SIGNOISE=50.
      DO 1000 I=1,ORDER
      DO 1000 J=1,ORDER
            PHI(I,J)=0.
            P(I,J)=0.
            IDN(I,J)=0.
            Q(I,J)=0.
1000  CONTINUE
      IDN(1,1)=1.
      IDN(2,2)=1
      IDN(3,3)=1
      IDNPZ(1,1)=1.
      P(1,1)=99999999999999.
      P(2,2)=99999999999999.
      P(3,3)=99999999999999.
      PPZ(1,1)=99999999999999.
      PHI(1,1)=1
      PHI(1,2)=TS
      PHI(1,3)=.5*TS*TS
      PHI(2,2)=1
```

(continued)

Listing 14.9 (*Continued*)

```
      PHI(2,3)=TS
      PHI(3,3)=1
      PHIPZ(1,1)=1
      DO 1100 I=1,ORDER
            H(1,1)=0.
1100  CONTINUE
      H(1,1)=1
      CALL MATTRN(H,1,ORDER,HT)
      R(1,1)=SIGNOISE**2
      CALL MATTRN(PHI,ORDER,ORDER,PHIT)
      CALL MATTRN(PHIPZ,ORDERPZ,ORDERPZ,PHITPZ)
      Q(1,1)=PHIS*TS**5/20
      Q(1,2)=PHIS*TS**4/8
      Q(1,3)=RHIS*TS**3/6
      Q(2,1)=Q(1,2)
      Q(2,2)=PHIS*TS**3/3
      Q(2,3)=PHIS*TS*TS/2
      Q(3,1)=Q(1,3)
      Q(3,2)=Q(2,3)
      Q(3,3)=PHIS*TS
      QPZ(1,1)=PHISPZ*TS
      DO 10 T=0.,200.,TS
            IF(T>XLOSE)THEN
                  R(1,1)=999999999999999.
            ENDIF
            CALL MATMUL(PHI,ORDER,ORDER,P,ORDER,ORDER,PHIP)
            CALL MATMUL(PHIP,ORDER,ORDER,PHIT,ORDER,ORDER,
              PHIPPHIT)
            CALL MATADD(PHIPPHIT,ORDER,ORDER,Q,M)
            CALL MATMUL(M,ORDER,ORDER,HT,ORDER,1,MHT)
            CALL MATMUL(H,1,ORDER,MHT,ORDER,1,HMHT)HMHTR(1,1)=
              HMHT(1,1)+R(1,1)HMHTRINV(1,1)=1./HMHTR(1,1)
            CALL MATMUL(MHT,ORDER,1,HMHTRINV,1,1,K)
            MATMUL(K,ORDER,1,1,H,1,ORDER,KH)
            CALL MATSUB(IDN,ORDER,ORDER,KH,IKH)
            CALL MATMUL(IKH,ORDER,ORDER, M,ORDER,ORDER,P)
            HPZ(1,1)=.5*T*T
            CALL MATTRN(HPZ,1,ORDERPZ,HTPZ)
            CALL MATMUL(PHIPZ,ORDERPZ,ORDERPZ,PPZ,ORDERPZ,
1             ORDERPZ,PHIPPZ)
            CALL MATMUL(PHIPPZ,ORDERPZ,ORDERPZ,PHITPZ,ORDERPZ,
1             ORDERPZ,PHIPPHITPZ)
            CALL MATADD(PHIPPHITPZ,ORDERPZ,ORDERPZ,QPZ,MPZ)
            CALL MATMUL(MPZ,ORDERPZ,ORDERPZ,HTPZ,ORDERPZ,1,
              MHTPZ)
            CALL MATMUL(HPZ,1,ORDERPZ,MHTPZ,ORDERPZ,1,HMHTPZ)
            HMHTRPZ(1,1)=HMHTPZ(1,1)+R(1,1)HMHTRINVPZ(1,1)=1./
              HMHTRPZ(1,1)
            CALL MATMUL(MHTPZ,ORDERPZ,1,HMHTRINVPZ,1,1,KPZ)
                                                    (continued)
```

Listing 14.9   (*Continued*)

```
        CALL  MATMUL(KPZ,ORDERPZ,1,HPZ,1,ORDERPZ,KHPZ)
        CALL  MATSUB(IDNPZ,ORDERPZ,ORDERPZ,KHPZ,IKHPZ)
        CALL  MATMUL(IKHPZ,ORDERPZ,ORDERPZ,MPZ,ORDERPZ,
1           ORDERPZ,PPZ)
        CALL  GAUSS(XNOISE,SIGNOISE)
        X=100.*T-20.*COS(W*T)/W+20./W
        XD=100.+20.*SIN(W*T)
        XDD=20.*W*COS(W*T)
        XS=X+XNOISE
        RES=XS-XH-TS*XDH-.5*TS*TS*XDDH
        XH=XH+XDH*TS+.5*TS*TS*XDDH+K(1,1)*RES
        XDH=XDH+XDDH*TS+K(2,1)*RES
        XDDH=XDDH+K(3,1)*RES
        XGPSS=XS
        XINSS=X-.5*BIAS*T*T
        XSPZ=XINSS-XGPSS
        RESPZ=XSPZ-.5*BIASH*T*T
        BIASH=BIASH+KPZ(1,1)*RESPZ
        IF(T>XLOSE)THEN
                XH=XINSS-.5*BIASH*T*T
        ENDIF
        SP11=SQRT(P(1,1))
        SP22=SQRT(P(2,2))
        SP33=SQRT(P(3,3))
        XHERR=X-XH
        XHDERR=XD-XDH
        XDDHERR=XDD-XDDH
        SP11PZ-SQRT(PPZ(1,1))
        BIASERR=BIAS-BIASH
        WRITE(9,*)T,X,XH,XD,XDH,XDD,XDDH,BIAS,BIASH
        WRITE(1,*)T,X,XH,XD,XDH,XDD,XDDH,BIAS,BIASH
        WRITE(2,*)T,XHERR,SP11,-SP11,XDHERR,SP22,-SP22,XDDHERR,
1           SP33,-SP33,BIASERR,SP11PZ,-SP11PZ
10  CONTINUE
    CLOSE(1)
    CLOSE(2)
    PAUSE

C SUBROUTINE GAUSS IS SHOWN IN LISTING 1.8
C SUBROUTINE MATTRN IS SHOWN IN LISTING 1.3
C SUBROUTINE MATMUL IS SHOWN IN LISTING 1.4
C SUBROUTINE MATADD IS SHOWN IN LISTING 1.1
C SUBROUTINE MATSUB IS SHOWN IN LISTING 1.2
```

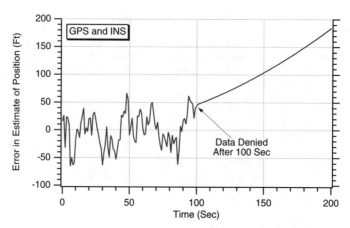

Fig. 14.52   Uncertainty in vehicle location has been dramatically reduced with aiding.

question. Finally, in the last section we discussed aiding. An example was chosen in which there were two sensors, both of which did not work satisfactorily by themselves, but when combined worked well under adverse circumstances.

References

[1]Maybeck, P. S., *Stochastic Models, Estimation, and Control*, Vol. 1, Academic International Press, New York, 1979, p. 229.

[2]Biezad, D. J., *Integrated Navigation and Guidance Systems*, AIAA, Reston, VA, 1999, pp. 104, 105.

[3]Merhav, S., *Aerospace Sensor Systems and Applications*, Springer–Verlag, New York, 1996, pp. 395–439.

Appendix: Fundamentals of Kalman-Filtering Software

Software Details

To facilitate learning, a CD that is formatted for both IBM- and Macintosh-compatible personal computers containing all of the text's FORTRAN source code listings is included. As a special feature for those who work in the MATLAB® or True BASIC languages, duplicate files, which are the MATLAB® and True BASIC equivalents of the FORTRAN listings, also can be found on the enclosed CD.

The FORTRAN source code should run as is with either Version 6.0 of the Absoft Power Macintosh Pro FORTRAN compiler or Version 4.3 of the Absoft FORTRAN compiler for IBM-compatible machines. Use of different FORTRAN compilers, in either the Macintosh- or IBM-compatible world, may require some slight modification of the source code. The MATLAB® source code should run as is using MATLAB® 5.2 on the Macintosh or Version 5.0 on IBM compatible computers. The True BASIC source code should run as is with the Bronze edition on the Macintosh or with the Silver edition on IBM-compatible computers.

The naming conventions for the source code files in each language are slightly different. The FORTRAN naming convention is CxLy.F, where x corresponds to chapter number and y corresponds to listing number. In other words, C4L2.F corresponds to FORTRAN Listing 4.2 of the text (i.e., Chapter 4, Listing 2). The MATLAB® naming convention is the same for both Macintosh- and IBM-compatible machines (C4L2.M). In addition, the True BASIC naming convention is also the same for both Macintosh- and IBM-compatible machines (C4L2.TRU).

Each of the source files on the enclosed CD has a few lines of extra code to make data files so that the user can plot or store the results after a run is made. The name of the generated data file is DATFIL for the Macintosh code and DATFIL.TXT for IBM-compatible code.

The data statements or definition of constants in each of the source code files correspond to those used in the numerical examples presented in the text. The user should first run the program of interest as is in order to verify that the data file generated corresponds to the appropriate figure in the text. Other cases of interest can be run by either changing constants and recompiling or modifying the source code to read input from the keyboard.

647

In the source code listings that make use of random numbers, use has been made of the FORTRAN uniform random number generators supplied by Absoft. If other FORTRAN compilers are used, the user will have to invoke the appropriate random-number language extension for the particular compiler or may have to write a random-number generator if it is not supplied by the compiler publisher. MATLAB® users have special statements for uniform and Gaussian distributions, whereas True BASIC offers a special statement for uniformly distributed random numbers.

MATLAB®

Many engineers prefer to work in MATLAB® because of its friendly environment and powerful plotting statements. In this Appendix some simple ways of converting the FORTRAN code to MATLAB® line by line are presented, and two examples taken from the text are used to illustrate the conversion algorithm. All of the text's source code has also been converted to MATLAB® M-files in this way and, as was already mentioned, is included on the enclosed CD. In addition, the MATLAB® code allows the reader to obtain the figures in the book directly without the need for porting the data to other plotting packages. The structure of the translation was originally provided by Dr Michael Dutton of the Weapons Systems Division, Defence Science and Technology Organisation (DSTO) in Salisbury, South Australia, for another book.[1]

The method of translation adopted was chosen so that the resultant M-files would closely resemble the original FORTRAN listings. This was done to facilitate a line-by-line conversion and to allow the code to run under earlier MATLAB® versions (including the student editions available on both PC and Macintosh platforms) without the need for invoking special library functions and various toolboxes.

MATLAB® is an interpretative language, and therefore its code, in general, will run much slower than compiled source code written in FORTRAN. In particular, the use of FOR or WHILE loops are normally avoided in MATLAB® because their use dramatically reduces program run-time performance. In this sense no serious attempt was made to optimize the running times of the M-files. However, most of the examples presented in the text took only a few seconds to run using MATLAB® on today's powerful desktop computers.

Unfortunately, if the MATLAB® code is to mirror closely the FORTRAN listings, the use of FOR and/or WHILE statements becomes necessary. This means that some of the programs will run more than an order of magnitude slower than their FORTRAN counterparts. In general those programs that run in the Monte Carlo mode or those simulations having a small integration interval or long operation times will have significantly longer MATLAB® running times. However, even in those cases the worst MATLAB® running times were still less than a minute on today's microcomputers.

As an example of how the FORTRAN source code was converted to MATLAB®, consider the case of Listing 4.4, which originally simulated in FORTRAN a first-order polynomial Kalman filter with gravity compensation. The MATLAB® equivalent of Listing 4.4 appears in Listing A.1. We can see that the MATLAB® listing is nearly identical to the FORTRAN listing, except that

Listing A.1   MATLAB® equivalent of FORTRAN Listing 4.4

```
TS=.1;
PHIS=0.;
A0=400000.;
A1=-6000.;
A2=-16.1;
XH=0.;
XDH=0.;
SIGNOISE=1000.;
ORDER=2;
T=0.;
S=0.;
H=.001;
PHI=[1 TS ;0 1];
P=[99999999 0;0 999999999];
IDNP=eye(ORDER);
Q=zeros(ORDER);
RMAT=SIGNOISE^2;
Q(1,1)=TS*TS*TS*PHIS/3.;
Q(1,2)=.5*TS*TS*PHIS;
Q(2,1)=Q(1,2);
Q(2,2)=PHIS*TS;
HMAT=[1 0];
HT=HMAT;
PHIT=PHI';
count=0;
for  T=0:TS:30
    PHIP=PHI*P;
    PHIPPHIT=PHIP*PHIT;
    M=PHIPPHIT+Q;
    HM=HMAT*M;
    HMH=HM*HT;
    HMHTR=HMHT+RMAT;
    HMHTRINV=inv(HMHTR);
    MHT=M*HT;
    K=MHT*HMHTRINV;
    KH=K*HMAT;
    IKH=IDNP-KH;
    P=IKH*M;
    XNOISE=SIGNOISE*randn;
    X=A0+A1*t+A2*T*T;
    XD=A1+2*A2*T;
    XS=X+XNOISE;
    RES=XS-XH-TS*XDH+16.1*TS*TS;
    XH=XH+XDH*TS-16.1*TS*TS+K(1,1)*RES;
    XDH=XDH-32.2*TS+K(2,1)*RES;
    SP11=sqrt(P(1,1));
    SP22=sqrt(P(2,2));
    XHERR=X-XH;
    XDHERR=XD-XDH;
```

(*continued*)

Listing A.1   (*Continued*)

```
        SP11P=-SP11;
        SP22P=-SP22;
        count=count+1;
        ArrayT(count)=T;
        ArrayX(count)=X;
        ArrayXH(count)=XH;
        ArrayXD(count)=XD;
        ArrayXDH(count)=XDH;
        ArrayXHERR(count)=XHERR;
        ArraySP11(count)=SP11P;
        ArrayXDHERR(count)=XDHERR;
        ArraySP22(count)=SP22;
        ArraySP22P(count)=SP22P;
end
figure
plot(ArrayT,ArrayXHERR,ArrayT,ArraySP11,ArrayT,ArarySP11P),grid
xlabel('Time  (Sec)')
ylabel('Error  in  Estimate  of  Altitude  (Ft)')
axis([0  30  -1500  1500])
figure
plot(ArrayT,ArrayXDHERR,ArrayT,ArraySP22,ArrayT,ArraySP22P),grid
xlabel('Time  (Sec)')
ylabel('Error  in  Estimate  of  Velocity  (Ft/Sec)')
axis([0  30  -200  200])
clc
output=[ArrayT',ArrayX',ArrayXH',ArrayXD',ArrayXDH'];
save datfil output -ascii
output=[ArrayT',ArrayXHERR|1,ArraySP11',ArraySP11P',ArrayXDHERR',ArraySP22',
    ArraySP2P'];
save covfil output -ascii
disp 'simulation  finished'
```

semicolons have been added to the end of each line and the output has been saved as individual arrays. At the end of the program, the resultant data are plotted by using MATLAB® statements, and the data are sent to ascii files called datfil and covfil. Use of special MATLAB® statements to simplify the writing of matrices has also been incorporated in the simulation. In addition, the matrix Riccati equations appear in a more efficient and compact form in the MATLAB® listing because subroutines for matrix algebra are no longer required. The FORTRAN Gaussian noise subroutine has been eliminated because MATLAB® Gaussian random numbers are automatically generated with the random statement.

The simulation of Listing A.1 was run, and data concerning both the estimates and errors in the estimates were automatically written to the files datfil and covfil. In addition, the plots of errors in the estimates of altitude and velocity were automatically generated with the preceding MATLAB® code and appear in Figs. A.1–A.2. Notice that Fig. A.1 is equivalent to FORTRAN-generated results of

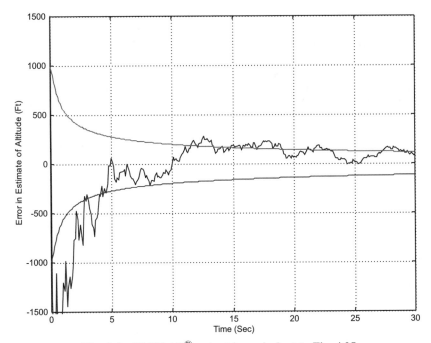

Fig. A.1   MATLAB® output is equivalent to Fig. 4.35.

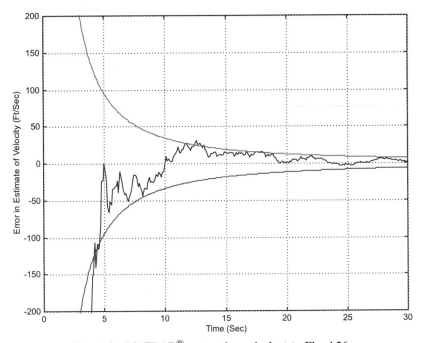

Fig. A.2   MATLAB® output is equivalent to Fig. 4.36.

Fig. 4.35, whereas Fig. A.2 is equivalent to the FORTRAN-generated results of Fig. 4.36.

For variety, another FORTRAN example was chosen for conversion because it contained a subroutine. Listing 7.5, which represented an extended Kalman filter with a special subroutine for state propagation, was converted from FORTRAN to MATLAB®. The main program that was converted to MATLAB® appears in Listing A.2, whereas the equivalent of the FORTRAN subroutine PROJECT appears as a separate M-file, known as PROJECT1.M, in Listing A.3.

The simulation of Listing A.2 was run, and data concerning the estimates and the errors in the estimates were automatically written to the files datfil and covfil. In addition, the few lines of extra code in Listing A.2 automatically generated the plots of Figs. A.3–A.4. Figure A.3 is equivalent to the FORTRAN-generated results of Fig. 7.21 whereas Fig. A.4 is equivalent to the FORTRAN-generated results of Fig. 7.22.

Listing A.2   MATLAB® equivalent of FORTRAN Listing 7.5

```
SIGNOISE=25.;
X=200000.;
XD=-6000.;
BETA=500.;
XH=200025.;
XDH=-6150.;
ORDER=2;
TS=.1;
TF=30.;
PHIS=0.;
T=0.;
S=0.;
H=.001;
PHI=zeros(ORDER,ORDER);
P=[SIGNOISE*SIGNOISE 0;0  20000.];
IDNP=eye(ORDER);
Q=zeros(ORDER,ORDER);
HMAT=[1  0];
HT=HMAT';
RMAT=SIGNOISE^2;
count=0;
while  T<=TF;
        XOLD=X;
        XDOLD=XD;
        XDD=.0034*32.2*XD*XD*exp(-X/22000.)/(2.*BETA)-32.2;
        X=X+H*XD;
        XD=XD+H*XDD;
        T=T+H;
        XDD=.0034*32.2*XD*XD*exp(-X/22000.)/(2.*BETA)-32.2;
        X=.5*(XOLD+X+H*XD);
        XD=.5*(XDOLD+XD+H*XDD);
        S=S+H;
        if  S>=(TS-.00001)
```

(continued)

Listing A.2   (*Continued*)

```
S=0;
RHOH=.0034*exp(-XH/22000.);
F21=-32.2*RHOH*XDH*XDH/(44000.*BETA);
F22=RHOH*32.2*XDH/BETA;
PHI(1,1)=1.;
PHI(1,2)=TS;
PHI(2,1)=F21*TS;
PHI(2,2)=1.+F22*TS;
Q(1,1)=PHIS*TS*TS*TS/3.;
Q(1,2)=PHIS*(TS*TS/2.+F22*TS*TS*TS/3.);
Q(2,1)=Q(1,2);
Q(2,2)=PHIS*(TS+F22*TS*TS+F22*F22*TS*TS*TS/3.);
PHIT=PHI';
PHIP=PHI*P;
PHIPPHIT=PHIP*PHIT;
M=PHIPPHIT+Q;
HM=HMAT*M;
HMHT=HM*HT;
HMHTR=HMHT+RMAT;
HMHTRINV=inv(HMHTR);
MHT=M*HT;
GAIN=HMHTRINV*MHT;
KH=GAIN*HMAT;
IKH=IDNP-KH;
P=IKH*M;
XNOISE=SIGNOISE*randn;
[XB,XDB,XDDB]=PROJECT(T,TS,XH,XDH,BETA);
RES=X+XNOISE-xb;
XH=XB+GAIN(1,1)*RES;
XDH=XDB+GAIN(2,1)*RES;
ERRX=X-XH;
SP11=sqrt(P(1,1));
ERRXD=XD-XDH;
SP22=sqrt(P(2,2));
SP11P=-SP11;
SP22P=SP22;
count=count+1;
ArrayT(count)=T;
ArrayX(count)=X;
ArrayXH(count)=XH;
ArrayXD(count)=XD;
ArrayXDH(count)=XDH;
ArrayERRX(count)=ERRX;
ArraySP11(count)=SP11;
ArraySP11P(count)=SP11P;
ArrayERRXD(count)=ERRXD;
ArraySP22(count)=SP22;
ArraySP22P(count)=SP22P;
end
```

(*continued*)

Listing A.2 *(Continued)*

```
end
figure
plot(ArrayT,ArrayERRX,ArrayT,ArraySP11,ArrayT,ArraySP11P),grid
xlabel('Time (Sec)')
ylabel('Error in Estimate of Altitude (Ft)')
axis([0 30 -20 20])
figure
plot(ArrayT,ArrayERRXD,ArrayT,ArraySP22,ArrayT,ArraySP22P),grid
xlabel('Time (Sec')
ylabel('Error in Estimate of Velocity (Ft/Sec)')
axis([0 30 -5 5])
clc
output=[ArrayT',ArrayX',ArrayXH',ArrayXD',ArrayXDH'];
save datfil output -ascii
output=[ArrayT',ArrayERRX',ArraySP11',ArraySP11P',ArrayERRXD',ArraySP22',
   ArraySP22P|1];
save covfil output -ascii
disp 'simulation finished'
```

True BASIC

To many older engineers BASIC is simply known as "FORTRAN without grief." BASIC was originally developed in 1963 at Dartmouth College by John Kemeny and Thomas Kurtz. Although many dialects of BASIC were originally implemented on microcomputers during the 1980s and early 1990s, the major non-object-oriented dialect to survive was the original, which is known commercially as True BASIC. True BASIC, like MATLAB®, is an interpretive language and is, therefore, much slower than a compiled language, such as FORTRAN.

Listing A.3 MATLAB® equivalent of FORTRAN subroutine PROJECT in Listing 7.5

```
function[XH,XDH,XDDH]=project1(TP,TS,XP,XDP,BETA)
T=0.;
X=XP;
XD=XDP;
H=.001;
while T<=(TS-.0001)
        XDD=.0034*32.2*XD*XD*exp(-X/22000.)/(2.*BETA)-32.2;
        XD=XD+H*XDD;
        X=X+H*XD;
        T=T+H;
end
XH=X;
XDH=XD;
XDDH=XDD;
```

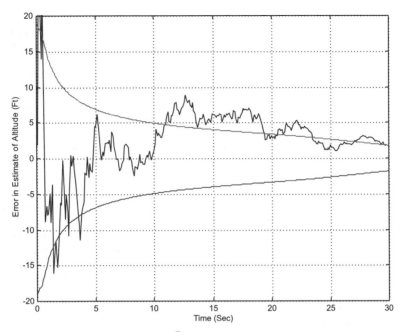

Fig. A.3 MATLAB® output is equivalent to Fig. 7.21.

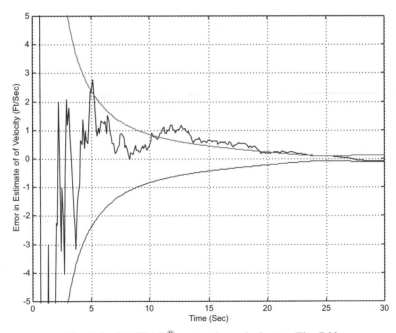

Fig. A.4 MATLAB® output is equivalent to Fig. 7.22.

However, as with MATLAB®, all of the programs in the text take only a few seconds to a minute of running time with True BASIC on today's powerful microcomputers. Its low cost (i.e., approximately $40 for the regular version and $20 for the student version) and ease of use also make it a natural choice for implementing the code in this text on personal computers or workstations. As with MATLAB®, True BASIC also has special functions for matrix manipulation. Therefore, many of the FORTRAN subroutines presented in the text are not required with True BASIC. With a few minor differences the True BASIC code is nearly identical to the FORTRAN code.

As an example of how the FORTRAN source code was converted to True BASIC, let us again consider the case of Listing 4.4, which simulated a first-order polynomial Kalman filter with gravity compensation. The True BASIC equivalent of Listing 4.4 appears in Listing A.4. We can see that the True BASIC listing is virtually identical to FORTRAN Listing 4.4. At the end of the program, the resultant data are sent to ascii files called DATFIL and COVFIL. Use of special True BASIC statements to simplify the writing of matrices also has been incorporated in the simulation. In addition, the matrix Riccati equations appear in a more efficient form the True BASIC listing because subroutines for matrix algebra are no longer required. Unlike MATLAB®, Gaussian random numbers are not automatically available with True BASIC. However, True BASIC provides a random-number generator yielding uniformly distributed numbers between 0 and 1 with the statement RND. A subroutine, similar to the FORTRAN subroutine GAUSS, makes use of the random-number generator in order to provide zero-mean Gaussian-distributed random numbers.

Listing A.4. True BASIC equivalent of Listing 4.4

```
OPTION NOLET
REM UNSAVE "DATFIL"
REM UNSAVE "COVFIL"
OPEN #1:NAME "DATFIL",ACCESS OUTPUT,CREATE NEW,ORGANIZATION
    TEXT
OPEN #2:NAME "COVFIL",ACCESS OUTPUT,CREATE NEW,ORGANIZATION
    TEXT
SET #1: MARGIN 1000
SET #2: MARGIN 1000
DIM P(2,2),Q(2,2),M(2,2),PHI(2,2),HMAT(1,2),HT(2,1),PHIT(2,2)
DIM RMAT(1,1),IDNP(2,2),PHIP(2,2),PHIPPHIT(2,2),HM(1,2)
DIM HMHT(1,1),HMHTR(1,1),HMHTRINV(1,1),MHT(2,1),K(2,1)
DIM KH(2,2),IKH(2,2)
TS=.1
PHIS=0.
A0=400000.
A1=-6000.
A2=-16.1
XH=0.
XDH=0.
SIGNOISE=1000.
ORDER=2
```

(continued)

Listing A.4   (*Continued*)

```
T=0.
S=0.
H=.001
MAT  PHI=ZER(ORDER,ORDER)
MAT  P=ZER(ORDER,ORDER)
MAT  IDNP=IDN(ORDER,ORDER)
MAT  Q=ZER(ORDER,ORDER)
RMAT(1,1)=SIGNOISE^2
P(1,1)=99999999999.
P(2,2)=99999999999.
PHI(1,1)=1.
PHI(1,2)=TS
PHI(2,2)=1.
Q(1,1)=TS*TS*TS*PHIS/3.
Q(1,2)=.5*TS*TS*PHIS
Q(2,1)=Q(1,2)
Q(2,2)=PHIS*TS
HMAT(1,1)=1.
HMAT(1,2)=0.
FOR T=0 TO 30 STEP TS
     MAT  PHIT=TRN(PHI)
     MAT  HT=TRN(HMAT)
     MAT  PHIP=PHI*P
     MAT  PHIPPHIT=PHIP*PHIT
     MAT  M=PHIPPHIT+Q
     MAT  HM=HMAT*M
     MAT  HMHT=HM*HT
     MAT  HMHTR=HMHT+RMAT
     HMHTRINV(1,1)=1./HMHTR(1,1)
     MAT  MHT=M*HT
     MAT  K=MHT*HMHTRINV
     MAT  KH=K*HMAT
     MAT  IKH=IDNP-KH
     MAT  P=IKH*M
     CALL  GAUSS(XNOISE,SIGNOISE)
     X=A0+A1*T+A2*T*T
     XD=A1+2*A2*T
     XS=X+XNOISE
     RES=XS-XH-TS*XDH+16.1*TS*TS
     XH=XH+XDH*TS-16.1*TS*TS+K(1,1)*RES
     XDH=XDH-32.2*TS+K(2,1)*RES
     SP11=SQR(P(1,1))
     SP22=SQR(P(2,2))
     XHERR=X-XH
     XDHERR=XD-XDH
     PRINT  T,X,XH,XD,XDH
     PRINT  #1:T,X,DH,XD,XDH
     PRINT  #2:T,XHERR,SP11,-SP11,XDHERR,SP22, − SP22
NEXT T
```

(*continued*)

Listing A.4   (*Continued*)

```
CLOSE #1
CLOSE #2
END

SUB  GAUSS(X,SIG)
LET  X=RND+RND+RND+RND+RND+RND-3
LET  X=1.414*X*SIG
END  SUB
```

True BASIC has a problem creating a new file if an old file with the same name already exists. Therefore, the user must be cautious when running True BASIC several times in a row when data files are automatically created. Either the user must manually delete previously created data files or have statements in the program automatically deleting the files. Listing A.4 has two statements at the beginning of the program to delete previously created files DATFIL and COVFIL. These statements are initially commented out by use of the REM statement. After the program is run once, the word REM can be deleted from the listing two times so that the files of the previous run will automatically be removed.

The True BASIC simulation of Listing A.4 was run, and data concerning the errors in the estimates were automatically written to the file COVFIL. The error in the estimates of altitude and velocity are displayed in Figs. A.5 and A.6, respectively. Figure A.5 is equivalent to the FORTRAN-generated results of Fig. 4.35, whereas Fig. A.6 is equivalent to the FORTRAN-generated results of Fig. 4.36. These results also agree with the MATLAB®-generated results of the preceding section (see Figs. A.1 and A.2).

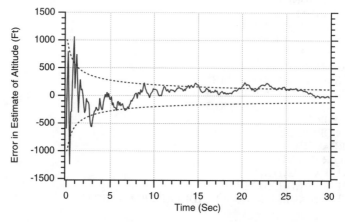

Fig. A.5   True BASIC output is equivalent to Fig. 4.35.

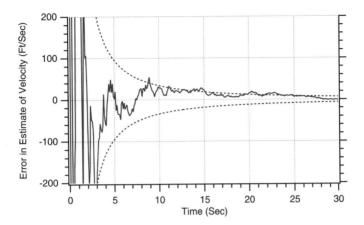

Fig. A.6   True BASIC output is equivalent to Fig. 4.36.

For completeness, Listing 7.5 was also converted to True BASIC and appears in Listing A.5. We can see that with the exception of the matrix operations the True BASIC listing appears to be nearly identical to the FORTRAN listing.

The simulation of Listing A.5 was run, and data concerning the errors in the estimates were automatically written to the file COVFIL. The information in this file was plotted, and the results appear in Figs. A.7 and A.8. Figure A.7 is equivalent to the FORTRAN-generated results of Fig. 7.21, whereas Fig. A.8 is equivalent to the FORTRAN-generated results of Fig. 7.22. These results are also equivalent to the MATLAB®-generated results of the preceding section (see Figs. A.3 and A.4).

Listing A.5   True BASIC equivalent of Listing 7.5

```
OPTION NOLET
REM UNSAVE "DATFIL"
REM UNSAVE "COVFIL"
OPEN #1:NAME "DATFIL",ACCESS OUTPUT,CREATE NEW,ORGANIZATION
    TEXT
OPEN #2:NAME "COVFIL",ACCESS OUTPUT,CREATE NEW,ORGANIZATION
    TEXT
SET #1: MARGIN 1000
SET #2: MARGIN 1000
DIM  PHI(2,2),P(2,2),M(2,2),PHIP(2,2),PHIPPHIT(2,2),GAIN(2,1)
DIM  Q(2,2),HMAT(1,2),HM(1,2),MHT(2,1)
DIM  PHIT(2,2)
DIM  HMHT(1,1),HT(2,1),KH(2,2),IDNP(2,2),IKH(2,2)
SIGNOISE=25.
X=200000.
XD=-6000.
BETA=500.
```

(continued)

Listing A.5 (*Continued*)

```
XH=200025.
XDH=-6150.
ORDER=2
TS=.1
TF=30.
PHIS=0.
T=0.
S=0.
H=.001
MAT PHI=ZER(ORDER,ORDER)
MAT P=ZER(ORDER,ORDER)
MAT IDNP=IDNP(ORDER,ORDER)
MAT Q=ZER(ORDER,ORDER)
P(1,1)=SIGNOISE*SIGNOISE
P(2,2)=20000.
MAT HMAT=ZER(1,ORDER)
MAT HT=ZER(ORDER,1)
HMAT(1,1)=1.
HT(1,1)=1.
DO WHILE T<=TF
        XOLD=X
        XDOLD=XD
        XDD=.0034*32.2*XD*XD*EXP(-X/22000.)/(2.*BETA)-32.2
        X=X+H*XD
        XD=XD+H*XDD
        T=T+H
        XDD=.0034*32.2*XD*XD*EXP(-X/22000.)/(2.*BETA)-32.2
        X=.5*(XOLD+X+H*XD)
        XD=.5*(XDOLD+XD+H*XDD)
        S=S+H
        IF S>=(TS-.00001) THEN
                S=0.
                RHOH=.0034*EXP(-XH/22000.)
                F21=-32.2*RHOH*XDH*XDH/(44000.*BETA)
                F22=RHOH*32.2*XDH/BETA
                PHI(1,1)=1.
                PHI(1,2)=TS
                PHI(2,1)=F21*TS
                PHI(2,2)=1.+F22*TS
                Q(1,1)=PHIS*TS*TS*TS/3.
                Q(1,2)=PHIS*(TS*TS/2.+F22*TS*TS*TS/3.)
                Q(2,1)=Q(1,2)
                Q(2,2)=PHIS*(TS+F22*TS*TS+F22*TS*TS*TS/3.)
                MAT PHIT=TRN(PHI)
                MAT PHIP=PHI*P
                MAT PHIPPHIT=PHIP*PHIT
                MAT M=PHIPPHIT+Q
                MAT HM=HMAT*M
                MAT HMHT=HM*HT
                HMHTR=HMHT(1,1)+SIGNOISE*SIGNOISE
```

(*continued*)

Listing A.5   *(Continued)*

```
                HMHTRINV=1./HMHTR
                MAT MHT=M*HT
                MAT GAIN=HMHTRINV*MHT
                MAT KH=GAIN*HMAT
                MAT IKH=IDNP=KH
                MAT P=IKH*M
                CALL GAUSS(XNOISE,SIGNOISE)
                CALL PROJECT(T,TS,XH,XDH,BETA,XB,XDB,XDDB)
                RES=X+XNOISE-XB
                XH=XB+GAIN(1,1)*RES
                XDH=XDB+GAIN(2,1)*RES
                ERRX=X-XH
                SP11=SQR(P(1,1))
                ERRXD=XD-XDH
                SP22=SQR(P(2,2))
                PRINT T,X,XH,XD,XDH
                PRINT #1:T,X,XH,XD,XDH
                PRINT #2:T,ERRX,SP11,-SP11,ERRXD,SP22,-SP22
        END IF
LOOP
CLOSE #1
CLOSE #2
END

SUB GAUSS(X,SIG)
LET X=RND+RND+RND+RND+RND+RND-3
LET X=1.414*X*SIG
END SUB
SUB PROJECT(TP,TS,XP,XDP,BETA,XH,XDH,XDDH)
T=0.
X=XP
XD=XDP
H=.001
DO WHILE T<=(TS-.0001)
        XDD=.0034*32.2*XD*XD*EXP(-X/22000.)/(2.*BETA)-32.2
        XD=XD+H*XDD
        X=X+H*XD
        T=T+H
LOOP
XH=X
XDH=XD
XDDH=XDD
END SUB
```

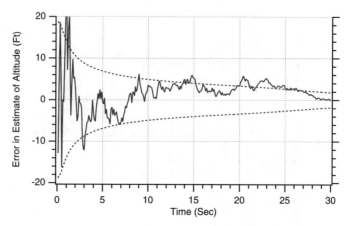

Fig. A.7   True BASIC output is equivalent to Fig. 7.21.

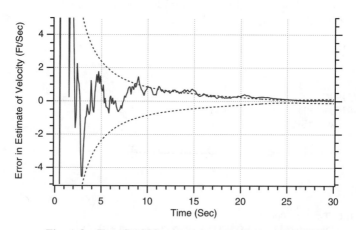

Fig. A.8   True BASIC output is equivalent to Fig. 7.22.

Reference

[1]Zarchan, P., *Tactical and Strategic Missile Guidance*, 3rd ed., Progress in Astronautics and Aeronautics, AIAA, Reston, VA, 1998, pp. 585–595.

Index

PROGRESS IN ASTRONAUTICS AND AERONAUTICS
SERIES VOLUMES

*Out of print.

*Out of print.

*40. Aerodynamics of Base Combustion (1976)
S. N. B. Murthy, Editor
J. R. Osborn,
Associate Editor
Purdue University
A. W. Barrows
J. R. Ward,
Associate Editors
Ballistics Research Laboratories
ISBN 0-915928-04-3

*41. Communications Satellite Developments: Systems (1976)
Gilbert E. LaVean
Defense Communications Agency
William G. Schmidt
CML Satellite Corp.
ISBN 0-915928-05-1

*42. Communications Satellite Developments: Technology (1976)
William G. Schmidt
CML Satellite Corp.
Gilbert E. LaVean
Defense Communications Agency
ISBN 0-915928-06-X

*43. Aeroacoustics: Jet Noise, Combustion and Core Engine Noise (1976)
Ira R. Schwartz, Editor
NASA Ames Research Center
Henry T. Nagamatsu,
Associate Editor
General Electric Research and Development Center
Warren C. Strahle,
Associate Editor
Georgia Institute of Technology
ISBN 0-915928-07-8

*44. Aeroacoustics: Fan Noise and Control; Duct Acoustics; Rotor Noise (1976)
Ira R. Schwartz, Editor
NASA Ames Research Center
Henry T. Nagamatsu,
Associate Editor
General Electric Research and Development Center
Warren C. Strahle,
Associate Editor
Georgia Institute of Technology
ISBN 0-915928-08-6

*45. Aeroacoustics: STOL Noise; Airframe and Airfoil Noise (1976)
Ira R. Schwartz, Editor
NASA Ames Research Center
Henry T. Nagamatsu,
Associate Editor
General Electric Research and Development Center
Warren C. Strahle,
Associate Editor
Georgia Institute of Technology
ISBN 0-915928-09-4

*46. Aeroacoustics: Acoustic Wave Propagation; Aircraft Noise Prediction; Aeroacoustic Instrumentation (1976)
Ira R. Schwartz, Editor
NASA Ames Research Center
Henry T. Nagamatsu,
Associate Editor
General Electric Research and Development Center
Warren C. Strahle,
Associate Editor
Georgia Institute of Technology
ISBN 0-915928-10-8

*47. Spacecraft Charging by Magnetospheric Plasmas (1976)
Alan Rosen
TRW Inc.
ISBN 0-915928-11-6

*48. Scientific Investigations on the Skylab Satellite (1976)
Marion I. Kent
Ernst Stuhlinger
NASA George C. Marshall Space Flight Center
Shi-Tsan Wu
University of Alabama
ISBN 0-915928-12-4

*49. Radiative Transfer and Thermal Control (1976)
Allie M. Smith
ARO Inc.
ISBN 0-915928-13-2

*50. Exploration of the Outer Solar System (1976)
Eugene W. Greenstadt
TRW Inc.
Murray Dryer
National Oceanic and Atmospheric Administration
Devrie S. Intriligator
University of Southern California
ISBN 0-915928-14-0

*51. Rarefied Gas Dynamics, Parts I and II (two volumes) (1977)
J. Leith Potter
ARO Inc.
ISBN 0-915928-15-9

*52. Materials Sciences in Space with Application to Space Processing (1977)
Leo Steg
General Electric Co.
ISBN 0-915928-16-7

---

*Out of print.

*53. Experimental
Diagnostics in Gas Phase
Combustion Systems
(1977)
Ben T. Zinn, Editor
*Georgia Institute of
Technology*
Craig T. Bowman,
Associate Editor
Stanford University
Daniel L. Hartley,
Associate Editor
Sandia Laboratories
Edward W. Price,
Associate Editor
*Georgia Institute of
Technology*
James G. Skifstad,
Associate Editor
Purdue University
ISBN 0-915928-18-3

*54. Satellite
Communication: Future
Systems (1977)
David Jarett
TRW Inc.
ISBN 0-915928-18-3

*55. Satellite
Communications:
Advanced Technologies
(1977)
David Jarett
TRW Inc.
ISBN 0-915928-19-1

*56. Thermophysics of
Spacecraft and Outer
Planet Entry Probes
(1977)
Allie M. Smith
ARO Inc.
ISBN 0-915928-20-5

*57. Space-Based
Manufacturing from
Nonterrestrial Materials
(1977)
Gerald K. O'Neill, Editor
Brian O'Leary,
Assistant Editor
Princeton University
ISBN 0-915928-21-3

*58. Turbulent
Combustion (1978)
Lawrence A. Kennedy
*State University of New
York at Buffalo*
ISBN 0-915928-22-1

*59. Aerodynamic
Heating and Thermal
Protection Systems (1978)
Leroy S. Fletcher
University of Virginia
ISBN 0-915928-23-X

*60. Heat Transfer and
Thermal Control Systems
(1978)
Leroy S. Fletcher
University of Virginia
ISBN 0-915928-24-8

*61. Radiation Energy
Conversion in Space
(1978)
Kenneth W. Billman
*NASA Ames Research
Center*
ISBN 0-915928-26-4

*62. Alternative
Hydrocarbon Fuels:
Combustion and Chemical
Kinetics (1978)
Craig T. Bowman
Stanford University
Jorgen Birkeland
Department of Energy
ISBN 0-915928-25-6

*63. Experimental
Diagnostics in Combustion
of Solids (1978)
Thomas L. Boggs
Naval Weapons Center
Ben T. Zinn
*Georgia Institute of
Technology*
ISBN 0-915928-28-0

*64. Outer Planet Entry
Heating and Thermal
Protection (1979)
Raymond Viskanta
Purdue University
ISBN 0-915928-29-9

*65. Thermophysics and
Thermal Control (1979)
Raymond Viskanta
Purdue University
ISBN 0-915928-30-2

*66. Interior Ballistics of
Guns (1979)
Herman Krier
*University of Illinois at
Urbana–Champaign*
Martin Summerfield
New York University
ISBN 0-915928-32-9

*67. Remote Sensing of
Earth from Space: Role of
"Smart Sensors" (1979)
Roger A. Breckenridge
*NASA Langley Research
Center*
ISBN 0-915928-33-7

*68. Injection and Mixing
in Turbulent Flow (1980)
Joseph A. Schetz
*Virginia Polytechnic
Institute and State
University*
ISBN 0-915928-35-3

*69. Entry Heating and
Thermal Protection (1980)
Walter B. Olstad
NASA Headquarters
ISBN 0-915928-38-8

*70. Heat Transfer,
Thermal Control, and
Heat Pipes (1980)
Walter B. Olstad
NASA Headquarters
ISBN 0-915928-39-6

*71. Space Systems and
Their Interactions with
Earth's Space
Environment (1980)
Henry B. Garrett
Charles P. Pike
Hanscom Air Force Base
ISBN 0-915928-41-8

*72. Viscous Flow Drag
Reduction (1980)
Gary R. Hough
*Vought Advanced
Technology Center*
ISBN 0-915928-44-2

*73. Combustion
Experiments in a Zero-
Gravity Laboratory (1981)
Thomas H. Cochran
*NASA Lewis Research
Center*
ISBN 0-915928-48-5

*74. Rarefied Gas
Dynamics, Parts I and II
(two volumes) (1981)
Sam S. Fisher
University of Virginia
ISBN 0-915928-51-5

*75. Gasdynamics of
Detonations and
Explosions (1981)
J. R. Bowen
*University of Wisconsin
at Madison*
N. Manson
Universite de Poitiers
A. K. Oppenheim
*University of California
at Berkeley*
R. I. Soloukhin
*Institute of Heat and Mass
Transfer, BSSR Academy
of Sciences*
ISBN 0-915928-46-9

*76. Combustion in
Reactive Systems (1981)
J. R. Bowen
*University of Wisconsin
at Madison*
N. Manson
Universite de Poitiers
A. K. Oppenheim
*University of California
at Berkeley*
R. I. Soloukhin
*Institute of Heat and Mass
Transfer, BSSR Academy
of Sciences*
ISBN 0-915928-47-7

*77.
Aerothermodynamics and
Planetary Entry (1981)
A. L. Crosbie
University of Missouri-Rolla
ISBN 0-915928-52-3

*78. Heat Transfer and
Thermal Control (1981)
A. L. Crosbie
University of Missouri-Rolla
ISBN 0-915928-53-1

*79. Electric Propulsion
and Its Applications to
Space Missions (1981)
Robert C. Finke
*NASA Lewis Research
Center*
ISBN 0-915928-55-8

*80. Aero-Optical
Phenomena (1982)
Keith G. Gilbert
Leonard J. Otten
*Air Force Weapons
Laboratory*
ISBN 0-915928-60-4

*81. Transonic
Aerodynamics (1982)
David Nixon
*Nielsen Engineering &
Research, Inc.*
ISBN 0-915928-65-5

*82. Thermophysics of
Atmospheric Entry (1982)
T. E. Horton
University of Mississippi
ISBN 0-915928-66-3

*83. Spacecraft Radiative
Transfer and
Temperature Control
(1982)
T. E. Horton
University of Mississippi
ISBN 0-915928-67-1

*84. Liquid-Metal Flows
and Magneto-
hydrodynamics (1983)
H. Branover
*Ben-Gurion University
of the Negev*
P. S. Lykoudis
Purdue University
A. Yakhot
*Ben-Gurion University
of the Negev*
ISBN 0-915928-70-1

---

*Out of print.

*85. Entry Vehicle Heating and Thermal Protection Systems: Space Shuttle, Solar Starprobe, Jupiter Galileo Probe (1983)
Paul E. Bauer
McDonnell Douglas Astronautics Co.
Howard E. Collicott
The Boeing Co.
ISBN 0-915928-74-4

*86. Spacecraft Thermal Control, Design, and Operation (1983)
Howard E. Collicott
The Boeing Co.
Paul E. Bauer
McDonnell Douglas Astronautics Co.
ISBN 0-915928-75-2

*87. Shock Waves, Explosions, and Detonations (1983)
J. R. Bowen
University of Washington
N. Manson
Universite de Poitiers
A. K. Oppenheim
University of California at Berkeley
R. I. Soloukhin
Institute of Heat and Mass Transfer, BSSR Academy of Sciences
ISBN 0-915928-76-0

*88. Flames, Lasers, and Reactive Systems (1983)
J. R. Bowen
University of Washington
N. Manson
Universite de Poitiers
A. K. Oppenheim
University of California at Berkeley
R. I. Soloukhin
Institute of Heat and Mass Transfer, BSSR Academy of Sciences
ISBN 0-915928-77-9

*89. Orbit-Raising and Maneuvering Propulsion: Research Status and Needs (1984)
Leonard H. Caveny
Air Force Office of Scientific Research
ISBN 0-915928-82-5

*90. Fundamentals of Solid-Propellant Combustion (1984)
Kenneth K. Kuo
Pennsylvania State University
Martin Summerfield
Princeton Combustion Research Laboratories, Inc.
ISBN 0-915928-84-1

91. Spacecraft Contamination: Sources and Prevention (1984)
J. A. Roux
University of Mississippi
T. D. McCay
NASA Marshall Space Flight Center
ISBN 0-915928-85-X

92. Combustion Diagnostics by Nonintrusive Methods (1984)
T. D. McCay
NASA Marshall Space Flight Center
J. A. Roux
University of Mississippi
ISBN 0-915928-86-8

93. The INTELSAT Global Satellite System (1984)
Joel Alper
COMSAT Corp.
Joseph Pelton
INTELSAT
ISBN 0-915928-90-6

94. Dynamics of Shock Waves, Explosions, and Detonations (1984)
J. R. Bowen
University of Washington
N. Manson
Universite de Poitiers
A. K. Oppenheim
University of California at Berkeley
R. I. Soloukhin
Institute of Heat and Mass Transfer, BSSR Academy of Sciences
ISBN 0-915928-91-4

95. Dynamics of Flames and Reactive Systems (1984)
J. R. Bowen
University of Washington
N. Manson
Universite de Poitiers
A. K. Oppenheim
University of California at Berkeley
R. I. Soloukhin
Institute of Heat and Mass Transfer, BSSR Academy of Sciences
ISBN 0-915928-92-2

96. Thermal Design of Aeroassisted Orbital Transfer Vehicles (1985)
H. F. Nelson
University of Missouri-Rolla
ISBN 0-915928-94-9

97. Monitoring Earth's Ocean, Land, and Atmosphere from Space—Sensors, Systems, and Applications (1985)
Abraham Schnapf
Aerospace Systems Engineering
ISBN 0-915928-98-1

---

*Out of print.

98. Thrust and Drag: Its Prediction and Verification (1985)
Eugene E. Covert
Massachusetts Institute of Technology
C. R. James
Vought Corp.
William F. Kimzey
Sverdrup Technology AEDC Group
George K. Richey
U.S. Air Force
Eugene C. Rooney
U.S. Navy Department of Defense
ISBN 0-930403-00-2

99. Space Stations and Space Platforms—Concepts, Design, Infrastructure, and Uses (1985)
Ivan Bekey
Daniel Herman
NASA Headquarters
ISBN 0-930403-01-0

100. Single- and Multi-Phase Flows in an Electromagnetic Field: Energy, Metallurgical, and Solar Applications (1985)
Herman Branover
Ben-Gurion University of the Negev
Paul S. Lykoudis
Purdue University
Michael Mond
Ben-Gurion University of the Negev
ISBN 0-930403-04-5

101. MHD Energy Conversion: Physiotechnical Problems (1986)
V. A. Kirillin
A. E. Sheyndlin
Soviet Academy of Sciences
ISBN 0-930403-05-3

102. Numerical Methods for Engine-Airframe Integration (1986)
S. N. B. Murthy
Purdue University
Gerald C. Paynter
Boeing Airplane Co.
ISBN 0-930403-09-6

103. Thermophysical Aspects of Re-Entry Flows (1986)
James N. Moss
NASA Langley Research Center
Carl D. Scott
NASA Johnson Space Center
ISBN 0-930430-10-X

*104. Tactical Missile Aerodynamics (1986)
M. J. Hemsch
PRC Kentron, Inc.
J. N. Nielson
NASA Ames Research Center
ISBN 0-930403-13-4

105. Dynamics of Reactive Systems Part I: Flames and Configurations; Part II: Modeling and Heterogeneous Combustion (1986)
J. R. Bowen
University of Washington
J.-C. Leyer
Universite de Poitiers
R. I. Soloukhin
Institute of Heat and Mass Transfer, BSSR Academy of Sciences
ISBN 0-930403-14-2

106. Dynamics of Explosions (1986)
J. R. Bowen
University of Washington
J.-C. Leyer
Universite de Poitiers
R. I. Soloukhin
Institute of Heat and Mass Transfer, BSSR Academy of Sciences
ISBN 0-930403-15-0

*107. Spacecraft Dielectric Material Properties and Spacecraft Charging (1986)
A. R. Frederickson
U.S. Air Force Rome Air Development Center
D. B. Cotts
SRI International
J. A. Wall
U.S. Air Force Rome Air Development Center
F. L. Bouquet
Jet Propulsion Laboratory, California Institute of Technology
ISBN 0-930403-17-7

*108. Opportunities for Academic Research in a Low-Gravity Environment (1986)
George A. Hazelrigg
National Science Foundation
Joseph M. Reynolds
Louisiana State University
ISBN 0-930403-18-5

109. Gun Propulsion Technology (1988)
Ludwig Stiefel
U.S. Army Armament Research, Development and Engineering Center
ISBN 0-930403-20-7

---

*Out of print.

110. Commercial Opportunities in Space (1988)
F. Shahrokhi
K. E. Harwell
University of Tennessee Space Institute
C. C. Chao
National Cheng Kung University
ISBN 0-930403-39-8

111. Liquid-Metal Flows: Magnetohydrodynamics and Application (1988)
Herman Branover
Michael Mond
Yeshajahu Unger
Ben-Gurion University of the Negev
ISBN 0-930403-43-6

112. Current Trends in Turbulence Research (1988)
Herman Branover
Micheal Mond
Yeshajahu Unger
Ben-Gurion University of the Negev
ISBN 0-930403-44-4

113. Dynamics of Reactive Systems Part I: Flames; Part II: Heterogeneous Combustion and Applications (1988)
A. L. Kuhl
R&D Associates
J. R. Bowen
University of Washington
J.-C. Leyer
Universite de Poitiers
A. Borisov
USSR Academy of Sciences
ISBN 0-930403-46-0

114. Dynamics of Explosions (1988)
A. L. Kuhl
R & D Associates
J. R. Bowen
University of Washington
J.-C. Leyer
Universite de Poitiers
A. Borisov
USSR Academy of Sciences
ISBN 0-930403-47-9

115. Machine Intelligence and Autonomy for Aerospace (1988)
E. Heer
Heer Associates, Inc.
H. Lum
NASA Ames Research Center
ISBN 0-930403-48-7

116. Rarefied Gas Dynamics: Space Related Studies (1989)
E. P. Muntz
University of Southern California
D. P. Weaver
U.S. Air Force Astronautics Laboratory (AFSC)
D. H. Campbell
University of Dayton Research Institute
ISBN 0-930403-53-3

117. Rarefied Gas Dynamics: Physical Phenomena (1989)
E. P. Muntz
University of Southern California
D. P. Weaver
U.S. Air Force Astronautics Laboratory (AFSC)
D. H. Campbell
University of Dayton Research Institute
ISBN 0-930403-54-1

118. Rarefied Gas Dynamics: Theoretical and Computational Techniques (1989)
E. P. Muntz
University of Southern California
D. P. Weaver
U.S. Air Force Astronautics Laboratory (AFSC)
D. H. Campbell
University of Dayton Research Institute
ISBN 0-930403-55-X

119. Test and Evaluation of the Tactical Missile (1989)
Emil J. Eichblatt Jr.
Pacific Missile Test Center
ISBN 0-930403-56-8

120. Unsteady Transonic Aerodynamics (1989)
David Nixon
Nielsen Engineering & Research, Inc.
ISBN 0-930403-52-5

121. Orbital Debris from Upper-Stage Breakup (1989)
Joseph P. Loftus Jr.
NASA Johnson Space Center
ISBN 0-930403-58-4

122. Thermal-Hydraulics for Space Power, Propulsion and Thermal Management System Design (1990)
William J. Krotiuk
General Electric Co.
ISBN 0-930403-64-9

---

*Out of print.

123. Viscous Drag Reduction in Boundary Layers (1990)
Dennis M. Bushnell
Jerry N. Hefner
NASA Langley Research Center
ISBN 0-930403-66-5

*124. Tactical and Strategic Missile Guidance (1990)
Paul Zarchan
Charles Stark Draper Laboratory, Inc.
ISBN 0-930403-68-1

125. Applied Computational Aerodynamics (1990)
P. A. Henne
Douglas Aircraft Company
ISBN 0-930403-69-X

126. Space Commercialization: Launch Vehicles and Programs (1990)
F. Shahrokhi
University of Tennessee Space Institute
J. S. Greenberg
Princeton Synergetics Inc.
T. Al-Saud
Ministry of Defense and Aviation Kingdom of Saudi Arabia
ISBN 0-930403-75-4

127. Space Commercialization: Platforms and Processing (1990)
F. Shahrokhi
University of Tennessee Space Institute
G. Hazelrigg
National Science Foundation
R. Bayuzick
Vanderbilt University
ISBN 0-930403-76-2

128. Space Commercialization: Satellite Technology (1990)
F. Shahrokhi
University of Tennessee Space Institute
N. Jasentuliyana
United Nations
N. Tarabzouni
King Abulaziz City for Science and Technology
ISBN 0-930403-77-0

*129. Mechanics and Control of Large Flexible Structures (1990)
John L. Junkins
Texas A&M University
ISBN 0-930403-73-8

130. Low-Gravity Fluid Dynamics and Transport Phenomena (1990)
Jean N. Koster
Robert L. Sani
University of Colorado at Boulder
ISBN 0-930403-74-6

131. Dynamics of Deflagrations and Reactive Systems: Flames (1991)
A. L. Kuhl
Lawrence Livermore National Laboratory
J.-C. Leyer
Universite de Poitiers
A. A. Borisov
USSR Academy of Sciences
W. A. Sirignano
University of California
ISBN 0-930403-95-9

132. Dynamics of Deflagrations and Reactive Systems: Heterogeneous Combustion (1991)
A. L. Kuhl
Lawrence Livermore National Laboratory
J.-C. Leyer
Universite de Poitiers
A. A. Borisov
USSR Academy of Sciences
W. A. Sirignano
University of California
ISBN 0-930403-96-7

133. Dynamics of Detonations and Explosions: Detonations (1991)
A. L. Kuhl
Lawrence Livermore National Laboratory
J.-C. Leyer
Universite de Poitiers
A. A. Borisov
USSR Academy of Sciences
W. A. Sirignano
University of California
ISBN 0-930403-97-5

134. Dynamics of Detonations and Explosions: Explosion Phenomena (1991)
A. L. Kuhl
Lawrence Livermore National Laboratory
J.-C. Leyer
Universite de Poitiers
A. A. Borisov
USSR Academy of Sciences
W. A. Sirignano
University of California
ISBN 0-930403-98-3

---

*Out of print.

135. Numerical Approaches to Combustion Modeling (1991)
Elaine S. Oran
Jay P. Boris
Naval Research Laboratory
ISBN 1-56347-004-7

136. Aerospace Software Engineering (1991)
Christine Anderson
U.S. Air Force Wright Laboratory
Merlin Dorfman
Lockheed Missiles & Space Company, Inc.
ISBN 1-56347-005-0

137. High-Speed Flight Propulsion Systems (1991)
S. N. B. Murthy
Purdue University
E. T. Curran
Wright Laboratory
ISBN 1-56347-011-X

138. Propagation of Intensive Laser Radiation in Clouds (1992)
O. A. Volkovitsky
Yu. S. Sedenov
L. P. Semenov
Institute of Experimental Meteorology
ISBN 1-56347-020-9

139. Gun Muzzle Blast and Flash (1992)
Günter Klingenberg
Fraunhofer-Institut für Kurzzeitdynamik, Ernst-Mach-Institut
Joseph M. Heimerl
U.S. Army Ballistic Research Laboratory
ISBN 1-56347-012-8

140. Thermal Structures and Materials for High-Speed Flight (1992)
Earl. A. Thornton
University of Virginia
ISBN 1-56347-017-9

141. Tactical Missile Aerodynamics: General Topics (1992)
Michael J. Hemsch
Lockheed Engineering & Sciences Company
ISBN 1-56347-015-2

142. Tactical Missile Aerodynamics: Prediction Methodology (1992)
Michael R. Mendenhall
Nielsen Engineering & Research, Inc.
ISBN 1-56347-016-0

143. Nonsteady Burning and Combustion Stability of Solid Propellants (1992)
Luigi De Luca
Politecnico di Milano
Edward W. Price
Georgia Institute of Technology
Martin Summerfield
Princeton Combustion Research Laboratories, Inc.
ISBN 1-56347-014-4

144. Space Economics (1992)
Joel S. Greenberg
Princeton Synergetics, Inc.
Henry R. Hertzfeld
HRH Associates
ISBN 1-56347-042-X

145. Mars: Past, Present, and Future (1992)
E. Brian Pritchard
NASA Langley Research Center
ISBN 1-56347-043-8

146. Computational Nonlinear Mechanics in Aerospace Engineering (1992)
Satya N. Atluri
Georgia Institute of Technology
ISBN 1-56347-044-6

147. Modern Engineering for Design of Liquid-Propellant Rocket Engines (1992)
Dieter K. Huzel
David H. Huang
Rocketdyne Division of Rockwell International
ISBN 1-56347-013-6

148. Metallurgical Technologies, Energy Conversion, and Magneto-hydrodynamic Flows (1993)
Herman Branover
Yeshajahu Unger
Ben-Gurion University of the Negev
ISBN 1-56347-019-5

149. Advances in Turbulence Studies (1993)
Herman Branover
Yeshajahu Unger
Ben-Gurion University of the Negev
ISBN 1-56347-018-7

150. Structural Optimization: Status and Promise (1993)
Manohar P. Kamat
Georgia Institute of Technology
ISBN 1-56347-056-X

151. Dynamics of Gaseous Combustion (1993)
A. L. Kuhl
Lawrence Livermore National Laboratory
J.-C. Leyer
Universite de Poitiers
A. A. Borisov
USSR Academy of Sciences
W. A. Sirignano
University of California
ISBN 1-56347-060-8

152. Dynamics of Heterogeneous Gaseous Combustion and Reacting Systems (1993)
A. L. Kuhl
Lawrence Livermore National Laboratory
J.-C. Leyer
Universite de Poitiers
A. A. Borisov
USSR Academy of Sciences
W. A. Sirignano
University of California
ISBN 1-56347-058-6

153. Dynamic Aspects of Detonations (1993)
A. L. Kuhl
Lawrence Livermore National Laboratory
J.-C. Leyer
Universite de Poitiers
A. A. Borisov
USSR Academy of Sciences
W. A. Sirignano
University of California
ISBN 1-56347-057-8

154. Dynamic Aspects of Explosion Phenomena (1993)
A. L. Kuhl
Lawrence Livermore National Laboratory
J.-C. Leyer
Universite de Poitiers
A. A. Borisov
USSR Academy of Sciences
W. A. Sirignano
University of California
ISBN 1-56347-059-4

155. Tactical Missile Warheads (1993)
Joseph Carleone
Aerojet General Corporation
ISBN 1-56347-067-5

156. Toward a Science of Command, Control, and Communications (1993)
Carl R. Jones
Naval Postgraduate School
ISBN 1-56347-068-3

*157. Tactical and Strategic Missile Guidance Second Edition (1994)
Paul Zarchan
Charles Stark Draper Laboratory, Inc.
ISBN 1-56347-077-2

158. Rarefied Gas Dynamics: Experimental Techniques and Physical Systems (1994)
Bernie D. Shizgal
University of British Columbia
David P. Weaver
Phillips Laboratory
ISBN 1-56347-079-9

159. Rarefied Gas Dynamics: Theory and Simulations (1994)
Bernie D. Shizgal
University of British Columbia
David P. Weaver
Phillips Laboratory
ISBN 1-56347-080-2

160. Rarefied Gas Dynamics: Space Sciences and Engineering (1994)
Bernie D. Shizgal
University of British Columbia
David P. Weaver
Phillips Laboratory
ISBN 1-56347-081-0

161. Teleoperation and Robotics in Space (1994)
Steven B. Skaar
University of Notre Dame
Carl F. Ruoff
Jet Propulsion Laboratory, California Institute of Technology
ISBN 1-56347-095-0

162. Progress in Turbulence Research (1994)
Herman Branover
Yeshajahu Unger
Ben-Gurion University of the Negev
ISBN 1-56347-099-3

163. Global Positioning System: Theory and Applications, Volume I (1996)
Bradford W. Parkinson
Stanford University
James J. Spilker Jr.
Stanford Telecom
Penina Axelrad,
Associate Editor
University of Colorado
Per Enge,
Associate Editor
Stanford University
ISBN 1-56347-107-8

164. Global Positioning System: Theory and Applications, Volume II (1996)
Bradford W. Parkinson
Stanford University
James J. Spilker Jr.
Stanford Telecom
Penina Axelrad,
Associate Editor
University of Colorado
Per Enge,
Associate Editor .
Stanford University
ISBN 1-56347-106-X

---

*Out of print.

---

*Out of print.